Lens Epithelium and Posterior Capsular Opacification

Shizuya Saika • Liliana Werner
Frank J. Lovicu
Editors

Lens Epithelium and Posterior Capsular Opacification

Springer

Editors
Shizuya Saika
Department of Ophthalmology
Wakayama Medical University
School of Medicine
Wakayama, Wakayama, Japan

Liliana Werner
John A. Moran Eye Center
University of Utah
Salt Lake City, Utah, USA

Frank J. Lovicu
School of Medical Sciences
Anatomy and Histology
Lens Research Laboratory
The University of Sydney
Sydney, New South Wales, Australia

ISBN 978-4-431-54299-5 ISBN 978-4-431-54300-8 (eBook)
DOI 10.1007/978-4-431-54300-8
Springer Tokyo Heidelberg New York Dordrecht London

Library of Congress Control Number: 2014942688

Printed on acid-free paper

Springer is part of Springer Science+Business Media (www.springer.com)

Foreword

This timely book, *Lens Epithelium and Posterior Capsular Opacification,* brings together an impressive list of contributions, all from leaders in the respective fields that relate to the chronic complication of cataract surgery, posterior capsular opacification (PCO). At the outset it is recognized that despite all the efforts towards its amelioration, PCO remains a significant eye health issue, with its most serious consequences impacting on paediatric cataract patients. This book begins with an excellent compendium of articles covering all the key aspects of lens epithelial cell biology, both in health and disease. In many cases the researchers have combined their interest in normal developmental processes with gaining an understanding of how cells go off track and progress down pathological pathways that lead to PCO. The contributors describe the great strides that have been made in identifying some of the key molecular triggers for these processes, as well as conditions that exacerbate PCO. The underlying principle here is that by furthering our understanding of how basic cell behaviour is regulated, this will provide the essentials for developing a molecular toolkit for alleviating this condition. This ties in very nicely with the analysis of the cellular types present in the capsular bag after surgery. In addition to all the latest information on the ubiquitous fibroblasts and myofibroblasts that contribute to the formation of the fibrotic plaques, as well as other features of PCO, it is refreshing to see some detail of the cells that make up Elschnig pearls and Soemmering ring. Cells within these discrete clusters exhibit some of the features of lens fibre cells. Given current knowledge about the factors that control the normal epithelial to fibre differentiation, future work directed at generating conditions that promote more normal (and less abnormal) lens cell behaviour could provide an important stepping stone towards devising strategies to regenerate lens structure and function after surgery, as is depicted in a number of experimental animal models.

A further clinical section details approaches that are aimed at preventing or reducing PCO by ridding the capsular bag of residual epithelial cells, or generating intraocular lenses (IOLs) or techniques that restrict these cells to domains of the capsular bag where they do little harm. So far, no treatment that eliminates these cells without resulting in more serious damaging consequences has been devised.

On a positive note, experimentation with IOL materials and IOL designs, in particular the inclusion of IOLs with sharp edges, has clearly provided some relief from the problem of PCO. However, as always with the march of progress, for example in the push to design IOLs that restore some accommodative function, this has led to other unwanted clinical outcomes. There is no doubt that because of the costs and commercial interests associated with these programs, this area will continue to receive much attention but will invariably throw up additional problems that will need to be countered.

As with all medical research, it is important to see that efforts aimed at alleviating PCO are being vigorously explored from all angles. In future, we can expect to see continued improvements in IOL designs and associated surgical procedures. In the longer term, it will be important to continue explorations into the possibility of regenerating a normal lens after cataract surgery. Surely this field must advance hand-in-hand with the exciting and burgeoning field of regenerative biology in general.

The editors should be congratulated for bringing together, in one volume, the leaders in all the fields relevant to PCO. The result provides a must-read text for all who wish to avail themselves of much of the latest knowledge on the behaviour of lens epithelial cells in health and disease.

John McAvoy, Ph.D.
Professor of Experimental Ophthalmology
Director of Laboratory Research
Save Sight Institute, University of Sydney
Sydney, NSW, Australia

Foreword

This is an impressive collection, notable for its thorough examination of the lens epithelium in lens development and growth and the essential contribution of the lens epithelium to posterior capsular opacification (PCO) following cataract surgery. The book guides the reader through the important functions of the lens epithelium in the intact lens and the biological mechanisms known to contribute to PCO. Integrated into this biological story is a thorough treatment of the surgical and design strategies that have been employed to reduce or prevent PCO.

The book starts by providing the developmental history of the lens progenitor cells and lens formation. This is followed by a novel assessment of the anatomy and normal functions of lens epithelial cells. This chapter has several beautiful figures illustrating the progression of lens epithelial cells to fiber cells and a cogent assessment of the events that occur during this process. Lens biology is further dissected to assess the factors that regulate lens growth and differentiation, which are obvious contributors to PCO. This section of the book is enhanced by thorough consideration of the structure and biology of the lens capsule, the substrate upon which PCO occurs, the transcription factors required for early lens development, and a relevant consideration of lens regeneration, a promising alternative to PCO. Understanding lens epithelial cell biology offers novel approaches to addressing PCO in the future.

The collection then moves on to careful consideration of the biological basis for PCO: wound healing and fibrosis. It also discusses the formation of Elschnig's pearls, which can degrade vision although representing a clear "attempt" by the lens to regenerate. Consideration is then given to the histology and frequency of PCO and the effects of PCO when occurring with intraocular lenses (IOLs) of different design. This provides a natural transition to the surgical methods that have been and are being used to reduce PCO. Included in this section are combination surgical and biological treatments (expose the posterior capsule to more aqueous humor by increasing the size of the anterior capsulorhexis or by keeping the capsular bag in a more open configuration) and strictly surgical innovations (the "bag-in-the-lens" approach and the use of capsular tensioning and bending rings to prevent the migration of lens epithelial cells into the visual axis). IOL design and materials play a large role in the incidence of PCO and, as expected, these receive comprehensive treatment. This section concludes with a forward-looking consideration of microincision IOLs and PCO and the special problems associated with pediatric cataract surgery, in which PCO is the major challenge. Not explicitly addressed are the future challenges posed by accommodating IOLs and other potential innovations. However, the in-depth treatment of the biological basis of PCO should provide valuable clues about how to address this pathology in whatever context it appears.

David C. Beebe, Ph.D.
Janet and Bernard Becker Professor of Ophthalmology and Visual Sciences
Professor of cell biology and physiology
Washington University School of Medicine
St. Louis, MO, USA

Preface

The crystalline lens, like the cornea, is one of the most transparent tissues dedicated to optical function. I, with my colleagues, have long been engaged in research on this remarkably unique tissue, in many diverse fields including embryology, physiology, pathology, and others. Disturbance of the optical nature of the crystalline lens by opacification caused by aging and/or other diseases directly impairs visual function. Treatment of the opacification of the tissue, primarily through cataract surgery, has developed into a very precise technique, along with continuous advances in medicine. Classical whole extraction of the opacified crystalline lens with an implantation of an iris-fixed intraocular lens (IOL) has long been replaced by the current modern procedure of extracapsular lens extraction, in association with a better understanding of the importance of the preservation of the lens capsule postoperatively. The design of IOLs has also progressively improved to suit this surgical procedure. After establishing the concept of phacoemulsification and aspiration of the cataractous lens, further innovation of surgical devices including the phaco machine, advanced microscopy, and improvement of biomaterials and the shape of IOLs encompass today's sophisticated cataract surgery. Modern phaco surgery has certainly achieved a dramatic recovery of patients' vision and provided an enormous contribution to healthy aging in humans.

Modern phaco cataract surgery involving IOL implantation is not yet free from complications, which still need to be addressed and overcome. Although postoperative infection or loss of accommodative vision occurs, one of the most common and to-be-prevented complications is posterior capsular opacification (PCO). PCO impairs the patient's vision and is associated with a requirement for further medical treatment, also disturbing the ability for ocular fundus examination by an ophthalmologist. Subsequent surgery or laser treatment of PCO also potentially increases the risk of rhegmatogenous retinal detachment of a "gloomy floater" and remnants of the PCO tissues.

The residual lens epithelial cells retained after extracapsular lens extraction are responsible for PCO. This cell is of ectodermal origin during embryonic development and lines the inner anterior surface of the lens capsule, the native basement membrane of the cell. The lens epithelial cell undergoes aberrant behavior post-phaco surgery. This epithelial cell can migrate posteriorly along the inner surface of the lens capsule, sometimes reaching the posterior lens pole in line with the optic portion of an IOL. The cells may then differentiate into lens-fiber-like tissues or de-differentiate into a mesodermally derived fibroblastic-like cell, accompanied by aberrant deposition of extracellular matrix, all of which may contribute to the development of PCO.

It is timely that we present this publication dedicated to our current knowledge and understanding of lens epithelial cell biology and PCO, focusing on the key issues with respect to the basic and clinical sciences. I am so lucky to have friends who are leading experts in this field to edit the chapters with me: Frank and Liliana. With their extreme energy and efforts, it is fortunate that the book can deliver 25 specialist chapters by a number of internationally distinguished investigators leading the research in the basic science aspects of lens cells behavior, as well as the clinical problems of lens cells leading to PCO. This volume is edited not only as a series of review articles on cutting-edge findings from researchers, but also as a primer of basic and clinical research on the lens epithelium and PCO for young research investigators, residents, and clinicians alike.

We would like to express our deepest appreciation to Makie Kambara and Mariko Kubota, Springer Japan, as well as our sincerest thanks to the many external reviewers who kindly spent their invaluable time reviewing the chapters.

Wakayama, Wakayama, Japan Shizuya Saika

Dr. Shizuya Saika is chairman/professor of the Department of Ophthalmology, Wakayama Medical University School of Medicine, Wakayama, Japan. Dr. Saika completed his M.D. in 1988 and his Ph.D. in 1993 from this university. He was also engaged in basic science research in the Department of Ophthalmology, University of Cincinnati School of Medicine, Ohio, USA, in 1998–1999. He has been engaged in basic science research in wound healing of lens and cornea besides his clinical work, especially of surgeries for cataract, glaucoma, and vitreoretinal disorders.

Preface

This book provides invaluable and updated information on posterior capsular opacification. It approaches the problem in a very comprehensive manner, from basic science to clinical outcomes and surgical or intraocular device-related methods currently used to prevent it. I was very pleased and honored when Shizuya invited me to join him and Frank in this project, to help develop the clinical/surgical section of the book represented by Parts II to VI.

Cataract surgical removal with intraocular lens implantation evolved into an extremely successful procedure. Yet, despite considerable advances in basic science research, as well as surgical techniques and intraocular device manufacturing, posterior capsular opacification remains the most frequent long-term postoperative complication after cataract surgery. Prevention of any form of opacification within the capsular bag has actually become a primary research goal, particularly with the development of specialized intraocular lenses. For example, accommodating lenses, which are generally designed to move within the capsular bag or have shape alterations upon efforts for accommodation, could have their functionality impaired postoperatively by cellular proliferation and fibrosis. It is only through the prevention of this complication that we will be able to enjoy the full potential of these modern devices.

I would like to thank Shizuya and Frank again for this collaborative opportunity, as well as Makie Kambara and Mariko Kubota of Springer Japan, for their hard work in bringing this project to term. I also would like to thank our international friends and colleagues who contributed with outstanding chapters and who are also all considered leaders in their respective fields. We sincerely hope that this comprehensive report will set the basis for increasing research efforts in the near future towards the complete eradication of posterior capsular opacification. To my colleagues and fellows at the John A. Moran Eye Center, as well as my family, especially my parents, Heron and Nilma, thank you so much for your constant professional and personal support!

Salt Lake City, Utah, USA Liliana Werner

Liliana Werner, M.D., Ph.D., is a tenured associate professor of ophthalmology and visual sciences and co-director of the Intermountain Ocular Research Center at the John A. Moran Eye Center, University of Utah, Salt Lake City, UT, USA. She has an M.D. degree from Brazil and a Ph.D. degree (biomaterials) from France. Dr. Werner's research is centered on the interaction between ocular tissues and different intraocular lens designs, materials, and surface modifications. These include intraocular lenses implanted after cataract surgery, and also phakic lenses for refractive surgery as well as ophthalmic implantable devices in general.

Preface

Since the turn of the last century, research on the ocular lens has flourished, with major advances in our understanding of its normal biology and pathology emerging, in particular over the last 30 years with the advent and development of powerful molecular tools. Although many candidate molecules have now been identified to play essential roles in the induction, growth, and maintenance of the ocular lens, much more needs to be understood in order to regulate many of the cellular processes that are essential for maintaining lens transparency and preventing cataract. Medical and surgical advances have also progressed with the years, and vision has been restored to many blind worldwide suffering from cataract, through what today seems to be a routine procedure of extracapsular extraction of the cataractous tissue and implantation of a prosthetic lens. This too, unfortunately, is not without its complications, one of the main being posterior capsular opacification (PCO). Much research on this secondary cataract has shown it to arise from lens epithelial cells remaining after surgery which undergo aberrant fiber cell differentiation and/or transformation into myofibroblast-like cells that form subcapsular plaques embedded in extracellular matrix. Although procedures have also been developed to deal with this secondary complication, much research has focused on its prevention. It is in this context of better understanding PCO and its cellular origins that this text was devoted specifically to the lens epithelium and PCO.

This text has allowed us to bring together, for the first time, an international panel of well-known lens cell biologists and clinicians currently working on research specifically related to lens induction, development, differentiation, growth, maintenance, pathology, and surgical practices. The first chapters provide a current view of basic research on the lens epithelium, serving as a foundation for the more clinical-based chapters that comprise the remainder of the text. These early chapters, where relevant, include discussions on lens pathology, such as processes leading to cataract and even PCO, to complement the basic lens research. From the establishment of the embryonic lens epithelium, its detailed morphology and association with its overlying basement membrane, the lens capsule, we cover the regulation of lens epithelial cell proliferation and their differentiation into fiber cells. From this we explore many of the key regulatory transcription factors implicated in congenital cataract before moving on to lens regeneration, a process unique to some lower vertebrates but with potential application to humans in the years to come. We conclude the basic science chapters highlighting different models investigating the molecules and fibrotic changes to lens epithelia leading to cataract as well as fibrosis associated with lens wound healing. The remainder of the text is dedicated to clinical outcomes of PCO, with a heavy focus on the different modes of its prevention. By showcasing these clinical issues we hope to better educate and bring together our basic lens researchers who have the knowledge and expertise to potentially address some of the problems preventing the progression of this specialist field of research.

This text is unique in its unsurpassed depth of information specifically focused on the lens epithelium and PCO, written by many of the very people who lead the field. It is an invaluable resource for those interested in lens biology and pathology, providing an entry point into the primary research literature. I am privileged to have taken part in this endeavor initiated by Professor Saika, and we hope that graduate students, residents, postdoctoral fellows, principal investigators, and clinicians alike will all enjoy reading and using this book as much as we have enjoyed editing it.

On a final note, I would like to sincerely thank all of the contributing authors who made this work possible, as well as the many other lens researchers who have kindly offered feedback revising the chapters. Last but not least, it has been a great honor working with Professors Saika and Werner. Their professionalism and work ethic have made this exercise a most rewarding and pleasurable experience. To my three sons, Christopher, Alexander, and Matthew Lovicu, thank you for always keeping me grounded and for giving me the time to work on this project.

Sydney, New South Wales, Australia Frank J. Lovicu

Dr. Frank J. Lovicu is a professor in the School of Medical Sciences, in the Discipline of Anatomy and Histology, Bosch Institute and Save Sight Institute, at the University of Sydney, NSW, Australia. He currently heads the Lens Research Laboratory in Anatomy and Histology, where he primarily studies the role of growth factor signaling in regulating lens cell behavior. His research promotes the understanding of the cellular and molecular processes leading to normal lens development and growth, as well as those leading to cataract. His research findings have the potential for development of new strategies to treat and/or prevent cataract.

Contents

Part I

Lens Epithelial Cell Biology

From Zygote to Lens: Emergence of the Lens Epithelium

1

Michael L. Robinson

Abstract

At the dawn of the twentieth century, Hans Spemann first discovered embryonic induction when he demonstrated that lenses failed to form in the frog, *Rana fusca*, following the destruction of the optic vesicle. Since that time, lens induction remains at the forefront of investigations of embryonic development. However, unlike in the early days of vertebrate experimental embryology, advanced genetic tools and techniques make it possible to paint a detailed molecular picture of how the lens comes into being. It is this picture, although not complete, that emerges in the following pages. Drawing from investigations using a number of different model systems and experimental approaches, this work traces the journey from fertilized zygote to the differentiation of the lens epithelium. These transitions from zygote to inner cell mass to epiblast to ectoderm to neural plate border to preplacodal ectoderm to lens placode to lens vesicle to lens epithelium occur through sequential paracrine signals that induce specific transcription factors leading to the production of proteins characterizing each stage. Understanding the paracrine signals that take place during normal embryonic development informs much of the current revolution fueled by the conversion of embryonic stem cells or induced pluripotent stem cells into differentiated cell types in vitro.

Keywords

Lens induction • Lens placode • Neural plate border • Preplacodal ectoderm

M.L. Robinson (✉)
Department of Biology, Miami University, Oxford, OH, USA
e-mail: Robinsm5@miamioh.edu

S. Saika et al. (eds.), *Lens Epithelium and Posterior Capsular Opacification*,
DOI 10.1007/978-4-431-54300-8_1, © Springer Japan 2014

1.1 In the Beginning

As all animal life begins with the union of sperm and egg, this seems an appropriate, but perhaps surprising, place to start describing the formation of the lens epithelium. Developmental and stem cell biologists stand at the precipice of fundamental understanding of how uncommitted cells receive and interpret instructions to differentiate into virtually any cell type. The lens, in fact, provided one of the first compelling examples of this with the creation of human lens progenitor cells from embryonic stem cells in 2010 [1]. The accomplishment of this feat entailed the sequential exposure of pluripotent cells to paracrine factors to mimic the sequence of events that lens progenitor cells normally experience during embryonic development. These paracrine factors likely initiate the epigenetic alterations in chromatin structure that ultimately direct cell fate decisions. Understanding the molecular control of cell type-specific chromatin structure remains the holy grail of developmental genetics.

The union of sperm and egg to form the totipotent (able to form all embryonic and extraembryonic cell types) zygote initiates the epic journey of development that, in the case of humans, results in the formation of over 200 distinct cell types. Much of the information currently known about early cell fate decisions in mammals comes from studies of mouse embryos. Although there are subtle differences in early mouse versus human development [2], these are far outweighed by the similarities, and given the difficulty of obtaining and observing early human embryos, the mouse makes a suitable human proxy. Following fertilization, the zygote initiates a number of cleavages that subdivide the early embryo into multiple cells without increasing the overall size of the embryo. In contrast to early cleavage-stage embryos in most animals, which rely on proteins translated from maternal mRNA transcripts, mammals initiate transcription of the embryonic genome (derived from the union of sperm and egg) very early in development. Active zygotic transcription initiates prior to the 2-cell stage in mice and between the 4- and 8-cell stages in human embryos (reviewed in [3]).

The lens epithelium, as well as all differentiated cell types, derives from the totipotent zygote, but the precise point at which cell differentiation begins in mammals remains a controversial topic. In many animals, differentiation begins immediately upon the initiation of cleavage, with divergent developmental fates imparted on each of the first two blastomeres. Oftentimes, asymmetric distribution of cytoplasmic components within the egg, or the point of sperm entry, dictates polarized development of nonmammalian cleavage-stage embryos. In contrast, the mouse oocyte lacks any obvious molecular polarity [4]. However, several studies suggested intrinsic polarity in the early mouse embryo, particularly in determining the embryonic–abembryonic axis (division of the blastocyst between embryo forming and extraembryonic tissue forming regions). One of these observed that the second polar body specifically localized to the embryonic–abembryonic boundary of the future blastocyst [5], and another suggested that the sperm entry point always coincided with this boundary in mice [6]. Yet, other investigations found no bias in the developmental fate of the first two mouse blastomeres [7]. Recently, sophisticated fate mapping experiments demonstrated that a proportion of mouse embryos exhibit nonrandom bias of 4-cell stage blastomeres with respect to contribution to embryonic or extraembryonic portions of the blastocyst [8].

Although there may be factors that bias the development of particular early blastomeres in the early mammalian embryo toward one developmental fate versus another, the first definitive differentiation event in mouse takes place subsequent to the third cleavage when the embryo, now known as a morula, consists of eight blastomeres. At this point, cells first develop apical–basal polarity and the embryo undergoes a process known as compaction in which the cluster of eight blastomeres go from resembling a cluster of marbles to a sphere with no obvious physical distinction between cells. Uvomorulin (also known as E-cadherin or CDH1), originally identified as playing an essential role in this compaction process [9], later distinguishes lens epithelium from differentiated lens fiber cells. Up to the compaction stage, all of the early mammalian blastomeres maintain totipotency. Compaction marks an end to this totipotent stage and sets the stage for the first differentiation event on the road to lens formation.

In the compacted mammalian embryo, some cells reside on the inside of the cell mass, being totally surrounded by other cells, and the remaining cells have a cell membrane exposed to the outer surface of the morula. Prior to compaction, all of the blastomeres co-express the transcription factor genes *Oct4* and *Cdx2*, but by the time the morula becomes a blastocyst, *Oct4* and *Cdx2* exhibit mutually exclusive expression in the inner cell mass and trophectoderm, respectively [10–12]. This distinction between cells destined for trophoblast and inner cell mass fates may be mediated by the Hippo (HPO) pathway [13, 14]. The cells lying interior, within the compacted morula, exhibit active HPO signaling. HPO signaling activates the Lats kinase which phosphorylates and thereby destabilizes the YAP protein and prevents YAP from entering the nucleus. However, in cells surrounding the inside morula cells, lower levels of HPO signaling lead to the nuclear accumulation of YAP where it acts as a cofactor for the TEAD4 transcription factor. Transcriptional upregulation of *Cdx2* requires theTEAD4/YAP complex and results in both the inhibition of *Oct4* expression and the transcriptional activation of trophoblast specifying genes including *Eomes*, *Psx1*, and *Hand1*. Likewise, *Oct4* expression in the inner cells suppresses *Cdx2* expression and activates the expression of *Sox2* and *Nanog* which, together with OCT4, maintain the pluripotency of the inner cell mass cells. The trophoblast cells go on to form most of the placenta but do not contribute to any tissues of the embryo proper.

The *Oct4* expressing inner cell mass (ICM) of the developing blastocyst provides the stem cells for the entire embryo, including the lens. These ICM cells subdivide into the primitive endoderm and epiblast prior to the implantation of the blastocyst into the uterus. FGF signaling within the inner cell mass pushes cells toward primitive endoderm fate [15], leading to the segregation of a layer of *Gata6* expressing primitive endoderm cells separating the epiblast cells from the blastocoel cavity of the blastocyst (see Fig. 1.1). The primitive endoderm cells will differentiate into the parietal and visceral endoderm of the yolk sac while the *Nanog*-expressing epiblast continues on the path leading to lens and other embryonic cell types.

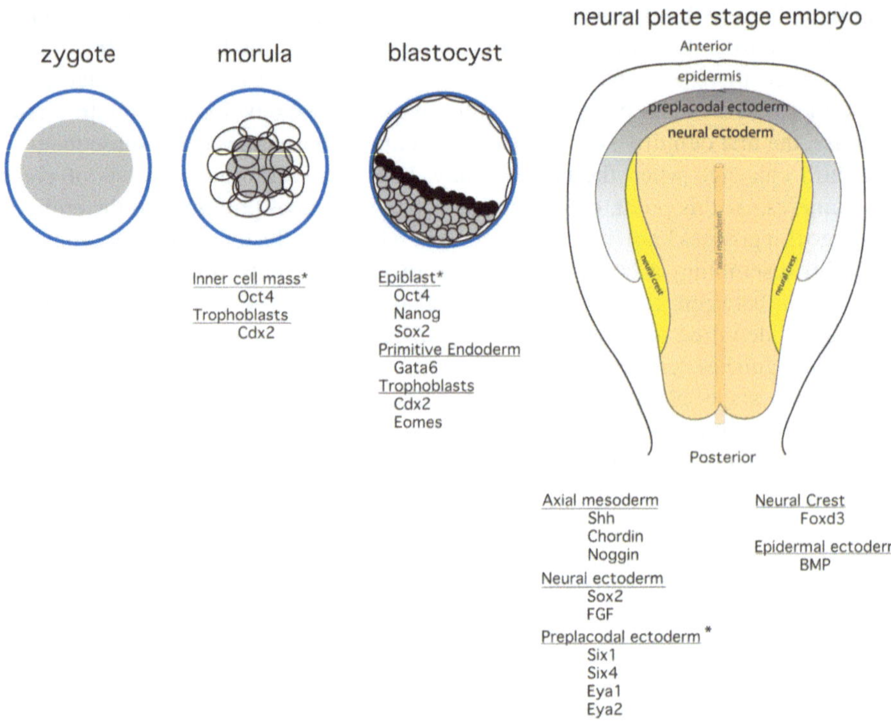

Fig. 1.1 Journey of the lens-forming ectoderm from zygote to neural plate. The zygote is pluripotent, capable of generating any cell type including lens. The first differentiation decision takes place in the morula when the inner cell mass cells (*gray*), from which the lens originates, are distinguished from the trophoblast lineage (*colorless*). At the blastocyst stage, the inner cell mass has differentiated into the primitive endoderm (*black*) and the epiblast (*gray*) which contains the future lens cells. After implantation, the embryo undergoes gastrulation. The axial mesoderm, underlying the neural ectoderm, secretes BMP inhibitors including CHORDIN, FOLLISTATIN, and NOGGIN that allows the overlying ectoderm to adopt a neural fate (*beige*). The neural crest cells (*yellow*) and the preplacodal ectoderm (*gray*) both receive an intermediate level of BMP signaling, but these cells are distinguished by Six1/4 and Eya1/2 expression in the preplacodal ectoderm and Foxd3 expression in the neural crest cells. The tissue expressing high levels of BMP becomes the epidermal ectoderm. The preplacodal ectoderm is later subdivided into the anterior preplacodal ectoderm (*darker gray*), from which the lens is derived, and the posterior preplacodal ectoderm (*lighter gray*). Underneath each diagram are listed genes or gene products distinguishing each tissue. Tissues from which the lens develops are indicated by *astricies*

After the implantation of the blastocyst into the uterus, the epiblast segregates into the amniotic ectoderm and the embryonic epiblast. The embryonic epiblast undergoes gastrulation to form the three primary germ layers of the embryo proper: ectoderm, mesoderm, and endoderm. The lens precursors derive from the embryonic ectoderm, that portion of the embryonic epiblast that avoids going through the primitive streak during gastrulation. The embryonic ectoderm that forms the lens originates in the head-forming region of the epiblast. Defining the head requires interaction of the embryonic epiblast with the underlying anterior visceral endoderm (AVE) derived from the primitive endoderm. The AVE defines the anterior of the embryo by secreting CERBERUS and LEFTY-1 which protects the overlying epiblast by antagonizing the WNT3a and NODAL paracrine signals emanating from the node at the posterior side of the embryo. Thus, initially the lens-forming ectoderm avoids WNT and NODAL patterning from the node that would otherwise activate mesodermal fate in this tissue. The foregut endoderm and anterior-most axial chordamesoderm invade the midline space between the epiblast and AVE where they secrete BMP antagonists including CHORDIN, NOGGIN, and FOLLISTATIN. The paracrine signals from the AVE, the foregut endoderm, and the axial mesoderm suppress BMP, NODAL, and WNT signaling in the epiblast fated to form the neural ectoderm.

1.2 Subdividing the Ectoderm: Formation of the Neural Plate Border

As the neural plate forms, the cells that ultimately give rise to the lens reside in a region of ectoderm between the neural and presumptive epidermal ectoderm known as the neural plate border (NPB) region. Although suppression of BMP signaling is required for neural ectoderm specification, the NPB, where the precursors of both the neural crest (NC) and the preplacodal ectoderm (PPE) reside, requires a low to intermediate level of BMP signaling. Active research continues to define the signals and transcription factors that pattern the NC and PPE, and the emerging story includes complications coming from different types of experiments and interspecies variations in the signals and transcription factors specifying these tissues.

In all vertebrates, induction of the PPE requires FGF signaling as well as attenuation of both WNT and BMP signaling [16–18]. The PPE, while initially indistinct, eventually separates into discrete regions where the ectoderm thickens and develops into specific sensory placodes that differ both in the transcription factors they express and their ultimate anatomical fate. The lens, olfactory, and anterior pituitary (adenohypophysis) placodes arise from the anterior PPE, while trigeminal, epibranchial, and otic placodes form from the posterior PPE. Neurons ultimately derive from all but the lens and adenohypophyseal placodes. In contrast, the NC develops from ectoderm medial and posterior to the PPE, ultimately giving rise to a wide variety of cell types including peripheral neurons and glia, smooth muscle cells, melanocytes, cells of the adrenal medulla, cardiac cells, and the cartilage and bones of the face.

Since both PPE and NC come from the NPB region, these cells may initially share a common potential to develop into either cell type, with subsequent divergence based on differences in paracrine signaling gradients. The finding that neural plate grafts into epidermal ectoderm induce the expression of the definitive placodal marker *Six1* on either side of the graft supports this model [17]. Evidence from chick explants suggests that the cells fated to become the NBP are specified as NC cells as early as the late blastula stage [19], but no cells of sensory placodal identity differentiate from these blastula cells cultured in isolation. An alternative model from *Xenopus* studies suggests that cells of the NBP consist of two distinct populations [16, 20]. One of these derives from the neural ectoderm and gives rise to the NC while the PPE originates exclusively from the non-neural ectoderm.

Although neural ectoderm requires a complete blockage of BMP signaling for early specification, BMP plays a complex role in patterning the remaining ectoderm. Complete blockage of BMP signals by 200 or 100 μM dorsomorphin in the zebrafish late blastula/early gastrula resulted in dorsalization with the complete loss of non-neural ectoderm, but 50 μM dorsomorphin caused a ventral expansion of NC ectoderm and loss of both PPE and epidermal ectoderm [21]. However, treatment of zebrafish embryos with 100 μM dorsomorphin 3 h later (7 h postfertilization) failed to prevent either NC or PPE formation, and in contrast to the ventral expansion of the NC with an early exposure to 50 μM dorsomorphin, no concentration of dorsomorphin expanded PPE at the expense of epidermal ectoderm. The PPE always formed in a domain stripe between the NC and the epidermal ectoderm [21]. In these experiments, the expression of the transcription factor genes, *AP2α*, *Foxi1*, *Gata3*, and *Dlx3*, required BMP signaling at an early stage, but became BMP independent shortly after their expression initiated. In zebrafish, both NC and PPE development require AP2α and DLX3, but only the PPE depends on FOXI1 and GATA3 for subsequent development. AP2α, FOXI1, and GATA3 work together in zebrafish to provide PPE competence, as placodal derivatives form in morpholino-mediated knockdown of any one of these transcription factors. Once the expression of *AP2α*, *Foxi1*, and *Gata3* initiates, these factors form a cross-regulatory network that maintains their expression in the absence of further BMP signaling [22]. In fact, the ectopic expression, in zebrafish, of any two of these transcription factors activated the expression of *Foxd3* and *Six4.1*, definitive markers of NC and PPE, respectively, in the absence of any BMP signaling [22]. In mice, the requirement of FOXI1 and GATA3 in global placodal development remains less clear, although mice lacking *Foxi1* demonstrate specific functional deficiencies in derivatives of the otic placode [23]. Likewise, null mutations of *Gata3* cause deafness in heterozygous mice, and homozygotes fail to develop the ear beyond the otocyst [24].

Whether or not the NPB initially consists of two distinct populations of cells, this region requires initial BMP signaling and expresses AP2α, FOXI, GATA, and DLX transcription factors, and these factors cooperate to initiate expression of *Six1* and *Eya1/2*. Precisely which *Foxi*, *Gata*, and *Dlx* genes are expressed in the NPB depends on the species. For example, chick *Foxi3* replaces zebrafish *Foxi1*, and while both zebrafish and *Xenopus* express *Dlx3* in the PPE, chick and mouse express *Dlx5*. In contrast to the transcription factors that confer PPE competence but display wide expression within the NPB region, only the PPE expresses *Eya1/2* and *Six1/4*, making the expression of these genes definitive for placodal precursors (reviewed in [25]).

1.3 Defining the Preplacodal Ectoderm

The *Drosophila* genome contains a single *eyes absent* (*Eya*) gene, but vertebrates possess four homologous *Eya* genes (*Eya1–4*). All animal *Eya* genes encode proteins with several distinct biochemical functions (reviewed in [26]). These include an N-terminal Pro/Ser/Thr-rich domain containing threonine phosphatase and transcriptional transactivation activity, and a highly conserved C-terminal EYA domain containing tyrosine phosphatase activity and a protein–protein interaction motif. To date, the variant histone, H2AX, represents the only validated substrate for EYA tyrosine phosphatase activity [27, 28], and substrates for EYA threonine phosphatases remain unrevealed. EYA proteins possess no DNA-binding affinity. Therefore, the potential of the transactivation domain depends entirely upon the interaction with another protein with an intrinsic DNA-binding domain. Both the homeodomain-containing SIX family and SOX2 transcription factors interact with EYA proteins and permit EYA-mediated transactivation of their respective DNA-binding target genes [29–32]. Interestingly, the transactivation activity of EYA proteins depends on intact EYA tyrosine phosphatase activity [33]. Of the vertebrate *Eya* genes, *Eya1* and *Eya2* specifically play roles in cranial placode development [34–38].

The Six family of transcription factors derive their name from the *si̲ne oculis* homeobo̲x gene of *Drosophila*. The *Drosophila* genome encodes three distinct SIX proteins: SINE OCULIS, OPTIX, and DSIX4. Likewise, vertebrate *Six* genes fall into three subfamilies based on similarities to these three Drosophila genes (reviewed in [39]). As their name suggests, all *Six* genes encode a DNA-binding homeodomain and additionally a SIX domain that mediates protein–protein interactions. Although SIX proteins may possess some intrinsic transcriptional activation activity [40], their main effect on transcription depends on interactions with transcriptional cofactors. The SIX domain of these transcription factors associates with members of the GROUCHO/TLE [30, 41–43] or DACHSHUND [33, 44] families of proteins to repress transcription or with members of the EYA protein family to activate transcription. In the PPE, SIX1 and SIX4 members of the sine oculis and DSIX4 subfamilies, respectively, likely regulate cell proliferation and survival. SIX3, a member of the OPTIX subfamily, expression initiates later during the specification of the lens placode.

The expression of *Six1* and *Eya1/2* defines the PPE. In the neighboring NC cells, AP2α and DLX transcription factors activate *Foxd3* expression. FOXD3, in turn, represses the expression of *Six1* and *Eya1/2*. In the neural tissue, SOX2 represses the expression of both *Six1* and *Eya1/2*. In the PPE SIX1, in conjunction with GROUCHO, may repress the expression of genes promoting neural or NC fates. Simultaneously, SIX1, associated with EYA1 or EYA2, may activate the transcription of genes required for placode development (see Fig. 1.1).

1.4 Distinguishing the Lens Placode

With the PPE specified, individual placodes must adopt their independent fate. The lens placode arises from the anterior PPE, along with the olfactory and adenohypophysis placodes, and some evidence suggests that the entire PPE adopts an initial anterior PPE character with lens formation being the default state for all placodes. Support for this notion comes from explant culture of isolated regions of chick PPE where both anterior and posterior regions autonomously expressed genes characteristic of lens placode including *Pax6*, *L-maf*, *δ-crystallin*, and *FoxC1* [45]. During gastrulation the anterior–posterior patterning of the neural ectoderm is accomplished, in part, by a gradient of WNT signaling. Secretion of Wnt antagonists by the anterior visceral endoderm, pharyngeal endoderm, and prechordal plate mesoderm results in a posterior to anterior gradient of WNT signaling. This gradient of WNT signals also distinguishes anterior PPE from posterior PPE, with the anterior region expressing *Pax6*, *Six3/6*, *Otx1/2*, and *Pitx1/2c/3* transcription factors and the posterior region expressing *Pax3/2/8*, *Irx1/2/3*, *Gbx2*, and *Msx1/2* transcription factors (reviewed in [46]).

Recent experiments in chick suggest that *Pax6* expression in the anterior PPE depends on somatostatin (SST) secreted from the underlying anterior mesendoderm during gastrulation [47]. The anterior PPE specifically expresses the somatostatin receptor, SSTR5, and ablation of the anterior mesendoderm, chemical antagonists of SST, and morpholinos to SSTR5 all reduce or eliminate the expression of *Pax6* in the anterior PPE. In addition to PAX6, SST signaling also initiates the expression of *nociceptin* in the anterior PPE and NOCICEPTIN (also known as ORPHANIN FQ) cooperates with SST to induce *Pax6* in lens and olfactory precursors in both chick and zebrafish [47]. These studies also demonstrated that signals from the posterior mesoderm including, but not limited to FGF, repress *nociceptin* and *Pax6* expression in the posterior PPE.

Distinguishing individual placodal identity within the anterior PPE also depends on local paracrine signaling. Hedgehog signaling from the axial mesoderm specifies the development of the adenohypophyseal placode in the medial anterior PPE. The cells destined to form the lens and olfactory placodes, though indistinguishable in the gastrula, begin the process of separation during neurulation. At the neural fold stage the lens and olfactory precursors commonly express *Pax6*, *Dlx5*, *Six1*, and *Sox2* transcription factor genes, but as these precursors separate, the lens placode cells maintain *Pax6* expression while turning off *Dlx5*, while the olfactory placode cells do the opposite. This specification of visual from chemosensory precursors recalls the similar situation in *Drosophila*, where the common eye/antennal imaginal disc diverges into *Eyeless* (the *Drosophila Pax6* homolog)-expressing eye precursors and *Distalless* (the Drosophila *Dlx* homolog)-expressing antenna precursors [48].

Although both olfactory and lens placode precursors require both BMP and FGF signaling prior to placode formation, the olfactory precursors receive more FGF signaling from the anterior neural ridge and this increased FGF signaling may distinguish olfactory from lens precursors. In fact, the induction of FOXG1, a transcription factor essential for olfactory development [49], requires FGF8 signaling from the anterior neural ridge [45]. However, exogenous FGF8 fails to induce the expression of olfactory-specific genes in lens precursors from neural stage embryos [45, 50]. Other studies suggest that increased BMP signaling in lens precursors distinguish these cells from those giving rise to olfactory fate [50]. In a recent review, Gunhaga suggests a model where FGFs and BMPs antagonistically pattern the neural plate border where FGFs from the neural tissue prevent this region from becoming epidermis while BMPs from the epidermal ectoderm specify lens/olfactory progenitor cells in the anterior PPE (reviewed in [51]).

In classical *Xenopus* embryonic transplantation experiments several different regions of head surface ectoderm retained competence for lens formation in response to optic vesicle induction during early development [52, 53]. These observations, in addition to experiments suggesting the default lens-forming fate of the PPE [45], imply that lens-forming potential in the ectoderm requires active restriction to ensure the correct placement of a single lens on each side of the head. At least some of the lens-forming restriction results from sonic hedgehog (SHH) signaling from the axial mesoderm. In fact, mutation of either *Shh* or the SHH effector *Gli2* results in the formation of a third midline lens in zebrafish that replaces the adenohypophysis [54]. Likewise, increased SHH signaling provides the mechanism for lens and eye degeneration in the blind cavefish, *Astyanax mexicanus*, which results in expanded expression of *Pax2* at the expense of *Pax6* [55].

Canonical WNT signaling also appears to restrict lens-forming potential in the head surface ectoderm. Although WNTs may induce several distinct signal transduction cascades, canonical WNT signaling leads to the stabilization of β-catenin, a molecule that functions both in the formation of adherens junctions between epithelial cells and as a transcriptional co-activator in combination with TCF/LEF transcription factors. The initial canonical WNT signaling complex consists of a WNT ligand, a FRIZZLED receptor, and an LDL-related coreceptor (either LRP5 or LRP6). This WNT/FRIZZLED/LRP5/6 complex activates DISHEVELLED (DVL) which subsequently inactivates the β-CATENIN phosphorylation/destruction complex. The genetic removal of *β-catenin* in lens cells disrupts lens morphogenesis [56–58], but at least some of this disruption results from the adhesion functions of β-CATENIN rather than its role as a transcriptional effector.

Indeed, abundant evidence suggests that lens induction requires a canonical WNT-free zone. A number of different canonical WNT signaling reporter strains fail to detect evidence for β-CATENIN-mediated transcriptional activity in the lens placode [56, 57, 59, 60–62]. In contrast, canonical WNT signaling characterizes both the ocular mesenchyme, consisting of a mixture of head mesoderm and neural crest cells, and the ectoderm surrounding the lens placode (particularly after invagination of the lens pit). Indeed, removal of *β-catenin* from the *Pax6* expressing surface ectoderm resulted in ectopic patches of lens formation, particularly in the ectoderm between the eye and nose [56, 57]. Furthermore, stabilization of β-CATENIN within the lens placode disrupts lens formation resulting in decreased expression of both *AP2α* and *Pax6* and increased expression of markers of neural ectoderm including *β-tubulin* [56]. The apposition of the growing optic vesicle with the overlying presumptive lens ectoderm excludes ocular mesenchyme which underlies the surrounding surface ectoderm. This contact between the optic vesicle and surface ectoderm contributes to the WNT-free zone that permits the upregulation of *Pax6* required for the final stage of lens induction.

Although the expression of *Pax6* occurs throughout the anterior PPE, *Pax6* expression dramatically increases specifically during formation of the lens placode. This two-phase expression of *Pax6* (*$Pax6^{preplacode}$* and *$Pax6^{placode}$*) plays an important role in the development of the lens (reviewed in [63]). Specifically, the increase in *Pax6* expression during lens placode formation marks definitive lens induction. As discussed above, the lens-forming ectoderm receives numerous inductive signals before lens placode formation that bias the ectoderm toward lens fate. While lens development requires these previous inductive signals, controversy continues as to whether signaling from the optic vesicle plays an essential role in lens induction. Much of this debate comes from the formation of lenses in the absence of retinal tissue, primarily in amphibians where the optic vesicle had been ablated. Lens formation in the absence of retina can also be induced in the anterior pituitary [54] or in the otic placode [64] by interfering with SHH signaling or ectopically expressing *Six3*, respectively. However, at least in mammals, definitive lens induction requires the optic vesicle.

Perhaps the strongest evidence for the requirement of the optic vesicle for mammalian lens induction comes from studies of mice and humans with mutations in the retinal homeobox gene *Rax* (formally known as *Rx*). In the absence of RAX, the optic vesicle fails to approach the presumptive lens ectoderm; neither morphological retinal development nor retina-specific gene expression occurs [65, 66]. Likewise, lens formation fails with RAX deficiency in both mice and humans [67]. The optic vesicle secretes BMP4 and lens placode formation fails in BMP4 null mouse embryos [68]. In RAX-deficient mouse embryos, while lens development fails, the development of the eyelids, conjunctiva, and lacrimal gland proceeds rather normally [69]. Although *$Pax6^{preplacode}$* expression remains unaltered in RAX mutant mice, the upregulation of *Pax6* (*$Pax6^{placode}$*) fails to occur in the absence of the optic vesicle. Therefore, $\text{Pax6}^{\text{preplacode}}$ expression suffices for the development of auxiliary eye structures (eyelids, conjunctiva, and lacrimal gland), but lens development requires higher levels of *Pax6* expression. Notably, deletion of *β-catenin* induced the upregulation of *Pax6* as well as the expression of definitive lens markers, *Foxe3* and α-CRYSTALLIN, in the presumptive lens ectoderm of RAX-deficient mice [69]. Consequently, in addition to providing paracrine signals supporting lens induction, the optic vesicle participates in preventing canonical WNT signaling in the lens-forming ectoderm.

The extracellular matrix between the presumptive lens surface ectoderm and the optic vesicle represents a likely mechanism by which lens placode formation initiates. Studies in the chick observed the deposition of extracellular matrix between the surface ectoderm and optic vesicle [70, 71] that preceded an increase in cellular density as placode formation progressed [72, 73]. These observations led to the hypothesis that the extracellular matrix between the surface ectoderm and optic vesicle restricted the lateral movement of ectodermal cells, resulting in a thickening of the ectoderm as cell proliferation continued in the area of contact [72]. A genetic mechanism by which this "restricted expansion hypothesis" of lens placode formation could occur was recently tested in mouse embryos [74]. Mice missing *Pax6* from the surface ectoderm fail to induce a lens placode [75, 76] and, in contrast to wild-type mice, the cell density of the *Pax6*-deficient surface ectoderm fails to increase relative to the surrounding surface ectoderm and spreads over the optic vesicle surface during development [74]. Microarray analysis revealed that several extracellular matrix molecules, including FIBRONECTIN, exhibit reduced expression in the *Pax6*-deficient surface ectoderm. Furthermore, depletion of *Fibronectin* expression from mouse embryos at E8.5 blocked lens placode formation and lens pit invagination despite the normal localization of F-ACTIN to the apical ends of the surface ectoderm cells overlying the optic vesicle [74]. These observations support the "restricted expansion hypothesis" and suggest that the invagination of the lens placode depends, at least in part, to the extracellular matrix that initially cements the presumptive lens ectoderm to the optic vesicle.

There is evidence supporting a continuous role for BMP signaling in the formation and invagination of the lens placode. As discussed previously, the optic vesicle secretes BMP4 and the presumptive lens ectoderm secretes both BMP4 and BMP7 [68, 77, 78]. The role of BMP4 in lens development may not be direct as exogenous BMP4 could not rescue lens development in the BMP4 null surface ectoderm in the absence of the optic vesicle [68]. Loss of BMP7 in mouse embryos results in a range of ocular phenotypes including anophthalmia, and in the most severely affected mouse embryos, *Pax6* expression disappears from the lens placode [77–79].

In 2009, the Beebe laboratory published the results of conditionally deleting type I BMP receptors, *Bmpr1a* and *Acvr1*, as well as BMP signaling SMAD proteins *Smad1*, *Smad5*, and *Smad4* [80]. These studies utilized the *Le-Cre* transgenic mouse line which expresses CRE recombinase in the ocular surface ectoderm at E9.0, just prior to the thickening of the lens placode, to specifically delete loxP-flanked alleles from the surface ectoderm-derived eye structures [75]. Although lens formation proceeded in the absence of either *Bmpr1a* or *Acvr1*, lens placodes from BMPR1a-deficient ectoderm exhibited a more than twofold increase in apoptosis while lens placodes from ACVR1-deficient ectoderm exhibited a significant decrease in the fraction of cells in S-phase of the cell cycle. Conditional deletion of both *Bmpr1a* and *Acvr1* in the lens-forming surface ectoderm reduced lens placode thickening and prevented lens placode invagination [80]. In addition, these double BMP receptor knockout lens placodes failed to induce the expression of αA-CRYSTALLIN or FOXE3 and exhibited reduced expression of SOX2. However, the double BMP receptor knockout ectoderm exhibited the normal upregulation of PAX6 expression that accompanies lens placode formation. Interestingly, only the full expression of SOX2 depended on BMP signaling through the common SMAD4 in these studies suggesting that BMP utilizes noncanonical, SMAD-independent signaling to influence lens formation [80]. The failure of lens placode invagination in *Bmpr1a*/*Acvr1* double knockout mice likely results from defective cytoskeletal reorganization [80].

Cytoskeletal rearrangements play an essential role in the invagination of the lens placode to form the lens pit, which subsequently pinches off from the overlying surface ectoderm to form the hollow lens vesicle. Apical constriction represents a major mechanism by which epithelial cells undergo morphogenetic movements that create tubes or pits from initially planar sheets. Apical constriction of lens placode cells requires SHROOM3, an actin-binding protein induced by PAX6 [81]. The apical localization of both F-ACTIN and MYOSIN II, mediators of the contractile apparatus leading to apical constriction, in the lens placode cells requires *Shroom3* expression. Actin cytoskeletal reorganization in cells depends on signaling through small Rho GTPase molecules including RHOA, RAC1, and CDC42. In fact, each of these molecules participate in various aspects of early lens morphogenesis. RAC1 drives lens placode cell elongation while RHOA mediates apical constriction, and the mutual antagonism of these activities ensures the proper shape of the lens pit [82]. SHROOM3-mediated apical constriction in the lens placode also requires RHOA, ROCK, and TRIO, a RHOA guanine nucleotide exchange factor [83]. However F-ACTIN-rich filopodia that connect the lens placode surface ectoderm to the underlying optic vesicle require CDC42 activity, and disruption of these filopodia reduces lens pit invagination [84].

In mice, different experimental approaches led to different conclusions as to the role of FGF receptor (FGFR) signaling in lens placode induction. In one set of experiments, E8.5 mouse embryo heads were bisected and cultured for 24 h with and without the presence of SU5957, a pharmacological inhibitor of FGFR activity. The SU5957-treated heads specifically expressed reduced levels of PAX6 protein and displayed reduced expression of a *Pax6* reporter gene, suggesting the FGFR activity participated in the increased expression of PAX6 associated with lens placode induction [85]. In other experiments where *Le-Cre* mice were used to delete *Fgfr1* and *Fgfr2* specifically from the lens placode-forming ectoderm, Pax6 levels in the FGFR-deficient lens placode remained normal at E9.5 but declined by E10.5. In this case, the FGFR-deficient lens placodes displayed normal proliferative activity but demonstrated massive increases in apoptosis, and in most cases very little lens tissue remained beyond E10.5 [86]. The deletion of *Frs2α*, a key player in FGFR signal transduction, in the lens placode also increased lens cell apoptosis without affecting the expression of Pax6 during later lens development, although the phenotype of deleting *Frs2α* is much less severe than deleting *Fgfr1* and *Fgfr2* [87]. In fact, recent reports question the relative importance of FRS2α in the actions of FGFR signaling with respect to lens development [88]. In any case, Pax6 expression continued in lenses upon deletion of *Fgfr1*, *Fgfr2*, and *Fgfr3* in the lens vesicle. Here too, removal of these three *Fgfr* genes in the lens cells dramatically increased apoptosis [89]. The requirement of FGFR activity for the survival of lens placode-derived cells remains a consistent theme through these analyses of genetically engineered mice, but direct involvement of FGFRs in lens placode formation remains controversial.

Interestingly, mutations in *Ndst1*, a gene encoding an enzyme required for the sulfation of heparin sulfate proteoglycans, caused variable, but severe deficits in early lens formation that sometimes included a reduction in the lens placodal upregulation of *Pax6* [90]. In particular, heparin sulfate proteoglycans mediate FGFR signaling, although they may function in BMP and WNT signaling as well. However, the lens defects observed in *Ndst1* mutant mice appeared to specifically associate with decreased FGF/FGFR signaling [90]. Specifically, the lens placodes and lens pits of *Ndst1* mutants exhibited a marked reduction in the phosphorylation of ERK1/2, a downstream effector of FGFR signaling, and a reduced expression of ETV5 (ERM), a transcription factor induced by FGFR signaling [90].

ERK1/2 phosphorylation likewise exhibits a complicated relationship with early lens formation. In fact, evidence exists that ERK1/2 phosphorylation must be downregulated during lens induction. Overactive ERK1/2 signaling provides an explanation behind variable aphakia seen in mice lacking *Nf1*, a gene encoding a GTPase-activating protein that counteracts RAS activation [91]. In the absence of NF1, an initial burst of ERK1/2 phosphorylation upregulates the expression of the RAS–ERK signaling antagonist SPRY2 which paradoxically leads to reduced ERK1/2 phosphorylation after lens induction that results in early lens degeneration [91]. Obviously, precise titration of ERK1/2 activation plays an important role in lens placode induction and subsequent stages of lens development.

The requirement for NOTCH signaling in lens placode induction may be species-specific. *Foxe3* expression follows the lens placodal upregulation of *Pax6* in vertebrates and therefore marks definitive lens placode induction. A 462 bp noncoding region, about 6 Kb upstream of the *Foxe3* gene in *Xenopus*, exhibits significant conservation with chicken, mouse, and human genomic sequences upstream of *Foxe3* [92]. An Rbpj (also known as Su(H))-binding motif (for NOTCH signaling) lies within this conserved noncoding region, and functional analyses in Xenopus, utilizing a GFP reporter, demonstrated that lens placode expression of the reporter required this motif. Furthermore, the PPE and the lens placode express *Notch2* while the underlying optic vesicle expresses the NOTCH ligand genes *Delta1* and *Delta2* in *Xenopus*. Injection of dominant negative mRNAs for Delta1, Delta2, or Rpbj into dorsal blastomeres of cleavage-stage Xenopus embryos eliminated or dramatically reduced *Foxe3* expression in the lens placode [92]. In contrast, genetic deletion of the NOTCH ligand *Jag1*, *Notch2*, or *Rbpj* from the lens placode or presumptive lens ectoderm in mice failed to reduce *Foxe3* expression in the lens placode or to inhibit lens placode invagination [93–96]. However, these mouse studies revealed a role for NOTCH signaling in the maintenance of the lens epithelium. In the absence of *Rbpj1* or *Jag1*, the anterior lens epithelium thins and specifically loses lens progenitor cells. During early lens development disrupted NOTCH signaling drives the entire lens epithelium to begin taking on characteristics of postmitotic transitional zone cells, with increased expression of p57KIP2 and PROX1, with eventual loss of epithelial characteristics including FOXE3 and E-CADHERIN, particularly in the *Jag1* conditional mutant lenses having the most severe phenotype [93, 95–97].

Also, loss of either *Rbpj1* or *Jag1* from the presumptive lens ectoderm led to a persistent lens stalk between the corneal epithelium and lens, characteristic of Peters' anomaly [93].

Failure to separate the lens from the surface ectoderm that will become the corneal epithelium occurs with several different mutations in mammals, including *Pax6* [98–100], *Foxe3* [101–104], *Foxc1* [105, 106], *Pitx2* [107], *Pitx3* [108], *Sox11* [109], *Sip1* [110], *Cyp1b1* [111], *AP2α* [112], *Cited2* [113], *Msx2* [114], *Spry1* and *Spry2* [115], *RXRα/RARγ* double knockouts [116], activating mutations of *Fgfr2* [117], and transgenic lenses ectopically expressing *Fgf3* [118]. Apoptosis normally removes the lens stalk that connects the lens vesicle to the overlying surface ectoderm in both mammals [119–124] and nonmammalian vertebrates [125, 126]. The precise mechanisms inducing apoptosis in the lens stalk remain unknown but must be tightly controlled to ensure the specific loss of stalk cells without compromising the survival of either the lens or corneal epithelium. The large number of genes whose mutation can result in Peters' anomaly suggests that this pathway will involve a number of different signal transduction cascades and a complex interplay of transcription factors. The relative frequency of Peters' anomaly in cases of anterior segment dysgenesis makes this an important area for elucidation.

Once the lens vesicle forms and separates from the surface ectoderm, the lens epithelium can differentiate from the cells comprising the anterior hemisphere of the vesicle. Just as the posterior lens vesicle cells differentiate through the process of fiber cell differentiation, the anterior vesicle cells must also mature to adopt their final form. Although initially the entire lens vesicle remains competent for cell proliferation, this characteristic quickly becomes the exclusive domain of the lens epithelial cells and ultimately proliferation in the lens epithelium narrows to a band of cells slightly anterior to the lens equator.

There is evidence that canonical WNT signaling participates in the development of the lens epithelium. Mice, homozygous for mutations in *Lrp6*, a coreceptor for canonical WNT signaling, exhibit defective lens epithelial cell differentiation and ectopically express β-crystallin, providing some evidence that lens epithelial cells inappropriately shift to a fiber cell fate [127]. Alternatively, constitutive activation of WNT/β-CATENIN signaling caused an expansion of lens epithelial cells with inhibited fiber cell differentiation [128]. As discussed previously, the initial formation of the lens placode requires a canonical WNT-free zone. The WNT-receptor antagonist, *Sfrp2*, is expressed early in lens placode formation [78]. This expression pattern led to the suggestion that SFRP2 may participate in suppressing canonical WNT signaling in the lens placode [129]. However, deletion of both *Sfrp2* and *Sfrp1* did not lead to increased β-CATENIN signaling in the lens placode but actually reduced canonical WNT signaling in the lens epithelium in E13.5 mouse embryos, demonstrating that these classical WNT antagonists maintain canonical WNT signaling in the lens epithelium [130].

So with the emergence of the lens epithelium from the lens vesicle, our journey from the fertilized zygote draws to an end. Despite the numerous paracrine signals, transcription factors and intracellular proteins essential for formation of the lens epithelium already identified, the story remains far from complete. In particular, understanding the epigenetic regulation of lens epithelial development, including the role of chromatin modifications, DNA methylation, microRNAs, and long noncoding RNAs, remains in its infancy. The recent examples of creating lens cells from embryonic stem cells and/or induced pluripotent stem cells derived from both mouse [131] and human [1, 132, 133] sources represent unique opportunities to define the temporal and chemical cues required to recapitulate or replace the normal process of lens generation during development. Perhaps someday iPS-derived lenses will routinely replace cataractous lenses or provide accommodating lenses to those of us past our youth suffering from presbyopia.

Acknowledgement I would like to acknowledge my friend and colleague Dr. Frank J. Lovicu for his patience and encouragement through the preparation of this chapter and my wife Julia E. Robinson for her assistance with the preparation of Fig. 1.1.

References

1. Yang C, Yang Y, Brennan L, Bouhassira EE, Kantorow M, Cvekl A (2010) Efficient generation of lens progenitor cells and lentoid bodies from human embryonic stem cells in chemically defined conditions. FASEB J 24(9):3274–3283. doi:10.1096/fj.10-157255
2. Niakan KK, Eggan K (2013) Analysis of human embryos from zygote to blastocyst reveals distinct gene expression patterns relative to the mouse. Dev Biol 375(1):54–64. doi:10.1016/j.ydbio.2012.12.008
3. Li L, Lu X, Dean J (2013) The maternal to zygotic transition in mammals. Mol Aspects Med 34(5):919–938. doi:10.1016/j.mam.2013.01.003
4. Rossant J, Tam PP (2004) Emerging asymmetry and embryonic patterning in early mouse development. Dev Cell 7(2):155–164. doi:10.1016/j.devcel.2004.07.012
5. Gardner RL (1997) The early blastocyst is bilaterally symmetrical and its axis of symmetry is aligned with the animal-vegetal axis of the zygote in the mouse. Development 124 (2):289–301
6. Piotrowska K, Zernicka-Goetz M (2001) Role for sperm in spatial patterning of the early mouse embryo. Nature 409(6819):517–521. doi:10.1038/35054069
7. Alarcon VB, Marikawa Y (2005) Unbiased contribution of the first two blastomeres to mouse blastocyst development. Mol Reprod Dev 72(3):354–361. doi:10.1002/mrd.20353
8. Tabansky I, Lenarcic A, Draft RW, Loulier K, Keskin DB, Rosains J, Rivera-Feliciano J, Lichtman JW, Livet J, Stern JN, Sanes JR, Eggan K (2013) Developmental bias in cleavage-stage mouse blastomeres. Curr Biol 23(1):21–31. doi:10.1016/j.cub.2012.10.054
9. Peyrieras N, Hyafil F, Louvard D, Ploegh HL, Jacob F (1983) Uvomorulin: a nonintegral membrane protein of early mouse embryo. Proc Natl Acad Sci U S A 80(20):6274–6277
10. Niwa H, Toyooka Y, Shimosato D, Strumpf D, Takahashi K, Yagi R, Rossant J (2005) Interaction between Oct3/4 and Cdx2 determines trophectoderm differentiation. Cell 123 (5):917–929. doi:10.1016/j.cell.2005.08.040
11. Dietrich JE, Hiiragi T (2007) Stochastic patterning in the mouse pre-implantation embryo. Development 134(23):4219–4231. doi:10.1242/dev.003798
12. Ralston A, Rossant J (2008) Cdx2 acts downstream of cell polarization to cell-autonomously promote trophectoderm fate in the early mouse embryo. Dev Biol 313(2):614–629. doi:10.1016/j.ydbio.2007.10.054
13. Nishioka N, Inoue K, Adachi K, Kiyonari H, Ota M, Ralston A, Yabuta N, Hirahara S, Stephenson RO, Ogonuki N, Makita R, Kurihara H, Morin-Kensicki EM, Nojima H,

Rossant J, Nakao K, Niwa H, Sasaki H (2009) The Hippo signaling pathway components Lats and Yap pattern Tead4 activity to distinguish mouse trophectoderm from inner cell mass. Dev Cell 16(3):398–410. doi:10.1016/j.devcel.2009.02.003
14. Cockburn K, Biechele S, Garner J, Rossant J (2013) The Hippo pathway member Nf2 is required for inner cell mass specification. Curr Biol 23(13):1195–1201. doi:10.1016/j.cub.2013.05.044
15. Yamanaka Y, Lanner F, Rossant J (2010) FGF signal-dependent segregation of primitive endoderm and epiblast in the mouse blastocyst. Development 137(5):715–724. doi:10.1242/dev.043471
16. Ahrens K, Schlosser G (2005) Tissues and signals involved in the induction of placodal Six1 expression in Xenopus laevis. Dev Biol 288(1):40–59. doi:10.1016/j.ydbio.2005.07.022
17. Litsiou A, Hanson S, Streit A (2005) A balance of FGF, BMP and WNT signalling positions the future placode territory in the head. Development 132(18):4051–4062. doi:10.1242/dev.01964
18. Brugmann SA, Pandur PD, Kenyon KL, Pignoni F, Moody SA (2004) Six1 promotes a placodal fate within the lateral neurogenic ectoderm by functioning as both a transcriptional activator and repressor. Development 131(23):5871–5881. doi:10.1242/dev.01516
19. Patthey C, Edlund T, Gunhaga L (2009) Wnt-regulated temporal control of BMP exposure directs the choice between neural plate border and epidermal fate. Development 136(1):73–83. doi:10.1242/dev.025890
20. Pieper M, Ahrens K, Rink E, Peter A, Schlosser G (2012) Differential distribution of competence for panplacodal and neural crest induction to non-neural and neural ectoderm. Development 139(6):1175–1187. doi:10.1242/dev.074468
21. Kwon HJ, Bhat N, Sweet EM, Cornell RA, Riley BB (2010) Identification of early requirements for preplacodal ectoderm and sensory organ development. PLoS Genet 6(9): e1001133. doi:10.1371/journal.pgen.1001133
22. Bhat N, Kwon HJ, Riley BB (2013) A gene network that coordinates preplacodal competence and neural crest specification in zebrafish. Dev Biol 373(1):107–117. doi:10.1016/j.ydbio.2012.10.012
23. Hulander M, Kiernan AE, Blomqvist SR, Carlsson P, Samuelsson EJ, Johansson BR, Steel KP, Enerback S (2003) Lack of pendrin expression leads to deafness and expansion of the endolymphatic compartment in inner ears of Foxi1 null mutant mice. Development 130 (9):2013–2025
24. Karis A, Pata I, van Doorninck JH, Grosveld F, de Zeeuw CI, de Caprona D, Fritzsch B (2001) Transcription factor GATA-3 alters pathway selection of olivocochlear neurons and affects morphogenesis of the ear. J Comp Neurol 429(4):615–630
25. Lleras-Forero L, Streit A (2012) Development of the sensory nervous system in the vertebrate head: the importance of being on time. Curr Opin Genet Dev 22(4):315–322. doi:10.1016/j.gde.2012.05.003
26. Tadjuidje E, Hegde RS (2013) The eyes absent proteins in development and disease. Cell Mol Life Sci 70(11):1897–1913. doi:10.1007/s00018-012-1144-9
27. Cook PJ, Ju BG, Telese F, Wang X, Glass CK, Rosenfeld MG (2009) Tyrosine dephosphorylation of H2AX modulates apoptosis and survival decisions. Nature 458(7238):591–596. doi:10.1038/nature07849
28. Krishnan N, Jeong DG, Jung SK, Ryu SE, Xiao A, Allis CD, Kim SJ, Tonks NK (2009) Dephosphorylation of the C-terminal tyrosyl residue of the DNA damage-related histone H2A. X is mediated by the protein phosphatase eyes absent. J Biol Chem 284(24):16066–16070. doi:10.1074/jbc.C900032200
29. Ohto H, Kamada S, Tago K, Tominaga SI, Ozaki H, Sato S, Kawakami K (1999) Cooperation of six and eya in activation of their target genes through nuclear translocation of Eya. Mol Cell Biol 19(10):6815–6824
30. Silver SJ, Davies EL, Doyon L, Rebay I (2003) Functional dissection of eyes absent reveals new modes of regulation within the retinal determination gene network. Mol Cell Biol 23 (17):5989–5999

31. Xu PX, Cheng J, Epstein JA, Maas RL (1997) Mouse Eya genes are expressed during limb tendon development and encode a transcriptional activation function. Proc Natl Acad Sci U S A 94(22):11974–11979
32. Zou D, Erickson C, Kim EH, Jin D, Fritzsch B, Xu PX (2008) Eya1 gene dosage critically affects the development of sensory epithelia in the mammalian inner ear. Hum Mol Genet 17(21):3340–3356. doi:10.1093/hmg/ddn229
33. Li X, Oghi KA, Zhang J, Krones A, Bush KT, Glass CK, Nigam SK, Aggarwal AK, Maas R, Rose DW, Rosenfeld MG (2003) Eya protein phosphatase activity regulates Six1-dach-Eya transcriptional effects in mammalian organogenesis. Nature 426(6964):247–254. doi:10.1038/nature02083
34. Sahly I, Andermann P, Petit C (1999) The zebrafish eya1 gene and its expression pattern during embryogenesis. Dev Genes Evol 209(7):399–410
35. David R, Ahrens K, Wedlich D, Schlosser G (2001) Xenopus Eya1 demarcates all neurogenic placodes as well as migrating hypaxial muscle precursors. Mech Dev 103(1–2):189–192
36. Ishihara T, Ikeda K, Sato S, Yajima H, Kawakami K (2008) Differential expression of Eya1 and Eya2 during chick early embryonic development. Gene Expr Patterns 8(5):357–367. doi:10.1016/j.gep.2008.01.003
37. Xu PX, Woo I, Her H, Beier DR, Maas RL (1997) Mouse Eya homologues of the Drosophila eyes absent gene require Pax6 for expression in lens and nasal placode. Development 124 (1):219–231
38. Mishima N, Tomarev S (1998) Chicken eyes absent 2 gene: isolation and expression pattern during development. Int J Dev Biol 42(8):1109–1115
39. Kumar JP (2009) The sine oculis homeobox (SIX) family of transcription factors as regulators of development and disease. Cell Mol Life Sci 66(4):565–583. doi:10.1007/s00018-008-8335-4
40. Weasner B, Salzer C, Kumar JP (2007) Sine oculis, a member of the SIX family of transcription factors, directs eye formation. Dev Biol 303(2):756–771. doi:10.1016/j.ydbio.2006.10.040
41. Zhu CC, Dyer MA, Uchikawa M, Kondoh H, Lagutin OV, Oliver G (2002) Six3-mediated auto repression and eye development requires its interaction with members of the Groucho-related family of co-repressors. Development 129(12):2835–2849
42. Lopez-Rios J, Tessmar K, Loosli F, Wittbrodt J, Bovolenta P (2003) Six3 and Six6 activity is modulated by members of the groucho family. Development 130(1):185–195
43. Kenyon KL, Li DJ, Clouser C, Tran S, Pignoni F (2005) Fly SIX-type homeodomain proteins Sine oculis and Optix partner with different cofactors during eye development. Dev Dyn 234 (3):497–504. doi:10.1002/dvdy.20442
44. Li X, Perissi V, Liu F, Rose DW, Rosenfeld MG (2002) Tissue-specific regulation of retinal and pituitary precursor cell proliferation. Science 297(5584):1180–1183. doi:10.1126/science.1073263
45. Bailey AP, Bhattacharyya S, Bronner-Fraser M, Streit A (2006) Lens specification is the ground state of all sensory placodes, from which FGF promotes olfactory identity. Dev Cell 11(4):505–517. doi:10.1016/j.devcel.2006.08.009
46. Schlosser G (2010) Making senses development of vertebrate cranial placodes. Int Rev Cell Mol Biol 283:129–234. doi:10.1016/S1937-6448(10)83004-7
47. Lleras-Forero L, Tambalo M, Christophorou N, Chambers D, Houart C, Streit A (2013) Neuropeptides: developmental signals in placode progenitor formation. Dev Cell 26(2):195–203. doi:10.1016/j.devcel.2013.07.001
48. Kumar JP, Moses K (2001) Eye specification in Drosophila: perspectives and implications. Semin Cell Dev Biol 12(6):469–474. doi:10.1006/scdb.2001.0270
49. Duggan CD, DeMaria S, Baudhuin A, Stafford D, Ngai J (2008) Foxg1 is required for development of the vertebrate olfactory system. J Neurosci 28(20):5229–5239. doi:10.1523/JNEUROSCI.1134-08.2008

50. Sjodal M, Edlund T, Gunhaga L (2007) Time of exposure to BMP signals plays a key role in the specification of the olfactory and lens placodes ex vivo. Dev Cell 13(1):141–149. doi:10.1016/j.devcel.2007.04.020
51. Gunhaga L (2011) The lens: a classical model of embryonic induction providing new insights into cell determination in early development. Philos Trans R Soc Lond B Biol Sci 366(1568):1193–1203. doi:10.1098/rstb.2010.0175
52. Grainger RM (1992) Embryonic lens induction: shedding light on vertebrate tissue determination. Trends Genet 8(10):349–355
53. Saha MS, Spann CL, Grainger RM (1989) Embryonic lens induction: more than meets the optic vesicle. Cell Differ Dev 28(3):153–171
54. Kondoh H, Uchikawa M, Yoda H, Takeda H, Furutani-Seiki M, Karlstrom RO (2000) Zebrafish mutations in Gli-mediated hedgehog signaling lead to lens transdifferentiation from the adenohypophysis anlage. Mech Dev 96(2):165–174
55. Yamamoto Y, Stock DW, Jeffery WR (2004) Hedgehog signalling controls eye degeneration in blind cavefish. Nature 431(7010):844–847. doi:10.1038/nature02864
56. Smith AN, Miller LA, Song N, Taketo MM, Lang RA (2005) The duality of beta-catenin function: a requirement in lens morphogenesis and signaling suppression of lens fate in periocular ectoderm. Dev Biol 285(2):477–489. doi:10.1016/j.ydbio.2005.07.019
57. Kreslova J, Machon O, Ruzickova J, Lachova J, Wawrousek EF, Kemler R, Krauss S, Piatigorsky J, Kozmik Z (2007) Abnormal lens morphogenesis and ectopic lens formation in the absence of beta-catenin function. Genesis 45(4):157–168. doi:10.1002/dvg.20277
58. Cain S, Martinez G, Kokkinos MI, Turner K, Richardson RJ, Abud HE, Huelsken J, Robinson ML, de Iongh RU (2008) Differential requirement for beta-catenin in epithelial and fiber cells during lens development. Dev Biol 321(2):420–433. doi:10.1016/j.ydbio.2008.07.002
59. Miller LA, Smith AN, Taketo MM, Lang RA (2006) Optic cup and facial patterning defects in ocular ectoderm beta-catenin gain-of-function mice. BMC Dev Biol 6:14. doi:10.1186/1471-213X-6-14
60. Zhou CJ, Molotkov A, Song L, Li Y, Pleasure DE, Pleasure SJ, Wang YZ (2008) Ocular coloboma and dorsoventral neuroretinal patterning defects in Lrp6 mutant eyes. Dev Dyn 237 (12):3681–3689. doi:10.1002/dvdy.21770
61. Song N, Schwab KR, Patterson LT, Yamaguchi T, Lin X, Potter SS, Lang RA (2007) Pygopus 2 has a crucial, Wnt pathway-independent function in lens induction. Development 134(10):1873–1885. doi:10.1242/dev.001495
62. Liu H, Thurig S, Mohamed O, Dufort D, Wallace VA (2006) Mapping canonical Wnt signaling in the developing and adult retina. Invest Ophthalmol Vis Sci 47(11):5088–5097. doi:10.1167/iovs.06-0403
63. Lang RA (2004) Pathways regulating lens induction in the mouse. Int J Dev Biol 48(8–9):783–791. doi:10.1387/ijdb.041903rl
64. Oliver G, Loosli F, Koster R, Wittbrodt J, Gruss P (1996) Ectopic lens induction in fish in response to the murine homeobox gene Six3. Mech Dev 60(2):233–239
65. Mathers PH, Grinberg A, Mahon KA, Jamrich M (1997) The Rx homeobox gene is essential for vertebrate eye development. Nature 387(6633):603–607. doi:10.1038/42475
66. Swindell EC, Bailey TJ, Loosli F, Liu C, Amaya-Manzanares F, Mahon KA, Wittbrodt J, Jamrich M (2006) Rx-Cre, a tool for inactivation of gene expression in the developing retina. Genesis 44(8):361–363. doi:10.1002/dvg.20225
67. Voronina VA, Kozhemyakina EA, O'Kernick CM, Kahn ND, Wenger SL, Linberg JV, Schneider AS, Mathers PH (2004) Mutations in the human RAX homeobox gene in a patient with anophthalmia and sclerocornea. Hum Mol Genet 13(3):315–322. doi:10.1093/hmg/ddh025
68. Furuta Y, Hogan BL (1998) BMP4 is essential for lens induction in the mouse embryo. Genes Dev 12(23):3764–3775
69. Swindell EC, Liu C, Shah R, Smith AN, Lang RA, Jamrich M (2008) Eye formation in the absence of retina. Dev Biol 322(1):56–64. doi:10.1016/j.ydbio.2008.07.009

70. Hendrix RW, Zwaan J (1974) Changes in the glycoprotein concentration of the extracellular matrix between lens and optic vesicle associated with early lens differentiation. Differentiation 2(6):357–362
71. Hendrix RW, Zwaan J (1975) The matrix of the optic vesicle-presumptive lens interface during induction of the lens in the chicken embryo. J Embryol Exp Morphol 33(4):1023–1049
72. Zwaan J, Hendrix RW (1973) Changes in cell and organ shape during early development of the ocular lens. Am Zool 13:1039–1049
73. McKeehan MS (1951) Cytological aspects of embryonic lens induction in the chick. J Exp Zool 117:31–64
74. Huang J, Rajagopal R, Liu Y, Dattilo LK, Shaham O, Ashery-Padan R, Beebe DC (2011) The mechanism of lens placode formation: a case of matrix-mediated morphogenesis. Dev Biol 355(1):32–42. doi:10.1016/j.ydbio.2011.04.008
75. Ashery-Padan R, Marquardt T, Zhou X, Gruss P (2000) Pax6 activity in the lens primordium is required for lens formation and for correct placement of a single retina in the eye. Genes Dev 14(21):2701–2711
76. Smith AN, Miller LA, Radice G, Ashery-Padan R, Lang RA (2009) Stage-dependent modes of Pax6-Sox2 epistasis regulate lens development and eye morphogenesis. Development 136 (17):2977–2985. doi:10.1242/dev.037341
77. Dudley AT, Robertson EJ (1997) Overlapping expression domains of bone morphogenetic protein family members potentially account for limited tissue defects in BMP7 deficient embryos. Dev Dyn 208(3):349–362. doi:10.1002/(SICI)1097-0177(199703)208:3<349::AID-AJA6>3.0.CO;2-I
78. Wawersik S, Purcell P, Rauchman M, Dudley AT, Robertson EJ, Maas R (1999) BMP7 acts in murine lens placode development. Dev Biol 207(1):176–188. doi:10.1006/dbio.1998.9153
79. Luo G, Hofmann C, Bronckers AL, Sohocki M, Bradley A, Karsenty G (1995) BMP-7 is an inducer of nephrogenesis, and is also required for eye development and skeletal patterning. Genes Dev 9(22):2808–2820
80. Rajagopal R, Huang J, Dattilo LK, Kaartinen V, Mishina Y, Deng CX, Umans L, Zwijsen A, Roberts AB, Beebe DC (2009) The type I BMP receptors, Bmpr1a and Acvr1, activate multiple signaling pathways to regulate lens formation. Dev Biol 335(2):305–316. doi:10.1016/j.ydbio.2009.08.027
81. Plageman TF Jr, Chung MI, Lou M, Smith AN, Hildebrand JD, Wallingford JB, Lang RA (2010) Pax6-dependent Shroom3 expression regulates apical constriction during lens placode invagination. Development 137(3):405–415. doi:10.1242/dev.045369
82. Chauhan BK, Lou M, Zheng Y, Lang RA (2011) Balanced Rac1 and RhoA activities regulate cell shape and drive invagination morphogenesis in epithelia. Proc Natl Acad Sci U S A 108 (45):18289–18294. doi:10.1073/pnas.1108993108
83. Plageman TF Jr, Chauhan BK, Yang C, Jaudon F, Shang X, Zheng Y, Lou M, Debant A, Hildebrand JD, Lang RA (2011) A Trio-RhoA-Shroom3 pathway is required for apical constriction and epithelial invagination. Development 138(23):5177–5188. doi:10.1242/dev.067868
84. Chauhan BK, Disanza A, Choi SY, Faber SC, Lou M, Beggs HE, Scita G, Zheng Y, Lang RA (2009) Cdc42- and IRSp53-dependent contractile filopodia tether presumptive lens and retina to coordinate epithelial invagination. Development 136(21):3657–3667. doi:10.1242/dev.042242
85. Faber SC, Dimanlig P, Makarenkova HP, Shirke S, Ko K, Lang RA (2001) Fgf receptor signaling plays a role in lens induction. Development 128(22):4425–4438
86. Garcia CM, Huang J, Madakashira BP, Liu Y, Rajagopal R, Dattilo L, Robinson ML, Beebe DC (2011) The function of FGF signaling in the lens placode. Dev Biol 351(1):176–185. doi:10.1016/j.ydbio.2011.01.001
87. Madakashira BP, Kobrinski DA, Hancher AD, Arneman EC, Wagner BD, Wang F, Shin H, Lovicu FJ, Reneker LW, Robinson ML (2012) Frs2alpha enhances fibroblast growth factor-mediated survival and differentiation in lens development. Development 139(24):4601–4612. doi:10.1242/dev.081737

88. Li H, Tao C, Cai Z, Hertzler-Schaefer K, Collins TN, Wang F, Feng GS, Gotoh N, Zhang X (2013) Frs2alpha and Shp2 signal independently of Gab to mediate FGF signaling in lens development. J Cell Sci. doi:10.1242/jcs.134478
89. Zhao H, Yang T, Madakashira BP, Thiels CA, Bechtle CA, Garcia CM, Zhang H, Yu K, Ornitz DM, Beebe DC, Robinson ML (2008) Fibroblast growth factor receptor signaling is essential for lens fiber cell differentiation. Dev Biol 318(2):276–288. doi:10.1016/j.ydbio.2008.03.028
90. Pan Y, Woodbury A, Esko JD, Grobe K, Zhang X (2006) Heparan sulfate biosynthetic gene Ndst1 is required for FGF signaling in early lens development. Development 133(24):4933–4944. doi:10.1242/dev.02679
91. Carbe C, Zhang X (2011) Lens induction requires attenuation of ERK signaling by Nf1. Hum Mol Genet 20(7):1315–1323. doi:10.1093/hmg/ddr014
92. Ogino H, Fisher M, Grainger RM (2008) Convergence of a head-field selector Otx2 and Notch signaling: a mechanism for lens specification. Development 135(2):249–258. doi:10.1242/dev.009548
93. Le TT, Conley KW, Mead TJ, Rowan S, Yutzey KE, Brown NL (2012) Requirements for Jag1-Rbpj mediated Notch signaling during early mouse lens development. Dev Dyn 241 (3):493–504. doi:10.1002/dvdy.23739
94. Saravanamuthu SS, Le TT, Gao CY, Cojocaru RI, Pandiyan P, Liu C, Zhang J, Zelenka PS, Brown NL (2012) Conditional ablation of the Notch2 receptor in the ocular lens. Dev Biol 362(2):219–229. doi:10.1016/j.ydbio.2011.11.011
95. Le TT, Conley KW, Brown NL (2009) Jagged 1 is necessary for normal mouse lens formation. Dev Biol 328(1):118–126. doi:10.1016/j.ydbio.2009.01.015
96. Rowan S, Conley KW, Le TT, Donner AL, Maas RL, Brown NL (2008) Notch signaling regulates growth and differentiation in the mammalian lens. Dev Biol 321(1):111–122. doi:10.1016/j.ydbio.2008.06.002
97. Jia J, Lin M, Zhang L, York JP, Zhang P (2007) The notch signaling pathway controls the size of the ocular lens by directly suppressing p57Kip2 expression. Mol Cell Biol 27(20):7236–7247. doi:10.1128/MCB.00780-07
98. Baulmann DC, Ohlmann A, Flugel-Koch C, Goswami S, Cvekl A, Tamm ER (2002) Pax6 heterozygous eyes show defects in chamber angle differentiation that are associated with a wide spectrum of other anterior eye segment abnormalities. Mech Dev 118(1–2):3–17
99. Collinson JM, Quinn JC, Buchanan MA, Kaufman MH, Wedden SE, West JD, Hill RE (2001) Primary defects in the lens underlie complex anterior segment abnormalities of the Pax6 heterozygous eye. Proc Natl Acad Sci U S A 98(17):9688–9693. doi:10.1073/pnas.161144098
100. Hanson IM, Fletcher JM, Jordan T, Brown A, Taylor D, Adams RJ, Punnett HH, van Heyningen V (1994) Mutations at the PAX6 locus are found in heterogeneous anterior segment malformations including Peters' anomaly. Nat Genet 6(2):168–173. doi:10.1038/ng0294-168
101. Ormestad M, Blixt A, Churchill A, Martinsson T, Enerback S, Carlsson P (2002) Foxe3 haploinsufficiency in mice: a model for Peters' anomaly. Invest Ophthalmol Vis Sci 43(5):1350–1357
102. Blixt A, Mahlapuu M, Aitola M, Pelto-Huikko M, Enerback S, Carlsson P (2000) A forkhead gene, FoxE3, is essential for lens epithelial proliferation and closure of the lens vesicle. Genes Dev 14(2):245–254
103. Brownell I, Dirksen M, Jamrich M (2000) Forkhead Foxe3 maps to the dysgenetic lens locus and is critical in lens development and differentiation. Genesis 27(2):81–93
104. Semina EV, Brownell I, Mintz-Hittner HA, Murray JC, Jamrich M (2001) Mutations in the human forkhead transcription factor FOXE3 associated with anterior segment ocular dysgenesis and cataracts. Hum Mol Genet 10(3):231–236
105. Honkanen RA, Nishimura DY, Swiderski RE, Bennett SR, Hong S, Kwon YH, Stone EM, Sheffield VC, Alward WL (2003) A family with Axenfeld-Rieger syndrome and Peters

anomaly caused by a point mutation (Phe112Ser) in the FOXC1 gene. Am J Ophthalmol 135 (3):368–375

106. Weisschuh N, Wolf C, Wissinger B, Gramer E (2008) A novel mutation in the FOXC1 gene in a family with Axenfeld-Rieger syndrome and Peters' anomaly. Clin Genet 74(5):476–480. doi:10.1111/j.1399-0004.2008.01025.x
107. Doward W, Perveen R, Lloyd IC, Ridgway AE, Wilson L, Black GC (1999) A mutation in the RIEG1 gene associated with Peters' anomaly. J Med Genet 36(2):152–155
108. Semina EV, Ferrell RE, Mintz-Hittner HA, Bitoun P, Alward WL, Reiter RS, Funkhauser C, Daack-Hirsch S, Murray JC (1998) A novel homeobox gene PITX3 is mutated in families with autosomal-dominant cataracts and ASMD. Nat Genet 19(2):167–170. doi:10.1038/527
109. Wurm A, Sock E, Fuchshofer R, Wegner M, Tamm ER (2008) Anterior segment dysgenesis in the eyes of mice deficient for the high-mobility-group transcription factor Sox11. Exp Eye Res 86(6):895–907. doi:10.1016/j.exer.2008.03.004
110. Yoshimoto A, Saigou Y, Higashi Y, Kondoh H (2005) Regulation of ocular lens development by Smad-interacting protein 1 involving Foxe3 activation. Development 132(20):4437–4448. doi:10.1242/dev.02022
111. Vincent A, Billingsley G, Priston M, Williams-Lyn D, Sutherland J, Glaser T, Oliver E, Walter MA, Heathcote G, Levin A, Heon E (2001) Phenotypic heterogeneity of CYP1B1: mutations in a patient with Peters' anomaly. J Med Genet 38(5):324–326
112. Pontoriero GF, Deschamps P, Ashery-Padan R, Wong R, Yang Y, Zavadil J, Cvekl A, Sullivan S, Williams T, West-Mays JA (2008) Cell autonomous roles for AP-2alpha in lens vesicle separation and maintenance of the lens epithelial cell phenotype. Dev Dyn 237 (3):602–617. doi:10.1002/dvdy.21445
113. Chen Y, Doughman YQ, Gu S, Jarrell A, Aota S, Cvekl A, Watanabe M, Dunwoodie SL, Johnson RS, van Heyningen V, Kleinjan DA, Beebe DC, Yang YC (2008) Cited2 is required for the proper formation of the hyaloid vasculature and for lens morphogenesis. Development 135(17):2939–2948. doi:10.1242/dev.021097
114. Zhao J, Kawai K, Wang H, Wu D, Wang M, Yue Z, Zhang J, Liu YH (2012) Loss of Msx2 function down-regulates the FoxE3 expression and results in anterior segment dysgenesis resembling Peters anomaly. Am J Pathol 180(6):2230–2239. doi:10.1016/j.ajpath.2012.02.017
115. Kuracha MR, Burgess D, Siefker E, Cooper JT, Licht JD, Robinson ML, Govindarajan V (2011) Spry1 and Spry2 are necessary for lens vesicle separation and corneal differentiation. Invest Ophthalmol Vis Sci 52(9):6887–6897. doi:10.1167/iovs.11-7531
116. Kastner P, Grondona JM, Mark M, Gansmuller A, LeMeur M, Decimo D, Vonesch JL, Dolle P, Chambon P (1994) Genetic analysis of RXR alpha developmental function: convergence of RXR and RAR signaling pathways in heart and eye morphogenesis. Cell 78 (6):987–1003
117. Okajima K, Robinson LK, Hart MA, Abuelo DN, Cowan LS, Hasegawa T, Maumenee IH, Jabs EW (1999) Ocular anterior chamber dysgenesis in craniosynostosis syndromes with a fibroblast growth factor receptor 2 mutation. Am J Med Genet 85(2):160–170
118. Robinson ML, Ohtaka-Maruyama C, Chan CC, Jamieson S, Dickson C, Overbeek PA, Chepelinsky AB (1998) Dysregulation of ocular morphogenesis by lens-specific expression of FGF-3/int-2 in transgenic mice. Dev Biol 198(1):13–31
119. Silver J, Hughes AF (1973) The role of cell death during morphogenesis of the mammalian eye. J Morphol 140(2):159–170. doi:10.1002/jmor.1051400204
120. Silver J, Hughes AF (1974) The relationship between morphogenetic cell death and the development of congenital anophthalmia. J Comp Neurol 157(3):281–301. doi:10.1002/cne.901570303
121. Bozanic D, Tafra R, Saraga-Babic M (2003) Role of apoptosis and mitosis during human eye development. Eur J Cell Biol 82(8):421–429. doi:10.1078/0171-9335-00328
122. Ozeki H, Ogura Y, Hirabayashi Y, Shimada S (2001) Suppression of lens stalk cell apoptosis by hyaluronic acid leads to faulty separation of the lens vesicle. Exp Eye Res 72(1):63–70. doi:10.1006/exer.2000.0923

123. Mohamed YH, Amemiya T (2003) Apoptosis and lens vesicle development. Eur J Ophthalmol 13(1):1–10
124. Laemle LK, Puszkarczuk M, Feinberg RN (1999) Apoptosis in early ocular morphogenesis in the mouse. Brain Res Dev Brain Res 112(1):129–133
125. Garcia-Porrero JA, Collado JA, Ojeda JL (1979) Cell death during detachment of the lens rudiment from ectoderm in the chick embryo. Anat Rec 193(4):791–804. doi:10.1002/ar.1091930404
126. Cole LK, Ross LS (2001) Apoptosis in the developing zebrafish embryo. Dev Biol 240(1):123–142. doi:10.1006/dbio.2001.0432
127. Stump RJ, Ang S, Chen Y, von Bahr T, Lovicu FJ, Pinson K, de Iongh RU, Yamaguchi TP, Sassoon DA, McAvoy JW (2003) A role for Wnt/beta-catenin signaling in lens epithelial differentiation. Dev Biol 259(1):48–61
128. Martinez G, Wijesinghe M, Turner K, Abud HE, Taketo MM, Noda T, Robinson ML, de Iongh RU (2009) Conditional mutations of beta-catenin and APC reveal roles for canonical Wnt signaling in lens differentiation. Invest Ophthalmol Vis Sci 50(10):4794–4806. doi:10.1167/iovs.09-3567
129. Machon O, Kreslova J, Ruzickova J, Vacik T, Klimova L, Fujimura N, Lachova J, Kozmik Z (2010) Lens morphogenesis is dependent on Pax6-mediated inhibition of the canonical Wnt/beta-catenin signaling in the lens surface ectoderm. Genesis 48(2):86–95. doi:10.1002/dvg.20583
130. Sugiyama Y, Shelley EJ, Wen L, Stump RJ, Shimono A, Lovicu FJ, McAvoy JW (2013) Sfrp1 and Sfrp2 are not involved in Wnt/beta-catenin signal silencing during lens induction but are required for maintenance of Wnt/beta-catenin signaling in lens epithelial cells. Dev Biol 384(2):181–193. doi:10.1016/j.ydbio.2013.10.008
131. Kawamorita M, Suzuki C, Saito G, Sato T, Sato K (2002) In vitro differentiation of mouse embryonic stem cells after activation by retinoic acid. Hum Cell 15(3):178–182
132. Qiu X, Yang J, Liu T, Jiang Y, Le Q, Lu Y (2012) Efficient generation of lens progenitor cells from cataract patient-specific induced pluripotent stem cells. PLoS One 7(3):e32612. doi:10.1371/journal.pone.0032612
133. Dincer Z, Piao J, Niu L, Ganat Y, Kriks S, Zimmer B, Shi SH, Tabar V, Studer L (2013) Specification of functional cranial placode derivatives from human pluripotent stem cells. Cell Rep. doi:10.1016/j.celrep.2013.10.048

Cell Biology of Lens Epithelial Cells

2

Steven Bassnett

Abstract

The epithelium of the vertebrate lens plays a critical role in tissue homeostasis and maintenance of lens clarity. The epithelium is the most metabolically active region of the lens and contains all the mitotically active cells. Cell division in the epithelium occurs exclusively in the germinative zone, a swathe of cells encircling the lens just above the equator. Fibroblast growth factor, a molecule with a demonstrated role in lens fiber cell differentiation, may promote epithelial cell division although other growth factors likely contribute. The organization of cells within the lens epithelium has often been likened to a cobblestone pattern. However, recent three-dimensional imaging studies have revealed that individual epithelial cells have a complex, polarized anatomy, with morphologically distinct apical and basolateral domains. The apical membrane is delineated by a hybrid junctional complex consisting of adherens junctions and tight junctions. Adherens junctions play a critical role in epithelial organization and loss of nectins or cadherins, two core components of adherens junctions, has catastrophic consequences for lens organization and transparency. Tight junctions, the apical-most junctional element, restrict the paracellular flow of ions into the lens but also serve as scaffolds for an assemblage of important polarity proteins. Targeted disruption of the partitioning defective (Par) family of polarity proteins results in loss of apical cell junctions and promotes epithelial-to-mesenchymal transition.

Keywords

Adherens junction • Germinative zone • Mitosis • Polarity

S. Bassnett (✉)
Department of Ophthalmology and Visual Sciences, Washington University School of Medicine, 660 S. Euclid Ave, Box 8096, St. Louis, MO 63117, USA
e-mail: bassnett@vision.wustl.edu

S. Saika et al. (eds.), *Lens Epithelium and Posterior Capsular Opacification*,
DOI 10.1007/978-4-431-54300-8_2, © Springer Japan 2014

2.1 Introduction

The anterior hemisphere of the vertebrate lens is covered by a continuous monolayer of cells—the lens epithelium. In adult mice, the epithelium consists of approximately 40,000 cells [1]. The epithelium of the much larger human lens is believed to contain more than ten times that number [2]. Histologically, the lens epithelium is classified as a *simple* epithelium (i.e., every cell is in direct contact with the overlying basement membrane and extends across the full thickness of the epithelial layer). In the young lens, the epithelial cells are generally described as *cuboidal* (meaning that they are about as high as they are wide). Later in life, when in many species the epithelium is thinner and the cells less densely packed, the lens epithelium might be better described as a simple *squamous* epithelium.

The cellular organization of the vertebrate lens has been the subject of many investigations over the past 100 years or more. Only recently, however, have techniques been developed that allow the morphology of living lens cells to be visualized. One such approach is the induced expression of fluorescent proteins in individual lens cells. Expressing cells can be optically isolated from their nonfluorescent neighbors by confocal microscopy. Their structure can then be imaged at high resolution and in three dimensions [3–5]. Applied to cells in the epithelium, this methodology has provided a new conception of lens cell organization. Viewed in situ, living lens epithelial cells are shown to have a surprisingly complex and dynamic morphology (Fig. 2.1a). Although the apical membranes of the epithelial cells are polygonal in shape, their basolateral membranes are highly folded and irregular. Numerous lamellipodia-like processes extend from the cell body. In the lens literature, the epithelium is commonly described as having a "cobblestone" appearance and individual cells as being polygonal in shape. The discrepancy between this description and the morphology of cells as revealed in fluorescent labeling studies (Fig. 2.1a) arises because the junctional complex at the apicolateral border of the cells (see below) is particularly prominent under conventional phase contrast microscopy. Since the junctional complex delineates the shape of the (largely polygonal) apical membrane, conventional microscopy gives the impression that epithelial cells have a more regular structure than is in fact the case.

At the edge of the epithelium, near the lens equator, epithelial cells terminally differentiate into fiber cells. Fiber cells account for almost all of the lens volume and, as their name implies, have a highly elongated form (Fig. 2.1b), quite unlike the epithelial cells from which they are derived. During the initial stages of terminal differentiation, the epithelial progenitor is first reshaped into a flat, ribbonlike cell [4]. As differentiation proceeds, cell length and volume increase and the lateral membranes of the fiber cells are extensively remodeled (Fig. 2.1b).

Although only accounting for a small fraction of the lens volume, the epithelial layer contains all the mitotically active cells in the tissue and can, therefore, be thought of as the "growth engine" of the lens. Cell division within the epithelium ultimately provides both the fiber cells to fill the tissue volume and the additional epithelial cells necessary to cover its expanding surface. The mitotic index varies with latitudinal position. Cells in the central region of the epithelium have a

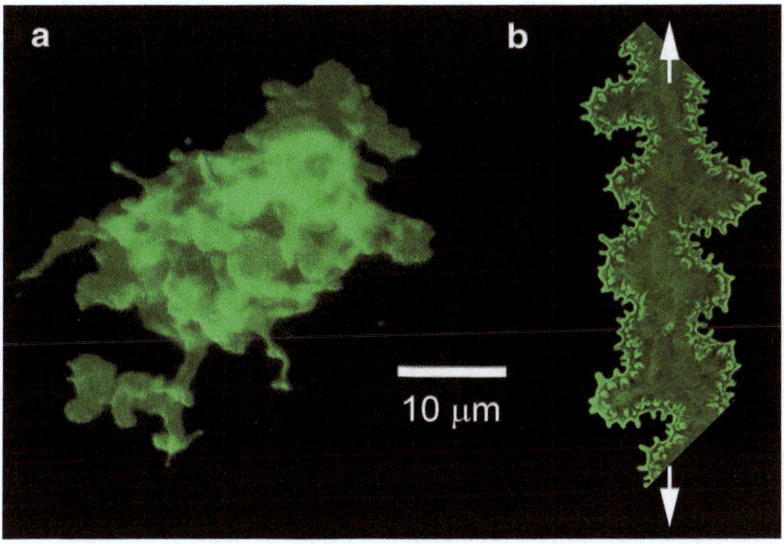

Fig. 2.1 Morphology of individual mouse lens cells as revealed by expression of green fluorescent protein (GFP). Epithelial cells (**a**, viewed from the capsular surface) have a complex and irregular morphology. At the lens equator, epithelial cells differentiate into fiber cells (**b**). The differentiation process involves a radical restructuring of cellular morphology. Fiber cells are physically interlinked via a system of tooth-like membrane processes protruding from the lateral surfaces. The cells are too long to visualize in their entirety at high magnification. At the scale shown, the cell shown in **b** would extend for several meters in either direction

relatively low mitotic index. This area lies within the pupillary space and is therefore exposed directly to light. Considerably higher rates of cell division are observed in the peripheral lens epithelium. In this location, the proliferating cells are sequestered in the shadow of the iris and thus protected from potentially mutagenic exposure to ultraviolet radiation [6].

As with all epithelia, the lens epithelium has an intimate association with its basement membrane. The basement membrane of the lens is called the lens capsule and completely envelopes the tissue. The lens capsule is synthesized by the epithelial cells themselves [7] and its detailed composition and properties will be discussed elsewhere in this volume. Here, we note merely that the capsule is particularly enriched in type IV collagen and contains components such as laminin and fibronectin that are common constituents of basement membranes throughout the body. Although its biochemical composition is not remarkable, the lens capsule is one of the thickest basement membranes in the body. A typical basement membrane might be 0.1 μm thick [8]. In contrast, the human lens capsule is approximately 10 μm thick [9]. Perhaps surprisingly, the lens capsule in mice is even thicker [10]. The elastic lens capsule helps mold the lens substance at rest and during accommodation. Near the equator, the capsule also serves as the anchor point for the ciliary zonule (Fig. 2.2), the rigging of fibrillin-rich fibrils that connects the lens to the adjacent

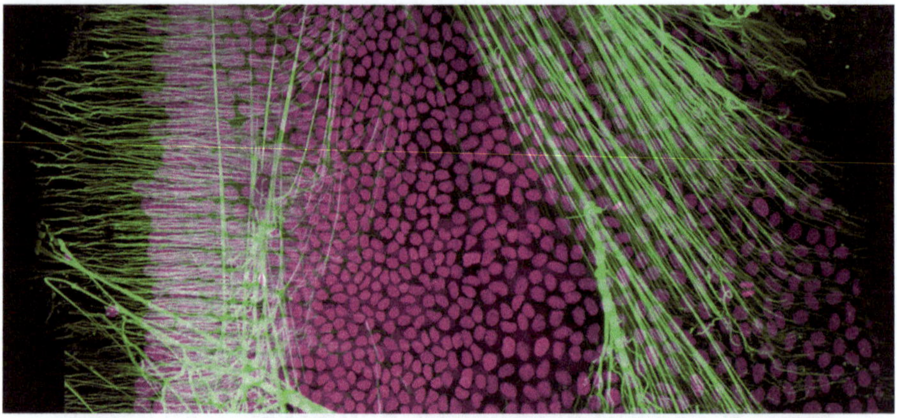

Fig. 2.2 Attachment of the ciliary zonule (*green*, visualized with anti-Magp1) at the lens equator. The arrangement of lens epithelial nuclei is visualized with Draq5 staining (*magenta*). On the left-hand side of the figure, cell nuclei are aligned in rows, signifying the onset of fiber cell differentiation. Mitotic figures are visible among the interphase epithelial nuclei

ciliary body and thus centers the lens in the eye [11]. The zonular fibers insert into the lens capsule and in accommodating species, such as humans, transmit the forces that flatten the lens when the eye focuses on distant objects.

2.2 Junctional Organization and Polarity in the Lens Epithelium

In common with all simple epithelia, the lens epithelium is highly polarized. As a result, the plasma membrane is partitioned into distinct apical and basolateral domains (asymmetry in the plane of the epithelium, so-called planar cell polarity, will be dealt with elsewhere). Because the lens epithelium arises from an invagination of the embryonic head ectoderm, epithelial cells are oriented inwards; their apical membranes apposed to the apical membranes of the underlying fiber cells. The highly folded basal membranes of the epithelial cells are in direct contact with the overlying capsule. The presence of an elaborate junctional complex at the apicolateral border of the cells allows the distinct compositions of the apical and basolateral membranes to be established and maintained. Examples of membrane proteins that show a segregated distribution in lens epithelial cell plasma membranes include ZO-1, which is restricted to the apical membrane domain [12, 13], and Cadm1, which is localized exclusively to the basolateral domain [14].

Contact points between neighboring cells in simple epithelia usually contain three distinct types of adhesive structures that can be recognized at the ultrastructural level. Proceeding from the apical to the basal surface these are tight junctions (zonula occludens), cadherin-based adhering junctions (zonula adherens), and desmosomes (macula adherens).

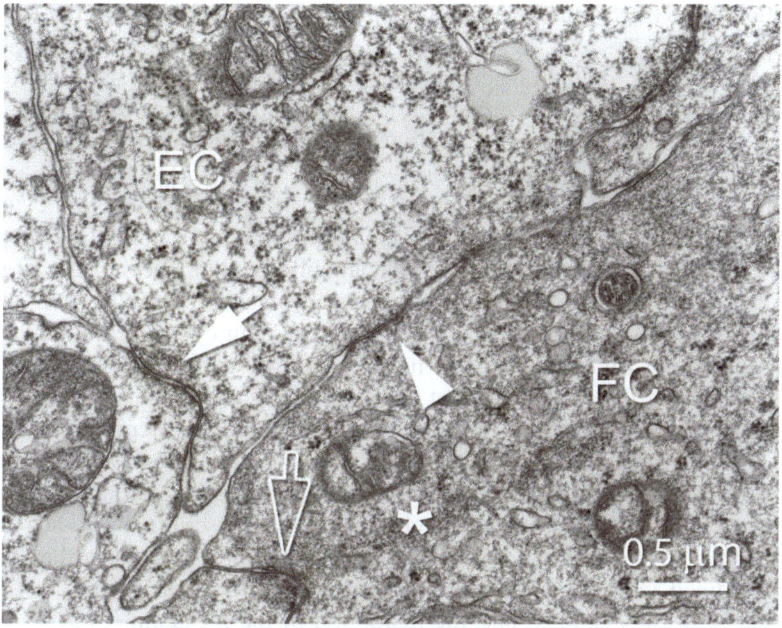

Fig. 2.3 Transmission electron micrograph of adhering junctions (AJs) between epithelial cells (*filled arrow*), between epithelial and fiber cells (*arrowhead*), and between fiber cells (*open arrow*) in the chicken lens. A band of actin microfilaments (*asterisk*) is associated with the fiber cell AJ. *EC* epithelial cell, *FC* fiber cell. Image courtesy of Ken Lo

Desmosomes, which impart tensile strength to epithelial sheets, are only infrequently observed in the lens epithelium [15], although they may be somewhat more common in the lenses of primates than in those of other species [16].

Transmission electron microcopy reveals that adhering junctions (AJs) are numerous between lens epithelial cells (Fig. 2.3) [17]. AJs are also found between epithelial cells and the underlying fiber cells [16], particularly near the edges of the epithelium [18]. The transmembrane adhesive core of AJs consists of a mosaic of clustered cadherin and nectin molecules [19]. Lens epithelial cells express both epithelial cadherin (E-cadherin; Cdh1) [20] and neuronal cadherin (N-cadherin; Cdh2) [21]. E-cadherin and N-cadherin are single-pass membrane proteins with five extracellular cadherin domains and are thus defined as "classical" cadherins. E-cadherin, as its name implies, is widely expressed in epithelial tissues. In the lens, its expression is restricted to the epithelium, the fiber cell membrane proteome containing only trace levels of E-cadherin [22]. The presence of N-cadherin in the lens epithelium is less expected. Elsewhere in the body, N-cadherin expression is restricted largely to neuronal cell types, the lens being one of the rare epithelial tissues to express this membrane protein [23]. Cadherins mediate calcium-*dependent* cell-cell adhesion through *trans*-cadherin interactions between neighboring cells. Conditional deletion of either E- or N-cadherin in the lens leads to profound cellular disruption, testifying to the key roles that AJs play in maintenance of the epithelial phenotype [24].

The cytoplasmic domain of cadherins contains a juxtamembrane region and a C-terminal, catenin-binding domain [25]. The juxtamembrane region binds p120 catenin (and associated proteins), while the catenin-binding domain interacts with β-catenin and γ-catenin (plakoglobin). p120 is thought to regulate cadherin stability. The pool of AJ-associated β-catenin exists in equilibrium with cytosolic, membrane, and nuclear pools, reflecting the multifunctional nature of the β-catenin protein. β-Catenin interacts with α-catenin, an obligate component of AJs. It seems likely that AJs, which are invariably positioned adjacent to the circumferential actin belt (see Fig. 2.3), are physically connected to the actin cytoskeleton. Such a link would generate a transcellular actin network, allowing mechanical forces to be distributed across the epithelial sheet as a whole. Because it is a bona fide actin-binding protein in vitro, it has long been suggested that α-catenin represents the physical linkage between cadherins in AJs and the actin cytoskeleton. However, the precise role of α-catenin remains controversial [26]. In addition to its role in regulating cell adhesion, β-catenin also functions as a transcriptional co-activator in the canonical Wnt signaling pathway [27]. The role of β-catenin in the lens has been examined in mice by conditionally disrupting the gene using the Cre-lox approach [28]. Interestingly, specific disruption of the β-catenin (*Ctnb1*) locus in lens fiber cells has comparatively little effect on the lens. In contrast, when *Ctnb1* is disrupted simultaneously in epithelial and fiber cells, the lens is profoundly disturbed, with loss of E-cadherin and disturbed apical/basal epithelial polarity. This observation underlines the importance of β-catenin in maintenance of the lens epithelial phenotype.

The second key group of adhesive proteins at AJs is the nectins. Nectins are members of the immunoglobulin superfamily and, in contrast to cadherins, mediate calcium-*independent* cell adhesion. Unlike the cadherins (which undergo only homophilic interactions in *trans*), nectins undergo both *trans*-homophilic and *trans*-heterophilic interactions at AJs. There are four members of the nectin family (nectin-1–4; encoded in humans by *PVRL1–4*), each of which have three extracellular domains, a transmembrane segment, and a cytoplasmic C-terminal domain that interacts with the PDZ domain of the scaffolding protein afadin. Afadin is a large, f-actin-binding protein that connects nectins to the actin cytoskeleton. The nectin–afadin complex is the first to be assembled at initial intercellular contact points and may help recruit cadherins to maturing AJs [29]. Both nectins and afadin are present in lens epithelial cell AJs [30]. Microarray analysis suggests that *PVRL3* (which encodes nectin-3) is the most abundantly expressed of the nectin genes in the eye [23]. In humans, mutations in *PVRL3* result in severe congenital cataracts [30]. Moreover, mice with targeted or spontaneous disruptions of the *Pvrl3* locus exhibit multiple lens defects. These observations underscore the importance of nectin-3 in lens morphogenesis and homeostasis.

AJs are the first elements of the epithelial junctional complex to be assembled at sites of intercellular contact. Their assembly triggers the activation of polarity complexes (see below) that help establish the asymmetric compositions of the apical and basolateral membranes. Surprisingly, the polarized organization of *cytoplasmic* structures, including the nucleus, Golgi apparatus, and centrosome, also appear to depend on AJ formation [31].

Tight junctions (TJs) are the apical-most components of the classical tripartite epithelial junctional complex. By transmission electron microscopy, TJs appear as close membrane appositions or "kissing points" between neighboring cells. Viewed in three dimensions, TJs appear as belt-like structures that extend in an anastomosing network around each cell, connecting cells to their neighbors and forming a tight seal across the extracellular space. Thus, the central role of TJs is to regulate paracellular permeability (diffusion, via the intercellular space from one side of the epithelium to the other). In addition to this barrier function, TJs have also long been thought to act as a "fence," preventing the intermingling of apical and basolateral components in the plane of the membrane, although recent evidence has cast some doubt on this latter role [32]. A great many proteins have been localized to the TJ, but the backbone of the TJ appears to be composed of claudins, a family of intrinsic membrane proteins. Hydropathy plots indicate that claudins have four transmembrane helices. Claudins interact with other claudins in the same cell through their N-terminal extracellular loops (*cis*-interactions) and with claudins in adjacent cells through their C-terminal extracellular loops (*trans*-interactions) [33]. Significantly, transfection of claudins into fibroblasts (a cell type that normally does not form TJs) is sufficient to trigger the formation of a network of TJ-like strands [34]. There are a number of other transmembrane protein components at TJs but these have less well-defined roles. This group includes occludin, tricellulin, JAM (junction adhesion molecule), and CAR (coxsackievirus and adenovirus receptor). At the cytoplasmic face of TJs, scaffolding proteins congregate, including ZO-1, ZO-2, and ZO-3.

For many years the existence of TJs in the lens epithelium was a contentious issue. Early freeze-fracture experiments gave conflicting results. The anastomosing strands indicative of TJs in freeze-fracture replicas were seen in some lens epithelia [35] but not others [36]. However, the use of electron dense tracers has since helped confirm the existence of TJs at the apicolateral border of lens epithelial cells [35]. Claudin-1 and occludin are expressed in lens epithelial cells [37]. Jam-1 is present in the lens, but mice deficient in *F11r* (the gene encoding Jam-1) do not have a lens phenotype [38]. In contrast, in humans, frameshift mutations in *JAM-3* result in intracranial hemorrhage and congenital cataracts [39]. Similarly, in mice, disruption of the *Jam-3* locus results in nuclear cataracts and other ocular pathology [40]. These findings are consistent with proteomic studies which show that Jam-3 is expressed particularly strongly in the lens [22].

Much of what is known about the establishment and maintenance of epithelial cell apical–basal polarity has come from studies of spindle orientation in *Drosophila* neuroblasts [41]. Remarkably, the protein complexes that determine the polarized distribution of cell fate determinants in dividing *Drosophila* neuroblasts are highly conserved and play equally pivotal roles in epithelial morphogenesis in vertebrates. Three distinct assemblages of polarity proteins are believed to control epithelial apical–basal polarity. These are the Crumbs complex (consisting of

Crumbs/Pals1/Patj), the partitioning defective (Par) complex (consisting of Par3/Par6/atypical protein kinase C (aPKC)/Cdc42), and the Scribble complex (consisting of Scrib/Dlg/Lgl). Crumbs is localized to the apical side of the AJ and is required for its formation. Par is associated with the tight junction. Together Par and Crumbs help specify the apical membrane domain. Scribble localizes to and helps define the basolateral membrane. In general, the complexes function to promote the establishment and expansion of the membrane domains with which they are associated.

The expression of polarity proteins in the lens epithelium has been examined in a number of studies. The Scribble complex members, Dlg-1 and Scrib, are widely expressed in the lens, often colocalizing with E- and N-cadherin at AJs [13]. Conditional deletion of Dlg-1 in the lens results in multilayering of the epithelium and redistribution of ZO-1 and E-cadherin [42].

The Par complex members aPKC, Par3, and Par6β colocalize with claudin-1 and occludin at lens epithelial cell TJs [37]. Conditional deletion of aPKC disrupts apical cell junctions and promotes epithelial-to-mesenchymal transition (EMT). At the edge of the lens epithelium, aPKC is also necessary for the formation of the "lens fulcrum," the region in which cells pivot through 180° as they begin the process of terminal cell differentiation [43].

During EMT, polarized epithelial cells adopt a fibroblastoid, motile phenotype. In the lens, this phenomenon is associated with posterior capsule opacification following extracapsular cataract surgery [44]. EMT involves loss of apical–basal polarity, reorganization of the cytoskeleton, and disassembly of the junctional complex. The Par polarity complex is targeted directly during TGFβ-induced EMT. During EMT, TGFβ receptors I and II (TGFβRI, II) associate with Par6 at TJs. Binding of TGFβ ligand to TGFβRII results in phophorylation of both TGFβRI and Par6. Phosphorylated Par6 mediates the destruction of RhoA at TJs and subsequent dissolution of the junctions [45]. Loss of E-cadherin, disintegration of AJs, and disappearance of the cortical actin ring are other characteristic early findings in EMT [46].

2.3 Proliferative Compartments in the Lens Epithelium

The lens increases in size and mass throughout life. In most species, growth is rapid initially but slows subsequently as, later in life, lens weight approaches some asymptotic maximum value [47]. Primate lenses appear to be an exception in that they follow a biphasic, growth pattern. During prenatal development in humans, for example, lens growth is rapid and asymptotic. After birth, however, growth is slow and linear across the remaining lifespan [48].

The ultimate driver of lens growth is cellular proliferation. Early investigators used the incorporation of tritiated thymidine by S-phase cells to visualize the distribution of mitoses in the lens. Such studies quickly established that mitotic cells are restricted to the lens epithelium [49–51]. In neonatal animals, S-phase cells are detected throughout the lens epithelium, but with age, cell division is

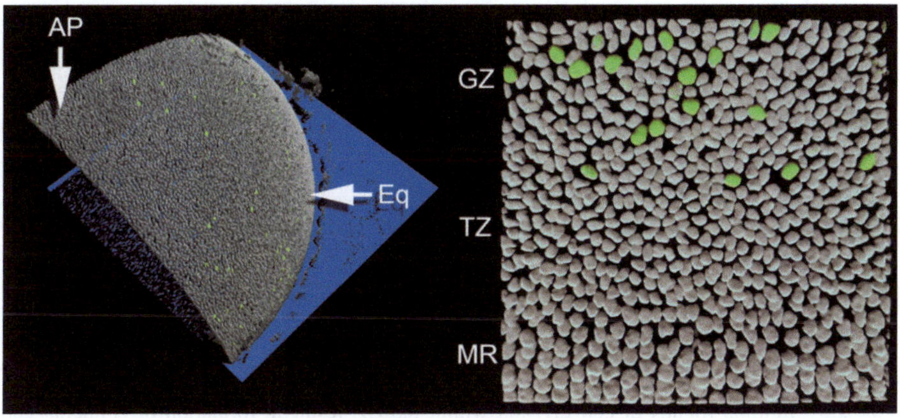

Fig. 2.4 Distribution of proliferating cells in 2-month-old mouse lens. S-phase cells (*green*) are labeled with the thymidine analog EdU. Mitosis is rare near the anterior pole (AP) of the lens but more common near the equator (Eq, *left panel*). High magnification, en face view of the lens equator (*right panel*) shows the proliferatively active germinative zone (GZ), the postmitotic transitional zone (TZ) at the epithelial margin, and the meridional rows (MR) formed as a result of nuclear alignment in differentiating fiber cells

increasingly concentrated in a band of epithelial cells that encircles the lens above the equator (Fig. 2.4). This region is called the proliferative or germinative zone (GZ) of the lens and contains most of the dividing cells. Between the germinative zone and the edge of the epithelium proper is a band of epithelial cells $\approx$ 10 cells wide. This transition zone (TZ) contains cells that have permanently withdrawn from the cell cycle. Epithelial cells in the TZ region express the cyclin-dependent kinase inhibitors $p27^{Kip1}$ and $p57^{Kip2}$ [52]. Together, Kip1 and Kip2 ensure timely exit of TZ epithelial cells from the cell cycle. Expression of both proteins persists in the fiber cell compartment, and combined knockout of *Kip1* and *2* in mice causes an over-proliferation defect in the lens and attendant apoptosis [53].

The striking, lifelong growth of the lens has led some authors to suggest that the epithelium may contain a contingent of tissue stem cells. A characteristic of many stem cells (e.g., limbal stem cells in the cornea) is that they divide only infrequently. When pre-labeled with BrdU, such cells therefore behave as “label-retaining cells” because the BrdU staining intensity is not diminished by multiple rounds of cell division. Cells in the central lens epithelium act as label-retaining cells, prompting the suggestion that this region harbors lens stem cells [54]. Other studies, however, have examined the expression of various stem cell markers and concluded that lens stem cells might instead reside anterior to [55] or within the GZ [56]. To date, no empirical studies have convincingly demonstrated the existence of a stem cell niche in the lens. It is equally plausible that lens epithelial cells are of a single type and that the proliferative behavior of a given cell is dictated entirely by its latitudinal position on the lens surface and its exposure to local growth factors. Interestingly, cells in the central epithelium (which are thought to be arrested in the G_0 phase of the cell cycle) reinitiate DNA synthesis following traumatic injury to the lens [57].

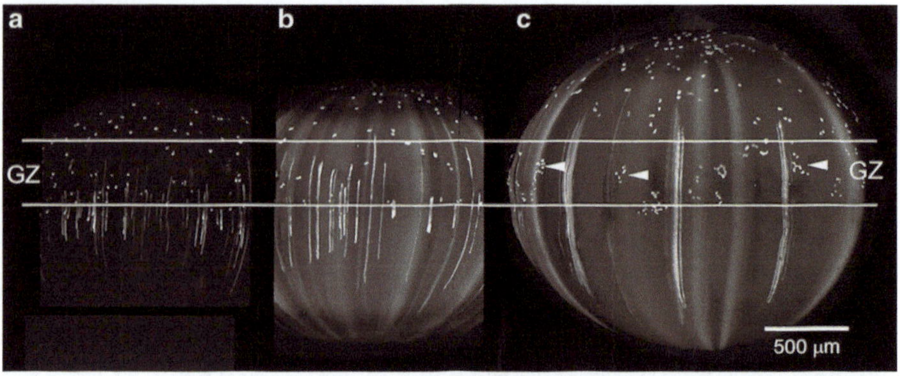

Fig. 2.5 Multiple rounds of cell division within the GZ result in formation of epithelial cell clones. GFP expression was induced in lenses of 3-week-old mice by tamoxifen treatment (see Shi and Bassnett [3]). Lenses were examined 3 days (**a**), 2 weeks (**b**), or 13 weeks (**c**) later. Note that although individual labeled epithelial cells are present initially, over time, only clusters of GFP-labeled cells (*arrowheads*) are detected within the GZ

Despite the significant increase in the anterior surface area of the lens that accompanies postnatal growth, there does not appear to be a corresponding increase in the number of epithelial cells [55]. Instead, the area covered by individual epithelial cells increases significantly. To retain epithelial cell constancy, therefore, cells lost from the epithelium due to fiber cell differentiation (or cell death) must be continuously replaced by epithelial cell division. Given that in adult lenses the central cells rarely if ever divide, new cells can only be introduced within the proliferatively active GZ. Addition of cells to this zone will inevitably result in posterior displacement of cells situated between the GZ and the lens equator. We have called this pattern of cellular displacement the "penny pusher" model of lens growth. Among other predictions, the penny pusher model suggests that the speed at which lens epithelial cells move will increase as cells approach the equator. Consider, for example, a cell situated near the anterior border of the GZ. It moves only slowly because the chance of cells dividing in more anterior regions (i.e., within the quiescent central epithelium) is very small. In contrast, for cells located near the equator, many mitoses are likely to occur in the GZ region that separates the equatorial cells from the central epithelium. Consequently, equatorial cell displacement will be much more rapid. A second important prediction of the penny pusher model is that cells are resident in the GZ long enough for several rounds of cell division to occur. If this is the case, then clones of cells should be produced as cells traverse the GZ. This prediction has been validated using induced expression of GFP in individual lens cell as a lineage tracer [5]. When GFP expression is triggered in random lens epithelial cells using the Cre-lox system, individual cells are labeled initially. However, in the succeeding weeks, clusters (clones) of GFP-expressing cells emerge in the GZ, and clone size gradually increases (Fig. 2.5). The synchronous differentiation of such clones results in the deposition of a cuneiform group of fluorescent fiber cells [5].

In young rodent lenses, the labeling index (percent of S-phase cells) in the GZ is 3–5 % [58, 59]. This is >20-fold higher than in the central epithelium. What accounts for the relatively high mitotic rate in the GZ? An attractive early hypothesis was that low concentrations of FGF might serve to stimulate epithelial cell proliferation in this zone [60]. The ability of *high* levels of FGF to stimulate lens fiber cell differentiation had been amply demonstrated in previous studies [61]. In vitro experiments suggested that at lower concentrations, FGF stimulates cell migration and proliferation rather than differentiation [60]. FGF levels are known to be higher in the vitreous humor than in the aqueous humor. Conceivably, therefore, a gradient of a single growth factor, FGF, might account for both the proliferative and differentiation behavior of cells in various locations on the lens surface. More recent studies have suggested that although FGF may indeed be implicated in the control of lens cell proliferation, it is unlikely to act in isolation. Work in several laboratories has established that lens epithelial cells express many growth factor receptors including the PDGF and EGF receptors. The distribution of some of these receptors parallels the proliferative behavior of the cells (i.e., the receptors are most abundant in the equatorial epithelium). Functional imaging studies have also shown that otherwise evenly distributed receptors often show enhanced responsiveness in the GZ [62]. Explanted lens epithelial cells retain the capacity to proliferate in response to treatment with FGF-depleted aqueous humor or with a range of growth factor ligands in vitro [63]. Based on these and other studies, it seems likely that no single growth factor is responsible for epithelial cell proliferation. Finally, recent studies have visualized the attachment of the ciliary zonule to the lens surface and correlated this with the distribution of S-phase cells in the underlying epithelium. Significantly, the GZ is located in the region spanned by the zonular fibers [11]. Cells that exist in physically dynamic environments often both sense and transduce the forces acting upon them. Thus mechanical stress can be transduced into signaling cascades that activate transcription factors and stimulate cell proliferation [64]. It would, perhaps, be surprising if cells near the lens equator were not sensitive to the forces generated by the zonular suspension system, particularly in species like humans, where the lens tissue is physically distorted each time we focus our eyes.

References

1. Bassnett S, Shi Y (2010) A method for determining cell number in the undisturbed epithelium of the mouse lens. Mol Vis 16:2294–2300
2. Kuszak JR, Costello MJ (2004) The structure of the vertebrate lens. In: Robinson ML, Lovicu FJ (eds) Development of the ocular lens. Cambridge University Press, Cambridge, pp 71–118
3. Shi Y, Bassnett S (2007) Inducible gene expression in the lens using tamoxifen and a GFP reporter. Exp Eye Res 85:732–737
4. Bassnett S (2005) Three-dimensional reconstruction of cells in the living lens: the relationship between cell length and volume. Exp Eye Res 81:716–723
5. Shi Y, Barton K, De Maria A, Petrash JM, Shiels A, Bassnett S (2009) The stratified syncytium of the vertebrate lens. J Cell Sci 122:1607–1615

6. Mallet JD, Rochette PJ (2011) Ultraviolet light-induced cyclobutane pyrimidine dimers in rabbit eyes. Photochem Photobiol 87:1363–1368
7. Haddad A, Bennett G (1988) Synthesis of lens capsule and plasma membrane glycoproteins by lens epithelial cells and fibers in the rat. Am J Anat 183:212–225
8. Halfter W, Candiello J, Hu H, Zhang P, Schreiber E, Balasubramani M (2013) Protein composition and biomechanical properties of in vivo-derived basement membranes. Cell Adh Migr 7:64–71
9. Ziebarth NM, Manns F, Uhlhorn SR, Venkatraman AS, Parel JM (2005) Noncontact optical measurement of lens capsule thickness in human, monkey, and rabbit postmortem eyes. Invest Ophthalmol Vis Sci 46:1690–1697
10. Danysh BP, Czymmek KJ, Olurin PT, Sivak JG, Duncan MK (2008) Contributions of mouse genetic background and age on anterior lens capsule thickness. Anat Rec (Hoboken) 291:1619–1627
11. Shi Y, Tu Y, De Maria A, Mecham RP, Bassnett S (2013) Development, composition, and structural arrangements of the ciliary zonule of the mouse. Invest Ophthalmol Vis Sci 54:2504–2515
12. Nielsen PA, Baruch A, Shestopalov VI et al (2003) Lens connexins alpha3Cx46 and alpha8Cx50 interact with zonula occludens protein-1 (ZO-1). Mol Biol Cell 14:2470–2481
13. Nguyen MM, Rivera C, Griep AE (2005) Localization of PDZ domain containing proteins discs large-1 and scribble in the mouse eye. Mol Vis 11:1183–1199
14. De Maria A, Shi Y, Luo X, Van Der Weyden L, Bassnett S (2011) Cadm1 expression and function in the mouse lens. Invest Ophthalmol Vis Sci 52:2293–2299
15. Kuwabara T (1975) The maturation of the lens cell: a morphologic study. Exp Eye Res 20:427–443
16. Lo WK (1988) Adherens junctions in the ocular lens of various species: ultrastructural analysis with an improved fixation. Cell Tissue Res 254:31–40
17. Lo WK, Shaw AP, Paulsen DF, Mills A (2000) Spatiotemporal distribution of zonulae adherens and associated actin bundles in both epithelium and fiber cells during chicken lens development. Exp Eye Res 71:45–55
18. Kuszak JR, Novak LA, Brown HG (1995) An ultrastructural analysis of the epithelial-fiber interface (EFI) in primate lenses. Exp Eye Res 61:579–597
19. Indra I, Hong S, Troyanovsky R, Kormos B, Troyanovsky S (2013) The adherens junction: a mosaic of cadherin and nectin clusters bundled by actin filaments. J Invest Dermatol 133:2546–2554
20. Nose A, Takeichi M (1986) A novel cadherin cell adhesion molecule: its expression patterns associated with implantation and organogenesis of mouse embryos. J Cell Biol 103:2649–2658
21. Hatta K, Takagi S, Fujisawa H, Takeichi M (1987) Spatial and temporal expression pattern of N-cadherin cell adhesion molecules correlated with morphogenetic processes of chicken embryos. Dev Biol 120:215–227
22. Bassnett S, Wilmarth PA, David LL (2009) The membrane proteome of the mouse lens fiber cell. Mol Vis 15:2448–2463
23. Lattin JE, Schroder K, Su AI et al (2008) Expression analysis of G protein-coupled receptors in mouse macrophages. Immunome Res 4:5
24. Pontoriero GF, Smith AN, Miller LA, Radice GL, West-Mays JA, Lang RA (2009) Co-operative roles for E-cadherin and N-cadherin during lens vesicle separation and lens epithelial cell survival. Dev Biol 326:403–417
25. Etienne-Manneville S (2011) Control of polarized cell morphology and motility by adherens junctions. Semin Cell Dev Biol 22:850–857
26. Brieher WM, Yap AS (2013) Cadherin junctions and their cytoskeleton(s). Curr Opin Cell Biol 25:39–46
27. Cadigan KM, Peifer M (2009) Wnt signaling from development to disease: insights from model systems. Cold Spring Harb Perspect Biol 1:a002881

28. Cain S, Martinez G, Kokkinos MI et al (2008) Differential requirement for beta-catenin in epithelial and fiber cells during lens development. Dev Biol 321:420–433
29. Takai Y, Ikeda W, Ogita H, Rikitake Y (2008) The immunoglobulin-like cell adhesion molecule nectin and its associated protein afadin. Annu Rev Cell Dev Biol 24:309–342
30. Lachke SA, Higgins AW, Inagaki M et al (2012) The cell adhesion gene PVRL3 is associated with congenital ocular defects. Hum Genet 131:235–250
31. Dupin I, Camand E, Etienne-Manneville S (2009) Classical cadherins control nucleus and centrosome position and cell polarity. J Cell Biol 185:779–786
32. Ikenouchi J, Suzuki M, Umeda K et al (2012) Lipid polarity is maintained in absence of tight junctions. J Biol Chem 287:9525–9533
33. Piontek J, Winkler L, Wolburg H et al (2008) Formation of tight junction: determinants of homophilic interaction between classic claudins. FASEB J 22:146–158
34. Furuse M, Sasaki H, Fujimoto K, Tsukita S (1998) A single gene product, claudin-1 or -2, reconstitutes tight junction strands and recruits occludin in fibroblasts. J Cell Biol 143:391–401
35. Lo WK (1987) In vivo and in vitro observations on permeability and diffusion pathways of tracers in rat and frog lenses. Exp Eye Res 45:393–406
36. Goodenough DA, Dick JS 2nd, Lyons JE (1980) Lens metabolic cooperation: a study of mouse lens transport and permeability visualized with freeze-substitution autoradiography and electron microscopy. J Cell Biol 86:576–589
37. Sugiyama Y, Prescott AR, Tholozan FM, Ohno S, Quinlan RA (2008) Expression and localisation of apical junctional complex proteins in lens epithelial cells. Exp Eye Res 87:64–70
38. Kang LI, Wang Y, Suckow AT et al (2007) Deletion of JAM-A causes morphological defects in the corneal epithelium. Int J Biochem Cell Biol 39:576–585
39. Mochida GH, Ganesh VS, Felie JM et al (2010) A homozygous mutation in the tight-junction protein JAM3 causes hemorrhagic destruction of the brain, subependymal calcification, and congenital cataracts. Am J Hum Genet 87:882–889
40. Daniele LL, Adams RH, Durante DE, Pugh EN Jr, Philp NJ (2007) Novel distribution of junctional adhesion molecule-C in the neural retina and retinal pigment epithelium. J Comp Neurol 505:166–176
41. Wodarz A (2005) Molecular control of cell polarity and asymmetric cell division in Drosophila neuroblasts. Curr Opin Cell Biol 17:475–481
42. Rivera C, Yamben IF, Shatadal S, Waldof M, Robinson ML, Griep AE (2009) Cell-autonomous requirements for Dlg-1 for lens epithelial cell structure and fiber cell morphogenesis. Dev Dyn 238:2292–2308
43. Sugiyama Y, Akimoto K, Robinson ML, Ohno S, Quinlan RA (2009) A cell polarity protein aPKClambda is required for eye lens formation and growth. Dev Biol 336:246–256
44. Eldred JA, Dawes LJ, Wormstone IM (2011) The lens as a model for fibrotic disease. Philos Trans R Soc Lond B Biol Sci 366:1301–1319
45. Viloria-Petit AM, Wrana JL (2010) The TGFbeta-Par6 polarity pathway: linking the Par complex to EMT and breast cancer progression. Cell Cycle 9:623–624
46. Tiwari N, Gheldof A, Tatari M, Christofori G (2012) EMT as the ultimate survival mechanism of cancer cells. Semin Cancer Biol 22:194–207
47. Augusteyn RC (2008) Growth of the lens: in vitro observations. Clin Exp Optom 91:226–239
48. Augusteyn RC (2010) On the growth and internal structure of the human lens. Exp Eye Res 90:643–654
49. Harding CV, Hughes WL, Bond VP, Schork P (1960) Autoradiographic localization of tritiated thymidine in wholemount preparations of lens epithelium. Arch Ophthalmol 63:58–65
50. Hanna C, O'Brien JE (1961) Cell production and migration in the epithelial layer of the lens. Arch Ophthalmol 66:103–107
51. Thomson DS, Pirie A, Overall M (1962) Autoradiography of lens epithelium after parenteral injection of tritiated thymidine. Arch Ophthalmol 67:464–469

52. Nagahama H, Hatakeyama S, Nakayama K, Nagata M, Tomita K (2001) Spatial and temporal expression patterns of the cyclin-dependent kinase (CDK) inhibitors p27Kip1 and p57Kip2 during mouse development. Anat Embryol (Berl) 203:77–87
53. Zhang P, Wong C, DePinho RA, Harper JW, Elledge SJ (1998) Cooperation between the Cdk inhibitors p27(KIP1) and p57(KIP2) in the control of tissue growth and development. Genes Dev 12:3162–3167
54. Zhou M, Leiberman J, Xu J, Lavker RM (2006) A hierarchy of proliferative cells exists in mouse lens epithelium: implications for lens maintenance. Invest Ophthalmol Vis Sci 47:2997–3003
55. Yamamoto N, Majima K, Marunouchi T (2008) A study of the proliferating activity in lens epithelium and the identification of tissue-type stem cells. Med Mol Morphol 41:83–91
56. Oka M, Toyoda C, Kaneko Y, Nakazawa Y, Aizu-Yokota E, Takehana M (2010) Characterization and localization of side population cells in the lens. Mol Vis 16:945–953
57. Harding CV, Donn A, Srinivasan BD (1959) Incorporation of thymidine by injured lens epithelium. Exp Cell Res 18:582–585
58. Mikulicich AG, Young RW (1963) Cell proliferation and displacement in the lens epithelium of young rats injected with tritiated thymidine. Invest Ophthalmol 2:344–354
59. Rafferty NS, Rafferty KA Jr (1981) Cell population kinetics of the mouse lens epithelium. J Cell Physiol 107:309–315
60. McAvoy JW, Chamberlain CG (1989) Fibroblast growth factor (FGF) induces different responses in lens epithelial cells depending on its concentration. Development 107:221–228
61. Lovicu FJ, McAvoy JW (2005) Growth factor regulation of lens development. Dev Biol 280:1–14
62. Maidment JM, Duncan G, Tamiya S, Collison DJ, Wang L, Wormstone IM (2004) Regional differences in tyrosine kinase receptor signaling components determine differential growth patterns in the human lens. Invest Ophthalmol Vis Sci 45:1427–1435
63. Iyengar L, Patkunanathan B, McAvoy JW, Lovicu FJ (2009) Growth factors involved in aqueous humour-induced lens cell proliferation. Growth Factors 27:50–62
64. Mendez MG, Janmey PA (2012) Transcription factor regulation by mechanical stress. Int J Biochem Cell Biol 44:728–732

The Lens Capsule: Synthesis, Remodeling, and MMPs

3

Judith A. West-Mays and Anna Korol

Abstract

The lens capsule is an amorphous, elastic structure that encapsulates the ocular lens. It is secreted by cells of the lens and is composed primarily of type IV collagen and laminin along with additional extracellular matrix (ECM) components such as entactin/nidogen, heparin sulfate proteoglycans (HSPG), and secreted protein acidic and rich in cysteine (SPARC), which act to stabilize the lens capsule structure. As the lens grows during development and with age, new capsular lamellae are synthesized, deposited, and organized by the lens epithelial and fiber cells. The main receptors of lens cells that adhere them to the ECM of the capsule are the integrins, heterodimeric transmembrane cell adhesion molecules. These adhesion molecules also act as bidirectional signaling molecules, mediating signals between the lens and the surrounding ocular media. The composition and arrangement of both the ECM of the lens capsule and the integrins are altered in fibrotic cataracts such as posterior capsule opacification (PCO) and anterior subcapsular cataract (ASC). This includes the aberrant deposition of ECM components not normally expressed in the lens capsule and a corresponding change in the profile of integrins expressed in the lens. The matrix metalloproteinases (MMPs), a family of matrix-degrading enzymes, have been shown to release growth factors from the lens capsule and activate receptors. They are also aberrantly expressed in PCO and ASC and their inhibition has been shown to suppress events involved in fibrotic cataract formation including lens epithelial cell migration, capsular contraction, and the transformation of cells into myofibroblasts.

Keywords

Extracellular matrix • Integrins • Lens capsule • Matrix metalloproteinases

J.A. West-Mays (✉) • A. Korol
Department of Pathology and Molecular Medicine, McMaster University Health Science Centre, Rm 1R10, 1200 Main St. West, Hamilton, ON, Canada L8N 3Z5
e-mail: westmayj@mcmaster.ca; korola3@mcmaster.ca

S. Saika et al. (eds.), *Lens Epithelium and Posterior Capsular Opacification*, DOI 10.1007/978-4-431-54300-8_3, © Springer Japan 2014

3.1 Lens Capsule Synthesis and Composition

3.1.1 Synthesis

As the lens pinches off from the overlying ectoderm during embryonic development, it becomes encapsulated by a basement membrane, known as the lens capsule. The lens capsule is an acellular, transparent elastic structure that surrounds the ocular lens, separating it from the surrounding ocular media. In addition to its function in compartmentalizing the lens, it is responsible for maintenance of the lens structures, control of fluid, and substrate exchange and regulation of normal lens epithelial cell growth and differentiation [1]. The capsule is formed through the deposition of multiple layers of basal lamina, predominantly composed of the extracellular matrix (ECM) components type IV collagen and laminin [2–8]. It is secreted by the lens epithelial and fiber cells, and as the lens grows rapidly during embryogenesis and throughout adult life, so must the capsule and this is accomplished by the ongoing deposition of matrix lamellae to its inner surface [9]. Once development is completed, the anterior lens epithelial cells continue to synthesize capsule matrix, whereas the posterior fiber cells are more limited in their ability to secrete new basal lamina. As a result, in the adult lens, the anterior capsule is thicker than the posterior capsule, the latter of which is thinnest at the posterior pole of the lens [9].

Lens epithelial cells and lens fiber cells in the outer cortical region remain in contact with the lens capsule and only when fibers cells reach the lens suture do they lose contact with it. Interestingly, as the lens differentiates, the matrix of the lens capsule also changes suggesting that it plays an active role in development and differentiation of the lens. For example, the ECM component fibronectin, which is known to control cell migration, is expressed during earlier stages of lens development but is absent in the lens during later stages of differentiation [7]. Cell–ECM interactions are known to be important for facilitating a variety of morphogenetic events during development and differentiation of tissues. In the lens it has been shown that ECM components of the capsule, such as laminin and nidogen, among others are required for normal lens development [10–12], while other ECM molecules such as type IV collagen are important for maintenance of the lens epithelial phenotype in the adult lens [1, 7, 13, 14]. Other matrix molecules are important for promoting lens cell differentiation, including laminin and heparan sulfate proteoglycan [15].

3.1.2 Composition

The lens capsule is a structurally complex ECM structure consisting of a meshwork of various glycoproteins and proteoglycans [16]. The ECM composition of the lens capsule has been determined for a number species and includes predominantly type IV collagen and laminin, as well as entactin/nidogen, heparin sulfate proteoglycans (HSPG), fibronectin, tenascin, and secreted protein acidic and rich in cysteine

(SPARC) [2–8, 17–19]. The mature lens consists of all of the α subunit chains of type IV collagen, which form a cross-linked network that provides stability to the capsule [20]. The importance of type IV collagen in maintaining lens capsule structure is demonstrated by the fact that patients with mutations in the α5 chain of type IV collagen have Alport syndrome and present with thinning of and fragility of the capsule, which can lead to capsular rupture [21]. Type IV collagen in the lens capsule has also been shown to be important in regulating lens epithelial signaling and cell survival. For example, primary LEC and lens cell lines cultured on type IV collagen have been shown to be protected against FAS-induced apoptosis [13].

Laminin is another major ECM component of the developing and adult lens capsule. It is a heterotrimeric protein comprised of α, β, and γ chains, and while all of these chains are expressed in the developing lens, the α1 chain has mainly been detected in the adult lens capsule [2, 6, 7, 22, 23]. Laminin, like type IV collagen, has been shown to promote and maintain lens epithelial cell survival and their normal phenotype. Indeed, mutations in the β2 chain of laminin in humans have been shown to cause Pierson syndrome in which ocular anomalies include posterior lenticonus [24, 25]. Studies in zebrafish have demonstrated that a knockdown of the α1 chain of laminin results in arrested lens development at the lens vesicle stage further suggesting that laminin is important in regulating lens differentiation [12].

Entactin (nidogen) is an additional constituent of the lens capsule and it binds to type IV collagen, laminin, and HSPG to help stabilize the supramolecular structure of the capsule [26, 27]. In particular, entactin 1 is the main type detected in the adult lens capsule. It has been shown that mice lacking entactin 1 exhibit a posterior lens capsule that is invaded by fiber cell processes, and the fiber suture is disrupted [10]. HSPG are lens capsule components that also bind to the other capsule molecules and also play a role in presenting growth factors, such as the fibroblast growth factors (FGFs), to lens epithelial and fiber cells [28].

SPARC is a more recently identified component of the capsule. It plays a role in ECM secretion and binds to ECM components such as collagen [17, 18]. SPARC has anti-adhesive properties reducing focal contacts and cell–cell and cell–matrix adhesion [29]. Its expression is highest in the peripheral lens epithelium, terminating in the equatorial region where cells differentiate into lens fiber cells [17, 19, 30–34]. In SPARC-null mice cataracts are evident by 3–4 months and the lens phenotypes consist of abnormal cell protrusions in the lens capsule as well as abnormal expression of laminin and type IV collagen [35].

Finally, the expression of additional ECM components, fibronectin, tenascin, and vitronectin is thought to occur in the developing lens capsule and not in the normal adult lens capsule [16]. Expression of these ECM constituents, however, is detected in the lens during cataract formation as discussed in a subsequent section (Sect. 1.3).

The lens capsule is not a static structure and is thought to undergo continuous remodeling. Matrix is produced anteriorly by the lens epithelial cells and posteriorly by the newly differentiated fiber cells [36]. As SPARC regulates the production of ECM proteins and binds to several collagens including the capsule type IV collagen, it is thought to be an important regulator of the organization and assembly of the components in the lens capsular matrix [37–42]. Less is known about the degradation and turnover of the ECM of the capsule. However, enzymes such as the matrix metalloproteinases (MMPs), which are known to remodel the ECM, are secreted constitutively by lens epithelial cells and further induced during lens fibrosis (see Sect. 1.4 for more on MMPs).

3.2 Cell–ECM Interactions of the Lens Capsule (Integrins)

Integrins are the main receptors for the ECM molecules in the lens capsule and have been associated with critical events in development and differentiation, morphogenesis, and migration [1, 16, 43]. They are a family of glycosylated, heterodimeric transmembrane cell adhesion molecules that form an interaction with their respective ECM ligands, enabling lens cells to adhere and migrate across the ECM. Each integrin is a heterodimer consisting of an α and β subunit and currently 18 α and 8 β subunits have been identified in mammals that together form 24 different integrin receptors, which bind to a specific ligand or set of ligands [1, 16, 43]. As signaling receptors they are considered to act bidirectionally due to their transmembrane nature and their interaction with the actin-based cytoskeleton. Thus, integrin signaling has been described as both "inside-out" (transmitting signals from within the cells to the integrin activity on the cell surface) and "outside-in" (transmitting extracellular signals into the cell) [1, 16, 43]. Based on the complex nature of the lens capsule matrix and the importance of this matrix in signaling it is not surprising that the lens expresses a number of different integrin receptors during development and in adulthood.

During development, the lens expresses an extensive array of integrins that corresponds to the ECM ligands that are present. Thus, it is likely that each integrin has a unique function in regulating lens differentiation. Embryonic mouse and chick lenses were shown to express α2, α6A, α3, and β1 subunits [44, 45]. More recent reports describe the expression of α9, α10, α11, αV, β5, and β8 subunits as well [16]. In the adult lens, integrins include the laminin receptors, α3β1, α6β1, and α6β4, type IV collagen receptors, α1β1 and α2β1, fibronectin receptor α5β1, and the tenascin receptor αVβ3 [16]. Microarray analyses of human lens epithelial cells revealed that the β1, β2, β3, β5, and β6 subunits are expressed with the β1 integrin subunit being the most abundantly expressed [46]. With regard to the α subunits, these cells expressed α1, α2, α3, α4, α5, α6, α7, α10, α11, αE, αM, and αV [46].

β1 integrins serve as heterodimeric partners for twelve different α integrin subunits and are expressed in many regions of the developing and adult lens. For example, they are found along the basal surfaces of lens epithelial and fiber cells where they contact ECM ligands in the capsule, as well as along the epithelial–fiber cell interface and in these cases likely involved in cell–cell interactions [47–50]. In cell cultures, function blocking antibodies to β1-integrin attenuate the ability of lens cells to bind to collagen and laminin suggesting that this molecule is important for lens cell–capsule communication [51]. However, β1-integrin is also found on the lateral and apical sides of lens cells in regions lacking lens capsule components [48, 49]. In the equatorial zone, β1-integrin's partner, α6, interacts with the IGF receptor and regulates ERK phosphorylation suggesting that integrin–growth factor receptor crosstalk is important for lens morphogenesis [23, 52].

The importance of integrins in lens development and maintenance of the capsule is further shown by studies of null mutations in mice. For example, mice with a double deletion of the α3 and α6 subunits ($\alpha3/\alpha6^{-/-}$) exhibit a lens phenotype consisting of a disrupted capsule and epithelium at E13.5, along with fiber cells extruding into the cornea [53]. In addition, conditional deletion of the β1 subunit in all cells of the lens at the vesicle stage of development results in a disorganized lens epithelium that expresses fiber cell markers, such as the β and γ crystallins [54]. The mutant lens epithelial cells also undergo epithelial-to-mesenchymal transition (EMT) as determined by the expression of α-smooth muscle actin (αSMA). These data demonstrated that β1-integrin is required to maintain the lens epithelial phenotype.

3.3 Altered ECM and Integrins of the Lens Capsule During PCO

3.3.1 Aberrant ECM in PCO

During lens pathologies, such as posterior capsular opacification (PCO) and anterior subcapsular cataract (ASC), lens cells begin to secrete aberrant amounts of ECM that are not normally detected in the lens capsule, including type I collagen, fibronectin, and vitronectin [55–59]. Both PCO and ASC are considered to be fibrotic cataracts with overlapping features yet distinct etiologies. PCO occurs when LECs remain within the capsule after cataract surgery and are triggered to proliferate and migrate to the posterior lens capsule [55, 57, 60, 61]. A proportion of these cells undergo a transition into myofibroblasts, through a phenomenon known as epithelial-to-mesenchymal transition (EMT). This is accompanied by the upregulation of filaments, such as α smooth muscle actin (αSMA) and desmin [58, 59]. As the myofibroblasts accumulate they deposit aberrant types and amounts of ECM and cause capsular wrinkling and as a result opacities develop. ASC is a primary cataract that develops following a pathological insult such as ocular trauma, surgery, or, systemically, as with diseases like atopic dermatitis and retinitis pigmentosa [1, 42]. Similar to PCO, in ASC, LECs are triggered to proliferate and transform into myofibroblasts, through EMT [55–57]. These myofibroblasts also deposit aberrant amounts and types of ECM and as a result, fibrotic plaques are formed directly beneath the anterior lens capsule, which cause light scattering.

The cytokine, transforming growth factor-β (TGFβ), is one of the most important modulators of these fibrotic cataracts and is also considered to be a critical modulator of the ECM. Indeed treatment of LECs with active TGFβ results in an aberrant production and deposition of matrix molecules such as collagen types I and III, fibronectin, and tenascin [62]. In turn, the accumulation of this aberrant matrix has also been shown to influence cell behavior, further promote EMT and cell migration and thereby exacerbate the fibrotic pathology. Thus, the preservation of the normal ECM environment for LECs, namely the ECM of the native lens capsule, is important for maintaining the normal lens phenotype and preventing LEC apoptosis and EMT. Classic experiments have demonstrated that when lens epithelial explants (lens cells remaining adhered to their capsule) were placed in a 3D type I collagen gel, the cells that made contact with the type I collagen underwent EMT, whereas those that remained in contact with their native basement membrane were protected and retained their epithelial phenotype [63]. Interestingly, the cells that underwent EMT no longer expressed laminin, a major component of the normal lens capsule.

A number of in vitro and in vivo models of PCO and ASC have been developed in order to understand the mechanism underlying these pathologies. Models for ASC have been developed in rats and mice, including rat lens explants treated with TGFβ, transgenic mice expressing active TGFβ in the lens (driven by a crystallin promoter), a lens puncture model, and adenoviral-mediated TGFβ overexpression in the lens of both rats and mice [64–68]. The subcapsular plaques that develop in these models closely resemble what has been reported for human ASC including the aberrant expression and elaboration of collagen types I and III, tenascin, and fibronectin [69]. Interestingly, the proteoglycan lumican is another ECM component implicated in promoting ASC formation, since lumican-null mice have delayed expression of αSMA, the marker of transdifferentiation [70]. Similarly, in PCO-related events, the ECM components vitronectin and fibronectin are deposited and have been shown to stimulate migration and induce αSMA expression. ECM players such as type I collagen have also been shown to provide a more rigid environment that provides further cues to promote fibrosis. For example, rigid microenvironments have been shown to cause myofibroblasts to release latent TGFβ from the ECM, demonstrating that mechanotransduction from the matrix contribute to the progression of lens fibrotic diseases like PCO [43].

3.3.2 Altered Integrins in PCO

Along with changes in the lens matrix that occur during PCO and ASC, corresponding changes in their receptors, the integrins, have also been well documented. In the classic study outlined earlier, in which lens epithelial cells are plated in an environment of type I collagen and undergo EMT, these cells also exhibit an altered profile of integrins: the laminin ligand α6 integrin is downregulated, while the α5 integrin, the ligand for fibronectin, a typical marker of mesenchymal cells, is induced [63]. Treatment of human lens epithelial cells with TGFβ resulted in increased expression of α5, α11, αV, and β5 integrin subunits [46]. Of particular interest, α5β1 integrin is upregulated by TGFβ in a human capsular bag model (PCO model) and a human lens cell line [46, 71]. This is not surprising since its ligand, fibronectin, is upregulated in these models and the interaction of this integrin with fibronectin is thought to contribute to αSMA expression and LEC transdifferentiation. Patients with ASC also exhibit a colocalized expression of α5β1 with fibronectin and αSMA in plaque cells [72]. The expression of β1-integrin is also upregulated during EMT occurring as a consequence of lens epithelial cell suspension in collagen I gels [63] and in response to TGFβ treatment. In fact, it has been proposed that interference with β1-integrin function may be clinically useful to block EMT of lens cells leading to PCO, a common side effect of modern cataract surgery [73, 74].

In other models of ASC and PCO, the αV integrins, which are subunits of integrin receptors for tenascin, vitronectin, and fibronectin, are also upregulated following TGFβ-induced EMT in the lens [43]. The αVβ5 integrin is thought to be important in the development of fibrotic pathologies due to its role in mechanotransduction signals from the ECM, which causes transdifferentiation of cells to myofibroblasts [75, 76]. Injury by mechanical trauma is thought to modulate expression of this integrin, which has relevance to the fibrosis that occurs after cataract surgery (PCO). In human capsular bags, αVβ6 integrin expression was increased compared to cultured, intact whole lenses that have not been injured [77]. Interestingly, αVβ6 is thought to activate TGFβ through its association with an RGD peptide in the latency-associated peptide [78]. Thus, both αVβ5 and αVβ6 may have important roles in sustaining TGFβ activation and further promoting fibrotic events like PCO.

Finally, integrin-linked kinase (ILK), a serine–threonine kinase that binds to the cytoplasmic tails of β1, β2, and β3 subunits, is weakly expressed in the lens, but has found to be upregulated in TGFβ transgenic lenses and correlated with LEC EMT [62, 79]. Also, ILK has been shown to colocalize with α5β1 and this was enhanced in the presence of fibronectin, suggesting that ILK may be involved in EMT via this interaction [79].

3.4 MMPs in Lens Capsule Remodeling and PCO

Beyond lens capsule-associated integrin signaling, lens cell proliferation and survival are also dependent upon the availability and presence of growth factors and cytokines. The lens capsule houses a multitude of growth factors and cytokines, which are normally ECM-sequestered in an inactive state and upon their release are free to activate signaling sequences that are propagated to the cell nucleus [80, 81]. The matrix metalloproteinases (MMPs) are a family of zinc-dependent endopeptidases responsible for the degradation of the structural components of the ECM, including that found in the lens capsule. From their initial discovery in the eye in corneal wound healing [82], MMPs have since been implicated in a number of ocular pathologies including proliferative retinopathies [83, 84], glaucomatous optic neuropathy [85], corneal disorders [86], scleritis [87], uveitis [88], pterygium [89], and macular degeneration [90]. More recently, MMPs have been studied in the lens and have been shown to play an integral role in the development of fibrotic cataracts, such as ASC and PCO [64, 91].

The MMP family consists of at least 23 structurally related members that can be classified as membrane-bound (MMP-14, MMP-15, MMP-16, MMP-17, MMP-24, MMP-25) or membrane-secreted proteins, which are further subdivided into four categories: collagenases (MMP-1, MMP-8, MMP-13), gelatinases (MMP-2, MMP-9), stromelysins (MMP-3, MMP-10, MMP-11, MMP-12, MMP-28), and matrilysins (MMP-7, MMP-26). MMPs are primarily regulated at the transcriptional level by growth factors, hormones, and cytokines, as well as cell–cell and cell–matrix interactions [92, 93]. Their activity is localized near the cell surface and depends upon the balance between MMP activators and inhibitors, such as tissue inhibitors of metalloproteinases (TIMPs), in the extracellular space [94, 95]. Each MMP selectively degrades components of the ECM leading to alterations in the surrounding microenvironment.

In the lens, multiple MMPs and TIMPs are constitutively expressed in a variety of species (Table 3.1) [91]. Human LEC lines exhibit endogenous expression of *MMP-2*, *MMP-9*, *MMP-14*, *TIMP-2*, and *TIMP-3* [46, 101, 108]. The role of constitutively expressed MMPs in the lens is not well understood; however, one study demonstrated that LECs grown on bovine lens capsules were protected from stress-related apoptosis due to MMP-2-dependent FGF-2 release from the lens capsule, facilitating cell survival [113]. This revealed MMP-2 as a potential survival factor and demonstrated a need for proteolytic processing of the capsule for the release of growth factors necessary for LEC viability and survival.

The majority of work on MMPs in the lens has focused on their induction following treatment with growth factors or in response to injury or stress (Table 3.1) [91]. For example, induced secretion of MMPs occurs from LECs following lens injury by ultraviolet irradiation, oxidative stress, or cataract surgery [91, 96, 97, 106, 110, 114, 115]. In addition, increased levels of MMPs are detected in the media of cultured cataractous lenses compared to normal non-cataractous lenses, as well as capsular bags following sham cataract surgery [110]. Following cataract surgery, the migration of residual LECs from the anterior lens to the posterior capsule requires ECM contraction that is triggered and mediated by MMPs. Correspondingly, inhibition of MMP activity results in reduced LEC migration and capsule contraction both in vitro and in in vivo capsular bag cultures [98, 107, 116–118].

The main structural components of the lens capsule, collagen IV and laminin, are substrates for the gelatinases, MMP-2 and MMP-9 [92, 119, 120] (Fig. 3.1). Thus, the gelatinases are the most widely studied MMPs in the lens, where normally they are expressed at low levels (Table 3.1) [100]. The induction of both MMP-2 and MMP-9 expression was demonstrated in primary chick lens cells as well as whole rat lenses following, and specific to, treatment with TGFβ [99, 111]. MMP-2 and MMP-9 are also induced by TGFβ (human) in lens capsular bags [59, 107]. To test whether MMPs promote PCO, a broad-spectrum MMP inhibitor, GM6001, was utilized on human donor lens capsules and revealed a significant inhibition in the migration of LECs [107]. A significant reduction in capsular contraction was also observed in the GM6001-treated capsular bags. In addition, MMP inhibitors have been shown to suppress ASC formation [64]. For example, using the excised rat lens model in which whole lenses were cultured with TGFβ, it was shown that co-treatment of the excised lenses with TGFβ and either GM6001, the broad MMP inhibitor (MMPI), or a MMP-2/MMP-9 specific inhibitor significantly suppressed the formation of ASC plaques [64]. Importantly, it was further shown that inhibition

Table 3.1 Expression of MMPs in the lens

Study (year)	Region of the lens/cell line	Species	Detection method	MMP/TIMP (constitutive (C) or induced (I) expression)
Dadoukis et al. (2013) [96]	Whole lens	Rabbit	Zymography	MMP-2[I] by radiation
Alapure et al. (2012) [97]	Anterior lens capsules	Human (postmortem)	RT-QPCR	MMP-2[I] MMP-9[I]
Eldred et al. (2012) [98]	FHL 124 cells	Human	ELISA Western blot	MMP-2[C] and [I] by TGFβ MT1-MMP[C] and [I] by TGFβ
Nathu et al. (2009) [99]	Lens cells laser captured	Rat	RT-QPCR	MMP-2 [C] and [I] by TGFβ MMP-9 [C] and [I] by TGFβ MT1-MMP [C] and [I] by TGFβ TIMP 1 [C] and [I] by TGFβ
Hodgkinson et al. (2007) [100]	Whole lens (regions)	Human (postmortem)	RT-QPCR	MMP-2 [C] MMP-9 [C] MMP-10 [C] MMP-11 [C] MMP-14 [C] MMP-15 [C] MMP-16 [C] MMP-17 [C] MMP-19 [C] MMP-21 [C] MMP-23 [C] MMP-24 [C] MMP-28 [C] TIMP-1 [C] TIMP-2 [C] TIMP-3 [C] TIMP-4 [C]

(continued)

Table 3.1 (continued)

Study (year)	Region of the lens/cell line	Species	Detection method	MMP/TIMP (constitutive (C) or induced (I) expression)
Chang et al. (2007) [101]	HLE cells	Human	RT-QPCR	MMP-2[I] by radiation MMP-3[I] by radiation MMP-9[I] by radiation
Chang et al. (2007) [101]	HLF cells	Human	RT-QPCR	MMP-2[I] by radiation MMP-3[I] by radiation MMP-9[I] by radiation
Dawes et al. (2007) [46]	FHL 124 cells	Human	RT-QPCR; microarray	MMP-2[C] and [I] by TGFβ MMP-14 [I] by TGFβ TIMP-3 [I] by TGFβ
Li et al. (2008) [102]	Capsular bag	Human	Immunolocalization ELISA assay	MMP-2[I] by sham operation MMP-9[I] by sham operation
Dwivedi et al. (2006) [64]	Whole lens	Rat	Western blot; zymography	MMP-9[C] and [I] by TGFβ MMP-2 [I] by TGFβ
Jiang et al. (2006) [103]	HLE cells	Human	Zymography	MMP-2[I] by EGF
Descamps et al. (2005) [104]	Whole lens	Mouse	Zymography	MMP-2[C] MMP-9[I] by sham operation
John et al. (2004) [105]	Whole lens Lens epithelium	Rat	Western blot Zymography Reverse transcription PCR	MMP-2[C] MMP-9[C] MMP-14[C] MMP-16[C] TIMP-1[C] TIMP-2[C] TIMP-3[C]
Sachdev et al. (2004) [106]	Whole lens	Human	Immunolocalization	MMP-1[C] and [I] by UV-B MMP-2[C] and [I] by TNFα/IL-1 MMP-3[C] and [I] by TNFα/IL-1 MMP-9[C] and [I] by TNFα/IL-1 TIMP-1[C]

				TIMP-2[C] TIMP-3[C]
Wong et al. (2004) [107]	Capsular bag	Human	ELISA	MMP-2[C] and [I] by sham operation MMP-9[C] and [I] by sham operation
Wormstone et al. (2002) [59]	Capsular bag	Human	Zymography	MMP-2[C] and [I] by TGFβ MMP-9[C] and [I] by TGFβ
Seomun et al. (2001) [108]	HLE-B3 cells	Human	Reverse transcription PCR	MMP-2[C] and [I] by TGFβ MMP-14[C] TIMP-2 [C]
Kawashima et al. (2000) [109]	Capsular bag	Human donor tissue (fibrous capsule)	Immunolocalization	MMP-1[C] MMP-2[C] MMP-3[C] MMP-9[C] TIMP-1[C] TIMP-2[C]
Tamiya et al. (2000) [110]	Whole	Porcine	Zymography	MMP-2[I] by H_2O_2
Tamiya et al. (2000) [110]	Capsular bag	Porcine	Zymography	MMP-9[I] by sham operation
Richiert and Ireland (1999) [111]	Annular pad cells	Chicken	Zymography	MMP-2 [I] by TGFβ/PDGF MMP-9 [I] by TGFβ/PDGF
Smine and Planter (1997) [112]	Whole lens	Human (postmortem)	Western blot	MMP 14[C]

LEC lens epithelial cells, *MMP* matrix metalloproteinase, *TIMP* tissue inhibitors of MMPs, *RT-QPCR* real-time quantitative PCR

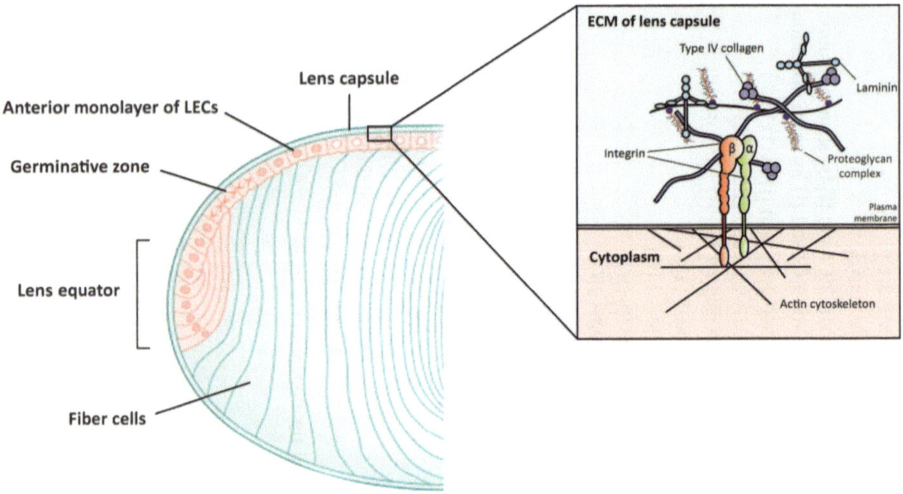

Fig. 3.1 Schematic diagram of the ocular lens. Integrins form a transmembrane heterodimer consisting of an **α** and **β** subunit responsible for adhering lens epithelial (LECs) to the extracellular matrix (ECM) of the lens capsule, which is composed primarily of collagen IV, laminin, and proteoglycans

of these MMPs prevented EMT of the LECs that is typically observed in the development of ASC as well as PCO [64].

A potential target of MMPs, and particularly MMP-2 and MMP-9, in TGF-β-induced EMT is the adherens junction constituent, E-cadherin, essential for proper formation and maintenance of epithelial cell junctions. Cleavage and loss of E-cadherin, and thus a loss of intercellular contacts, correlate with EMT, and recent studies are focusing on the proteolytic activity of MMPs to explain this correlation [121]. The loss of E-cadherin associated with TGFβ-mediated EMT has been demonstrated using mouse and rat lens epithelial explants, in which TGFβ treatment induced the loss of E-cadherin expression and its delocalization from the cell junctions [122, 123]. MMPs are able to cleave the N-terminal extracellular domain of E-cadherin releasing a unique fragment ranging in size from 50 to 84 kDa, a phenomenon known as E-cadherin shedding [64, 124–126]. Using the whole lens model, it was further shown that TGFβ causes E-cadherin disruption resulting in E-cadherin shedding, and extracellular E-cadherin fragments were detected in the conditioned media [64]. Interestingly, co-treatment with MMPIs reduced the appearance of the E-cadherin fragments suggesting that this may be the mechanism by which MMPIs suppress ASC formation.

MMPs are a diverse family with multiple roles in matrix remodeling, cell invasion, and migration as well as cellular transformation and evidence suggests that they are involved in promoting fibrotic cataracts such as PCO and ASC. However, further studies are required to determine the mechanism(s) by which MMPs participate in lens capsule remodeling and lens fibrosis.

3.5 Summary

The lens capsule is an avascular basement membrane that surrounds the lens and is important in regulating its development and differentiation. In this chapter we have provided an overview of the literature regarding lens capsule synthesis, composition, and its receptors, the integrins. Importantly, we also discuss how the capsule is not a static structure and during lens fibrotic disease including PCO and ASC is both altered and actively participates in disease progression. Finally, remodeling enzymes, such as the MMPs, have been shown to release growth factors and activate receptors of the lens capsule and participate in the development of PCO and ASC.

References

1. Duncan MK (2004) In: Lovicu FJ, Robinson ML (eds) Development of the ocular lens. Cambridge University Press, Cambridge/New York, pp xv, 398p
2. Cammarata PR, Cantu-Crouch D, Oakford L, Morrill A (1986) Macromolecular organization of bovine lens capsule. Tissue Cell 18(1):83–97
3. Cammarata PR, Smith JY (1987) Colocalization of laminin and fibronectin in bovine lens epithelial cells in vitro. In Vitro Cell Dev Biol 23(9):611–620
4. Kaplony A, Zimmermann DR, Fischer RW, Imhof BA, Odermatt BF, Winterhalter KH, Vaughan L (1991) Tenascin Mr 220,000 isoform expression correlates with corneal cell migration. Development 112(2):605–614
5. Kelley PB, Sado Y, Duncan MK (2002) Collagen IV in the developing lens capsule. Matrix Biol 21(5):415–423
6. Parmigiani C, McAvoy J (1984) Localisation of laminin and fibronectin during rat lens morphogenesis. Differentiation 28(1):53–61
7. Parmigiani CM, McAvoy JW (1991) The roles of laminin and fibronectin in the development of the lens capsule. Curr Eye Res 10(6):501–511
8. Schulz MW, Chamberlain CG, McAvoy JW (1997) Binding of FGF-1 and FGF-2 to heparan sulphate proteoglycans of the mammalian lens capsule. Growth Factors 14(1):1–13
9. Krag S, Andreassen TT (2003) Mechanical properties of the human posterior lens capsule. Invest Ophthalmol Vis Sci 44(2):691–696
10. Dong L, Chen Y, Lewis M, Hsieh JC, Reing J, Chaillet JR, Howell CY, Melhem M, Inoue S, Kuszak JR, DeGeest K, Chung AE (2002) Neurologic defects and selective disruption of basement membranes in mice lacking entactin-1/nidogen-1. Lab Invest 82(12):1617–1630
11. Gross JM, Perkins BD, Amsterdam A, Egana A, Darland T, Matsui JI, Sciascia S, Hopkins N, Dowling JE (2005) Identification of zebrafish insertional mutants with defects in visual system development and function. Genetics 170(1):245–261. doi:10.1534/genetics.104.039727
12. Zinkevich NS, Bosenko DV, Link BA, Semina EV (2006) Laminin alpha 1 gene is essential for normal lens development in zebrafish. BMC Dev Biol 6:13. doi:10.1186/1471-213X-6-13
13. Futter CE, Crowston JG, Allan BD (2005) Interaction with collagen IV protects lens epithelial cells from Fas-dependent apoptosis by stimulating the production of soluble survival factors. Invest Ophthalmol Vis Sci 46(9):3256–3262. doi:10.1167/iovs.05-0086
14. Olivero DK, Furcht LT (1993) Type IV collagen, laminin, and fibronectin promote the adhesion and migration of rabbit lens epithelial cells in vitro. Invest Ophthalmol Vis Sci 34 (10):2825–2834

15. Lovicu FJ, Chamberlain CG, McAvoy JW (1995) Differential effects of aqueous and vitreous on fiber differentiation and extracellular matrix accumulation in lens epithelial explants. Invest Ophthalmol Vis Sci 36(7):1459–1469
16. Wederell ED, de Iongh RU (2006) Extracellular matrix and integrin signaling in lens development and cataract. Semin Cell Dev Biol 17(6):759–776. doi:10.1016/j.semcdb.2006.10.006
17. Bassuk JA, Birkebak T, Rothmier JD, Clark JM, Bradshaw A, Muchowski PJ, Howe CC, Clark JI, Sage EH (1999) Disruption of the Sparc locus in mice alters the differentiation of lenticular epithelial cells and leads to cataract formation. Exp Eye Res 68(3):321–331. doi:10.1006/exer.1998.0608
18. Norose K, Lo WK, Clark JI, Sage EH, Howe CC (2000) Lenses of SPARC-null mice exhibit an abnormal cell surface-basement membrane interface. Exp Eye Res 71(3):295–307. doi:10.1006/exer.2000.0884
19. Yan Q, Clark JI, Sage EH (2000) Expression and characterization of SPARC in human lens and in the aqueous and vitreous humors. Exp Eye Res 71(1):81–90. doi:10.1006/exer.2000.0853
20. Ohkubo S, Takeda H, Higashide T, Ito M, Sakurai M, Shirao Y, Yanagida T, Oda Y, Sado Y (2003) Immunohistochemical and molecular genetic evidence for type IV collagen alpha5 chain abnormality in the anterior lenticonus associated with Alport syndrome. Arch Ophthalmol 121(6):846–850. doi:10.1001/archopht.121.6.846
21. Colville DJ, Savige J (1997) Alport syndrome. A review of the ocular manifestations. Ophthalmic Genet 18(4):161–173
22. Cammarata PR, Spiro RG (1985) Identification of noncollagenous components of calf lens capsule: evaluation of their adhesion-promoting activity. J Cell Physiol 125(3):393–402. doi:10.1002/jcp.1041250306
23. Walker JL, Menko AS (1999) alpha6 Integrin is regulated with lens cell differentiation by linkage to the cytoskeleton and isoform switching. Dev Biol 210(2):497–511. doi:10.1006/dbio.1999.9277
24. Pierson M, Cordier J, Hervouuet F, Rauber G (1963) An unusual congenital and familial congenital malformative combination involving the eye and kidney. J Genet Hum 12:184–213
25. Zenker M, Aigner T, Wendler O, Tralau T, Muntefering H, Fenski R, Pitz S, Schumacher V, Royer-Pokora B, Wuhl E, Cochat P, Bouvier R, Kraus C, Mark K, Madlon H, Dotsch J, Rascher W, Maruniak-Chudek I, Lennert T, Neumann LM, Reis A (2004) Human laminin beta2 deficiency causes congenital nephrosis with mesangial sclerosis and distinct eye abnormalities. Hum Mol Genet 13(21):2625–2632. doi:10.1093/hmg/ddh284
26. Durkin ME, Wewer UM, Chung AE (1995) Exon organization of the mouse entactin gene corresponds to the structural domains of the polypeptide and has regional homology to the low-density lipoprotein receptor gene. Genomics 26(2):219–228
27. Fox JW, Mayer U, Nischt R, Aumailley M, Reinhardt D, Wiedemann H, Mann K, Timpl R, Krieg T, Engel J et al (1991) Recombinant nidogen consists of three globular domains and mediates binding of laminin to collagen type IV. EMBO J 10(11):3137–3146
28. Schulz MW, Chamberlain CG, de Iongh RU, McAvoy JW (1993) Acidic and basic FGF in ocular media and lens: implications for lens polarity and growth patterns. Development 118(1):117–126
29. Brekken RA, Sage EH (2000) SPARC, a matricellular protein: at the crossroads of cell-matrix. Matrix Biol 19(7):569–580
30. Gilbert RE, Cox AJ, Kelly DJ, Wilkinson-Berka JL, Sage EH, Jerums G, Cooper ME (1999) Localization of secreted protein acidic and rich in cysteine (SPARC) expression in the rat eye. Connect Tissue Res 40(4):295–303
31. Kantorow M, Horwitz J, Carper D (1998) Up-regulation of osteonectin/SPARC in age-related cataractous human lens epithelia. Mol Vis 4:17

32. Kantorow M, Huang Q, Yang XJ, Sage EH, Magabo KS, Miller KM, Horwitz J (2000) Increased expression of osteonectin/SPARC mRNA and protein in age-related human cataracts and spatial expression in the normal human lens. Mol Vis 6:24–29
33. Maillard C, Malaval L, Delmas PD (1992) Immunological screening of SPARC/Osteonectin in nonmineralized tissues. Bone 13(3):257–264
34. Sawhney RS (1995) Identification of SPARC in the anterior lens capsule and its expression by lens epithelial cells. Exp Eye Res 61(5):645–648
35. Yan Q, Perdue N, Blake D, Sage EH (2005) Absence of SPARC in murine lens epithelium leads to increased deposition of laminin-1 in lens capsule. Invest Ophthalmol Vis Sci 46 (12):4652–4660. doi:10.1167/iovs.05-0460
36. Johnson MC, Beebe DC (1984) Growth, synthesis and regional specialization of the embryonic chicken lens capsule. Exp Eye Res 38(6):579–592
37. Kupprion C, Motamed K, Sage EH (1998) SPARC (BM-40, osteonectin) inhibits the mitogenic effect of vascular endothelial growth factor on microvascular endothelial cells. J Biol Chem 273(45):29635–29640
38. Maurer P, Hohenadl C, Hohenester E, Gohring W, Timpl R, Engel J (1995) The C-terminal portion of BM-40 (SPARC/osteonectin) is an autonomously folding and crystallisable domain that binds calcium and collagen IV. J Mol Biol 253(2):347–357. doi:10.1006/jmbi.1995.0557
39. Mayer U, Aumailley M, Mann K, Timpl R, Engel J (1991) Calcium-dependent binding of basement membrane protein BM-40 (osteonectin, SPARC) to basement membrane collagen type IV. Eur J Biochem 198(1):141–150
40. Raines EW, Lane TF, Iruela-Arispe ML, Ross R, Sage EH (1992) The extracellular glycoprotein SPARC interacts with platelet-derived growth factor (PDGF)-AB and -BB and inhibits the binding of PDGF to its receptors. Proc Natl Acad Sci U S A 89(4):1281–1285
41. Sage H, Vernon RB, Funk SE, Everitt EA, Angello J (1989) SPARC, a secreted protein associated with cellular proliferation, inhibits cell spreading in vitro and exhibits Ca + 2-dependent binding to the extracellular matrix. J Cell Biol 109(1):341–356
42. Sasaki T, Hohenester E, Gohring W, Timpl R (1998) Crystal structure and mapping by site-directed mutagenesis of the collagen-binding epitope of an activated form of BM-40/SPARC/osteonectin. EMBO J 17(6):1625–1634. doi:10.1093/emboj/17.6.1625
43. Walker J, Menko AS (2009) Integrins in lens development and disease. Exp Eye Res 88 (2):216–225. doi:10.1016/j.exer.2008.06.020
44. Thorsteinsdottir S, Roelen BA, Freund E, Gaspar AC, Sonnenberg A, Mummery CL (1995) Expression patterns of laminin receptor splice variants alpha 6A beta 1 and alpha 6B beta 1 suggest different roles in mouse development. Dev Dyn 204(3):240–258. doi:10.1002/aja.1002040304
45. Wu JE, Santoro SA (1994) Complex patterns of expression suggest extensive roles for the alpha 2 beta 1 integrin in murine development. Dev Dyn 199(4):292–314. doi:10.1002/aja.1001990405
46. Dawes LJ, Elliott RM, Reddan JR, Wormstone YM, Wormstone IM (2007) Oligonucleotide microarray analysis of human lens epithelial cells: TGFbeta regulated gene expression. Mol Vis 13:1181–1197
47. Bassnett S, Missey H, Vucemilo I (1999) Molecular architecture of the lens fiber cell basal membrane complex. J Cell Sci 112(Pt 13):2155–2165
48. Duncan MK, Kozmik Z, Cveklova K, Piatigorsky J, Cvekl A (2000) Overexpression of PAX6 (5a) in lens fiber cells results in cataract and upregulation of (alpha)5(beta)1 integrin expression. J Cell Sci 113(Pt 18):3173–3185
49. Menko AS, Philip NJ (1995) Beta 1 integrins in epithelial tissues: a unique distribution in the lens. Exp Cell Res 218(2):516–521. doi:10.1006/excr.1995.1186
50. Wederell ED, Brown H, O'Connor M, Chamberlain CG, McAvoy JW, de Iongh RU (2005) Laminin-binding integrins in rat lens morphogenesis and their regulation during fibre differentiation. Exp Eye Res 81(3):326–339. doi:10.1016/j.exer.2005.02.005

51. Nishi O, Nishi K, Mano C, Ichihara M, Honda T, Saitoh I (1997) Inhibition of migrating lens epithelial cells by blocking the adhesion molecule integrin: a preliminary report. J Cataract Refract Surg 23(6):860–865
52. Walker JL, Zhang L, Zhou J, Woolkalis MJ, Menko AS (2002) Role for alpha 6 integrin during lens development: Evidence for signaling through IGF-1R and ERK. Dev Dyn 223 (2):273–284. doi:10.1002/dvdy.10050
53. De Arcangelis A, Mark M, Kreidberg J, Sorokin L, Georges-Labouesse E (1999) Synergistic activities of alpha3 and alpha6 integrins are required during apical ectodermal ridge formation and organogenesis in the mouse. Development 126(17):3957–3968
54. Simirskii VN, Wang Y, Duncan MK (2007) Conditional deletion of beta1-integrin from the developing lens leads to loss of the lens epithelial phenotype. Dev Biol 306(2):658–668. doi:10.1016/j.ydbio.2007.04.004
55. Font RL, Brownstein S (1974) A light and electron microscopic study of anterior subcapsular cataracts. Am J Ophthalmol 78(6):972–984
56. Hay ED (1995) An overview of epithelio-mesenchymal transformation. Acta Anat 154 (1):8–20
57. Novotny GE, Pau H (1984) Myofibroblast-like cells in human anterior capsular cataract. Virchows Arch A Pathol Anat Histopathol 404(4):393–401
58. Saika S (2004) Relationship between posterior capsule opacification and intraocular lens biocompatibility. Prog Retin Eye Res 23(3):283–305. doi:10.1016/j.preteyeres.2004.02.004
59. Wormstone IM, Tamiya S, Anderson I, Duncan G (2002) TGF-beta2-induced matrix modification and cell transdifferentiation in the human lens capsular bag. Invest Ophthalmol Vis Sci 43(7):2301–2308
60. Eldred JA, Dawes LJ, Wormstone IM (2011) The lens as a model for fibrotic disease. Philos Trans R Soc Lond B Biol Sci 366(1568):1301–1319. doi:10.1098/rstb.2010.0341
61. Kappelhof JP, Vrensen GF (1992) The pathology of after-cataract. A minireview. Acta ophthalmol Suppl (205):13–24
62. de Iongh RU, Wederell E, Lovicu FJ, McAvoy JW (2005) Transforming growth factor-beta-induced epithelial-mesenchymal transition in the lens: a model for cataract formation. Cells Tissues Organs 179(1–2):43–55. doi:10.1159/000084508
63. Zuk A, Hay ED (1994) Expression of beta 1 integrins changes during transformation of avian lens epithelium to mesenchyme in collagen gels. Dev Dyn 201(4):378–393. doi:10.1002/aja.1002010409
64. Dwivedi DJ, Pino G, Banh A, Nathu Z, Howchin D, Margetts P, Sivak JG, West-Mays JA (2006) Matrix metalloproteinase inhibitors suppress transforming growth factor-beta-induced subcapsular cataract formation. Am J Pathol 168(1):69–79
65. Godin D, Ivan E, Johnson C, Magid R, Galis ZS (2000) Remodeling of carotid artery is associated with increased expression of matrix metalloproteinases in mouse blood flow cessation model. Circulation 102(23):2861–2866
66. Hales AM, Chamberlain CG, Dreher B, McAvoy JW (1999) Intravitreal injection of TGFbeta induces cataract in rats. Invest Ophthalmol Vis Sci 40(13):3231–3236
67. Hales AM, Chamberlain CG, McAvoy JW (1995) Cataract induction in lenses cultured with transforming growth factor-beta. Invest Ophthalmol Vis Sci 36(8):1709–1713
68. Robertson JV, Nathu Z, Najjar A, Dwivedi D, Gauldie J, West-Mays JA (2007) Adenoviral gene transfer of bioactive TGFbeta1 to the rodent eye as a novel model for anterior subcapsular cataract. Mol Vis 13:457–469
69. Lovicu FJ, Schulz MW, Hales AM, Vincent LN, Overbeek PA, Chamberlain CG, McAvoy JW (2002) TGFbeta induces morphological and molecular changes similar to human anterior subcapsular cataract. Br J Ophthalmol 86(2):220–226
70. Saika S, Miyamoto T, Tanaka S, Tanaka T, Ishida I, Ohnishi Y, Ooshima A, Ishiwata T, Asano G, Chikama T, Shiraishi A, Liu CY, Kao CW, Kao WW (2003) Response of lens epithelial cells to injury: role of lumican in epithelial-mesenchymal transition. Invest Ophthalmol Vis Sci 44(5):2094–2102
71. Dawes LJ, Eldred JA, Anderson IK, Sleeman M, Reddan JR, Duncan G, Wormstone IM (2008) TGF beta-induced contraction is not promoted by fibronectin-fibronectin receptor

interaction, or alpha SMA expression. Invest Ophthalmol Vis Sci 49(2):650–661. doi:10.1167/iovs.07-0586
72. Yoshino M, Kurosaka D, Obazawa M, Takayama F (2001) Presence of alpha 5 beta 1 integrin and fibronectin in the anterior subcapsular cataract. Nippon Ganka Gakkai Zasshi 105 (2):83–87
73. Kim JT, Lee DH, Chung KH, Kang IC, Kim DS, Joo CK (2002) Inhibitory effects of salmosin, a disintegrin, on posterior capsular opacification in vitro and in vivo. Exp Eye Res 74(5):585–594. doi:10.1006/exer.2001.1150
74. Palmade F, Sechoy-Chambon O, Regnouf de Vains JB, Coquelet C, Bonne C (1994) Inhibition of cell adhesion to lens capsule by LCM 1910, an RGD-derived peptide. J Ocul Pharmacol 10(4):623–632
75. Lygoe KA, Norman JT, Marshall JF, Lewis MP (2004) AlphaV integrins play an important role in myofibroblast differentiation. Wound Repair Regen 12(4):461–470. doi:10.1111/j.1067-1927.2004.12402.x
76. Wipff PJ, Rifkin DB, Meister JJ, Hinz B (2007) Myofibroblast contraction activates latent TGF-beta1 from the extracellular matrix. J Cell Biol 179(6):1311–1323. doi:10.1083/jcb.200704042
77. Sponer U, Pieh S, Soleiman A, Skorpik C (2005) Upregulation of alphavbeta6 integrin, a potent TGF-beta1 activator, and posterior capsule opacification. J Cataract Refract Surg 31 (3):595–606. doi:10.1016/j.jcrs.2004.05.058
78. Sheppard D (2004) Roles of alphav integrins in vascular biology and pulmonary pathology. Curr Opin Cell Biol 16(5):552–557. doi:10.1016/j.ceb.2004.06.017
79. Weaver MS, Toida N, Sage EH (2007) Expression of integrin-linked kinase in the murine lens is consistent with its role in epithelial-mesenchymal transition of lens epithelial cells in vitro. Mol Vis 13:707–718
80. Kessenbrock K, Plaks V, Werb Z (2010) Matrix metalloproteinases: regulators of the tumor microenvironment. Cell 141(1):52–67. doi:10.1016/j.cell.2010.03.015
81. Sivak JM, Fini ME (2002) MMPs in the eye: emerging roles for matrix metalloproteinases in ocular physiology. Prog Retin Eye Res 21(1):1–14
82. Slansky HH, Freeman MI, Itoi M (1968) Collagenolytic activity in bovine corneal epithelium. Arch Ophthalmol 80(4):496–498
83. Brown MM, Brown GC, Duker JS, Tasman WS, Augsburger JJ (1994) Exudative retinopathy of adults: a late sequela of retinopathy of prematurity. Int Ophthalmol 18(5):281–285
84. Das A, McGuire PG, Eriqat C, Ober RR, DeJuan E Jr, Williams GA, McLamore A, Biswas J, Johnson DW (1999) Human diabetic neovascular membranes contain high levels of urokinase and metalloproteinase enzymes. Invest Ophthalmol Vis Sci 40(3):809–813
85. Yan X, Tezel G, Wax MB, Edward DP (2000) Matrix metalloproteinases and tumor necrosis factor alpha in glaucomatous optic nerve head. Arch Ophthalmol 118(5):666–673
86. Fini ME, Parks WC, Rinehart WB, Girard MT, Matsubara M, Cook JR, West-Mays JA, Sadow PM, Burgeson RE, Jeffrey JJ, Raizman MB, Krueger RR, Zieske JD (1996) Role of matrix metalloproteinases in failure to re-epithelialize after corneal injury. Am J Pathol 149 (4):1287–1302
87. Di Girolamo N, Lloyd A, McCluskey P, Filipic M, Wakefield D (1997) Increased expression of matrix metalloproteinases in vivo in scleritis tissue and in vitro in cultured human scleral fibroblasts. Am J Pathol 150(2):653–666
88. El-Shabrawi YG, Christen WG, Foster SC (2000) Correlation of metalloproteinase-2 and -9 with proinflammatory cytokines interleukin-1b, interleukin-12 and the interleukin-1 receptor antagonist in patients with chronic uveitis. Curr Eye Res 20(3):211–214
89. Di Girolamo N, Wakefield D, Coroneo MT (2000) Differential expression of matrix metalloproteinases and their tissue inhibitors at the advancing pterygium head. Invest Ophthalmol Vis Sci 41(13):4142–4149
90. Plantner JJ, Jiang C, Smine A (1998) Increase in interphotoreceptor matrix gelatinase A (MMP-2) associated with age-related macular degeneration. Exp Eye Res 67(6):637–645. doi:10.1006/exer.1998.0552

91. West-Mays JA, Pino G (2007) Matrix metalloproteinases as mediators of primary and secondary cataracts. Expert Rev Ophthalmol 2(6):931–938. doi:10.1586/17469899.2.6.931
92. Fini ME, Girard MT, Matsubara M, Bartlett JD (1995) Unique regulation of the matrix metalloproteinase, gelatinase B. Invest Ophthalmol Vis Sci 36(3):622–633
93. Nagase H, Woessner JF Jr (1999) Matrix metalloproteinases. J Biol Chem 274 (31):21491–21494
94. Brew K, Dinakarpandian D, Nagase H (2000) Tissue inhibitors of metalloproteinases: evolution, structure and function. Biochim Biophys Acta 1477(1–2):267–283
95. Gomez DE, Alonso DF, Yoshiji H, Thorgeirsson UP (1997) Tissue inhibitors of metalloproteinases: structure, regulation and biological functions. Eur J Cell Biol 74(2):111–122
96. Dadoukis P, Klagas I, Komnenou A, Karakiulakis G, Karoutis A, Karampatakis V, Papakonstantinou E (2013) Infrared irradiation alters the expression of matrix metalloproteinases and glycosaminoglycans in the cornea and crystalline lens. Graefe's archive for clinical and experimental ophthalmology = Albrecht von Graefes Archiv fur klinische und experimentelle Ophthalmologie. doi:10.1007/s00417-013-2349-9
97. Alapure BV, Praveen MR, Gajjar DU, Vasavada AR, Parmar TJ, Arora AI (2012) Matrix metalloproteinase-2 and -9 activities in the human lens epithelial cells and serum of steroid induced posterior subcapsular cataracts. Mol Vis 18:64–73
98. Eldred JA, Hodgkinson LM, Dawes LJ, Reddan JR, Edwards DR, Wormstone IM (2012) MMP2 activity is critical for TGFbeta2-induced matrix contraction–implications for fibrosis. Invest Ophthalmol Vis Sci 53(7):4085–4098. doi:10.1167/iovs.12-9457
99. Nathu Z, Dwivedi DJ, Reddan JR, Sheardown H, Margetts PJ, West-Mays JA (2009) Temporal changes in MMP mRNA expression in the lens epithelium during anterior subcapsular cataract formation. Exp Eye Res 88(2):323–330. doi:10.1016/j.exer.2008.08.014
100. Hodgkinson LM, Duncan G, Wang L, Pennington CJ, Edwards DR, Wormstone IM (2007) MMP and TIMP expression in quiescent, dividing, and differentiating human lens cells. Invest Ophthalmol Vis Sci 48(9):4192–4199. doi:10.1167/iovs.06-1371
101. Chang PY, Bjornstad KA, Rosen CJ, Lin S, Blakely EA (2007) Particle radiation alters expression of matrix metalloproteases resulting in ECM remodeling in human lens cells. Radiat Environ Biophys 46(2):187–194. doi:10.1007/s00411-006-0087-7
102. Li JH, Wang NL, Wang JJ (2008) Expression of matrix metalloproteinases of human lens epithelial cells in the cultured lens capsule bag. Eye 22(3):439–444. doi:10.1038/sj.eye.6702735
103. Jiang Q, Zhou C, Bi Z, Wan Y (2006) EGF-induced cell migration is mediated by ERK and PI3K/AKT pathways in cultured human lens epithelial cells. J Ocul Pharmacol Ther 22(2):93–102. doi:10.1089/jop.2006.22.93
104. Descamps FJ, Martens E, Proost P, Starckx S, Van den Steen PE, Van Damme J, Opdenakker G (2005) Gelatinase B/matrix metalloproteinase-9 provokes cataract by cleaving lens betaB1 crystallin. FASEB J 19(1):29–35. doi:10.1096/fj.04-1837com
105. John M, Jaworski C, Chen Z, Subramanian S, Ma W, Sun F, Li D, Spector A, Carper D (2004) Matrix metalloproteinases are down-regulated in rat lenses exposed to oxidative stress. Exp Eye Res 79(6):839–846. doi:10.1016/j.exer.2004.08.025
106. Sachdev NH, Di Girolamo N, Nolan TM, McCluskey PJ, Wakefield D, Coroneo MT (2004) Matrix metalloproteinases and tissue inhibitors of matrix metalloproteinases in the human lens: implications for cortical cataract formation. Invest Ophthalmol Vis Sci 45 (11):4075–4082. doi:10.1167/iovs.03-1336
107. Wong TT, Daniels JT, Crowston JG, Khaw PT (2004) MMP inhibition prevents human lens epithelial cell migration and contraction of the lens capsule. Br J Ophthalmol 88(7):868–872. doi:10.1136/bjo.2003.034629
108. Seomun Y, Kim J, Lee EH, Joo CK (2001) Overexpression of matrix metalloproteinase-2 mediates phenotypic transformation of lens epithelial cells. Biochem J 358(Pt 1):41–48
109. Kawashima Y, Saika S, Miyamoto T, Yamanaka O, Okada Y, Tanaka S, Ohnishi Y (2000) Matrix metalloproteinases and tissue inhibitors of metalloproteinases of fibrous humans lens capsules with intraocular lenses. Curr Eye Res 21(6):962–967

110. Tamiya S, Wormstone IM, Marcantonio JM, Gavrilovic J, Duncan G (2000) Induction of matrix metalloproteinases 2 and 9 following stress to the lens. Exp Eye Res 71(6):591–597. doi:10.1006/exer.2000.0916
111. Richiert DM, Ireland ME (1999) Matrix metalloproteinase secretion is stimulated by TGF-beta in cultured lens epithelial cells. Curr Eye Res 19(3):269–275
112. Smine A, Plantner JJ (1997) Membrane type-1 matrix metalloproteinase in human ocular tissues. Curr Eye Res 16(9):925–929
113. Tholozan FM, Quinlan RA (2007) Lens cells: more than meets the eye. Int J Biochem Cell Biol 39(10):1754–1759. doi:10.1016/j.biocel.2007.06.021
114. Hodgkinson LM, Wang L, Duncan G, Edwards DR, Wormstone IM (2010) ADAM and ADAMTS gene expression in native and wound healing human lens epithelial cells. Mol Vis 16:2765–2776
115. Pladzyk A, Reddy AB, Yadav UC, Tammali R, Ramana KV, Srivastava SK (2006) Inhibition of aldose reductase prevents lipopolysaccharide-induced inflammatory response in human lens epithelial cells. Invest Ophthalmol Vis Sci 47(12):5395–5403. doi:10.1167/iovs.06-0469
116. Awasthi N, Wang-Su ST, Wagner BJ (2008) Downregulation of MMP-2 and -9 by proteasome inhibition: a possible mechanism to decrease LEC migration and prevent posterior capsular opacification. Invest Ophthalmol Vis Sci 49(5):1998–2003. doi:10.1167/iovs.07-0624
117. Morarescu D, West-Mays JA, Sheardown HD (2010) Effect of delivery of MMP inhibitors from PDMS as a model IOL material on PCO markers. Biomaterials 31(8):2399–2407. doi:10.1016/j.biomaterials.2009.11.108
118. Hazra S, Guha R, Jongkey G, Palui H, Mishra A, Vemuganti GK, Basak SK, Mandal TK, Konar A (2012) Modulation of matrix metalloproteinase activity by EDTA prevents posterior capsular opacification. Mol Vis 18:1701–1711
119. Mackay AR, Hartzler JL, Pelina MD, Thorgeirsson UP (1990) Studies on the ability of 65-kDa and 92-kDa tumor cell gelatinases to degrade type IV collagen. J Biol Chem 265 (35):21929–21934
120. McCawley LJ, Matrisian LM (2001) Matrix metalloproteinases: they're not just for matrix anymore! Curr Opin Cell Biol 13(5):534–540
121. Zheng G, Lyons JG, Tan TK, Wang Y, Hsu TT, Min D, Succar L, Rangan GK, Hu M, Henderson BR, Alexander SI, Harris DC (2009) Disruption of E-cadherin by matrix metalloproteinase directly mediates epithelial-mesenchymal transition downstream of transforming growth factor-beta1 in renal tubular epithelial cells. Am J Pathol 175 (2):580–591. doi:10.2353/ajpath.2009.080983
122. Korol A, Pino G, Dwivedi D, Robertson JV, Deschamps PA, West-Mays JA (2014) Matrix metalloproteinase-9 null mice are resistant to TGF-beta-induced anterior subcapsular cataract formation. Am J Pathol. doi:10.1016/j.ajpath.2014.03.013
123. Banh A, Deschamps PA, Vijayan MM, Sivak JG, West-Mays JA (2007) The role of Hsp70 and Hsp90 in TGF-beta-induced epithelial-to-mesenchymal transition in rat lens epithelial explants. Mol Vis 13:2248–2262
124. George SJ, Dwivedi A (2004) MMPs, cadherins, and cell proliferation. Trends Cardiovasc Med 14(3):100–105. doi:10.1016/j.tcm.2003.12.008
125. Ho AT, Voura EB, Soloway PD, Watson KL, Khokha R (2001) MMP inhibitors augment fibroblast adhesion through stabilization of focal adhesion contacts and up-regulation of cadherin function. J Biol Chem 276(43):40215–40224. doi:10.1074/jbc.M101647200
126. Mei JM, Borchert GL, Donald SP, Phang JM (2002) Matrix metalloproteinase(s) mediate (s) NO-induced dissociation of beta-catenin from membrane bound E-cadherin and formation of nuclear beta-catenin/LEF-1 complex. Carcinogenesis 23(12):2119–2122

Lens Epithelial Cell Proliferation

4

F.J. Lovicu, L. Iyengar, L.J. Dawes, and J.W. McAvoy

Abstract

Cell proliferation in the vertebrate ocular lens is essential for its establishment, development and growth. Lens cell proliferation features very early in its morphogenesis and results in rapid tissue growth, but becomes increasingly restricted, both spatially and temporally, with age. As the lens is established, so are defined growth zones characterised by polarised regions of cell proliferation and subsequent fibre cell differentiation. These growth zones are tightly regulated by growth factors in the surrounding ocular environment. Although lens epithelial cell proliferation persists throughout life, albeit at a markedly reduced rate with increased age, the majority of epithelial cells in the adult lens remain quiescent. If perturbed, as a result of various ocular pathologies, normal lens cell proliferation is deregulated as epithelial cells re-enter the cell cycle; such cellular hyperplasia often compromises lens function and subsequently results in cataract formation. Identifying the key ocular factors, and understanding the underlying mechanisms regulating lens cell proliferation, will further advance our understanding of the aetiology of cataracts that are characterised by aberrant lens cell proliferation.

F.J. Lovicu (✉)
Discipline of Anatomy and Histology, Bosch Institute,
University of Sydney, Sydney, NSW 2006, Australia

Save Sight Institute, University of Sydney,
Sydney, NSW 2006, Australia
e-mail: frank.lovicu@sydney.edu.au

L. Iyengar
Discipline of Anatomy and Histology, Bosch Institute,
University of Sydney, Sydney, NSW 2006, Australia

L.J. Dawes • J.W. McAvoy
Save Sight Institute, University of Sydney,
Sydney, NSW 2006, Australia

S. Saika et al. (eds.), *Lens Epithelium and Posterior Capsular Opacification*,
DOI 10.1007/978-4-431-54300-8_4, © Springer Japan 2014

Keywords
Cataract • ERK1/2 signalling • Growth factors • Lens cell proliferation

4.1 The Cell Cycle

Cell proliferation underlies the growth and development of all living organisms [1] and is involved in embryogenesis, tissue formation and patterning, postnatal growth and maintenance as well as replacing ageing or damaged cells [2]. Many aspects of abnormal development, disease, healing and ageing are related to changes in the abilities of cells to divide [2]. Cell division is mediated via the cell cycle [1], with both entry into the cell cycle from a state of quiescence and exit from the cell cycle requiring precise regulation for normal tissue growth and differentiation to be maintained (see [3]).

The mammalian cell cycle is divided into four phases, G1, S, G2 and M (see [3]). DNA synthesis and replication occur in the S-phase of the cell cycle before segregation into two daughter cells during mitosis, the M-phase of the cell cycle. DNA synthesis and mitosis are separated by two gap phases, G1 and G2. In multicellular organisms, most differentiated cells remain in the G1 phase [2] but can also enter a G0 state, where they become quiescent. Cell proliferation is primarily regulated at G1 prior to entry into S-phase and at G2 prior to entry into mitosis [4]. Commitment to cell division or exiting the cell cycle is decided at the "restriction point" in G1 [5]. Exit from G0 and the majority of G1 are growth factor-dependent phases [4]. Once this point is passed, cells are committed to the S-phase and the remainder of the cell cycle can progress in the absence of growth stimuli [5]. Thus, the restriction point is the primary event controlling cell proliferation (see [3, 4]).

Progression through different cell cycle phases is monitored by protein complexes, the cyclins and cyclin-dependent kinases (cdks; [6]). Cyclins are considered "periodic" proteins, synthesised during defined cell cycle movements and then quickly degraded [7]. In contrast, cdks are regulatory proteins, constitutively expressed and active only when cyclin-bound [8]. When activated by their cognate cyclin, the cdk is responsible for phosphorylating, and thereby inactivating, specific targets such as the retinoblastoma (Rb) protein that is considered a critical molecular factor in controlling the restriction point in the G1 phase of the cell cycle [8]. Acting in opposition to the cyclins as negative cell cycle regulators are the cyclin-dependent kinase inhibitors (CKIs), including members of the p21CIP1 family, such as p27KIP1 and p57KIP2, that function to inhibit cdks involved in G1 to S transition (see [3]).

4.2 Lens Epithelial Cell Proliferation

Cell proliferation is an important cellular process that regulates the growth and architecture of the ocular lens. As described in Chap. 1, the lens placode invaginates as a result of inductive interactions with the optic vesicle neuroectoderm. During this time, the majority of lens precursor cells actively proliferate as the lens pit invaginates and ultimately separates from the overlying ectoderm to form the lens vesicle [9–13]. High levels of proliferation occur throughout the lens vesicle, but as development progresses, proliferative activity becomes compartmentalised; cells in the anterior layer of the lens vesicle (facing the presumptive cornea and aqueous humour) continue to maintain their high proliferative capacity, as they differentiate into a cuboidal epithelial monolayer, whereas cells in the posterior half of the vesicle (facing the presumptive retina and vitreous humour) display a reduced level of cell proliferation, as they withdraw from the cell cycle and are driven to elongate and differentiate into primary lens fibres [9–11, 14, 15]. In this way, the lens acquires its distinct polarity, which is maintained throughout life.

4.2.1 The Ocular Environment

The importance of the ocular environment for determining and maintaining lens polarity is best illustrated by the pioneering work of Coulombre and Coulombre [16]. In this study they surgically inverted the lens of a 5-day-old chick embryo so that the epithelial monolayer that normally faces the presumptive cornea now faced the presumptive retina. In this new environment, the lens repolarised, forming a new epithelial sheet over its now new anterior surface, while the original central epithelial cells that now faced the presumptive retina elongated and differentiated into a new fibre mass [16]. This experiment not only highlights the importance of the ocular environment in regulating lens cell behaviour, but that epithelial cell proliferation is maintained and regulated by the anterior compartment, laying the foundation for future studies on the influence of ocular growth factors in lens biology.

4.2.2 Lens Growth Patterns

As embryonic development proceeds, mitotic activity in the lens progressively decreases and during postnatal growth the pattern of cell proliferation is further modified, becoming restricted to a narrow band of epithelial cells localised above the lens equator, in the germinative zone (Figs. 4.1 and 4.2) [12, 13, 15]. While most of these studies were conducted on rats and mice, the human lens appears to follow a similar pattern of proliferative activity [17]. The estimated cell cycle time is shortest in the germinative zone while cells in the central epithelium remain relatively quiescent throughout life [12]. The progeny of these cell divisions at the lens equator migrate and/or are displaced posteriorly, into a region known as the transitional zone, where they exit the cell cycle and initiate the fibre differentiation process (Fig. 4.2) [14, 15, 18].

The expression pattern of cell cycle regulators is consistent with lens polarity, with positive cell cycle regulators, the cyclins and their corresponding cdk counterparts, restricted to the proliferative epithelial cells [19–23], while negative

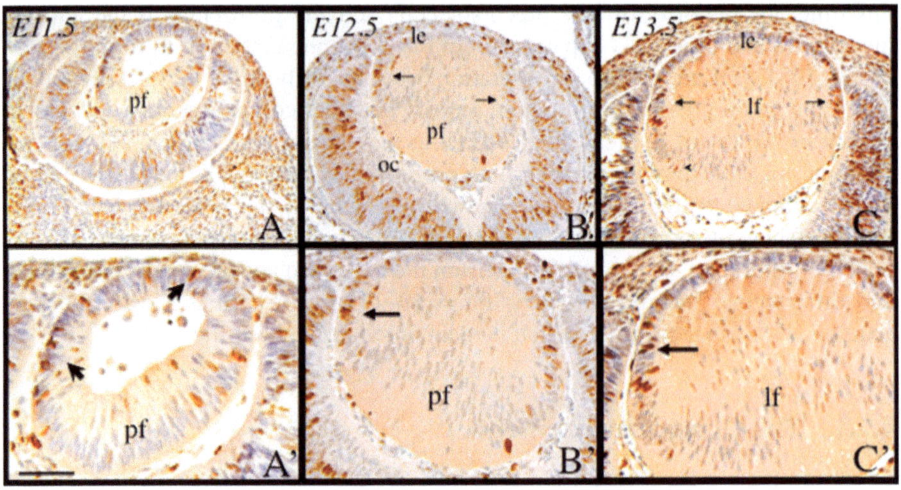

Fig. 4.1 Patterns of epithelial cell proliferation during murine lens morphogenesis. Midsagittal eye sections labelled for BrdU incorporation, counterstained with haematoxylin. BrdU labelling (*brown nuclei*) was used to mark cells in the S-phase of the cell cycle during lens development, from early embryonic day 11.5 (**a**, **a′**), 12.5 (**b**, **b′**) and E13.5 (**c**, **c′**). In the early lens vesicle (**a**, **a′**), as primary fibre (pf) cells begin to elongate, most lens cell proliferations are observed in the anterior lens vesicle cells (**a′**, *arrows*). **b**, **b′** Once the primary fibre cells make contact with the overlying epithelium (le), and the lens vesicle lumen is lost, cell proliferation is evident throughout the epithelium with more marked expression at the lens equator (*arrows*), adjacent to the anterior margin of the optic cup (oc). **c**, **c′** As the lens continues to grow, the peripheral cells of the epithelium still display the most pronounced cell proliferation (*arrows*), and this gives rise to the germinative zone. *lf* lens fibre cells. *Scale bar*: **a–c**, 200 μm, **a′–c′**, 100 μm (adapted from Kallifatidis [15])

cell cycle regulators such as CKIs are restricted to the post-mitotic cells in the transitional zone that are undergoing early fibre differentiation changes [14, 21].

4.2.3 Regulation of Lens Cell Proliferation

Cell proliferation in the lens occurs throughout embryogenesis, into postnatal growth, and continues throughout life. The tight regulation of this cellular process is critical for proper lens functioning as aberrant levels and patterns of cell proliferation can perturb lens transparency, resulting in cataract (see Sect. 4.6). Normal patterns of lens cell proliferation are thought to be orchestrated by the ocular environment, namely the aqueous humour that bathes the lens epithelium (see [24]). The ocular environment is a rich source of growth and regulatory factors (see [25, 26]). Although the factor(s) in the vitreous humour that stimulate fibre cell elongation and differentiation has been well documented (see [26]), the specific factors in the aqueous humour considered to promote and maintain lens cell proliferation in situ are not as well defined. In vitro, a multitude of growth factors

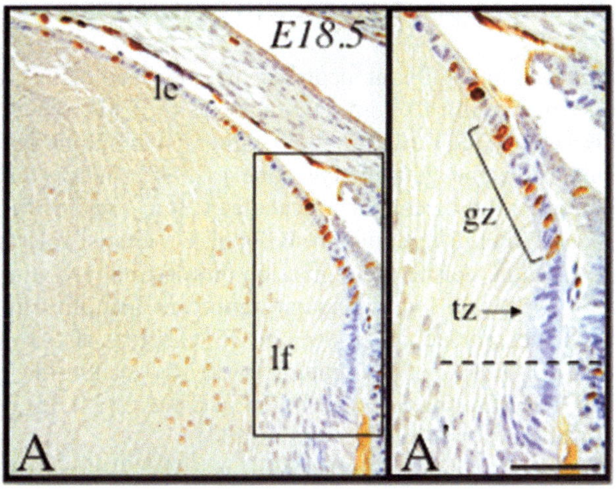

Fig. 4.2 Pattern of lens epithelial cell proliferation in the murine foetal lens. Midsagittal eye sections labelled for BrdU incorporation, counterstained with haematoxylin. (**a**) At murine embryonic day 18.5, BrdU-labelled (*brown*) cells are localised to the lens epithelium (le). At higher magnification (**a'**), above the lens equator (*broken line*), proliferating (BrdU-labelled) epithelial cells are localised to the germinative zone (gz), followed more posteriorly by another defined region of epithelial cells, the transitional zone (tz), that have exited the cell cycle as they begin to elongate and differentiate into secondary fibre cells (lf). *Scale bar*: **a**, 200 μm, **a'**, 100 μm (adapted from Kallifatidis [15])

have been shown to affect lens growth, but given the potential redundancy in their activities, it is not clear if some or all of these ocular-derived mitogens are involved in aqueous-induced lens epithelial cell proliferation.

4.3 The Aqueous Humour and Aqueous-Derived Mitogens (Growth Factors)

Growth factors within the aqueous and vitreous humours regulate lens growth. The primary source of the aqueous humour is the blood flowing through the ciliary arterial system, which allows small amounts of plasma proteins (including growth factors) to permeate into the ocular anterior chamber [27], making the aqueous humour functionally and biochemically distinct from the plasma [28]. The growth factors found in the aqueous humour of different species are largely conserved [29, 30] and have been shown to directly influence lens cell proliferation (see [25]). Some of these include members of the insulin and insulin-like growth factor (IGF), platelet-derived growth factor (PDGF), epidermal growth factor (EGF, including its related ligand transforming growth factor-α, TGFα), hepatocyte growth factor (HGF) and the fibroblast growth factor (FGF) families. Other molecules in the aqueous such as transforming growth factor-β (TGFβ) have also been shown to be involved in epithelial maintenance and growth (see [26]).

4.3.1 Insulin and Insulin-Like Growth Factor (IGF)

Members of the insulin family of peptides include insulin, insulin-like growth factor-I (IGF-I) and insulin-like growth factor-II (IGF-II). There are two IGF receptors, IGFR-I and IGFR-II, that are both expressed in mitotically active regions of the lens epithelium, the germinative zone [31, 32]. While insulin is mainly involved in the regulation of cellular metabolism, IGF-I and IGF-II are involved in promoting growth and differentiation. IGFs play a crucial role during embryogenesis and are essential during the regulation of postnatal growth (see [33]). The primary source of IGFs is the liver, although significant quantities are also produced in extra-hepatic tissues, including the eye [34, 35]. There is also a well-established association of IGFs with specific binding proteins (insulin-like growth factor binding proteins; IGFBPs) that are also found in the eye [34, 35], which function to stabilise IGF and deliver it to target tissues [34].

Radioimmunoassays indicate that the IGF-I and IGF-II concentrations in aqueous and vitreous are in the sub-nanomolar range [34] but are sufficient to allow binding to ocular IGFBPs and IGF-I receptors on lens epithelial cells [31, 34]. Interestingly, the levels of IGF-I and IGF-II in the aqueous are approximately twice those found in the vitreous [34], suggesting that IGF may play a more prominent role in the anterior chamber, bathing the epithelial cells. In vitro studies have indicated that IGF can induce proliferation in a range of lens culture systems including lens epithelial explants [36, 37], whole lens cultures [38, 39] and lens cell lines [40]. When combined with other growth factors found in the eye, such as EGF [39] or FGF [36], IGF can synergise with these factors to produce an enhanced lens cell proliferation response and in some instances (with FGF) fibre cell differentiation [36]. In vivo studies using transgenic mice that overexpressed IGF-I in the lens showed increased proliferation in the germinative zone and an expansion of the transitional zone posteriorly toward the retina [41]. This led to the hypothesis that the distribution of IGF in the eye may provide a spatial cue that defines the extent of the germinative zone and the location of the transitional zone, thus regulating the patterns of lens cell proliferation in vivo [41].

4.3.2 Platelet-Derived Growth Factor (PDGF)

PDGF is a potent mitogen released most prominently by platelets but also present to a lesser extent in other cell types [42]. The PDGF family is composed of four ligands, PDGFA-D. All PDGF isoforms form homodimers, but PDGF-A and PDGF-B can also be secreted as a heterodimer (PDGF-AB; [42]). The PDGF receptor consists of α and β subunits that dimerize to form three distinct receptor combinations ($\alpha\alpha$, $\alpha\beta$, $\beta\beta$), differing in their ability to bind the various PDGF isoforms (see [42]). The importance of PDGF signalling during development is highlighted by the *patch* mutation in mice (which harbours a deletion of the PDGF receptor α gene), resulting in a range of anomalies, including perturbed lens development [43, 44]. The pattern of lens cell proliferation in situ correlates with

PDGFα receptor and PDGF-A ligand expression [45]. PDGF is strongly expressed in the iris and the ciliary body, ocular structures juxtaposed to the germinative zone. During embryogenesis, PDGFα receptor expression is detected throughout the lens epithelium [45]. In the postnatal lens, as the central epithelial cells reach mitotic quiescence, PDGFα receptor expression is lost in these cells and becomes restricted to the mitotically active lens epithelium of the germinative zone [45].

The first in vitro studies with PDGF showed that it could maintain lens growth and transparency in whole lens cultures when delivered in a pulsatile fashion [46]; however, insulin, which can independently stimulate lens growth, was also present in the culture medium. More recently, in studies with rat lens epithelial explants, PDGF-A was shown to promote cell proliferation [37, 47], as well as potentiate FGF-induced fibre differentiation [47, 48], similar to IGF. Overexpression of PDGF-A in lenses of transgenic mice increased the levels of DNA synthesis in lens epithelial cells [45], supporting a mitogenic role for PDGF. More recent studies by Ray and colleagues [49] have shown a role for PDGF-D in lens cell proliferation as an antibody that sequestered PDGF-D could strongly inhibit lens cell proliferation in anterior segment organ culture, whereas other antibodies for PDGF-A, PDGF-B or FGF-2 could not [49].

4.3.3 Fibroblast Growth Factor (FGF)

FGF has been shown to play an important role in regulating many aspects of growth and development [50]. There are 22 known members of the FGF family (FGF1-FGF23, with mouse FGF15 being the ortholog of human FGF19; [51] that are highly conserved [52, 53] and their distribution in the eye has been studied extensively [54–57]. Cellular responses to FGF are mediated via a family of four receptor genes (*fgfr1-4*: [53, 58, 59]) with lens cells shown to express all four FGFR genes [60–63]. Over the years, both in vitro and in vivo models support a pivotal role for FGF in regulating lens fibre differentiation (see [26]). The development of the rat lens epithelial explant system [64, 65] was central to the identification of the dose-dependent responsiveness of lens epithelia to FGFs; with higher doses of FGF (100 ng/mL) serving as a potent morphogen for lens fibre differentiation, with much lower doses (1–5 ng/mL) sufficient for its mitogenic ability on these same cells [37, 66].

4.3.4 Epidermal Growth Factor (EGF)

Epidermal growth factor (EGF) influences development of the eye through its ability to induce early eye opening in mice [67] and has been localised to several ocular tissues [68]. The EGF family of receptor tyrosine kinases consists of four receptors: EGFR (ErbB-1, HER-1), HER-2/neu (c-erbB-2), HER-3 (c-erbB-3) and HER-4 (c-erbB-4). Fifteen members of the endogenous EGF ligand family have been identified, including amphiregulin, betacellulin, biregulin, EGF, epiregulin, heparin binding-EGF (HB-EGF), heregulin α/β, neuregulin (NRG) 1α/1β/2α/2β/3/4 and TGFα [69–71]. While the ligands overlap with respect to binding to the various EGF receptors, they have their own specificities and affinities for their respective receptors [72]. Many of the EGF ligands are present in the ocular media surrounding the lens [73] and lens cells of a number of species have been shown to express functionally active EGF receptors [38, 74–76]. Studies have indicated that EGF can induce proliferation in whole lens cultures [39], as well as cultures of various transformed epithelial cell lines [77, 78]. As mentioned earlier, EGF has also been reported to act synergistically with other growth factors such as IGF, to produce an enhanced mitotic response in whole lens cultures [39].

In addition to EGF, its related mitogen TGFα binds and mediates its effects specifically through the EGF receptor. TGFα shares a high degree of structural homology to EGF due to conservation of the cysteine residues involved in receptor binding [79]. In lens explants, TGFα and EGF have been shown to induce proliferation and migration of epithelial cells [37, 80]. Furthermore, EGF may play a significant role in contributing to posterior capsular opacification (PCO), a common problem associated with cataract surgery/trauma that results from the abnormal proliferation of lens epithelial cells as they undergo a wound healing response [81]; see also Sect. 1.6 and Chaps. 8 and 9).

4.3.5 Hepatocyte Growth Factor (HGF)

Hepatocyte growth factor has recently been shown to induce lens epithelial cell proliferation and migration in human lens capsular bags (a model system used to study PCO) and rabbit cells [81, 82] by activating its receptor c-met, which has been shown to be expressed in both human and rabbit lens epithelial cells [75, 83]. It has been found in abundance in human capsular bag cultures suggesting that it may contribute to the development of aberrant cell proliferation associated with PCO [81].

4.3.6 Transforming Growth Factor-β (TGFβ)

TGFβ has been shown to induce aberrant growth and differentiation of the lens epithelial cells causing them to undergo an epithelial-mesenchymal transition (EMT: see Chaps. 8 and 9), characteristic of the changes seen in the cataractous lens [84, 85]. TGFβ1, 2 and 3 are present throughout the lens epithelium during embryonic and postnatal development [86, 87]; however, although little reactivity for TGFβ receptor (TGFβRI and RII) expression is detected in newborn rats (postnatal day 3, P3), there is a substantial increase in immunoreactivity for the receptors in weanling rats (P21; [87]). As TGFβR expression inversely correlates with patterns of cell proliferation in the lens, it has been suggested that TGFβ signalling may act as a negative regulator of lens growth, similar to other cell types (see [88, 89]).

4.4 Cellular Signalling Pathways

Cell proliferation is a complex developmental process that undoubtedly requires numerous signals, and a cascade of signalling molecules are involved in this process. With respect to cell proliferation, the signal transduction pathways that have arguably been best studied in the lens include the MAPK pathway and the PI3-K pathway.

4.4.1 Mitogen-Activated Protein Kinase (MAPK) Signalling

Mitogen-activated protein kinases control many cellular events, from complex programmes such as embryogenesis, cell differentiation, cell proliferation and cell death, as well as short-term changes required for homeostasis and acute hormonal responses [90]. The MAPK family consists of four distinct signalling modules: the extracellular signal-regulated kinase (ERK1/2) pathway, the c jun-N terminal kinase (JNK) pathway also as known as the stress-activated protein kinase (SAPK) signalling pathway, the p38 pathway and ERK5, otherwise known as the big MAPK 1 (BMK1) signalling module. While the ERK1/2 signalling pathway is primarily associated with the mitogenic responses of a cell, being strongly activated in response to growth factor stimulation, the p38 pathway and JNK/SAPK pathway are more commonly recruited in response to cellular stress, inflammation and apoptosis, activated by interleukins, cytokines and other such inflammatory molecules [91]. ERK5 signalling has been shown to play a role in cell proliferation and survival although the exact role of this kinase is yet to be elucidated [92]. Although each MAP kinase has unique characteristics, a feature shared by the MAP kinase pathways studied to date is the fact that the terminal kinases are activated by protein kinase cascades that contain at least two upstream kinases that serve to ensure signal amplification and regulation [93]. Interestingly, ERK1/2 are also the most abundant MAPKs in lenses of a number of species [94].

There are numerous downstream targets of ERK1/2. Over 150 substrates of ERK1/2 have been reported so far, which are involved in several distinct functions. These include transcription factors, protein kinases and phosphatases, cytoskeletal and scaffold proteins, receptors, signalling molecules and others [95]. Interestingly, about 50 % of these proteins are localised in the nucleus, with the others found in the cytosol, plasma membrane and within cellular organelles, supporting a role for ERK1/2 in the regulation of both cytosolic and nuclear processes [95]. Within a few minutes of ERK1/2 activation and translocation to the nucleus, the most significant effect of ERK1/2 activation is immediate early gene (IEG) expression (e.g. c-fos, c-myc; see [96]). The protein products of these genes are involved in multiple aspects of cell proliferation, growth and differentiation. The transcription of IEGs is regulated by serum response elements (SRE), activated by transcription factors (e.g. serum response factor; SRF) that are direct substrates of the ERK1/2-signalling cascade (see [96]). Other targets include some of the better characterised transcription factors known as ternary complex factors (TCFs), including Elk-1 which is directly phosphorylated by ERK1/2 on multiple sites [96]. Upon complex formation with SRF, phosphorylated TCFs transcriptionally activate numerous mitogen-inducible genes regulated by these serum response elements [96]. ERK1/2 signalling has also been shown to directly link growth factor signalling to ribosome biogenesis by activating ribosomal S6-kinase-1 (RSK1), suggesting a role in protein synthesis which is necessary for proper cellular growth and proliferation [97, 98].

4.4.2 Akt/PI3-Kinase Signalling

Similar to MAPK/ERK signalling, the PI3-K signalling pathway has received considerable attention with respect to cell proliferation, survival and growth (see [99]). Unlike the ERK1/2 signalling pathway whose primary role is to regulate cell proliferation and differentiation, PI3-K has additional roles in maintaining cellular homeostasis, including insulin metabolism, and glycogen, fatty-acid and nitric oxide syntheses [99]. PI3-K is a heterodimeric molecule composed of a binding subunit (p85) and a catalytic subunit (p110). As a consequence of growth factor stimulation, activation of P13-K results in the generation of lipid second messengers, which serve as docking sites for proteins that harbour pleckstrin homology domains, notably Akt. Phosphorylated Akt can modulate the function of numerous substrates involved in the regulation of cell proliferation, such as glycogen synthase kinase-3 (GSK3), mammalian target of rapamycin (mTOR) and Cdc25B. Akt can also enhance cell cycle progression by blocking the activity of negative cell cycle regulators such as cyclin-dependent kinase inhibitors, p21/Waf1/Cip1 and p57/Kip2, and directly stimulate genes involved in cell proliferation [99].

4.5 Regulation of Cell Proliferation in the Lens

Cell proliferation is a tightly controlled process in the developing and adult vertebrate lens. While bone morphogenetic protein (BMP)-/activin-mediated signalling has been reported to be required for cell proliferation during lens morphogenesis [100], in vitro findings emphasise an important role for multiple growth factors, with particular emphasis on a key role for FGF signalling in lens cell proliferation. Although many growth factors have been shown to induce lens cell proliferation in vitro, little is known about how they relate to aqueous-induced lens cell proliferation. Interpretation and synthesis of data from numerous studies has been complicated by the fact they have been conducted in diverse culture systems from different species. In an earlier study, in the embryonic chick lens (prior to the formation of the aqueous humour), Hyatt and Beebe [101] demonstrated that serum proteins readily enter the anterior chamber of the eye and regulate lens cell proliferation. In this extensive study, it was shown that not one particular growth factor or combination of growth factors could replace the high proliferative capacity of embryonic serum [101]. Since this finding, the growth factors involved in aqueous-induced lens cell proliferation have been explored, including the signalling pathways involved in mediating this growth. As mentioned, an earlier study has shown a role for PDGF-D in the regulation of lens cell proliferation in the postnatal lens, as an antibody that sequestered PDGF-D could strongly inhibit lens epithelial cell proliferation in anterior segment organ culture [49]. Given that a multitude of growth factors are detected in the eye and have been shown to play a role in lens cell proliferation [39, 41, 45, 47, 73, 102], this prompted further investigations into cell proliferation in the postnatal lens. By comparing the signalling pathways induced by different ocular growth factors with those induced by aqueous, more recent studies have identified key aqueous-derived lens mitogens and their associated signalling mechanisms involved in mediating lens growth [37, 103, 104].

4.5.1 ERK1/2 and Akt Signalling in Lens Proliferation

Since the seminal experiments of the Coulombres [16, 105, 106], although the aqueous was considered to be the endogenous source of lens mitogens, this was only recently demonstrated in vitro [37]. By applying aqueous directly onto lens epithelial explants, and using the bromodeoxyuridine (BrdU)-incorporation assay, aqueous-induced proliferation of lens epithelial cells was shown to be similar to that induced by various growth factors. Both aqueous-induced proliferation and that induced by a number of growth factors were shown to be dependent on both MAPK/ERK1/2 and PI3-K/Akt signalling [37, 107, 108]. By utilising specific inhibitors for these signalling molecules, a strong correlation was observed between cell proliferation and both PI3-K/Akt- and ERK1/2-signalling activity, with ERK1/2 signalling likely being the key downstream mediator of this activity [37]. More recent studies in situ, examining mice with conditional deletions of ERK1/2 in the lens, demonstrate that phosphorylation of ERK2 (MAPK1) is required for lens epithelial cell proliferation in the germinative zone, with both ERK1 and ERK2 being required for proliferation of central epithelia [108]. This is consistent with increased levels of phosphorylated ERK1/2 in equatorial lens epithelial cells (most likely ERK2), associated with regions of higher levels of proliferation [74, 107]. Interestingly, although a number of mitogens could induce cell proliferation via ERK1/2 signalling in lens epithelial explants, they each differentially impacted on the phosphorylation of ERK1/2 (Fig. 4.3). While FGF typically showed induction of ERK1/2 phosphorylation for up to 6 h, similar to aqueous, other mitogens tested, including PDGF, IGF or EGF, only individually stimulated ERK1/2 phosphorylation for up to 1 h (see Fig. 4.3; [109]). By specifically blocking any one or a combination of high-affinity receptors for FGF, PDGF, EGF and IGF, in aqueous-treated lens epithelial explants, it was shown that aqueous-induced cell proliferation also appears to be due to a combination of growth factor signals, of which FGF plays a central role [109]. Moreover, there appears to be at least two phases of ERK1/2 phosphorylation, with an early phase dependent on IGF and/or PDGF and the later phase dependent on FGF [109]. By blocking FGF signalling, aqueous

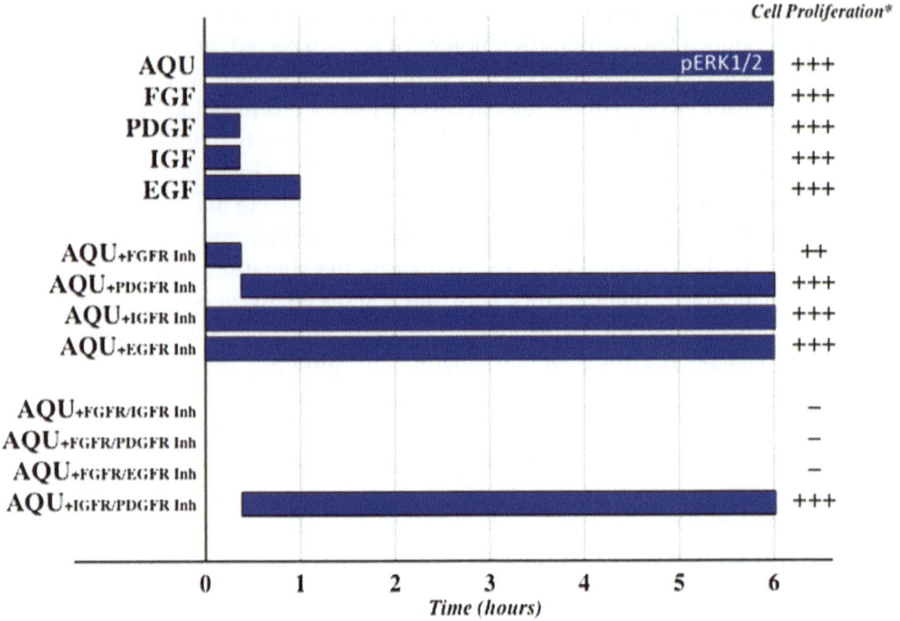

Fig. 4.3 Contribution of growth factor-stimulated phosphorylation of ERK1/2 in aqueous (AQU)-induced cell proliferation in rat lens epithelial explants. All growth factors examined induced lens cell proliferation comparable to aqueous humour; however, only FGF induced the phosphorylation of ERK1/2 (*blue bars*) for up to 6 h, comparable to aqueous. In the presence of any one of the selective receptor tyrosine kinase (RTK) inhibitors (Inh) for the respective growth factors: FGFR Inh (SU5402), PDGFR Inh (AG1296), IGFR Inh (AG1024) or EGFR Inh (PD153035), not one could completely block the ability of aqueous to stimulate cell proliferation, although ERK1/2 phosphorylation was compromised with the PDGFR inhibitor (early phase only) and more completely with the FGFR inhibitor. When different combinations of inhibitors were applied, aqueous-induced ERK1/2 phosphorylation and cell proliferation were completely blocked, providing that the FGFR inhibitor was included. All other inhibitor combinations did not perturb aqueous-induced lens cell proliferation, for example, IGFR- and PDGFR-inhibitor combinations. (*Asterisk*) Cell proliferation was assessed using BrdU incorporation and/or tritiated thymidine incorporation. This figure was compiled from data obtained from [37, 104]

could still induce lens cell proliferation; however, aqueous-induced ERK1/2 signalling was altered, now only transiently activated for less than 1 h (see Fig. 4.3). In contrast, blocking PDGFR signalling could block this initial phase of ERK1/2 phosphorylation induced by aqueous, but not proliferation (Fig. 4.3), highlighting the notion that growth factors such as PDGF and FGF may act together as primary mediators of aqueous-induced ERK1/2 signalling. Overall, different combinations of inhibitors selective for high-affinity growth factor receptors could block aqueous-induced cell proliferation, provided that FGF receptor signalling was one of those abrogated (Fig. 4.3; [109]). It can thus be hypothesised that while FGF and PDGF signalling may render cells competent in order to initiate a proliferative response, IGF and EGF may act as progression growth factors, allowing these "competent" cells to progress through the cell cycle, supporting a role for multiple growth factors in aqueous-induced lens cell proliferation in situ [110].

Consistent with FGFs being required for lens proliferation in situ, a more recent study has shown that conditional ablation of Shp2, an adaptor protein critical for FGF receptor signalling, resulted in a decrease in both ERK1/2 phosphorylation and epithelial cell proliferation in murine embryonic lenses, as well as increased apoptosis in these cells [111]. As multiple growth factors are involved in regulating lens cell proliferation, further insight into the role of other signalling pathways is imperative to gaining a better understanding of lens cell proliferation. Some of the more notable pathways include mammalian target of rapamycin (mTOR), phospholipase Cγ (PLCγ), PKC, 12(S)-HETE, 12-lipoxygenase, JAK-STAT, Wnt, Hedgehog and Notch signalling, as well as cPLA2α reactive oxygen species (ROS) generation from NADPH oxidase [77, 78, 112–116]. For example, conditional knockout of the Notch signalling regulator, the DNA-binding protein RPB-Jk, results in premature exit of lens epithelial cells from the cell cycle and a reduced number of epithelial progenitor cells for secondary fibre differentiation [117]. Evidence from several studies now indicates that this regulation is mediated by unidirectional Notch signalling, activated by Jag-1-expressing fibre cells [118–120].

Given the different roles of these functionally distinct growth factors, a better understanding of growth factor signalling at different stages of the cell cycle will no doubt shed light on the factors regulating lens cell proliferation in situ.

4.6 Cell Proliferation in Cataract

The majority of epithelial cells are quiescent in the adult vertebrate lens, dividing infrequently or when perturbed [121]. Cataract can occur as a result of deregulated proliferation, resulting from aberrant growth factor-mediated signalling. Elevated levels of many ocular growth factors are found in the aqueous in various ocular pathologies suggesting that growth factor-mediated signalling may play a significant role in aberrant growth that compromises lens function and results in cataract [30, 122–125]. Abnormal proliferation of the lens epithelium is most commonly associated with posterior subcapsular cataract (PSC) [126]. Steroid use is the primary risk factor for PSC; however diabetes and therapeutic radiation treatment have also been implicated [126]. The defining characteristics of steroid-induced PSC include an association only with steroids that possess glucocorticoid activity and the posterior migration of aberrant lens epithelial cells from the lens equator to the posterior lens pole [127]. Histologic studies of steroid cataracts describe the presence, at the posterior lens pole, of nucleated cells with the characteristics of epithelial cells, together with rounded Wedl or bladder cells, that contain degenerating nuclei [128, 129]. The presence of these cells is consistent with disruption to normal equatorial lens epithelial cell proliferation and differentiation. Although the cellular and molecular mechanisms of PSC formation are not well defined, recent investigations have focused toward discerning how glucocorticoid

(GC) steroids affect the regulation of lens proliferation. In a recent study utilising a rodent lens in vitro model, the GC dexamethasone (DEX), enhanced FGF-induced cell proliferation and coverage of the lens capsule [130]. Other studies on primary cultures of human lens epithelial cells and on a human lens cell line have also described changes in gene expression induced by DEX [131, 132]. The analysis indicated regulators of cell proliferation and apoptosis and components of signalling pathways such as MAPK/ERK were markedly changed by GC receptor activation. As discussed earlier in this chapter, ERK1/2 has been shown to be a key regulator of FGF- and aqueous-induced lens epithelial cell proliferation [104]; thus, one possible mechanism whereby GCs may alter regulation of lens cell proliferation is through modulation of ERK1/2 activity.

Cataracts characterised by aberrant proliferation of the lens epithelium are also associated with several human pathological syndromes. Most commonly, anterior polar cataract (APC) is associated with congenital eye diseases including Aniridia and Peters' anomaly and may present in certain systemic diseases such as Alport syndrome [133]. It has been postulated that APC is a product of abnormal lens epithelial proliferation causing the formation of a mass in the region of the anterior pole; however, imperfect separation of the lens from surface ectoderm during embryonic development may also contribute to its aetiology [133]. APC is recognised as a varying sized, white plaque at the anterior pole of the lens; it may project forward into the anterior chamber forming a pyramidal cataract and may further extend into the lens where degenerated cortical fibres are observed [134]. A histologic examination of human pyramidal opacities following cataract surgery showed lens epithelial hyperplasia, with some cells showing a fibroblast-like appearance that appeared embedded in dense collagenous connective tissue [135]. It is not clear as to what promotes the exacerbated proliferative response of lens epithelial cells to form polar opacities. In a rodent APC model, it was observed that degeneration of cortical lens cells preceded proliferation and invasion of epithelial cells into the degenerated anterior cortex, leading the investigator to hypothesise that proliferation of the epithelium is stimulated by degeneration of adjacent lens cells [135]. Both clinically and histologically, APC appears similar to anterior subcapsular cataract (ASC). Anterior subcapsular cataract is often associated with the ocular trauma that accompanies an impact injury and the inflammation or irritation of the eye that is characteristic of atopic dermatitis [136]. Following trauma to the anterior pole, epithelial cells become necrotic, after which adjacent cells migrate into the subcapsular area and proliferate to form an epithelial plaque. The proliferative lens epithelial cells undergo an epithelial to mesenchymal transition (EMT) to form myofibroblasts along with an excessive production and deposition of extracellular matrix proteins collagen I, III and fibronectin [137]. The hyper-proliferation of lens epithelial cells that precedes fibrotic metaplasia in ASC is also a key characteristic of posterior capsule opacification (or secondary cataract), the most common complication of cataract surgery [138].

Following cataract surgery residual lens epithelial cells attached to the anterior capsule hyper-proliferate onto the previously cell-free posterior capsule underlying the intraocular lens and encroach onto the visual axis; subsequent fibrotic changes at the posterior capsule lead to opacification [138]. Key in vitro investigations using BrdU labelling and detection of cells in the process of DNA synthesis revealed that once the human lens capsule has been breached and the fibres removed, the mitotic index is dramatically increased [139]. Moreover, disruption to postmortem capsular bags (generated following cataract surgery) by removal of a prosthetic intraocular lens (IOL) and fibres that form Soemmering's ring led to a significant increase in cell division [140]. Rakic and colleagues hypothesised that the restoration of proliferative activity that follows removal of fibres is likely due to differentiated fibres suppressing the proliferative activity of lens epithelial cells. However, more recent findings suggest unidirectional signalling from fibre cells to the overlying epithelial cells of the germinative zone permits proliferation [120]. To date it is unclear as to whether differentiated fibres that form in PCO have a key role in regulating lens cell proliferation. What is most apparent after cataract surgery is a phase of intensive lens cell proliferation and migration that decreases over time. Therefore, as a means to prevent PCO formation, several drugs that block lens cell proliferation and induce apoptosis have been investigated [141, 142]; however, the risk of toxic effects on surrounding tissues has inhibited their potential for human clinical trials. To date there is no effective biological agent that is used clinically to prevent the excessive lens cell growth that follows cataract surgery. In order to selectively inhibit proliferation post-cataract surgery, an understanding of those proteins and signalling mechanisms that promote the proliferative response is vital. During the surgical removal of a cataract, the blood-aqueous barrier is breached which can result in increased levels of proteins in the aqueous humour such as inflammatory mediators, cytokines and growth factors [143]. In particular, levels of thrombin [144], FGF [145] and HGF [81], are likely to increase following damage to the blood-aqueous barrier and induce lens cell proliferation. Much evidence also suggests autocrine signalling to be important in regulating the proliferative events that give rise to PCO particularly as the level of protein in the aqueous humour slowly returns to basal levels after cataract surgery [143]. A key investigation by Wormstone and colleagues demonstrated that after cataract extraction in vitro, adult human lens epithelial cells were able to proliferate in a medium without serum or added growth factors [146]. Protein expression analysis of the human native lens epithelium and the in vitro human capsular bag culture system has identified autocrine signalling systems for FGF, HGF and EGF (reviewed by [138]). Most importantly, in a capsular bag model, lens cell proliferation was significantly reduced following the application of receptor inhibitors to the growth factors and inhibition of their downstream signalling components ERK, p38 MAPK and JNK [74, 82, 147, 148]. Consistent with earlier in vitro reports, these studies highlight that multiple signalling pathways are responsible for promoting proliferative activity of lens epithelial cells following cataract surgery. Future investigations focused on inhibiting either multiple signalling components or regulators common to signalling pathways may be of importance in ablating lens cell proliferation and, as a result, prevent the development of PCO.

References

1. Nurse P (2002) Cyclin dependent kinases and cell cycle control (Nobel lecture). Chembiochem 3:596–603
2. Knox B, Ladiges P, Evans B (1994) Biology. McGraw Hill, Sydney
3. Griep AE, Zhang P (2004) Lens cell proliferation: the cell cycle. In: Lovicu FJ, Robinson ML (eds) Development of the ocular lens. Cambridge University Press, Cambridge, pp 191–211
4. Kazlauskas A (2005) The priming/completion paradigm to explain growth factor-dependent cell cycle progression. Growth Factors 23:203–210
5. Pardee AB (1989) G1 events and regulation of cell proliferation. Science 246:603–608
6. Sanchez I, Dynlacht BD (2005) New insights into cyclins. CDKs and cell cycle control. Semin Cell Dev Biol 16:311–321
7. Evans T, Rosenthal ET, Youngblom J, Distel D, Hunt T (1983) Cyclin: a protein specified by maternal RNA in sea urchin eggs that is destroyed at each cleavage division. Cell 33:389–396
8. Vincenzi B, Schiavon G, Silletta M, Santini D, Perrone G, Di Marino M, Angeletti S, Baldi A, Tonini G (2006) Cell cycle alterations and lung cancer. Histol Histopathol 21:423–435
9. Hanna C, O'Brien JE (1961) Cell production and migration in the epithelial layer of the lens. Arch Ophthalmol 66:103–107
10. Mikulicich AG, Young RW (1963) Cell proliferation and displacement in the lens epithelium of young rats injected with tritiated thymidine. Invest Ophthalmol Vis Sci 2:344–354
11. Modak SP, Morris G, Yamada T (1968) DNA synthesis and mitotic activity during early development of chick lens. Dev Biol 17:544–561
12. McAvoy JW (1978) Cell division, cell elongation and distribution of α-, β- and γ-crystallins in the rat lens. J Embryol Exp Morphol 44:149–165
13. McAvoy JW (1978) Cell division, cell elongation and the coordination of crystallin gene expression during lens morphogenesis in the rat. J Embryol Exp Morphol 45:271–281
14. Lovicu FJ, McAvoy JW (1999) Spatial and temporal expression of p57 (KIP2) during murine lens development. Mech Dev 86:165–169
15. Kallifatidis G, Boros J, Shin EH, McAvoy JW, Lovicu FJ (2011) The fate of dividing cells during lens morphogenesis, differentiation and growth. Exp Eye Res 92:502–511
16. Coulombre JL, Coulombre AJ (1963) Lens development: fiber elongation and lens orientation. Science 142:1489–1490
17. Bozanic D, Saraga-Babic M (2004) Cell proliferation during the early stages of human eye development. Anat Embryol (Berl) 208:381–388
18. Gao CY, Rampalli AM, Cai HC, He HY, Zelenka PS (1999) Changes in cyclin dependent kinase expression and activity accompanying lens fiber cell differentiation. Exp Eye Res 69:695–703
19. Fromm L, Overbeek PA (1996) Regulation of cyclin and cyclin-dependent kinase gene expression during lens differentiation requires the retinoblastoma protein. Oncogene 12:69–75
20. Gao CY, Zelenka PS (1997) Cyclins, cyclin-dependent kinases and differentiation. Bioessays 19:307–315
21. Zhang P, Wong C, DePinho RA, Harper JW, Elledge SJ (1998) Cooperation between the Cdk inhibitors p27(KIP1) and p57(KIP2) in the control of tissue growth and development. Genes Dev 12:3162–3167
22. Rampalli AM, Gao CY, Chauthaiwale VM, Zelenka PS (1998) pRb and p107 regulate E2F activity during lens fiber cell differentiation. Oncogene 16:399–408
23. Hyde RK, Griep AE (2002) Unique roles for E2F1 in the mouse lens in the absence of functional pRB proteins. Invest Ophthalmol Vis Sci 43:1509–1516
24. Lovicu FJ, Robinson ML (2004) Early lens development. In: Lovicu FJ, Robinson ML (eds) Development of the ocular lens. Cambridge University Press, Cambridge, pp 3–23
25. McAvoy JW, Chamberlain CG (1990) Growth factors in the eye. Prog Growth Factor Res 2:29–43

26. Lovicu FJ, McAvoy JW (2005) Growth factor regulation of lens development. Dev Biol 280:1–14
27. Segal MB (1992) Barriers and fluids of the eye and brain. CRC Press, Boca Raton, FL
28. Davson H (1962) The eye. Academic, New York
29. Tripathi RC, Millard CB, Tripathi BJ (1989) Protein composition of human aqueous humor: SDS-PAGE analysis of surgical and post-mortem samples. Exp Eye Res 48:117–130
30. Welge-Lussen U, May CA, Neubauer AS, Prilinger S (2001) Role of tissue growth factors in aqueous humor homeostasis. Curr Opin Ophthalmol 12:94–99
31. Alemany J, Girbau M, Bassas L, de Pablo F (1990) Insulin receptors and insulin-like growth factor I receptors are functional during organogenesis of the lens. Mol Cell Endocrinol 74:155–162
32. Bassnett S, Beebe DC (1991) Coincident loss of mitochondria and nuclei during lens fibre differentiation. Int Rev Exp Pathol 32:223–254
33. Le Roith D, Butler AA (1999) Insulin-like growth factors in pediatric health and disease. J Clin Endocrinol Metab 84:4355–4361
34. Arnold DR, Moshayedi P, Schoen TJ, Jones BE, Chader GJ, Waldbillig RJ (1993) Distribution of IGF-I and -II, IGF binding proteins (IGFBPs) and IGFBP mRNA in ocular fluids and tissues: potential sites of synthesis of IGFBPs in aqueous and vitreous. Exp Eye Res 56:555–565
35. Burren CP, Berka JL, Edmondson SR, Werther GA, Batch JA (1996) Localization of mRNAs for insulin-like growth factor-I (IGF-I), IGF-I receptor, and IGF binding proteins in rat eye. Invest Ophthalmol Vis Sci 37:1459–1468
36. Liu J, Chamberlain CG, McAvoy JW (1996) IGF enhancement of FGF-induced fibre differentiation and DNA synthesis in lens explants. Exp Eye Res 63:621–629
37. Iyengar L, Patkunanathan B, Lynch OT, McAvoy JW, Rasko JE, Lovicu FJ (2006) Aqueous humour- and growth factor-induced lens cell proliferation is dependent on MAPK/ERK1/2 and Akt/PI3-K signalling. Exp Eye Res 83:667–678
38. Hollenberg MD (1975) Receptors for insulin and epidermal growth factor: relation to synthesis of DNA in cultured rabbit lens epithelium. Arch Biochem Biophys 171:371–377
39. Reddan JR, Wilson-Dziedzic DC (1983) Insulin growth factor and epidermal growth factor trigger mitosis in lenses cultured in a serum-free medium. Invest Ophthalmol Vis Sci 24:409–416
40. Ibaraki N, Lin LR, Reddy VN (1996) A study of growth factor receptors in human lens epithelial cells and their relationship to fiber differentiation. Exp Eye Res 63:683–692
41. Shirke S, Faber SC, Hallem E, Makarenkova HP, Robinson ML, Overbeek PA, Lang RA (2001) Misexpression of IGF-1 in the mouse lens expands the transitional zone and perturbs lens polarization. Mech Dev 101:167–174
42. Shih AH, Holland EC (2006) Platelet-derived growth factor (PDGF) and glial tumorigenesis. Cancer Lett 232:139–147
43. Morrison-Graham K, Schatteman GC, Bork T, Bowen-Pope DF, Weston JA (1992) A PDGF receptor mutation in the mouse (Patch) perturbs the development of a non-neuronal subset of neural crest-derived cells. Development 115:133–142
44. Soriano P (1997) The PDGF alpha receptor is required for neural crest cell development and for normal patterning of the somites. Development 124:2691–2700
45. Reneker LW, Overbeek PA (1996) Lens-specific expression of PDGF-A alters lens growth and development. Dev Biol 180:554–565
46. Brewitt B, Clark JI (1988) Growth and transparency in the lens, an epithelial tissue, stimulated by pulses of PDGF. Science 242:777–779
47. Kok A, Lovicu FJ, Chamberlain CG, McAvoy JW (2002) Influence of platelet-derived growth factor on lens epithelial cell proliferation and differentiation. Growth Factors 20:27–34
48. Wang Q, McAvoy JW, Lovicu FJ (2010) Growth factor signaling in vitreous humor-induced lens fiber differentiation. Invest Ophthalmol Vis Sci 51:3599–3610

49. Ray S, Gao C, Wyatt K, Fariss RN, Bundek A, Zelenka P, Wistow G (2005) Platelet-derived growth factor D, tissue specific expression in the eye and a key role in control of lens epithelial cell proliferation. J Biol Chem 280:8494–8502
50. Dawid IB, Taira M, Good PJ, Rebagliati MR (1992) The role of growth factors in embryonic induction in Xenopus laevis. Mol Reprod Dev 32:136–144
51. Wright TJ, Ladher R, McWhirter J, Murre C, Schoenwolf GC, Mansour JL (2004) Mouse FGF15 is the ortholog of human and chick FGF19 but is not uniquely required for otic induction. Dev Biol 269:264–275
52. Powers CJ, McLeskey SW, Wellstein A (2000) Fibroblast growth factors, their receptors and signaling. Endocr Relat Cancer 7:165–197
53. Ornitz DM, Itoh N (2001) Fibroblast growth factors. Genome Biol 2:3005
54. Lovicu FJ, de Iongh RU, McAvoy JW (1997) Expression of FGF-1 and FGF-2 mRNA during lens morphogenesis, differentiation and growth. Curr Eye Res 16:222–230
55. Chamberlain CG, McAvoy JW (1997) Fibre differentiation and polarity in the mammalian lens: a key role for FGF. Prog Ret Eye Res 16:443–478
56. Govindarajan V, Overbeek PA (2001) Secreted FGFR3, but not FGFR1, inhibits lens fiber differentiation. Development 128:1617–1627
57. Zhao S, Hung FC, Colvin JS, White A, Dai W, Lovicu FJ, Ornitz DM, Overbeek PA (2001) Patterning the optic neuroepithelium by FGF signaling and Ras activation. Development 128:5051–5060
58. Ornitz DM (2000) FGFs, heparan sulfate and FGFRs: complex interactions essential for development. Bioessays 22:108–112
59. Bansal R (2002) Fibroblast growth factors and their receptors in oligodendrocyt development: implications for demyelination and remyelination. Dev Neurosci 24:35–46
60. de Iongh RU, Lovicu FJ, Hanneken A, Baird A, McAvoy JW (1996) FGF receptor-1 (flg) expression is correlated with fibre differentiation during rat lens morphogenesis and growth. Dev Dyn 206:412–426
61. de Iongh RU, Lovicu FJ, Chamberlain CG, McAvoy JW (1997) Differential expression of fibroblast growth factor receptors during rat lens morphogenesis and growth. Invest Ophthalmol Vis Sci 38:1688–1699
62. Marcelle C, Eichmann A, Halevy O, Breant C, Douarin NL (1994) Distinct developmental expression of a new avian FGFR. Development 120:683–694
63. Kurose H, Okamoto M, Shimizu H, Bito T, Marcelle C, Woji S, Ohuchi H (2005) FGF19-FGFR4 signalling elaborates lens induction with the FGF8-L-Maf cascade in the chick embryo. Dev Growth Diff 47:213–223
64. McAvoy JW (1980) Beta- and gamma-crystallin synthesis in rat lens epithelium explanted with neural retinal. Differentiation 17:85–91
65. West-Mays JA, Pino G, Lovicu FJ (2010) Development and use of the lens epithelial explant system to study lens differentiation and cataractogenesis. Prog Retin Eye Res 29:135–143
66. Chamberlain CG, McAvoy JW (1989) Induction of lens fibre differentiation by acidic and basic fibroblast growth factor (FGF). Growth Factors 1:125–134
67. Cohen S (1962) Isolation of a mouse submaxillary gland protein accelerating incisor eruption and eyelid opening in the new-born animal. J Biol Chem 237:1555–1562
68. Lakshmanan J, Perheentupa J, Hoath SB, Kim H, Gruters A, Odell C, Fisher DA (1985) Epidermal growth factor in mouse ocular tissue: effects of thyroxine and exogenous epidermal growth factor. Pediatr Res 19:315–319
69. Jones JT, Akita RW, Sliwkowski MX (1999) Binding specificities and affinities of egf domains for ErbB receptors. FEBS Lett 447:227–231
70. Olayioye MA, Neve RM, Lane HA, Hynes NE (2000) The ErbB signaling network: receptor heterodimerization in development and cancer. EMBO J 19:3159–3167
71. Yarden Y, Sliwkowski MX (2001) Untangling the ErbB signalling network. Nat Rev Mol Cell Biol 2:127–137

72. Oda K, Matsuoka Y, Funahashi A, Kitano H (2005) A comprehensive pathway map of epidermal growth factor receptor signaling. Mol Syst Biol 1:2005.0010
73. Majima K (1995) Human lens epithelial cells proliferate in response to exogenous EGF and have EGF and EGF receptor. Ophthalmic Res 27:356–365
74. Maidment JM, Duncan G, Tamiya S, Collison DJ, Wang L, Wormstone IM (2004) Regional differences in tyrosine kinase receptor signalling components determine differential growth patterns in the human lens. Invest Ophthalmol Vis Sci 45:1427–1435
75. Weng J, Liang Q, Mohan RR, Li Q, Wilson SE (1997) Hepatocyte growth factor, keratinocyte growth factor, and other growth factor-receptor systems in the lens. Invest Ophthalmol Vis Sci 38:1543–1554
76. Ireland ME, Mrock LK (2000) Differentiation of chick lens epithelial cells: involvement of the epidermal growth factor receptor and endogenous ligand. Invest Ophthalmol Vis Sci 41:183–190
77. Arora JK, Lysz TW, Zelenka PS (1996) A role for 12(S)- HETE in the response of human lens epithelial cells to epidermal growth factor and insulin. Invest Ophthalmol Vis Sci 37:1411–1418
78. Haque MS, Arora JK, Dikdan G, Lysz TW, Zelenka PS (1999) The rabbit lens epithelial cell line N/N1003A requires 12-lipoxygenase activity for DNA synthesis in response to EGF. Mol Vis 15:5–8
79. Marquardt H, Hunkapiller MW, Hood LE, Todaro GJ (1984) Rat transforming growth factor type 1: structure and relation to epidermal growth factor. Science 223:1079–1082
80. Wunderlich K, Knorr M (1994) Effect of platelet-derived growth factor PDGF on replication of cultured bovine lens epithelial cells. Opthalmologe 91:98–102
81. Wormstone IM, Tamiya S, Marcantonio JM, Reddan JR (2000) Hepatocyte growth factor function and c-Met expression in human lens epithelial cells. Invest Ophthalmol Vis Sci 41:4216–4222
82. Choi J, Park SY, Joo CK (2004) Hepatocyte growth factor induces proliferation of lens epithelial cells through activation of ERK1/2 and JNK/SAPK. Invest Ophthalmol Vis Sci 45:2696–2704
83. Fleming TP, Song Z, Andley UP (1998) Expression of growth control and differentiation genes in human lens epithelial cells with extended life span. Invest Ophthalmol Vis Sci 39:1387–1398
84. Hales AM, Chamberlain CG, McAvoy JW (1995) Cataract induction in lenses cultured with transforming growth factor-beta. Invest Ophthalmol Vis Sci 36:1709–1713
85. Lovicu FJ, Schulz MW, Hales AM, Vincent LN, Overbeek PA, Chamberlain CG, McAvoy JW (2002) TGFbeta induces morphological and molecular changes similar to human anterior subcapsular cataract. Br J Ophthalmol 86:220–226
86. Gordon-Thomson C, de Iongh RU, Hales AM, Chamberlain CG, McAvoy JW (1998) Differential cataractogenic potency of TGF-beta1, -beta2, and -beta3 and their expression in the postnatal rat eye. Invest Ophthalmol Vis Sci 39:1399–1409
87. de Iongh RU, Gordon-Thomson C, Chamberlain CG, Hales AM, McAvoy JW (2001) Tgfbeta receptor expression in lens: implications for differentiation and cataractogenesis. Exp Eye Res 72:649–659
88. Lang RA, McAvoy JW (2004) Growth factors in lens development. In: Lovicu FJ, Robinson ML (eds) Development of the ocular lens. Cambridge University Press, New York
89. Akhurst RJ, Derynck R (2001) TGF-beta signaling in cancer - a double-edged sword. Trends Cell Biol 11:S44–S51
90. Chen Z, Gibson TB, Robinson F, Silvestro L, Person G, Xu B, Wright A, Vanderbilt C, Cobb MH (2001) MAP kinases. Chem Rev 101:2449–2476
91. Kyriakias JM, Avruch J (1996) Sounding the alarm: protein kinase cascades activated by stress and inflammation. J Biol Chem 271:24313–24316
92. Hayashi M, Lee JD (2004) Role of the BMK1/ERK5 signaling pathway: lessons from knockout mice. J Mol Med 82:800–808

93. Yoon S, Seger R (2006) The extracellular signal-regulated kinase: multiple substrates regulate diverse cellular functions. Growth Factors 24:21–44
94. Li DW, Liu JP, Wang J, Mao YW, Hou LH (2003) Expression and activity of the signaling molecules for mitogen-activated protein kinase pathways in human, bovine, and rat lenses. Invest Ophthalmol Vis Sci 44:5277–5286
95. Seger R, Krebs EG (1995) The MAPK signaling cascade. FASEB J 9:726–735
96. Treisman R (1996) Regulation of transcription by MAP kinase cascades. Curr Opin Cell Biol 8:205–215
97. Stefanovsky VY, Pelletier G, Hannan R, Gagnon-Kugler T, Rothblum LI, Moss T (2001) An immediate response of ribosomal transcription to growth factor stimulation in mammals is mediated by ERK phosphorylation of UBF. Mol Cell 8:1063–1073
98. Felton-Edkins ZA, Fairley JA, Graham EL, Johnston IM, White RJ, Scott PH (2003) The mitogen-activated protein (MAP) kinase ERK induces tRNA synthesis by phosphorylating TFIIIB. EMBO J 22:2422–2432
99. Testa JR, Bellacosa A (2001) AKT plays a central role in tumorigenesis. Proc Natl Acad Sci 98:10983–10985
100. Rajagopal R, Huang J, Dattilo LK, Kaartinen V, Mishina Y, Deng CX, Umans L, Zwijsen A, Roberts AB, Beebe DC (2009) The type I BMP receptors, Bmpr1a and Acvr1, activate multiple signaling pathways to regulate lens formation. Dev Biol 335:305–316
101. Hyatt GA, Beebe DC (1993) Regulation of lens cell growth and polarity by an embryo-specific growth factor and by inhibitors of lens cell proliferation and differentiation. Development 117:701–709
102. McAvoy JW, Chamberlain CG (1989) Fibroblast growth factor (FGF) induces different responses in lens epithelial cells depending upon its concentration. Development 107:221–228
103. Iyengar L, Wang Q, Rasko JE, McAvoy JW, Lovicu FJ (2007) Duration of ERK1/2 phosphorylation induced by FGF or ocular media determines lens cell fate. Differentiation 75:662–668
104. Iyengar L, Patkunanathan B, McAvoy JW, Lovicu FJ (2009) Growth factors involved in aqueous humour-induced lens cell proliferation. Growth Factors 27:50–62
105. Coulombre JL, Coulombre AJ (1969) Lens development. IV. Size, shape, and orientation. Invest Ophthalmol 8:251–257
106. Coulombre JL, Coulombre AJ (1971) Lens development. V. Histological analysis of the mechanism of lens reconstitution from implants of lens epithelium. J Exp Zool 176:15–24
107. Lovicu FJ, McAvoy JW (2001) FGF-induced lens cell proliferation and differentiation is dependent on MAPK (ERK1/2) signalling. Development 128:5075–5084
108. Le AC, Musil LS (2001) FGF signalling in chick lens development. Dev Biol 233:394–411
109. Upadhya D, Ogata M, Reneker LW (2013) MAPK1 is required for establishing the pattern of cell proliferation and for cell survival during lens. Development 140:1573–1582
110. Stiles CD, Isberg RR, Pledger WJ, Antoniades HN, Scher CD (1979) Control of the Balb/c-3 T3 cell cycle by nutrients and serum factors: analysis using platelet-derived growth factor and platelet-poor plasma. J Cell Physiol 99:395–405
111. Pan Y, Carbe C, Powers A, Feng GS, Zhang X (2010) Sprouty2-modulated Kras signaling rescues Shp2 deficiency during lens and lacrimal gland development. Development 137:1085–1093
112. Janus A, Robak T, Smolewski P (2005) The mammalian target of the rapamycin (mTOR) kinase pathway: its role in tumourigenesis and targeted antitumour therapy. Cell Mol Biol Lett 10:479–498
113. Thisse B, Thisse C (2005) Functions and regulations of fibroblast growth factor signaling during embryonic development. Dev Biol 287:390–402
114. de Iongh RU, Abud HE, Hime GR (2006) WNT/Frizzled signaling in eye development and disease. Front Biosci 11:2442–2464

115. Kerr CL, Huang J, Williams T, West-Mays JA (2012) Activation of the hedgehog signaling pathway in the developing lens stimulates ectopic FoxE3 expression and disruption in fiber cell differentiation. Invest Ophthalmol Vis Sci 53:3316–3330
116. Wang Y, Xing K-Y, Lou M (2011) Regulation of cytosolic phospholipase A2 ($cPLA_2\alpha$) and its association with cell proliferation in human lens epithelial cells. Invest Ophthalmol Vis Sci 52:8231–8240
117. Jia J, Lin M, Zhang L, York JP, Zhang P (2007) The Notch signaling pathway controls the size of the ocular lens by directly suppressing p57Kip2 expression. Mol Cell Biol 27:7236–7247
118. Rowan S, Conley KW, Le TT, Donner AL, Maas RL, Brown NL (2008) Notch signaling regulates growth and differentiation in the mammalian lens. Dev Biol 321:111–122
119. Le TT, Conley KW, Brown NL (2009) Jagged 1 is necessary for normal lens formation. Dev Biol 328:118–126
120. Saravanamuthu SS, Gao CY, Zelenka PS (2009) Notch signaling is required for lateral induction of Jagged1 during FGF-induced lens fiber differentiation. Dev Biol 332:166–176
121. Zhou M, Leiberman J, Xu J, Lavker RM (2006) A hierarchy of proliferative cells exists in mouse lens epithelium: implications for lens maintenance. Invest Ophthalmol Vis Sci 47:2997–3003
122. Majima K, Kojima Y, Ouhashi F (1998) Cell biological analysis with respect to cause of fibrous opacification of the anterior capsule after cataract extraction. Ophthalmologica 212:364–368
123. Shigemitsu T, Ishiguro K, Shimizu Y, Horiguchi M, Kasahara M, Arakaki S (1999) Immunocytochemical features of lens after cataract tissue–signalling molecules (growth factors, cytokines, other signalling molecules), cytoskeleton proteins, cellular and extracellular matrix proteins. Int Ophthalmol 23:137–144
124. Shentu X, Yao K, Sun C, Xu W, Wu R (2002) Expression and effect of basic fibroblast growth factor on human cataract lens epithelial cells. Chin Med J (Engl) 115:268–271
125. Saika S, Miyamoto T, Ishida I, Shirai K, Ohnishi Y, Ooshima A, McAvoy JW (2002) TGF beta-Smad signalling in postoperative human lens epithelial cells. Br J Ophthalmol 86:1428–1433
126. Robman L, Taylor H (2005) External factors in the development of cataract. Eye 19:1074–1082
127. James E (2007) The etiology of steroid cataract. J Ocul Pharmacol Ther 23:403–420
128. Eshaghian J, Streeten B (1980) Human posterior subcapsular cataract. An ultrastructural study of the posteriorly migrating cells. Arch Ophthalmol 98:134–143
129. Greiner J, Chylack L (1979) Posterior subcapsular cataracts: histopathologic study of steroid-associated cataracts. Arch Ophthalmol 97:135–144
130. Wang C, Dawes LJ, Liu Y, Wen L, Lovicu FJ, McAvoy JW (2013) Dexamethasone influences FGF-induced responses in lens epithelial explants and promotes the posterior capsule coverage that is a feature of glucocorticoid-induced cataract. Exp Eye Res 111:79–87
131. James ER, Fresco VM, Robertson LL (2005) Glucocorticoid-induced changes in the global gene expression of lens epithelial cells. J Ocul Pharmacol Ther 21:11–27
132. Gupta V, Awasthi N, Wagner BJ (2007) Specific activation of the glucocorticoid receptor and modulation of signal transduction pathways in human lens epithelial cells. Invest Ophthalmol Vis Sci 48:1724–1734
133. Moross T, Vaithilingam SS, Styles S, Gardner HA (1984) Autosomal dominant anterior polar cataracts associated with a familial 2;14 translocation. J Med Genet 21:52–53
134. Wheeler DT, Mullaney PB, Awad A, Zwaan J (1999) Pyramidal anterior polar cataracts. Ophthalmology 106:2362–2367
135. Drenckhahn D (1978) Anterior polar cataract and lysosomal alterations in the lens of rats treated with the amphiphilic lipidosis-inducing drugs chloroquine and chlorphentermine. Virchows Arch B Cell Pathol 27:255–266

136. Eldred JA, Dawes LJ, Wormstone IM (2011) The lens as a model for fibrotic disease. Philos Trans R Soc Lond B Biol Sci 366:1301–1319
137. Lee EH, Joo CK (1999) Role of transforming growth factor-beta in transdifferentiation and fibrosis of lens epithelial cells. Invest Ophthalmol Vis Sci 40:2025–2032
138. Wormstone IM, Wang L, Liu CS (2009) Posterior capsule opacification. Exp Eye Res 88:257–269
139. Rakic JM, Galand A, Vrensen GF (1997) Separation of fibres from the capsule enhances mitotic activity of human lens epithelium. Exp Eye Res 64:67–72
140. Rakic JM, Galand A, Vrensen GF (2000) Lens epithelial cell proliferation in human posterior capsule opacification specimens. Exp Eye Res 71:489–494
141. Behar-Cohen FF, David T, D'Hermies F, Pouliquen YM, Buechler Y, Nova MP, Houston LL, Courtois Y (1995) In vivo inhibition of lens regrowth by fibroblast growth factor 2-saporin. Invest Ophthalmol Vis Sci 36:2434–2448
142. Duncan G, Wormstone IM, Liu CS, Marcantonio JM, Davies PD (1997) Thapsigargin-coated intraocular lenses inhibit human lens cell growth. Nat Med 3:1026–1028
143. Pande M, Spalton DJ, Marshall J (1996) Continuous curvilinear capsulorhexis and intraocular lens biocompatibility. J Cataract Refract Surg 22:89–97
144. James C, Collison DJ, Duncan G (2005) Characterization and functional activity of thrombin receptors in the human lens. Invest Ophthalmol Vis Sci 46:925–932
145. Namiki M, Tagami Y, Yamamoto M, Yamanaka A, Itoh M, Kanoh M (1992) Presence of human epidermal growth factor (hEGF), basic fibroblast growth factor (bFGF) in human aqueous. Nihon Ganka Gakkai Zasshi 96:652–656
146. Wormstone IM, Liu CS, Rakic JM, Marcantonio JM, Vrensen GF, Duncan G (1997) Human lens epithelial cell proliferation in a protein-free medium. Invest Ophthalmol Vis Sci 38:396–404
147. Wormstone IM, Del Rio-Tsonis K, McMahon G, Tamiya S, Davies PD, Marcantonio JM, Duncan G (2001) FGF: an autocrine regulator of human lens cell growth independent of added stimuli. Invest Ophthalmol Vis Sci 42:1305–1311
148. Dawes LJ, Duncan G, Wormstone IM (2013) Age-related differences in signaling efficiency of human lens cells underpin differential wound healing response rates following cataract surgery. Invest Ophthalmol Vis Sci 54:333–342

Growth Factor Signaling in Lens Fiber Differentiation

5

Robb U. de Iongh and Melinda K. Duncan

Abstract

Since the first reports of experimental lens inversion causing changes in lens cell fate in the 1960s, various factors including the mitogens IGF, EGF, PDGF, as well as Wnts, Notch, and BMPs have been implicated in the regulation of lens fiber cell differentiation, although the preponderance of evidence suggests that FGFs provide the major signal for this process. It is becoming increasingly apparent that complex interactions between the signaling pathways activated by these factors are required to generate functional fiber cells. Disruption of these signaling cascades usually results in the formation of congenital lens defects, including cataracts, and it is likely that these signaling cascades also regulate lens "regenerative" processes that follow extracapsular cataract extraction and lead to Soemmering's ring and Elschnig's pearls. Improved knowledge of the signaling processes that occur in normal lens development as well as in the lens epithelial cells (LECs) remaining in the capsular bag following cataract surgery may not only improve the outcomes of cataract treatment but may also eventually lead to the capacity to regenerate functional lenses either from "induced pluripotent stem cells" or from the patient's own remnant LECs.

Keywords

Bone morphogenetic proteins • Elschnig's pearl • Fibroblast growth factors • Heparan sulfate proteoglycans • Lens differentiation • Posterior capsule opacification • Soemmering's ring

---

R.U. de Iongh (✉)
Department of Anatomy and Neuroscience, The University of Melbourne,
Parkville, VIC 3010, Australia
e-mail: r.deiongh@unimelb.edu.au

M.K. Duncan
Department of Biological Sciences, University of Delaware, Newark, DE 19716, USA
e-mail: duncanm@udel.edu

S. Saika et al. (eds.), *Lens Epithelium and Posterior Capsular Opacification*,
DOI 10.1007/978-4-431-54300-8_5, © Springer Japan 2014

Abbreviations

BMP	Bone morphogenetic protein
Dvl	Dishevelled
EGF	Epidermal growth factor
EMT	Epithelial–mesenchymal transition
ERK	Extracellular-regulated kinase
FGF	Fibroblast growth factor
FGFR	Fibroblast growth factor receptor
Fzd	Frizzled
HSPG	Heparan sulfate proteoglycan
IGF	Insulin-like growth factor
JAK	Janus kinase
LEC	Lens epithelial cell
MAPK	Mitogen-activated protein kinase
NICD	Notch intracellular domain
PCO	Posterior capsule opacification
PCP	Planar cell polarity
PDGF	Platelet-derived growth factor
PI3K	Phosphatidylinositol-3-kinase
PLC	Phospholipase C
STAT	Signal transducer and activator of transcription
TGF	Transforming growth factor
TNF	Tumor necrosis factor

5.1 Introduction

The ocular lens is a transparent, cellular tissue that has two cell types: an anterior epithelial monolayer and the highly elongated lens fibers, which comprise the bulk of the lens. The lens is enclosed by a thick basement membrane, the lens capsule (see Chap. 3), which serves as the basal anchor for both lens epithelial and cortical fiber cells [1]. During embryonic development, the ocular lens arises from the head ectoderm which, under the influence of the neuroepithelium of the optic vesicle, invaginates and forms a spheroidal vesicle that is positioned within the anterior margins of the optic cup. Within the vesicle, the cells that face the optic cup undergo extensive elongation to form the primary lens fiber cells. By contrast, the vesicle cells that face the overlying ectoderm (the presumptive corneal epithelium) form the cuboidal lens epithelial cells (LEC). This allocation of different fates to anterior and posterior lens vesicle cells gives rise to the distinct polarity of the embryonic lens [2].

During subsequent embryonic and postnatal development, and indeed throughout life, the lens continues to grow by distinct patterns of cell proliferation,

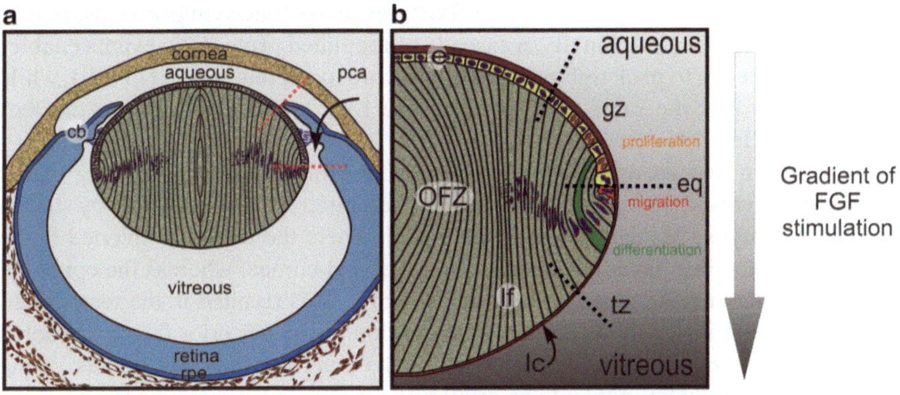

Fig. 5.1 The vertebrate eye and lens. (**a**) The highly polarized lens has two cell types: epithelial (e) and fiber cells (lf) contained with a thick basement membrane, the lens capsule (lc). It is suspended between the cornea and retina by collagenous zonular fibers that arise from the ciliary body (cb) and bathed by the aqueous and vitreous humors, including aqueous of the posterior chamber (pca). (**b**) The anterior epithelium (e), bathed by aqueous humor, is relatively quiescent; cell proliferation occurs mainly in the germinative zone (gz). Daughter cells contribute to the epithelium or, if they are displaced across the equator (eq), undergo extensive elongation and differentiation into secondary fiber cells in the transitional zone (tz), under the influence of the vitreous humor. Central lens fiber cells lose organelles and nuclei to form an organelle-free zone (OFZ). An anterior–posterior gradient of FGF stimulation, combined with other growth factors in the ocular media, is involved in regulating lens cell behavior. Adapted from Martinez et al. [4]

migration, and differentiation. Essentially all of the proliferative potential of lens cells resides in the lens epithelium, while lens fibers exhibit terminal cell cycle arrest [3] (Fig. 5.1). During embryonic development, cell proliferation is detected throughout the lens epithelium; however, from perinatal stages to the adult, rapidly dividing epithelial cells become restricted to a region just above the lens equator known as the "germinative zone" [5–7]. However, there is evidence for a population of slowly dividing "stem-like cells" in the central epithelium [8]. Daughter cells arising from divisions in the germinative zone migrate or are displaced below the equator into a region called the "transitional zone" where they withdraw from the cell cycle and start to differentiate into secondary fiber cells. As additional fiber cells elongate and differentiate, the earlier generated fibers are buried beneath them so that the oldest lens fibers are found at the center of the lens [2]. Differentiation of fiber cells is characterized by dramatic changes in length with concomitant increases in membrane area and protein synthesis (particularly soluble crystallins, specific membrane, and intermediate filament proteins that contribute to lens clarity). Fiber cells also acquire membrane specializations (e.g., ball and socket junctions, membrane protrusions, and furrowed membranes) that ensure the close association and packing of the fiber cells required to establish the internal circulation of the lens, necessary for the homeostasis of this avascular tissue [9]. At later stages of differentiation, fiber cells destroy their cell nuclei and other organelles, forming an organelle-free zone in the central region of the lens to reduce light scatter [10]. Finally, a cascade of regulated proteolytic events enables the lens fiber cells to pack tighter and the lens core to exclude water [11–19], while fiber cells within the same growth shell fuse [20, 21], contributing to the lens refractive index.

The notion that growth factors from the ocular media and surrounding ocular tissues are involved in regulating lens fiber differentiation was reinforced by studies in chick embryos [22] and postnatal mice [23], where the lens was inverted in the eye so that the fibers faced the anterior chamber and cornea, whereas the epithelial cells faced the vitreous chamber and neural retina. This resulted in the reestablishment of an anterior epithelium and differentiation of the posterior facing epithelial cells into elongating fiber cells. Since then, numerous in vitro studies, using cultured lens epithelial explants or primary cells isolated from rodents and chick lenses, combined with transgenic mouse studies, have identified several growth factor and cell–cell signaling pathways that are involved in lens fiber differentiation, including the IGFs, FGFs, BMPs, TGFβ, TNF, Notch, and Wnts [24]. While each of these has been shown to contribute to the fiber differentiation process, the FGFs are the only factors that seem to be essential and sufficient to initiate lens fiber differentiation in mammals [25]. Indeed, many of the other factors shown to affect fiber differentiation appear to interact with FGF signaling pathways and are particularly required to fine-tune the intracellular signaling pathways activated by FGF to ensure precise regulation of fiber cell differentiation [24] (Fig. 5.2).

5.2 The Fibroblast Growth Factor (FGF) Family

The FGF family comprises 22 structurally related polypeptides that can function as secreted paracrine (*Fgf1–10*, *Fgf16–18*, *Fgf20*, *Fgf22*) and endocrine signals (*Fgf15/19*, *Fgf21*, *Fgf23*) or as intracellular molecules (*Fgf11–14*) that are not secreted and function independently of FGF receptors [26]. Paracrine FGFs have high affinity for and require heparan sulfate proteoglycans (HSPGs) for optimum receptor binding. The endocrine FGFs, unlike paracrine FGFs, have a low affinity for HSPG and thus can diffuse through the extracellular matrix more easily. As they have a low binding affinity for the cognate FGF receptors, they rely on co-receptors of the Klotho family to activate FGFR signaling [26, 27]. Various members of the FGF family have been found to be expressed in the vertebrate eye during embryonic development or in the adult, including *Fgf1*, *Fgf2* [28], *Fgf3* [29], *Fgf5* [30], *Fgf6* [31, 32], *Fgf8*, *Fgf9*, and *Fgf15/19* [33–35]. A recent screen of the family, using real-time PCR, identified mRNA expression of all FGF family members in the adult mouse eye except *Fgf4*, *Fgf5*, *Fgf15*, *Fgf16*, *Fgf21*, and *Fgf23* [36]. Members of the intracellular FGF11 subfamily (*Fgf11–14*) were also detected; however, the functions of these molecules in the eye or in the lens have not been investigated.

There are five mammalian FGF receptor (FGFR) genes (*Fgfr1–Fgfr5*), of which *Fgfr1–Fgfr4* encode receptor tyrosine kinases, with an extracellular ligand-binding domain comprising three immunoglobulin-like domains (I, II, and III),

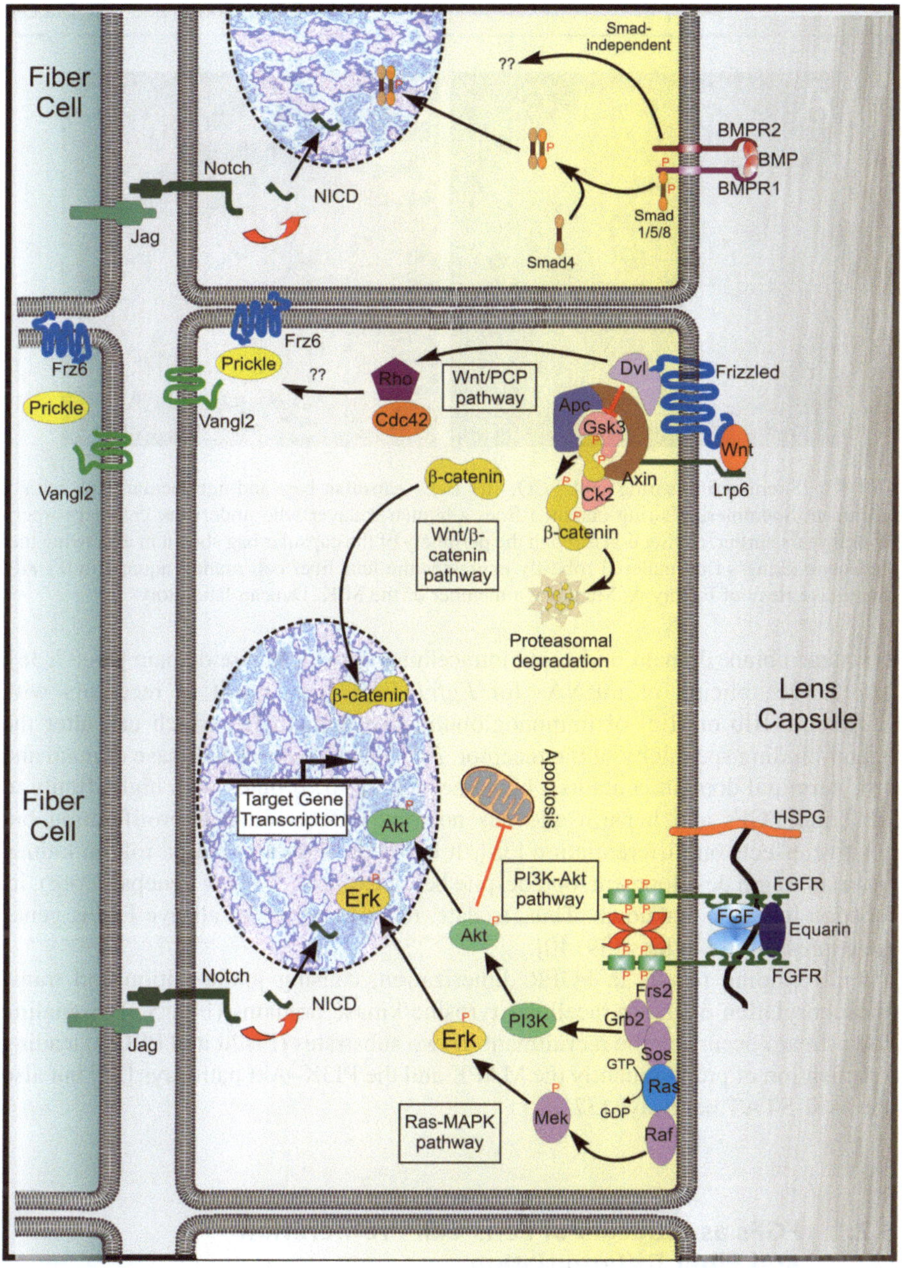

Fig. 5.2 Signaling pathways in lens fiber differentiation. Diagram of epithelial cells at the lens equator, highlighting key growth factor and cell–cell signaling pathways implicated in fiber cell differentiation. These include the FGFs acting via FGFRs to activate the MAPK and Ras–MAPK pathways, Wnts activating Frizzleds to activate the β-catenin or PCP pathways, BMPs engaging specific receptors to activate Smad-dependent and Smad-independent pathways, and activation of the Notch pathway resulting in release of the Notch intracellular domain (NICD). The outputs of these pathways often involve transcriptional changes but also affect cytoskeletal organization and membrane protein localization. See text for details. Adapted from Martinez et al. [4]

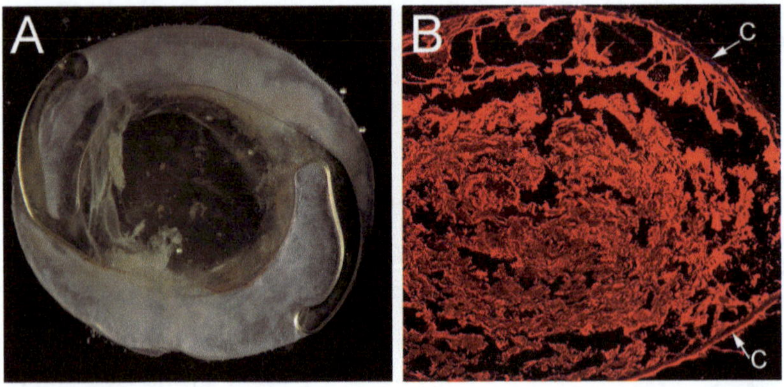

Fig. 5.3 Soemmering's ring and PCO. (**a**) Lens capsular bag and intraocular lens with a prominent Soemmering's ring obtained from a human cadaver who underwent cataract surgery several years earlier. (**b**) Section through the periphery of the capsular bag shown in **a** showing that the Soemmering's ring material robustly expresses the lens fiber cell marker aquaporin 0 (*red*). Images courtesy of Fahmy A. Mamuya, a member of the M.K. Duncan laboratory

a transmembrane domain, and a split intracellular tyrosine kinase domain [26, 37, 38]. Alternative splicing of mRNAs for *Fgfr1–Fgfr3* can result in receptors with variations (IIIb or IIIc) of immunoglobulin-like domain III, which can alter the ligand-binding specificity of the receptor. *Fgfr5* lacks a tyrosine kinase domain and its C-terminal domain contains a histidine-rich motif. It binds with high affinity to paracrine FGFs and heparin and has negative effects on cell proliferation but positive effects on differentiation [38]. It has been shown to play a role in kidney and diaphragm development, but, despite being expressed (www.genepaint.org), its function in lens is unknown. Thus, to date, representatives of all five FGFR genes are expressed in the lens [39, 40].

FGF binding results in FGFR dimerization, causing juxtaposition and trans-phosphorylation of the intracellular tyrosine kinase domains (Fig. 5.3). Signaling downstream occurs by the recruitment of two substrates (Frs2α and PLCγ), leading to activation of predominantly the MAPK and the PI3K–Akt pathways [24] but also the JAK–STAT pathways [37, 41].

5.2.1 FGFs as Inducers of Lens Cell Proliferation and Fiber Differentiation

Proliferative and fiber cell-inducing activity was first identified in extracts of retina and in retina-conditioned medium in assays using rodent lens epithelial explants [42] and dissociated bovine lens cells [43, 44]. However, the pivotal finding that this activity corresponded to fibroblast growth factor (FGF) [45–47], which induced concentration-dependent lens cell responses (proliferation, migration, and differentiation) [48], led to the proposal of an anterior–posterior gradient of FGF stimulation within the eye that controls these lens cell behaviors in vivo [49] (Fig. 5.1). Similar FGF concentration-dependent effects on chick lens cell proliferation and differentiation have been documented [50]. The FGF-gradient hypothesis posits that very low levels of FGF signals in the aqueous of the anterior chamber are sufficient to maintain epithelial cell survival, whereas increasing levels in the posterior aqueous chamber initiate cell proliferation in the germinative zone and that cells in the transition zone are exposed to high levels of FGF that initiate fiber cell differentiation [48]. Support for this FGF gradient comes from various studies, demonstrating different levels of FGF in the ocular media and lens capsule [51, 52], expression of FGF receptors in lens [40, 53, 54], and various FGFs in ocular tissues, particularly the retina [28, 33, 52, 55, 56], as well as transgenic studies, which disrupted the normal levels of FGF by overexpression and induced aberrant fiber differentiation in the epithelium [57–59]. Finally, the requirement for FGF in fiber differentiation was demonstrated by transgenic studies in which expression of dominant-negative receptors [60] or deletion of multiple FGF receptor genes [39] disrupted or inhibited fiber cell differentiation. There is a considerable redundancy of FGFR function in the lens, with only null mutation of *Fgfr2* resulting in a phenotype [39, 61], characterized by small lenses with impaired cell survival, cell cycle regulation, and fiber differentiation. By contrast, mice with null mutations of *Fgfr1* [62], *Fgfr3* [63], and *Fgfr4* [64] show normal lens development and differentiation. Indeed, lenses lacking two of the four FGFRs (*Fgfr1*/*Fgfr3* or *Fgfr3*/*Fgfr4*) showed normal lens development [64, 65], and it was not until three of the FGFRs (*Fgfr1–3*) were deleted that profound effects were observed on lens development [65].

The identification of numerous FGFs expressed in the eye has made it difficult to ascertain which FGFs regulate lens differentiation. Consistent with the idea that FGFs act redundantly, most knockouts of FGF gene show no phenotype (reviewed in [39]). However, studies of mice [66, 67] with *Fgf9* mutations showed attenuated fiber differentiation, suggesting that FGF9 is a key contributor to murine lens fiber differentiation. In zebrafish, knockdown of *Fgf19* attenuated fiber cell differentiation [68], but loss of its ortholog, *Fgf15*, in mice had no apparent effects on lens development [69]. Moreover in chick, *Fgf19* has been implicated in a negative feedback loop during early eye induction [34], suggesting that there are species differences in the way FGFs are utilized during ocular development.

5.2.2 Requirement for Heparan Sulfate Proteoglycans (HSPGs) for FGF Activity in Lens

Paracrine FGFs have a high affinity for heparin and heparan sulfate proteoglycans (HSPGs), a feature that was used in their initial isolation [70], and FGF binding to HSPG modulates its activity [71]. The lens capsule is rich in HSPGs [72, 73], which bind and co-localize with FGFs [55, 73, 74] and are required for presentation of FGF to receptors on the surface of lens cells, resulting in subsequent receptor dimerization and internalization [75]. The importance of HSPGs for lens differentiation is highlighted by studies showing disrupted fiber differentiation in lenses of mice that lack the heparan sulfate side chains on a major form of HSPG, perlecan [76], or that lack the enzymes (*Ndst1* and *Ndst2*) required to add sulfate groups to HSPG core proteins [77, 78].

HSPGs and other proteoglycans have also been detected in human samples of posterior capsule opacification (PCO), suggesting a possible role for FGF signaling during development of PCO [79]. Consistent with this, human capsular bags contain FGF [80, 81], either sequestered in the capsule or produced by the epithelial cells, that is sufficient to promote cell migration across the posterior capsule [81].

5.2.3 Other FGF Co-receptors

Recently, a novel secreted molecule, equarin (CCDC80 in mammals), has been implicated as a modulator of FGF signaling and cell adhesion during chick lens development by its interaction with HSPG [82–84]. Equarin binds heparin and HSPG and is expressed at the lens equator of the chick lens, where cells exit the cell cycle and initiate fiber differentiation. Exogenous equarin promotes fiber cell differentiation by increasing FGF-mediated activation of the ERK–MAPK pathway, whereas knockdown of equarin inhibits fiber differentiation [82]. Intriguingly, equarin is also responsive to BMP signaling (see below), and it has been proposed that a balance of BMP and FGF signals regulates proliferation and cell cycle exit via equarin expression at the chick lens equator [83].

The Klotho family of transmembrane glycoproteins has been identified as co-receptors for endocrine FGFs. One member, the lactase-like (*Lctl*) gene, is highly expressed in the adult eye but has not yet been detected in lens *per se*. Intriguingly, *Lctl*, which is a membrane-bound glycosidase-like protein, appears to function as a co-receptor for FGF15/19 with FGFRs and can activate ERK signaling pathways in cells [36]. While FGF15/19 is known to be involved in early lens differentiation in zebrafish [68] and chicken [34], it remains to be determined if *Lctl* is involved.

5.2.4 FGF Signaling Pathways in Lens

Investigation of the FGF downstream signaling pathways in the lens has provided strong evidence for the activation of the MAPK–ERK [50, 85–90], PI3K–Akt [87, 88, 91–93], and STAT [94, 95] pathways during fiber differentiation. Moreover, negative regulation of these pathways in the lens epithelium by intracellular (Sprouty) or transmembrane (Sefs) antagonists of FGF signaling plays an important role in maintaining the anterior epithelium and preventing inappropriate fiber differentiation [96–98]. Experiments using specific small molecule antagonists of the MAPK pathway in rat lens explants [85, 87] indicate that activation of ERK via the Ras–MAPK pathway is required for epithelial cell proliferation and fiber cell elongation but not for fiber-specific crystallin expression. Consistent with this, studies in transgenic mice using dominant-negative [99] and oncogenic [100] forms of Ras as well as conditional deletion of Mapk1 in the lens [101] show that the Ras–MAPK pathway is sufficient and required for epithelial cell proliferation but not for fiber differentiation.

Studies using small molecule antagonists of the PI3K–Akt pathway in the lens show that inhibition of signaling via this pathway blocks cell proliferation [102] and fiber differentiation, including the expression of β- and γ-crystallins [87].

Intriguingly, the activation profiles of pERK and pAkt in response to aqueous and vitreous humors are similar to low and high concentrations of FGF and correlated with proliferation and fiber differentiation responses, respectively [86, 87]. This suggested that the duration of pathway activation may determine whether cells proliferate or differentiate. However, further studies using antagonists for various receptors known to be present in the lens show that while various factors (IGF1, EGF, PDGF-A) contribute to pathway activation, FGF signaling was required for differentiation responses, notwithstanding any extended duration of ERK or Akt phosphorylation that could be achieved by other factors [88].

While the STAT pathways are activated in cultured lens cells following various growth factor (FGF, PDGF-A, IGF1) treatments [94], deletion of either of the *Stat1* and *Stat3* genes in mice suggests either that the JAK–STAT pathway is not essential for lens development or that there is functional redundancy among STAT genes [95].

5.3 TGFβ Superfamily Signaling in Lens Fiber Differentiation

The transforming growth factor superfamily comprises more than 30 secreted growth factors, including the prototypic TGFβ and bone morphogenetic protein (BMP) families. These factors elicit a plethora of responses and are involved in an extraordinarily diverse range of cellular functions and processes during embryonic development as well as adult tissue homeostasis and wound repair. While ligands and receptors are remarkably promiscuous, the mechanism of signaling is highly conserved and involves ligand dimers binding to type I and type II receptors, with type II receptors phosphorylating type I receptors (Fig. 5.3). Receptor-specific Smads (Smads2/3 for TGFβs or Smads1/5/8 for BMPs) bind to and are phosphorylated by the activated type I receptor leading to the formation of Smad heterodimers with the ubiquitous common co-Smad (Smad4), which then enter the nucleus to activate specific gene transcription [103–106]. In addition to Smad-dependent signaling, there are Smad-independent pathways that utilize small GTPases, the p38-MAPK, ERK–MAPK, and the PI3K pathways, and also apicobasal polarity proteins such as Par6 [106].

The ectopic expression of dominant-negative TGFβ receptors (*Tgfbr2* and *Tgfbr1*) in lens fibers of the murine lens abrogated Smad2 activation and indicated a requirement for TGFβ signaling during lens terminal differentiation [107]. Remarkably though, null mutations of various TGFbeta family receptors or of Smad4 and Smad2 have not recapitulated this phenotype [7, 108, 109]. Given the promiscuity of ligands and receptors [105], it is plausible that the dominant-negative receptors effectively block all signaling (Smad dependent and nondependent); however, the possibility that overexpression of mutant receptors within the secretory pathway caused nonspecific fiber cell death, perhaps via an unfolded protein response [110, 111], cannot be discounted.

A requirement for BMPs in lens induction and morphogenesis was originally demonstrated in mice that lack BMP7 or BMP4 (reviewed in [112]). Since then, several studies have demonstrated a requirement for BMP signaling during lens fiber differentiation in vitro and in vivo (reviewed in [24]). More recently, cross talk between FGF and BMP signals has been demonstrated, particularly in differentiating fiber cells, with BMPs contributing to the activity of vitreous-induced fiber cell differentiation, initiating the expression of key lens fiber differentiation marker genes, and regulating the gap junction-mediated coupling between fiber cells [83, 113, 114]. BMPs also combine with FGFs at the lens equator to regulate equarin expression and mediate cell cycle exit as epithelial cells differentiate into fiber cells [83]. Analysis of mice lacking Smads 1, 5, and 4 and various type I receptors indicates that BMPs mediate cell proliferation and probably crystallin expression by Smad-dependent signaling but that cell cycle exit and cytoskeletal changes during lens development are mediated via Smad-independent pathways [7, 109].

TGFβ signaling via both the canonical Smad-dependent and noncanonical Smad-independent pathways has been shown to mediate epithelial–mesenchymal transition (EMT) changes in the lens during formation of anterior subcapsular cataracts, in PCO and in response to injury [115] (see Chaps. 8 and 9). By contrast, adenoviral induction of BMP7 in lens cells suppresses EMT induced by injury [116], suggesting that BMP signaling antagonizes the EMT-promoting effects of TGFβ. However, the contribution of BMP (or TGFβ) signaling to aberrant fiber differentiation that occurs during the formation of Soemmering's rings or Elschnig's pearls is not known.

5.4 Wnt Signaling in Lens Fiber Differentiation

Wnts are a family of 19 highly glycosylated, secreted growth factors that have been shown to have a broad range of roles in embryonic development and in particular regulate stem cell populations and cell fate decisions as well as cancer cell biology. Wnts bind to seven-pass transmembrane proteins, called Frizzleds (Fzds), and the Lrp5/6 co-receptors. Recently, the transmembrane Lgr4–6 family of proteins, which complex with R-spondin, has been shown to facilitate and amplify Wnt/Frizzled/Lrp signaling in stem cells [117, 118]. Depending on which Wnt and Frizzled proteins are present in the membrane complex, different domains of the cytoplasmic scaffold protein, Dishevelled (Dvl), are utilized to activate either the canonical Wnt/β-catenin pathway or the noncanonical planar cell polarity (PCP) pathway (Fig. 5.3). In the canonical pathway, the Wnt/Fzd/Dvl complex inhibits a complex of cytosolic proteins that functions to phosphorylate and degrade β-catenin. Accumulated and stabilized β-catenin can then enter the nucleus to activate gene transcription. The Wnt/PCP pathway is less well understood but is known to utilize a different Dvl domain to activate small GTPases, such as Rho and Cdc42, and mediate changes in the cytoskeleton and asymmetric localization of various proteins (Fzd6, Vangl, Strabismus, Prickle) to different lateral membrane domains in epithelial cells [119–121]. Both pathways have been shown to play roles during lens development. While antagonism of the canonical Wnt/β-catenin pathway is required for lens induction [122, 123], the pathway appears to at least be transiently active during differentiation of the embryonic lens and be involved in regulating epithelial cell proliferation and modulating crystallin expression during early primary fiber differentiation [124–126]. However, it appears to not have a role in later differentiation of secondary fibers [24, 127]. Components of the Wnt/PCP pathway are expressed in the developing lens at various stages [128–130], and a recent report indicates that Wnt signaling, particularly the PCP pathway, is regulated by FGF in lens explants and that inhibition of Wnt signaling by the secreted extracellular antagonist Sfrp2 or by a small molecule inhibitor (IWP-2), which blocks cellular Wnt production, blocked FGF-induced crystallin expression [127].

There is also increasing evidence that aberrant Wnt signaling may be involved in cataract formation. TGFβ signaling, which is a known inducer of LEC fibrosis (see Chap. 8), can also upregulate Wnt pathway expression [131]. Moreover, overactivation of the Wnt pathway in transgenic mouse lens [132] and in lens cells in vitro [133] results in epithelial–mesenchymal transition (EMT) of LEC. It remains to be determined how Wnt signaling contributes to aberrant fiber differentiation and/or fibrosis in human lens capsules postsurgery, although in other systems, canonical Wnt and TGFβ signaling can cooperate to drive fibrosis [134–136].

5.5 Notch Signaling in Lens Fiber Differentiation

Notch proteins are transmembrane receptors that regulate cellular identity during development [137] and disease [138, 139]. Canonically, Notch receptors bind transmembrane ligands (Jagged or Delta-like) expressed by adjacent cells, resulting in Notch proteolysis (Fig. 5.3). The released Notch intracellular domain (NICD) then translocates to the nucleus to form a complex with Rbpj and Maml, which activates target gene transcription [140]. Notch activity can set up feedback loops leading to either lateral inhibition [137] (ligand concentration in the Notch-expressing cell is maintained at low levels) or lateral induction (ligand expression is activated in the Notch-expressing cell [141] leading to *cis*-inhibition of Notch activity [142]). These feedback loops often regulate the boundaries between proliferative (or stem cell) cell populations and those fated to differentiate [137]. In the lens, Notch 1 and 2 are expressed in the lens epithelium and engage with Jagged1, which is expressed by lens fibers [143, 144]. In the central epithelium, Notch signaling maintains cell proliferation during development and prevents LECs from transitioning to fibers [145, 146]. However, at the lens equator, where FGFRs are activated, Jagged1 levels increase due to cross talk between Notch and FGFR signaling. This leads to Notch 2 and Jagged1 co-expression in transition zone cells, a process necessary for lens fiber cell differentiation [147]. Notably, since the anatomical division of the eye into anterior and posterior chambers occurs much later than the initial segregation of lens epithelial and fiber cell fates in the lens vesicle, the FGF gradient is unlikely to be sharp during early development. Since FGF receptors are expressed in all cells of the lens vesicle, including early fibers [53, 54], it is likely that cross talk between Notch and FGFR signaling is crucial to fine-tune the position of the border between differentiating lens fibers and cells fated to be LECs.

5.6 The Role of Other Mitogens in Fiber Differentiation

Various other growth factors (PDGF, EGF, IGF), mitogenic for lens cells, have been identified in ocular media/tissues and been implicated in fiber cell differentiation (see [24] for review). While experiments in various systems indicate that they are mitogenic for lens epithelial cells (see Chap. 4), their role in fiber differentiation appears to be predominantly synergistic with FGF signaling (see above and [24]), particularly via the ERK–MAPK and Akt pathways. A key point that this highlights is that while FGFs may be the preeminent differentiation factor, it is the combined action of numerous factors that fine-tune the signaling pathways that control fiber differentiation [88], and their relative contributions to such fine-tuning appear to vary between species.

Historically IGF1/insulin or "lentropin" (an IGF-like molecule), rather than the FGFs, was considered to be the fiber-inducing factor(s) for the chick lens. This stemmed from several studies that documented an early cell elongation response (after 3–5 h) in lens explants and induction of δ-crystallin expression in response to IGF or insulin [148–151]. As a result, it was long considered that there were distinct differences in the factors that regulate fiber differentiation in avians as opposed to mammals. Recent studies comparing FGF2 and IGF1 responses in chick explants indicated that while IGF1 and insulin induced fiber cell differentiation marker (CP49 and δ-crystallin) expression, the cell fiber elongation response in 6-day cultures was variable. Moreover, the fiber differentiating activity of chick and bovine vitreous on chick explants was attributed to an FGF-like activity [50]. Similarly, rat lens explants have subpopulations of IGF1-responsive epithelial cells that express β- and γ-crystallin [152, 153] but do not elongate unless FGF is present. It thus appears that IGFs have the capacity to induce a limited set of fiber differentiation events in vitro and that the effects are more pronounced in chick than in rat. Intriguingly, in rat explants, insulin or IGF, in combination with FGF, alters the expression ratio of the α-, β-, and γ-crystallin genes, compared to FGF alone [154]. Thus, one way that IGFs may fine-tune FGF-induced fiber differentiation in mammals is to regulate crystallin expression and thus could be involved in the massive shifts in the ratio of the different crystallins produced by lens fiber cells at different stages of lens development [19, 155]. The evolutionary switch in birds to the use of δ-crystallin rather than β- and γ-crystallins in differentiating fibers may explain the different sensitivity of chick cells to IGFs and insulin. However, mouse transgenic studies indicate that IGF1 and insulin are insufficient to induce fiber differentiation and that insulin may actually inhibit fiber differentiation [156, 157]. In the case of ectopic expression of IGF1, the transgenic lenses showed expansion of the epithelium posteriorly into the transitional zone and there was increased proliferation, but not differentiation, of the epithelium [157].

PDGF has been implicated in lens growth and maintenance of lens clarity in vitro culture of whole lenses [158]. Like IGF1, PDGF on its own induces cell proliferation in rat lens epithelial explants [159] and induces a restricted repertoire of β- and γ-crystallin isoform expression in a subpopulation of cells within explants [160] but potentiates low doses of FGF to cause epithelial cells to proliferate and differentiate [159] via activation of ERK–MAPK and PI3K–Akt pathways [88]. Consistent with this, transgenic expression of PDGF-A in mouse lenses results in increased cell proliferation as well as a limited elongation response and β-crystallin expression [161]. The effects of PDGF on chick lens cells have been reported to stimulate proliferation [162] and activate JAK–STAT signaling but have no effect on δ-crystallin expression, despite activation of the PI3K–Akt pathway [92]. To our knowledge, interactions between FGF and PDGF in chick lens fiber differentiation have not been investigated.

5.7 Clinical Implications for PCO

During the most commonly performed cataract operation, extracapsular lens extraction (ECCE), the central lens capsule and the attached epithelium are removed by capsulorhexis, and the lens fibers are removed by phacoemulsification. While every attempt is made to remove all residual cellular material from the capsular bag by "polishing," the peripheral LECs are nearly impossible to remove due to their location and tight adherence to the lens capsule. As discussed above, these cells have the highest proliferative and differentiation potential and, in the intact lens, normally respond to ocular FGFs and other growth factors, by proliferating and differentiating into lens fiber cells.

Consistent with this, it was first reported in the nineteenth century that morphologically recognizable lenses can reform from the remaining capsule and epithelial cells, following lens fiber cell removal in mammals, if the capsule does not collapse and inflammation is minimized [163]. In the twentieth century, these findings were confirmed in mice [164], monkeys, and rabbits [163], and, in the twenty-first century, with refinements of surgical technique and use of biodegradable hyaluronic acid or refractive silicone scaffolds, combined with hyaluronic acid, almost complete lens regeneration, with a well-organized transparent cortex, has been achieved [165]. The use of such "natural regeneration" of lenses has not yet been applied clinically, but has been proposed as a potential method to restore accommodation after cataract surgery [165].

In current clinical practice, the implantation of an intraocular lens in the capsular bag during cataract surgery is often associated with the subsequent proliferation and differentiation of remnant LECs into lens fibers, presumably in response to FGFs and other ocular growth factors. This results in the formation of Soemmering's ring, when the cells remain at the periphery of the capsular bag), or Elschnig's "pearl-type" PCO, when the cells escape from this area [166]. However, in contrast to the transparent "regenerated" cortical fibers obtained in the rabbit experiments [165], Soemmering's rings (Fig. 5.1a) are usually opaque, despite the robust expression of fiber cell-specific genes such as aquaporin 0 (Fig. 5.1b). Similarly, while Elschnig's pearls are ultrastructurally similar to fiber cells [167], these structures also scatter light, probably due to a lack of the strict organization necessary for transparency [166]. Some reports propose that "pearl-type" PCO is more destructive to patient vision than "fibrotic" PCO due to its ability to robustly scatter light [168, 169], although it appears that Elschnig's pearl morphology is very dynamic in PCO [170, 171] and complete spontaneous regression of subclinical Elschnig's pearl-type PCO has been reported [172] (see Chap. 12).

Few studies have investigated specifically the molecular mechanisms driving Soemmering's ring/Elschnig's "pearl-type" PCO formation, although it is likely that it occurs when remnant LECs encounter not only the growth factors within the eye that drive fiber cell differentiation in the intact lens (see above) but also cytokines that are activated postsurgery. The fibers that form in the capsular bag after cataract are often misaligned or fail to elongate properly and are thus not transparent, likely due to a lack of appropriate directional cues. Investigations using rat lens epithelial explants have shown that when LECs are sandwiched between two layers of lens capsule, they are able to elongate and organize into a parallel alignment similar to the intact lens. However, when the apical surfaces of these cells are directly exposed to culture media containing FGF, they are more globular and irregularly arranged although they still express lens fiber cell markers. Moreover, explants exposed to culture media, containing 50 % vitreous humor v/v, differentiated into lens fiber cells and maintained a cell population resembling the lens epithelium, leading to a transparent structure morphologically similar to a normal lens [173, 174]. This maintenance of an epithelium in apposition to the lens fibers is likely to be critical to this cellular organization as it is clear that lens fibers require signals from the lens epithelium including planar cell polarity [175] and Notch signaling [146, 147] for their correct orientation and function.

Notably, this robust fiber cell differentiation response following cataract surgery is commonly accompanied by the epithelial–mesenchymal transition (EMT) of other LECs to myofibroblasts to form "fibrotic" PCO (see Chaps. 8, 9, and 10) leading to mixed populations of cells expressing either fiber cell or myofibroblast markers in the capsular bag. While little is known about how different fates are acquired by remnant lens cells, this may reflect intrinsic differences between LEC populations [8, 176, 177] and their stochastic responses to the cytokine/growth factor milieu present in the eye after cataract surgery, or both. Further improvements in the long-term visual outcomes of cataract surgery, including the prevention of PCO and the reestablishment of accommodation following cataract surgery, will likely require either that the phenotypic changes in the lens epithelium after surgery be blocked (which would allow "lens refilling" strategies for refractive replacement [178]) or that the lens epithelium is directed to undergo a regulated differentiation process that is compatible with long-term stabilization of an IOL and maintenance of a clear visual axis (as proposed in [165]). Both of these goals will require an improved understanding of the normal regulation of lens fiber cell differentiation and how this response is both enhanced and inhibited after cataract surgery.

References

1. Danysh BP, Duncan MK (2009) The lens capsule. Exp Eye Res 88(2):151–164. doi:10.1016/j.exer.2008.08.002
2. Piatigorsky J (1981) Lens differentiation in vertebrates: a review of cellular and molecular features. Differentiation 19:134–152
3. Rampalli AM, Gao CY, Chauthaiwale VM, Zelenka PS (1998) pRb and p107 regulate E2F activity during lens fiber cell differentiation. Oncogene 16(3):399–408. doi:10.1038/sj.onc.1201546
4. Martinez G, de Iongh RU (2010) The lens epithelium in ocular health and disease. Int J Biochem Cell Biol 42(12):1945–1963. doi:10.1016/j.biocel.2010.09.012
5. Sellitto C, Li L, White TW (2004) Connexin50 is essential for normal postnatal lens cell proliferation. Invest Ophthalmol Vis Sci 45(9):3196–3202. doi:10.1167/iovs.04-0194
6. White TW, Gao Y, Li L, Sellitto C, Srinivas M (2007) Optimal lens epithelial cell proliferation is dependent on the connexin isoform providing gap junctional coupling. Invest Ophthalmol Vis Sci 48(12):5630–5637. doi:10.1167/iovs.06-1540
7. Rajagopal R, Dattilo LK, Kaartinen V, Deng CX, Umans L, Zwijsen A, Roberts AB, Bottinger EP, Beebe DC (2008) Functions of the type 1 BMP receptor Acvr1 (Alk2) in lens development: cell proliferation, terminal differentiation, and survival. Invest Ophthalmol Vis Sci 49(11):4953–4960. doi:10.1167/iovs.08-2217
8. Zhou M, Leiberman J, Xu J, Lavker RM (2006) A hierarchy of proliferative cells exists in mouse lens epithelium: implications for lens maintenance. Invest Ophthalmol Vis Sci 47 (7):2997–3003. doi:10.1167/iovs.06-0130
9. Bassnett S, Shi Y, Vrensen GF (2011) Biological glass: structural determinants of eye lens transparency. Philos Trans R Soc Lond B Biol Sci 366(1568):1250–1264. doi:10.1098/rstb.2010.0302
10. Bassnett S, Beebe DC (2004) Lens fiber differentiation. In: Lovicu FJ, Robinson ML (eds) Development of the ocular lens. Cambridge University Press, New York, pp 214–244
11. Korlimbinis A, Berry Y, Thibault D, Schey KL, Truscott RJ (2009) Protein aging: truncation of aquaporin 0 in human lens regions is a continuous age-dependent process. Exp Eye Res 88 (5):966–973. doi:10.1016/j.exer.2008.12.008
12. Kumari SS, Eswaramoorthy S, Mathias RT, Varadaraj K (2011) Unique and analogous functions of aquaporin 0 for fiber cell architecture and ocular lens transparency. Biochim Biophys Acta 1812(9):1089–1097. doi:10.1016/j.bbadis.2011.04.001
13. Liu J, Xu J, Gu S, Nicholson BJ, Jiang JX (2011) Aquaporin 0 enhances gap junction coupling via its cell adhesion function and interaction with connexin 50. J Cell Sci 124(Pt 2):198–206. doi:10.1242/jcs.072652
14. Gonen T, Cheng Y, Kistler J, Walz T (2004) Aquaporin-0 membrane junctions form upon proteolytic cleavage. J Mol Biol 342(4):1337–1345. doi:10.1016/j.jmb.2004.07.076

15. Werten PJ, Vos E, De Jong WW (1999) Truncation of betaA3/A1-crystallin during aging of the bovine lens; possible implications for lens optical quality. Exp Eye Res 68(1):99–103. doi:10.1006/exer.1998.0584
16. Lampi KJ, Ma Z, Hanson SR, Azuma M, Shih M, Shearer TR, Smith DL, Smith JB, David LL (1998) Age-related changes in human lens crystallins identified by two-dimensional electrophoresis and mass spectrometry. Exp Eye Res 67(1):31–43. doi:10.1006/exer.1998.0481
17. Ma Z, Hanson SR, Lampi KJ, David LL, Smith DL, Smith JB (1998) Age-related changes in human lens crystallins identified by HPLC and mass spectrometry. Exp Eye Res 67(1):21–30. doi:10.1006/exer.1998.0482
18. Lampi KJ, Shih M, Ueda Y, Shearer TR, David LL (2002) Lens proteomics: analysis of rat crystallin sequences and two-dimensional electrophoresis map. Invest Ophthalmol Vis Sci 43 (1):216–224
19. Ueda Y, Duncan MK, David LL (2002) Lens proteomics: the accumulation of crystallin modifications in the mouse lens with age. Invest Ophthalmol Vis Sci 43(1):205–215
20. Shestopalov VI, Bassnett S (2000) Expression of autofluorescent proteins reveals a novel protein permeable pathway between cells in the lens core. J Cell Sci 113(Pt 11):1913–1921
21. Shestopalov VI, Bassnett S (2003) Development of a macromolecular diffusion pathway in the lens. J Cell Sci 116(Pt 20):4191–4199. doi:10.1242/jcs.00738
22. Coulombre JL, Coulombre AJ (1963) Lens development : fiber elongation and lens orientation. Science 142(3598):1489–1490
23. Yamamoto Y (1976) Growth of lens and ocular environment: role of neural retina in the growth of mouse lens as revealed by an implantation experiment. Dev Growth Differ 18 (3):273–278. doi:10.1111/j.1440-169X.1976.00273.x
24. Lovicu FJ, McAvoy JW, de Iongh RU (2011) Understanding the role of growth factors in embryonic development: insights from the lens. Philos Trans R Soc Lond B Biol Sci 366 (1568):1204–1218. doi:10.1098/rstb.2010.0339
25. Lovicu FJ, McAvoy JW (2005) Growth factor regulation of lens development. Dev Biol 280 (1):1–14. doi:10.1016/j.ydbio.2005.01.020
26. Itoh N, Ornitz DM (2011) Fibroblast growth factors: from molecular evolution to roles in development, metabolism and disease. J Biochem 149(2):121–130. doi:10.1093/jb/mvq121
27. Beenken A, Mohammadi M (2012) The structural biology of the FGF19 subfamily. Adv Exp Med Biol 728:1–24. doi:10.1007/978-1-4614-0887-1_1
28. Lovicu FJ, de Iongh RU, McAvoy JW (1997) Expression of FGF-1 and FGF-2 mRNA during lens morphogenesis, differentiation and growth. Curr Eye Res 16(3):222–230
29. Wilkinson DG, Bhatt S, McMahon AP (1989) Expression pattern of the FGF-related proto-oncogene int-2 suggests multiple roles in fetal development. Development 105(1):131–136
30. Gelfman CM, Kelleher CM, Hjelmeland LM (1998) Differentiation of retinal pigment epithelial cells in vitro uncovers silencer activity in the FGF-5 gene promoter. Exp Eye Res 67(2):151–162. doi:10.1006/exer.1998.0506
31. Weng J, Liang Q, Mohan RR, Li Q, Wilson SE (1997) Hepatocyte growth factor, keratinocyte growth factor, and other growth factor-receptor systems in the lens. Invest Ophthalmol Vis Sci 38(8):1543–1554
32. Wilson SE, Walker JW, Chwang EL, He YG (1993) Hepatocyte growth factor, keratinocyte growth factor, their receptors, fibroblast growth factor receptor-2, and the cells of the cornea. Invest Ophthalmol Vis Sci 34(8):2544–2561
33. Kurose H, Bito T, Adachi T, Shimizu M, Noji S, Ohuchi H (2004) Expression of Fibroblast growth factor 19 (Fgf19) during chicken embryogenesis and eye development, compared with Fgf15 expression in the mouse. Gene Expr Patterns 4(6):687–693. doi:10.1016/j.modgep.2004.04.005
34. Kurose H, Okamoto M, Shimizu M, Bito T, Marcelle C, Noji S, Ohuchi H (2005) FGF19-FGFR4 signaling elaborates lens induction with the FGF8-L-Maf cascade in the chick embryo. Dev Growth Differ 47(4):213–223. doi:10.1111/j.1440-169X.2005.00795.x

35. Cinaroglu A, Ozmen Y, Ozdemir A, Ozcan F, Ergorul C, Cayirlioglu P, Hicks D, Bugra K (2005) Expression and possible function of fibroblast growth factor 9 (FGF9) and its cognate receptors FGFR2 and FGFR3 in postnatal and adult retina. J Neurosci Res 79(3):329–339. doi:10.1002/jnr.20363
36. Fon Tacer K, Bookout AL, Ding X, Kurosu H, John GB, Wang L, Goetz R, Mohammadi M, Kuro-o M, Mangelsdorf DJ, Kliewer SA (2010) Research resource: comprehensive expression atlas of the fibroblast growth factor system in adult mouse. Mol Endocrinol (Baltimore, MD) 24(10):2050–2064. doi:10.1210/me.2010-0142
37. Brooks AN, Kilgour E, Smith PD (2012) Molecular pathways: fibroblast growth factor signaling: a new therapeutic opportunity in cancer. Clin Cancer Res 18(7):1855–1862. doi:10.1158/1078-0432.ccr-11-0699
38. Trueb B (2011) Biology of FGFRL1, the fifth fibroblast growth factor receptor. Cell Mol Life Sci 68(6):951–964. doi:10.1007/s00018-010-0576-3
39. Robinson ML (2006) An essential role for FGF receptor signaling in lens development. Semin Cell Dev Biol 17(6):726–740. doi:10.1016/j.semcdb.2006.10.002
40. Garcia CM, Huang J, Madakashira BP, Liu Y, Rajagopal R, Dattilo L, Robinson ML, Beebe DC (2011) The function of FGF signaling in the lens placode. Dev Biol 351(1):176–185. doi:10.1016/j.ydbio.2011.01.001
41. Eswarakumar VP, Lax I, Schlessinger J (2005) Cellular signaling by fibroblast growth factor receptors. Cytokine Growth Factor Rev 16(2):139–149. doi:10.1016/j.cytogfr.2005.01.001
42. Campbell MT, McAvoy JW (1984) Onset of fibre differentiation in cultured rat lens epithelium under the influence of neural retina-conditioned medium. Exp Eye Res 39(1):83–94
43. Barritault D, Arruti C, Courtois Y (1981) Is there a ubiquitous growth factor in the eye? Proliferation induced in different cell types by eye-derived growth factors. Differentiation 18 (1):29–42
44. Arruti C, Cirillo A, Courtois Y (1985) An eye-derived growth factor regulates epithelial cell proliferation in the cultured lens. Differentiation 28(3):286–290
45. Chamberlain CG, McAvoy JW (1987) Evidence that fibroblast growth factor promotes lens fibre differentiation. Curr Eye Res 6(9):1165–1169
46. Chamberlain CG, McAvoy JW (1989) Induction of lens fibre differentiation by acidic and basic fibroblast growth factor (FGF). Growth Factors 1(2):125–134
47. Moenner M, Chevallier B, Badet J, Barritault D (1986) Evidence and characterization of the receptor to eye-derived growth factor I, the retinal form of basic fibroblast growth factor, on bovine epithelial lens cells. Proc Natl Acad Sci U S A 83(14):5024–5028
48. McAvoy JW, Chamberlain CG (1989) Fibroblast growth factor (FGF) induces different responses in lens epithelial cells depending on its concentration. Development 107 (2):221–228
49. McAvoy JW, Chamberlain CG, de Iongh RU, Richardson NA, Lovicu FJ (1991) The role of fibroblast growth factor in eye lens development. Ann N Y Acad Sci 638:256–274
50. Le AC, Musil LS (2001) FGF signaling in chick lens development. Dev Biol 233(2):394–411. doi:10.1006/dbio.2001.0194
51. Schulz MW, Chamberlain CG, de Iongh RU, McAvoy JW (1993) Acidic and basic FGF in ocular media and lens: implications for lens polarity and growth patterns. Development 118 (1):117–126
52. Caruelle D, Groux-Muscatelli B, Gaudric A, Sestier C, Coscas G, Caruelle JP, Barritault D (1989) Immunological study of acidic fibroblast growth factor (aFGF) distribution in the eye. J Cell Biochem 39(2):117–128. doi:10.1002/jcb.240390204
53. de Iongh RU, Lovicu FJ, Chamberlain CG, McAvoy JW (1997) Differential expression of fibroblast growth factor receptors during rat lens morphogenesis and growth. Invest Ophthalmol Vis Sci 38(9):1688–1699
54. de Iongh RU, Lovicu FJ, Hanneken A, Baird A, McAvoy JW (1996) FGF receptor-1 (flg) expression is correlated with fibre differentiation during rat lens morphogenesis and growth.

Dev Dyn 206(4):412–426. doi:10.1002/(SICI)1097-0177(199608)206:4<412::AID-AJA7>3.0.CO;2-L
55. de Iongh R, McAvoy JW (1992) Distribution of acidic and basic fibroblast growth factors (FGF) in the foetal rat eye: implications for lens development. Growth Factors 6(2):159–177
56. de Iongh R, McAvoy JW (1993) Spatio-temporal distribution of acidic and basic FGF indicates a role for FGF in rat lens morphogenesis. Dev Dyn 198(3):190–202. doi:10.1002/aja.1001980305
57. Lovicu FJ, Overbeek PA (1998) Overlapping effects of different members of the FGF family on lens fiber differentiation in transgenic mice. Development 125(17):3365–3377
58. Robinson ML, Ohtaka-Maruyama C, Chan CC, Jamieson S, Dickson C, Overbeek PA, Chepelinsky AB (1998) Disregulation of ocular morphogenesis by lens-specific expression of FGF-3/int-2 in transgenic mice. Dev Biol 198(1):13–31
59. Robinson ML, Overbeek PA, Verran DJ, Grizzle WE, Stockard CR, Friesel R, Maciag T, Thompson JA (1995) Extracellular FGF-1 acts as a lens differentiation factor in transgenic mice. Development 121(2):505–514
60. Robinson ML, MacMillan-Crow LA, Thompson JA, Overbeek PA (1995) Expression of a truncated FGF receptor results in defective lens development in transgenic mice. Development 121(12):3959–3967
61. Garcia CM, Yu K, Zhao H, Ashery-Padan R, Ornitz DM, Robinson ML, Beebe DC (2005) Signaling through FGF receptor-2 is required for lens cell survival and for withdrawal from the cell cycle during lens fiber cell differentiation. Dev Dyn 233(2):516–527. doi:10.1002/dvdy.20356
62. Zhao H, Yang Y, Partanen J, Ciruna BG, Rossant J, Robinson ML (2006) Fibroblast growth factor receptor 1 (Fgfrl) is not essential for lens fiber differentiation in mice. Mol Vis 12:15–25
63. Deng C, Wynshaw-Boris A, Zhou F, Kuo A, Leder P (1996) Fibroblast growth factor receptor 3 is a negative regulator of bone growth. Cell 84(6):911–921
64. Weinstein M, Xu X, Ohyama K, Deng CX (1998) FGFR-3 and FGFR-4 function cooperatively to direct alveogenesis in the murine lung. Development 125(18):3615–3623
65. Zhao H, Yang T, Madakashira BP, Thiels CA, Bechtle CA, Garcia CM, Zhang H, Yu K, Ornitz DM, Beebe DC, Robinson ML (2008) Fibroblast growth factor receptor signaling is essential for lens fiber cell differentiation. Dev Biol 318(2):276–288. doi:10.1016/j.ydbio.2008.03.028
66. Zhao S, Hung FC, Colvin JS, White A, Dai W, Lovicu FJ, Ornitz DM, Overbeek PA (2001) Patterning the optic neuroepithelium by FGF signaling and Ras activation. Development 128(24):5051–5060
67. Puk O, Moller G, Geerlof A, Krowiorz K, Ahmad N, Wagner S, Adamski J, de Angelis MH, Graw J (2011) The pathologic effect of a novel neomorphic Fgf9(Y162C) allele is restricted to decreased vision and retarded lens growth. PLoS One 6(8):e23678. doi:10.1371/journal.pone.0023678
68. Nakayama Y, Miyake A, Nakagawa Y, Mido T, Yoshikawa M, Konishi M, Itoh N (2008) Fgf19 is required for zebrafish lens and retina development. Dev Biol 313(2):752–766. doi:10.1016/j.ydbio.2007.11.013
69. Wright TJ, Ladher R, McWhirter J, Murre C, Schoenwolf GC, Mansour SL (2004) Mouse FGF15 is the ortholog of human and chick FGF19, but is not uniquely required for otic induction. Dev Biol 269(1):264–275. doi:10.1016/j.ydbio.2004.02.003
70. Baird A, Esch F, Gospodarowicz D, Guillemin R (1985) Retina- and eye-derived endothelial cell growth factors: partial molecular characterization and identity with acidic and basic fibroblast growth factors. Biochemistry 24(27):7855–7860
71. Ornitz DM (2000) FGFs, heparan sulfate and FGFRs: complex interactions essential for development. Bioessays 22(2):108–112. doi:10.1002/(sici)1521-1878(200002)22:2<108::aid-bies2>3.0.co;2-m

72. Mohan PS, Spiro RG (1991) Characterization of heparan sulfate proteoglycan from calf lens capsule and proteoglycans synthesized by cultured lens epithelial cells. Comparison with other basement membrane proteoglycans. J Biol Chem 266(13):8567–8575
73. Schulz MW, Chamberlain CG, McAvoy JW (1997) Binding of FGF-1 and FGF-2 to heparan sulphate proteoglycans of the mammalian lens capsule. Growth Factors 14(1):1–13
74. Lovicu FJ, McAvoy JW (1993) Localization of acidic fibroblast growth factor, basic fibroblast growth factor, and heparan sulphate proteoglycan in rat lens: implications for lens polarity and growth patterns. Invest Ophthalmol Vis Sci 34(12):3355–3365
75. Mascarelli F, Fuhrmann G, Courtois Y (1993) aFGF binding to low and high affinity receptors induces both aFGF and aFGF receptors dimerization. Growth Factors 8(3):211–233
76. Rossi M, Morita H, Sormunen R, Airenne S, Kreivi M, Wang L, Fukai N, Olsen BR, Tryggvason K, Soininen R (2003) Heparan sulfate chains of perlecan are indispensable in the lens capsule but not in the kidney. EMBO J 22(2):236–245. doi:10.1093/emboj/cdg019
77. Pan Y, Woodbury A, Esko JD, Grobe K, Zhang X (2006) Heparan sulfate biosynthetic gene Ndst1 is required for FGF signaling in early lens development. Development 133 (24):4933–4944. doi:10.1242/dev.02679
78. Qu X, Hertzler K, Pan Y, Grobe K, Robinson ML, Zhang X (2011) Genetic epistasis between heparan sulfate and FGF-Ras signaling controls lens development. Dev Biol 355(1):12–20. doi:10.1016/j.ydbio.2011.04.007
79. Ishibashi T, Araki H, Sugai S, Tawara A, Inomata H (1995) Detection of proteoglycans in human posterior capsule opacification. Ophthalmic Res 27(4):208–213
80. Hayashi N, Kato H, Kiyosawa T, Hayashi H, Oshima K, Yamaoka M (1991) The change in immunohistochemical localization of basic fibroblast growth factor (b-FGF) around the lens capsule after extracapsular extraction. Nihon Ganka Gakkai Zasshi 95(7):621–624
81. Wormstone IM, Del Rio-Tsonis K, McMahon G, Tamiya S, Davies PD, Marcantonio JM, Duncan G (2001) FGF: an autocrine regulator of human lens cell growth independent of added stimuli. Invest Ophthalmol Vis Sci 42(6):1305–1311
82. Song X, Sato Y, Felemban A, Ito A, Hossain M, Ochiai H, Yamamoto T, Sekiguchi K, Tanaka H, Ohta K (2012) Equarin is involved as an FGF signaling modulator in chick lens differentiation. Dev Biol 368(1):109–117. doi:10.1016/j.ydbio.2012.05.029
83. Jarrin M, Pandit T, Gunhaga L (2012) A balance of FGF and BMP signals regulates cell cycle exit and Equarin expression in lens cells. Mol Biol Cell 23(16):3266–3274. doi:10.1091/mbc.E12-01-0075
84. Song X, Sato Y, Sekiguchi K, Tanaka H, Ohta K (2013) Equarin is involved in cell adhesion by means of heparan sulfate proteoglycan during lens development. Dev Dyn 242(1):23–29. doi:10.1002/dvdy.23902
85. Lovicu FJ, McAvoy JW (2001) FGF-induced lens cell proliferation and differentiation is dependent on MAPK (ERK1/2) signalling. Development 128(24):5075–5084
86. Iyengar L, Wang Q, Rasko JE, McAvoy JW, Lovicu FJ (2007) Duration of ERK1/2 phosphorylation induced by FGF or ocular media determines lens cell fate. Differentiation 75 (7):662–668. doi:10.1111/j.1432-0436.2007.00167.x
87. Wang Q, Stump R, McAvoy JW, Lovicu FJ (2009) MAPK/ERK1/2 and PI3-kinase signalling pathways are required for vitreous-induced lens fibre cell differentiation. Exp Eye Res 88 (2):293–306. doi:10.1016/j.exer.2008.08.023
88. Wang Q, McAvoy JW, Lovicu FJ (2010) Growth factor signaling in vitreous humor-induced lens fiber differentiation. Invest Ophthalmol Vis Sci 51(7):3599–3610. doi:10.1167/iovs.09-4797
89. Le AC, Musil LS (2001) A novel role for FGF and extracellular signal-regulated kinase in gap junction-mediated intercellular communication in the lens. J Cell Biol 154(1):197–216
90. Boswell BA, Le AC, Musil LS (2009) Upregulation and maintenance of gap junctional communication in lens cells. Exp Eye Res 88(5):919–927. doi:10.1016/j.exer.2008.11.031

91. Zatechka SD Jr, Lou MF (2002) Studies of the mitogen-activated protein kinases and phosphatidylinositol-3 kinase in the lens. 1. The mitogenic and stress responses. Exp Eye Res 74(6):703–717
92. Chandrasekher G, Sailaja D (2003) Differential activation of phosphatidylinositol 3-kinase signaling during proliferation and differentiation of lens epithelial cells. Invest Ophthalmol Vis Sci 44(10):4400–4411
93. Gotoh N, Ito M, Yamamoto S, Yoshino I, Song N, Wang Y, Lax I, Schlessinger J, Shibuya M, Lang RA (2004) Tyrosine phosphorylation sites on FRS2alpha responsible for Shp2 recruitment are critical for induction of lens and retina. Proc Natl Acad Sci U S A 101 (49):17144–17149. doi:10.1073/pnas.0407577101
94. Ebong S, Yu CR, Carper DA, Chepelinsky AB, Egwuagu CE (2004) Activation of STAT signaling pathways and induction of suppressors of cytokine signaling (SOCS) proteins in mammalian lens by growth factors. Invest Ophthalmol Vis Sci 45(3):872–878
95. Ebong S, Chepelinsky AB, Robinson ML, Zhao H, Yu CR, Egwuagu CE (2004) Characterization of the roles of STAT1 and STAT3 signal transduction pathways in mammalian lens development. Mol Vis 10:122–131
96. Boros J, Newitt P, Wang Q, McAvoy JW, Lovicu FJ (2006) Sef and Sprouty expression in the developing ocular lens: implications for regulating lens cell proliferation and differentiation. Semin Cell Dev Biol 17(6):741–752. doi:10.1016/j.semcdb.2006.10.007
97. Newitt P, Boros J, Madakashira BP, Robinson ML, Reneker LW, McAvoy JW, Lovicu FJ (2010) Sef is a negative regulator of fiber cell differentiation in the ocular lens. Differentiation 80(1):53–67. doi:10.1016/j.diff.2010.05.005
98. Pan Y, Carbe C, Powers A, Feng GS, Zhang X (2010) Sprouty2-modulated Kras signaling rescues Shp2 deficiency during lens and lacrimal gland development. Development 137 (7):1085–1093. doi:10.1242/dev.042820
99. Xie L, Overbeek PA, Reneker LW (2006) Ras signaling is essential for lens cell proliferation and lens growth during development. Dev Biol 298(2):403–414. doi:10.1016/j.ydbio.2006.06.045
100. Reneker LW, Xie L, Xu L, Govindarajan V, Overbeek PA (2004) Activated Ras induces lens epithelial cell hyperplasia but not premature differentiation. Int J Dev Biol 48(8–9):879–888. doi:10.1387/ijdb.041889lr
101. Upadhya D, Ogata M, Reneker LW (2013) MAPK1 is required for establishing the pattern of cell proliferation and for cell survival during lens development. Development 140 (7):1573–1582. doi:10.1242/dev.081042
102. Iyengar L, Patkunanathan B, Lynch OT, McAvoy JW, Rasko JE, Lovicu FJ (2006) Aqueous humour- and growth factor-induced lens cell proliferation is dependent on MAPK/ERK1/2 and Akt/PI3-K signalling. Exp Eye Res 83(3):667–678. doi:10.1016/j.exer.2006.03.008
103. Wu MY, Hill CS (2009) Tgf-beta superfamily signaling in embryonic development and homeostasis. Dev Cell 16(3):329–343. doi:10.1016/j.devcel.2009.02.012
104. Beyer TA, Narimatsu M, Weiss A, David L, Wrana JL (2013) The TGFbeta superfamily in stem cell biology and early mammalian embryonic development. Biochim Biophys Acta 1830(2):2268–2279. doi:10.1016/j.bbagen.2012.08.025
105. Mueller TD, Nickel J (2012) Promiscuity and specificity in BMP receptor activation. FEBS Lett 586(14):1846–1859. doi:10.1016/j.febslet.2012.02.043
106. Weiss A, Attisano L (2013) The TGFbeta superfamily signaling pathway. Wiley Interdiscip Rev Dev Biol 2(1):47–63. doi:10.1002/wdev.86
107. de Iongh RU, Lovicu FJ, Overbeek PA, Schneider MD, Joya J, Hardeman ED, McAvoy JW (2001) Requirement for TGFbeta receptor signaling during terminal lens fiber differentiation. Development 128(20):3995–4010
108. Beebe D, Garcia C, Wang X, Rajagopal R, Feldmeier M, Kim JY, Chytil A, Moses H, Ashery-Padan R, Rauchman M (2004) Contributions by members of the TGFbeta superfamily to lens development. Int J Dev Biol 48(8–9):845–856. doi:10.1387/ijdb.041869db

109. Rajagopal R, Huang J, Dattilo LK, Kaartinen V, Mishina Y, Deng CX, Umans L, Zwijsen A, Roberts AB, Beebe DC (2009) The type I BMP receptors, Bmpr1a and Acvr1, activate multiple signaling pathways to regulate lens formation. Dev Biol 335(2):305–316. doi:10.1016/j.ydbio.2009.08.027
110. Reneker LW, Chen H, Overbeek PA (2011) Activation of unfolded protein response in transgenic mouse lenses. Invest Ophthalmol Vis Sci 52(5):2100–2108. doi:10.1167/iovs.10-5650
111. Firtina Z, Danysh BP, Bai X, Gould DB, Kobayashi T, Duncan MK (2009) Abnormal expression of collagen IV in lens activates unfolded protein response resulting in cataract. J Biol Chem 284(51):35872–35884. doi:10.1074/jbc.M109.060384
112. Gunhaga L (2011) The lens: a classical model of embryonic induction providing new insights into cell determination in early development. Philos Trans R Soc Lond B Biol Sci 366 (1568):1193–1203. doi:10.1098/rstb.2010.0175
113. Boswell BA, Lein PJ, Musil LS (2008) Cross-talk between fibroblast growth factor and bone morphogenetic proteins regulates gap junction-mediated intercellular communication in lens cells. Mol Biol Cell 19(6):2631–2641. doi:10.1091/mbc.E08-02-0124
114. Boswell BA, Overbeek PA, Musil LS (2008) Essential role of BMPs in FGF-induced secondary lens fiber differentiation. Dev Biol 324(2):202–212. doi:10.1016/j.ydbio.2008.09.003
115. Eldred JA, Dawes LJ, Wormstone IM (2011) The lens as a model for fibrotic disease. Philos Trans R Soc Lond B Biol Sci 366(1568):1301–1319. doi:10.1098/rstb.2010.0341
116. Saika S, Ikeda K, Yamanaka O, Flanders KC, Ohnishi Y, Nakajima Y, Muragaki Y, Ooshima A (2006) Adenoviral gene transfer of BMP-7, Id2, or Id3 suppresses injury-induced epithelial-to-mesenchymal transition of lens epithelium in mice. Am J Physiol Cell Physiol 290(1):C282–C289. doi:10.1152/ajpcell.00306.2005
117. Holland JD, Klaus A, Garratt AN, Birchmeier W (2013) Wnt signaling in stem and cancer stem cells. Curr Opin Cell Biol 25(2):254–264. doi:10.1016/j.ceb.2013.01.004
118. Kim W, Kim M, Jho EH (2013) Wnt/beta-catenin signalling: from plasma membrane to nucleus. Biochem J 450(1):9–21. doi:10.1042/bj20121284
119. Roszko I, Sawada A, Solnica-Krezel L (2009) Regulation of convergence and extension movements during vertebrate gastrulation by the Wnt/PCP pathway. Semin Cell Dev Biol 20 (8):986–997. doi:10.1016/j.semcdb.2009.09.004
120. Strutt D (2003) Frizzled signalling and cell polarisation in Drosophila and vertebrates. Development 130(19):4501–4513. doi:10.1242/dev.00695
121. Gao B (2012) Wnt regulation of planar cell polarity (PCP). Curr Topics Dev Biol 101:263–295. doi:10.1016/b978-0-12-394592-1.00008-9
122. Smith AN, Miller LA, Song N, Taketo MM, Lang RA (2005) The duality of beta-catenin function: a requirement in lens morphogenesis and signaling suppression of lens fate in periocular ectoderm. Dev Biol 285(2):477–489. doi:10.1016/j.ydbio.2005.07.019
123. Machon O, Kreslova J, Ruzickova J, Vacik T, Klimova L, Fujimura N, Lachova J, Kozmik Z (2010) Lens morphogenesis is dependent on Pax6-mediated inhibition of the canonical Wnt/beta-catenin signaling in the lens surface ectoderm. Genesis 48(2):86–95. doi:10.1002/dvg.20583
124. Cain S, Martinez G, Kokkinos MI, Turner K, Richardson RJ, Abud HE, Huelsken J, Robinson ML, de Iongh RU (2008) Differential requirement for beta-catenin in epithelial and fiber cells during lens development. Dev Biol 321(2):420–433. doi:10.1016/j.ydbio.2008.07.002
125. Lyu J, Joo CK (2004) Wnt signaling enhances FGF2-triggered lens fiber cell differentiation. Development 131(8):1813–1824. doi:10.1242/dev.01060
126. Stump RJ, Ang S, Chen Y, von Bahr T, Lovicu FJ, Pinson K, de Iongh RU, Yamaguchi TP, Sassoon DA, McAvoy JW (2003) A role for Wnt/beta-catenin signaling in lens epithelial differentiation. Dev Biol 259(1):48–61
127. Dawes LJ, Sugiyama Y, Tanedo AS, Lovicu FJ, McAvoy JW (2013) Wnt-frizzled signaling is part of an FGF-induced cascade that promotes lens fiber differentiation. Invest Ophthalmol Vis Sci 54(3):1582–1590. doi:10.1167/iovs.12-11357

128. Sugiyama Y, Lovicu FJ, McAvoy JW (2011) Planar cell polarity in the mammalian eye lens. Organogenesis 7(3):191–201. doi:10.4161/org.7.3.18421
129. Chen Y, Stump RJ, Lovicu FJ, McAvoy JW (2006) A role for Wnt/planar cell polarity signaling during lens fiber cell differentiation? Semin Cell Dev Biol 17(6):712–725. doi:10.1016/j.semcdb.2006.11.005
130. Chen Y, Stump RJ, Lovicu FJ, Shimono A, McAvoy JW (2008) Wnt signaling is required for organization of the lens fiber cell cytoskeleton and development of lens three-dimensional architecture. Dev Biol 324(1):161–176. doi:10.1016/j.ydbio.2008.09.002
131. Chong CC, Stump RJ, Lovicu FJ, McAvoy JW (2009) TGFbeta promotes Wnt expression during cataract development. Exp Eye Res 88(2):307–313. doi:10.1016/j.exer.2008.07.018
132. Martinez G, Wijesinghe M, Turner K, Abud HE, Taketo MM, Noda T, Robinson ML, de Iongh RU (2009) Conditional mutations of beta-catenin and APC reveal roles for canonical Wnt signaling in lens differentiation. Invest Ophthalmol Vis Sci 50(10):4794–4806. doi:10.1167/iovs.09-3567
133. Bao XL, Song H, Chen Z, Tang X (2012) Wnt3a promotes epithelial-mesenchymal transition, migration, and proliferation of lens epithelial cells. Mol Vis 18:1983–1990
134. Akhmetshina A, Palumbo K, Dees C, Bergmann C, Venalis P, Zerr P, Horn A, Kireva T, Beyer C, Zwerina J, Schneider H, Sadowski A, Riener MO, MacDougald OA, Distler O, Schett G, Distler JH (2012) Activation of canonical Wnt signalling is required for TGF-beta-mediated fibrosis. Nat Commun 3:735. doi:10.1038/ncomms1734
135. Beyer C, Schramm A, Akhmetshina A, Dees C, Kireva T, Gelse K, Sonnylal S, de Crombrugghe B, Taketo MM, Distler O, Schett G, Distler JH (2012) beta-catenin is a central mediator of pro-fibrotic Wnt signaling in systemic sclerosis. Ann Rheum Dis 71(5):761–767. doi:10.1136/annrheumdis-2011-200568
136. Guo Y, Xiao L, Sun L, Liu F (2012) Wnt/beta-catenin signaling: a promising new target for fibrosis diseases. Physiol Res 61(4):337–346
137. Fior R, Henrique D (2009) "Notch-Off": a perspective on the termination of Notch signalling. Int J Dev Biol 53(8–10):1379–1384. doi:10.1387/ijdb.072309rf
138. Dang TP (2012) Notch, apoptosis and cancer. Adv Exp Med Biol 727:199–209. doi:10.1007/978-1-4614-0899-4_15
139. Hu YY, Zheng MH, Zhang R, Liang YM, Han H (2012) Notch signaling pathway and cancer metastasis. Adv Exp Med Biol 727:186–198. doi:10.1007/978-1-4614-0899-4_14
140. Chillakuri CR, Sheppard D, Lea SM, Handford PA (2012) Notch receptor-ligand binding and activation: insights from molecular studies. Semin Cell Dev Biol 23(4):421–428. doi:10.1016/j.semcdb.2012.01.009
141. Savill NJ, Sherratt JA (2003) Control of epidermal stem cell clusters by Notch-mediated lateral induction. Dev Biol 258(1):141–153
142. del Alamo D, Rouault H, Schweisguth F (2011) Mechanism and significance of cis-inhibition in Notch signalling. Curr Biol 21(1):R40–R47. doi:10.1016/j.cub.2010.10.034
143. Le TT, Conley KW, Brown NL (2009) Jagged 1 is necessary for normal mouse lens formation. Dev Biol 328(1):118–126. doi:10.1016/j.ydbio.2009.01.015
144. Saravanamuthu SS, Le TT, Gao CY, Cojocaru RI, Pandiyan P, Liu C, Zhang J, Zelenka PS, Brown NL (2012) Conditional ablation of the Notch2 receptor in the ocular lens. Dev Biol 362(2):219–229. doi:10.1016/j.ydbio.2011.11.011
145. Jia J, Lin M, Zhang L, York JP, Zhang P (2007) The Notch signaling pathway controls the size of the ocular lens by directly suppressing p57Kip2 expression. Mol Cell Biol 27 (20):7236–7247. doi:10.1128/mcb.00780-07
146. Rowan S, Conley KW, Le TT, Donner AL, Maas RL, Brown NL (2008) Notch signaling regulates growth and differentiation in the mammalian lens. Dev Biol 321(1):111–122. doi:10.1016/j.ydbio.2008.06.002
147. Saravanamuthu SS, Gao CY, Zelenka PS (2009) Notch signaling is required for lateral induction of Jagged1 during FGF-induced lens fiber differentiation. Dev Biol 332 (1):166–176. doi:10.1016/j.ydbio.2009.05.566

148. Beebe DC, Silver MH, Belcher KS, Van Wyk JJ, Svoboda ME, Zelenka PS (1987) Lentropin, a protein that controls lens fiber formation, is related functionally and immunologically to the insulin-like growth factors. Proc Natl Acad Sci U S A 84(8):2327–2330
149. Alemany J, Zelenka P, Serrano J, de Pablo F (1989) Insulin-like growth factor I and insulin regulate delta-crystallin gene expression in developing lens. J Biol Chem 264 (29):17559–17563
150. Alemany J, Borras T, de Pablo F (1990) Transcriptional stimulation of the delta 1-crystallin gene by insulin-like growth factor I and insulin requires DNA cis elements in chicken. Proc Natl Acad Sci U S A 87(9):3353–3357
151. Milstone LM, Piatigorsky J (1977) Delta-crystallin gene expression in embryonic chick lens epithelia cultured in the presence of insulin. Exp Cell Res 105(1):9–14
152. Klok E, Lubsen NH, Chamberlain CG, McAvoy JW (1998) Induction and maintenance of differentiation of rat lens epithelium by FGF-2, insulin and IGF-1. Exp Eye Res 67 (4):425–431
153. Richardson NA, Chamberlain CG, McAvoy JW (1993) IGF-1 enhancement of FGF-induced lens fiber differentiation in rats of different ages. Invest Ophthalmol Vis Sci 34(12): 3303–3312
154. Civil A, van Genesen ST, Klok EJ, Lubsen NH (2000) Insulin and IGF-I affect the protein composition of the lens fibre cell with possible consequences for cataract. Exp Eye Res 70 (6):785–794. doi:10.1006/exer.2000.0846
155. Hejtmancik JF, Beebe DC, Ostrer H, Piatigorsky J (1985) Delta- and beta-crystallin mRNA levels in the embryonic and posthatched chicken lens: temporal and spatial changes during development. Dev Biol 109(1):72–81
156. Xie L, Chen H, Overbeek PA, Reneker LW (2007) Elevated insulin signaling disrupts the growth and differentiation pattern of the mouse lens. Mol Vis 13:397–407
157. Shirke S, Faber SC, Hallem E, Makarenkova HP, Robinson ML, Overbeek PA, Lang RA (2001) Misexpression of IGF-I in the mouse lens expands the transitional zone and perturbs lens polarization. Mech Dev 101(1–2):167–174
158. Brewitt B, Clark JI (1990) A new method for study of normal lens development in vitro using pulsatile delivery of PDGF or EGF in HL-1 serum-free medium. In Vitro Cell Dev Biol 26 (3 Pt 1):305–314
159. Kok A, Lovicu FJ, Chamberlain CG, McAvoy JW (2002) Influence of platelet-derived growth factor on lens epithelial cell proliferation and differentiation. Growth Factors 20 (1):27–34
160. Vinader LM, van Genesen ST, de Jong WW, Lubsen NH (2003) Influence of hormones and growth factors on lens protein composition: the effect of dexamethasone and PDGF-AA. Mol Vis 9:723–729
161. Reneker LW, Overbeek PA (1996) Lens-specific expression of PDGF-A alters lens growth and development. Dev Biol 180(2):554–565. doi:10.1006/dbio.1996.0328
162. Potts JD, Kornacker S, Beebe DC (1998) Activation of the Jak-STAT-signaling pathway in embryonic lens cells. Dev Biol 204(1):277–292. doi:10.1006/dbio.1998.9077
163. Gwon A (2006) Lens regeneration in mammals: a review. Surv Ophthalmol 51(1):51–62. doi:10.1016/j.survophthal.2005.11.005
164. Call MK, Grogg MW, Del Rio-Tsonis K, Tsonis PA (2004) Lens regeneration in mice: implications in cataracts. Exp Eye Res 78(2):297–299
165. Gwon A, Gruber L (2010) Engineering the crystalline lens with a biodegradable or non-degradable scaffold. Exp Eye Res 91(2):220–228. doi:10.1016/j.exer.2010.05.011
166. Apple DJ, Escobar-Gomez M, Zaugg B, Kleinmann G, Borkenstein AF (2011) Modern cataract surgery: unfinished business and unanswered questions. Surv Ophthalmol 56 (6 Suppl):S3–S53. doi:10.1016/j.survophthal.2011.10.001
167. Sveinsson O (1993) The ultrastructure of Elschnig's pearls in a pseudophakic eye. Acta Ophthalmol 71(1):95–98

168. Moreno-Montanes J, Alvarez A, Maldonado MJ (2005) Objective quantification of posterior capsule opacification after cataract surgery, with optical coherence tomography. Invest Ophthalmol Vis Sci 46(11):3999–4006. doi:10.1167/iovs.04-1531
169. Cheng CY, Yen MY, Chen SJ, Kao SC, Hsu WM, Liu JH (2001) Visual acuity and contrast sensitivity in different types of posterior capsule opacification. J Cataract Refract Surg 27 (7):1055–1060
170. Findl O, Neumayer T, Hirnschall N, Buehl W (2010) Natural course of Elschnig pearl formation and disappearance. Invest Ophthalmol Vis Sci 51(3):1547–1553. doi:10.1167/iovs.09-3989
171. Neumayer T, Findl O, Buehl W, Georgopoulos M (2006) Daily changes in the morphology of Elschnig pearls. Am J Ophthalmol 141(3):517–523. doi:10.1016/j.ajo.2005.11.019
172. Yoneya H, Kato S, Sugita G, Oshika T (2007) Spontaneous regression of Elschnig pearl posterior capsule opacification. J Cataract Refract Surg 33(5):913–914. doi:10.1016/j.jcrs.2006.12.028
173. O'Connor MD, Wederell ED, de Iongh R, Lovicu FJ, McAvoy JW (2008) Generation of transparency and cellular organization in lens explants. Exp Eye Res 86(5):734–745. doi:10.1016/j.exer.2008.01.020
174. O'Connor MD, McAvoy JW (2007) In vitro generation of functional lens-like structures with relevance to age-related nuclear cataract. Invest Ophthalmol Vis Sci 48(3):1245–1252. doi:10.1167/iovs.06-0949
175. Sugiyama Y, Stump RJ, Nguyen A, Wen L, Chen Y, Wang Y, Murdoch JN, Lovicu FJ, McAvoy JW (2010) Secreted frizzled-related protein disrupts PCP in eye lens fiber cells that have polarised primary cilia. Dev Biol 338(2):193–201. doi:10.1016/j.ydbio.2009.11.033
176. Walker JL, Zhai N, Zhang L, Bleaken BM, Wolff I, Gerhart J, George-Weinstein M, Menko AS (2010) Unique precursors for the mesenchymal cells involved in injury response and fibrosis. Proc Natl Acad Sci U S A 107(31):13730–13735. doi:10.1073/pnas.0910382107
177. Yamamoto N, Majima K, Marunouchi T (2008) A study of the proliferating activity in lens epithelium and the identification of tissue-type stem cells. Med Mol Morphol 41(2):83–91. doi:10.1007/s00795-008-0395-x
178. Nishi Y, Mireskandari K, Khaw P, Findl O (2009) Lens refilling to restore accommodation. J Cataract Refract Surg 35(2):374–382. doi:10.1016/j.jcrs.2008.10.054

Lens-Specific Transcription Factors and Their Roles in Diagnosis and Treatment of Human Congenital Cataract

6

Ales Cvekl, Ilana B. Friedman, and Elena V. Semina

Abstract

Specific DNA-binding transcription factors (TFs) are critical components of gene regulatory networks (GRNs) that govern lens growth and differentiation. Lens placode formation is controlled by Pax6, Six3, and Sox2. Lens vesicle formation and its separation from the surface ectoderm are regulated by AP-2α, Foxe3, and retinoic acid/retinoid-activated nuclear receptors. The lens precursor cells exit cell cycle via a coordinated action of Gata3, Pitx3, and Prox1. Lens fiber cell differentiation is controlled by ATF4/CREB2, c-Maf, Gata3, Hsf4, Pax6, Pitx3, Prox1, and Sox1. Lens morphogenesis requires intricate temporal and spatial control of gene expression that is executed through multiple links between TFs and extracellular signaling, including BMP, FGF, Notch, and Wnt pathways. Disrupted function of lens-preferred TFs results in multiple lens developmental abnormalities that range from the lack of lens formation, lost of primordial lens (*aphakia*), incomplete separation of the lens vesicle from the surface ectoderm (corneal-lenticular stalk), disrupted lens fiber cell differentiation, and retention of subcellular organelles including the nuclei. These developmental abnormalities are not compatible with lens transparency and result in

A. Cvekl (✉)
The Departments Ophthalmology and Visual Sciences, Albert Einstein College of Medicine, Bronx, NY 10461, USA

The Departments of Genetics, Albert Einstein College of Medicine, Bronx, NY 10461, USA
e-mail: ales.cvekl@einstein.yu.edu

I.B. Friedman
The Departments Ophthalmology and Visual Sciences, Albert Einstein College of Medicine, Bronx, NY 10461, USA
e-mail: eyindoc@yahoo.com

E.V. Semina
Division of Developmental Biology, Medical College of Wisconsin, Milwaukee, WI 53226, USA

Department of Pediatrics, Medical College of Wisconsin, Milwaukee, WI 53226, USA
e-mail: esemina@mcw.edu

S. Saika et al. (eds.), *Lens Epithelium and Posterior Capsular Opacification*,
DOI 10.1007/978-4-431-54300-8_6, © Springer Japan 2014

congenital cataracts. These cataracts affect 1:2,500 individuals and are difficult to manage due to the age of the child and eye growth considerations, possible associated microcornea or microphthalmia, and risks of glaucoma and amblyopia. For example, up to ~40 % of children with bilateral cataracts develop glaucoma following surgery, particularly when cataracts are removed early in life. This review describes normal and abnormal lens formation through the perspective of TFs preferentially expressed in the lens, their contributions to the human pediatric cataracts, and current strategies in human genetic and clinical studies to identify novel cataract-associated genes to better understand molecular basis of these defects and its treatments.

Keywords
Cataract • FOXE3 • HSF4 • MAF • PAX6 • PITX3

6.1 Introduction

The function of ocular lens is to focus and transmit light onto the retina. In contrast to the lens made from polished glass or other man-made materials, the ocular lens is exclusively comprised from highly specialized cells that appear and function like the "biological glass" [1]. The biological lens is more than an engineered transparent/refractive material. Lens cellular design enables multiple additional functions, including accommodation and correction of spherical aberration, through the use of a gradient of refractive index and flexibility of the lens surface. Lens evolution has to follow basic laws of optics. The critical lens property is to prevent light scattering. The majority of biological tissues scatter and/or absorb light: the transparent tissues are limited to the cornea, lens, and retina. Transparency of lens cytosol is a result of short-range order interactions between the proteins dissolved [1]. However, this is not sufficient as the light-scattering organelles in lens fibers are eliminated from this compartment [2]. In addition, scattering at the cell borders is abolished through tight alignments of lens fiber cells and minimal space between them.

Cataract is defined as opacification or cloudiness of the ocular lens [3]. Breakdown of the lens microarchitecture, including formation of protein precipitates and damage of lens membranes, leads to fluctuation of the optical density that results in light scattering and these processes ultimately obstruct vision. Congenital cataracts are a special subgroup of cataracts that scatter light already at birth or shortly afterwards. Although this condition occurs with a relatively low incidence rate of 1.8 cases per 10,000 births [4], the majority of instances are isolated abnormalities. Juvenile cataracts are identified within the first decade of life. The genotypic basis as well as the phenotype can vary and can have varying degrees of impact on the visual outcome for the child.

Congenital cataract is the leading cause of treatable childhood blindness in the world. It accounts for between 5 and 20 % of blindness in children worldwide. The prevalence of cataracts has been studied and varies by geography with the highest percentage of blindness occurring in the developing world. This numbers about 1–4 per 10,000 children [5, 6]. In Sweden, the incidence of congenital cataract is 36 per 100,000 births, similar to that in the UK [6]. In Australia, Wirth et al. report an incidence of 2.2 per 10,000 births [7].

Genetic studies of congenital cataract already identified 38 genes that carry specific mutations and seven additional genetic loci, with the number constantly growing [3]. A significant portion of these genes involves either lens regulatory or fiber cell-specific structural proteins. In addition, studies in mouse and rat models have identified mutations in similar types of genes further demonstrating evolutionary conservation of lens-specific genetic programs and show that these model organisms are relevant to studies of human congenital cataract. The focus of the present review is to summarize our current knowledge about the lens regulatory genes and their role in developmental cataract formation and to explain lens abnormalities that lead to lens opacities. We provide an updated list of transcription factors and discuss possible downstream molecular mechanisms that induce these cataracts. Next, we summarize clinical aspects of the pediatric cataract, strategies to ameliorate this problem, and current approaches to map the mutations. Finally, we discuss future research in the identification of candidate genes, perspectives of genetic screening, and strategies to correct these early cataracts in young patients.

6.2 Embryonic Lens Formation, Lens Developmental Abnormalities, and Cataract Classification

6.2.1 Embryonic Lens Formation and Lens Developmental Abnormalities

Congenital cataracts can be also termed "developmental" cataracts as they are formed during lens embryonic development. In human, the beginning of lens formation occurs during the third week of gestation. By the end of the fourth week, lens nucleus is already formed. Mouse is considered as an excellent model to study genetic, epigenetic, and molecular mechanisms of human lens development [8–17]. In this book, Chaps. 4 and 5 include detailed descriptions of normal mammalian lens formation and differentiation. In this chapter we are focused on the lens developmental abnormalities, identified in human and mouse, and how these defects are linked to cataract formation.

The first stage of lens formation is the formation of lens progenitor cells resulting in the formation of the lens placode. Failure of lens placode formation results in severe eye abnormalities and congenital blindness and is referred as congenital anophthalmia. Pax6 (see Sect. 6.3.1) is a gene that is essential for lens placode formation and multiple subsequent steps in lens development [18]. The lens placode can be normally formed; however, during the process of lens vesicle formation, disrupted function of multiple genes, including Six3 (see Sect. 6.3.2) and Mab21l1, induces programmed cell death of the cells forming the lens vesicle leading to its regression [19, 20]. Lens vesicle is formed through the reciprocal invagination of the lens placode and optic vesicle/optic cup. The completion of the lens vesicle formation requires its detachment from the surface ectoderm, the prospective corneal epithelium. Incomplete separation of these cell layers, through tightly controlled apoptosis, results in the formation of corneal-lenticular stalk. Although this structure obstructs vision, it cannot be classified as cataract. In human patients, this condition is referred as Peters' anomaly (see Sects. 6.3.1 and 6.3.3). Specific mutations linked to Peters' anomaly have been identified in a number of human regulatory genes, including AP-2α (Sect. 6.3.3), FOXE3 (Sect. 6.3.3), PAX6 (Sect. 6.3.1), and PITX3 (Sect. 6.3.4). Gene targeting in mouse also produced corneal-lenticular stalk in additional genes, including Notch2, RBP-J/CBF1/Su (H), and Sox11 [21–23].

It is also possible that following the normal lens vesicle formation, this structure is subsequently lost and this condition is known as lens aphakia. The model locus (*ak*) first identified and linked to this condition contains Pitx3 gene (see Sect. 6.3.4). A range of lens abnormalities can form during the primary and secondary lens fiber cell differentiation. These differences include formation of lens vacuoles. These vacuoles represent a space between the lens fibers or inside of the cells. The lens fiber cells may not form proper sutures and "ball-and-socket" contacts between the adjacent lens fibers [24]. At the molecular level, crystallin gene expression is often disrupted in mouse models of these genes [25]. From these studies it is inferred that crystallins are also deregulated in human lens. In addition, it is likely that other lens structural proteins, including lens membrane protein MIP/aquaporin 0, and connexins, including Cx45, Cx43, and Cx47, are not properly expressed. In mouse, these lens differentiation defects are caused by mutations in AP-2α, c-Maf, Hsf4, Pax6, Prox1, RARβ/RXRγ, Sox1, and other genes [16]. A special case of cataracts can be caused by disrupted and/or incomplete degradation of lens fiber cell nuclei and possibly other subcellular organelles. These conditions were found in mouse models of Brg1/Smarca4, Dnase2b, Ncoa6, Notch2, and others [22, 26–28]. These abnormalities are not compatible with lens transparency and are direct causes of lens opacities as described earlier.

A "secondary" aphakia may occur as apparently normal lens differentiation can pass through the primary lens fiber cell differentiation and reach the phase of secondary lens formation. Lens degeneration becomes evident prior the onset of the lens fiber cell denucleation, the lens fiber cell compartment will display signs of apoptosis, and this condition is caused by loss of ATF4/CREB2 [29]. Over the time, lens is completely destroyed in these mutants. Reduced lens size is typically associated with the reduced size of the eye ("small eye," *Sey*), and this condition is also called microphthalmia. Smaller lenses often exhibit signs of cataract formation. The prototypic gene of *Sey* mutation is Pax6 (see Sect. 6.3.1).

Taken together, abnormal lens development often leads to changes in the lens fiber cell compartment that produce opacities and cataract. Mouse genetic studies generated a growing list of mutants with many of the identified at the molecular level [30–32]. Human genetic studies identified many of these genes as causative genes of congenital cataracts as described below (see Chap. 3). Identification of novel cataract-causing genes is described in Sect. 6.3.5.

6.2.2 Cataract Classification Based on Opacity Position

There have been many attempts to classify congenital cataracts, and many systems have been developed; however, many of them include cataracts that are associated with systemic disorders as opposed to purely heritable forms. In addition, many classifications have been named for the ophthalmologist or families in which the cataracts were first described. Because of this, there is no universal classification system, which may limit the ability of investigators to compare phenotypes directly, as well as cause difficulty in predicting prognosis. This may hamper genetic counseling for families in whom these lens changes are prevalent [33]. One method, which will be used here, is to discuss the phenotypes based on the position of the opacity.

The anterior polar cataract (APC) is situated in the anterior pole of the lens and is usually a discrete lesion seen in both eyes. These lesions may vary among family members, and the visual prognosis is usually consistent within a family but may differ among pedigrees. Because of the location in the visual axis, and their distance from the retina, they often do not have a significant impact on vision and may be followed for many years prior to needing extraction.

The posterior cataracts are located in the posterior pole of the lens and may be both static or progressive. They are usually easily identified at birth and are generally symmetrical. The progressive form is usually associated with the need for surgical extraction. These can be associated with persistent fetal vasculature, and the affected eye is usually microphthalmic. In the case of posterior lenticonus, the cataract usually develops after the critical period of visual development and is usually unilateral. Nuclear cataracts refer to opacification within the embryonal or fetal nuclei of the lens. They can vary in appearance from small dustlike (pulverulent), punctate dots, cerulean (blue dot), or complete nuclear opacification. These cataracts are generally vision threatening, nonprogressive, present at birth or early infancy, and require prompt removal in order to increase the chances of useful vision. Morphologically, the eyes are usually smaller than normal, and the inherited forms are usually autosomal dominant.

In the lamellar cataract, the lamellae are formed from concentric layers of secondary lens fiber deposition around the embryonal nucleus. These opacities are generally discrete and are surrounded by normal lamellae, which is suggestive of a transient disruption of normal lens development during gestation. The surrounding cortex and the embryonal nucleus are clear. The later in development the insult occurs, the larger the lamellar cataract is found. The majority of these cataracts, though already present at birth, may not obstruct visual development immediately. In many cases, they may be followed by serial examination and only require removal if vision deteriorates. They are generally bilateral, autosomal dominant, and the eyes are of normal size.

Pulverulent cataract has both an autosomal dominant and autosomal recessive form of inheritance. The recessive form is usually associated with later onset of opacities. The cataracts appear as powdery opacities that are present throughout the lens. There are some families in which they may be confined to one area of the lens and may be classified as nuclear or lamellar, respectively, based on their location.

The lens changes in cerulean cataract are distinctive in both appearance and prognosis. They are not present at birth, but rather appear in childhood and progress, not requiring surgery until adulthood in many cases. The appearance is that of discrete pinhead-shaped blue and white dots that are located throughout the lens, with a predilection for the cortex.

In total (complete) cataract, complete opacification of the fetal nucleus is found at birth, followed by opacification of the cortex, shortly after birth. They require surgery early in infancy to attempt to restore visual development. They are rarely autosomal dominant. A rare form is cortical cataract, in which opacification occurs in the outer cortex adjacent to the capsule. The other regions of the lens are clear, including the visual axis, and as such the prognosis is excellent, rarely requiring removal. Finally, sutural cataracts usually manifest as prominent opacification of the anterior and posterior sutures. They are most often associated with other forms of inherited cataract such as nuclear, cerulean, pulverulent, and lamellar.

6.3 Transcription Factors and Congenital Cataract

This section relates to nine DNA-binding TFs (FOXE3, HSF4, MAF, OTX2, PAX6, PITX3, SIX3, SOX2, and TFAP2A/AP-2α) (Table 6.1) and three transcriptional coactivators, including phosphatase EYA1 (Table 6.1), and two histone lysine acetyltransferases CBP/KAT3A and p300/EP300/KAT3B (see Sect. 6.3.4). When these genes are linked together via molecular biology studies, evidence exists that they regulate the expression of crystallin genes (see Sect. 6.3.6). For the purpose of this review, we define these proteins as "lens specific," as they play significant roles in lens formation and cataract. However, their expression should be rather called "lens-preferred" as they are expressed outside of the lens, in many cases in the retina, central nervous system, as well as in other organs and tissues. Correspondingly, mutations in these genes cause isolated defects in the lens (e.g., HSF4 and FOXE3) and/or more pleiotropic effects (e.g., PAX6 and AP-2α).

These proteins can be organized into smaller groups depending on their structure/function and molecular mechanisms. Herein, we used a classification based on similarity of defects and molecular mechanisms of cataract formation. This corresponds to earlier Sect. 6.2 and the aforementioned lens developmental defects. Three important features pertinent to their genes involve the gene dosage of effect, i.e., "haploinsufficiency," "dominant" effects, and nonsense RNA-mediated decay. Haploinsufficiency is defined as a type of gene dosage effect in which loss of function of one allele such as complete gene deletion is displayed by mild abnormalities, and loss of both alleles results in more severe phenotype. Dominant effects are mostly linked to gain-of-function missense mutations that increase activity of the protein. Nonsense RNA-mediated decay relates to nonsense mutations followed by degradation of mRNA without the synthesis of truncated proteins that could potentially act as dominant-negative repressors.

Table 6.1 Transcription factors and human cataract

Gene	MIM	Locus	Cataract type/ locus/AD, AR	Other eye phenotype	References
EYA1	601653	8q13.3	Congenital	ASD	[34]
FOXE3	601094	1p33	Congenital	ASD, aphakia, sclerocornea	[35]
HSF4	602438	16q22.1	Congenital, age-related, progressive/ cataract 5	–	[36]
MAF	177075	16q22	Congenital, juvenile/ cataract 21	Microcornea, coloboma	[37]
OTX2	600037	14q22.3	Congenital	Microphthalmia, coloboma	[38]
PAX6	607108	11p13	Congenital, presenile	Aniridia, ASD, foveal hypoplasia, keratitis, corneal dystrophy, optic nerve coloboma/hypoplasia	[39]
PITX3	602669	10q24.3	Congenital/ cataract 11	ASD	[40]
TFAP2A	107580	6p24.3	Congenital	Microphthalmia, coloboma, ASD	[41]
SIX3	603714	2p21	Presenile	Microphthalmia, coloboma, strabismus	[42]
SOX2	184429	3q26.3	Congenital	Microphthalmia, coloboma, sclerocornea	[43]

The inheritance is AD for all genes, except of HSF4 (both AD and AR; see text)
AD autosomal dominant, *AR* autosomal recessive, *ASD* anterior segment dysgenesis

6.3.1 PAX6

The DNA-binding TF PAX6 is a central regulator of eye development [18] and a focus of human genetic studies of congenital ocular abnormalities [44, 45]. Pax6 regulates all critical stages of lens development starting with the regulation of lens placode formation and culminating with the regulation of terminal differentiation and organelle degradation in the differentiating lens fiber cells [9, 18]. PAX6 ranks among the best-studied genes through human genetic studies as well as in model organism studies. The majority of PAX6 mutations cause aniridia, a panocular disease that is manifested by reduced iris, lens cataract, corneal and retinal abnormalities, glaucoma, and nystagmus.

PAX6 encodes a 422 amino acid (aa) protein (Table 6.1). PAX6 binds to DNA via 128 amino acid paired domain and an internal 60 amino acid homeodomain [46]. A major splice variant, PAX6(5a) is 14 amino acids longer due to an insertion in the paired domain [47]. The PAX6/PAX6(5a) ratio in lens is approximately 8:1. Pax6 proteins recognize 13–20 bp sequences on the DNA [25, 46, 47]. In mouse, *small eye* (*Sey*) alleles [48] carry mutations in Pax6 gene [31]. Several hundreds of mutations have been discovered in human PAX6 coding and noncoding sequences [49, 50]. The majority of nonsynonymous mutations include nonsense mutations that predict premature protein termination; however, these mutated mRNAs are likely degraded via the nonsense-mediated decay [45]. It has been shown recently in mouse that through a nonsense mutation suppression strategy using a topical drug formulation, including Ataluren, it is possible to reverse the *Sey* phenotype to nearly normal eye [51]. The "minor" missense mutations are mostly found in the paired domain and typically cause moderate eye defects, including cataract (mutations: G18W, A33P, and G64V) [52, 53]. In the majority of patients, cataract develops in teens or early adulthood [54]. The noncoding regulatory mutations (i.e., deletions) in *PAX6* locus were identified in the 3′-distal regions [55, 56]. Interestingly, an increased dosage/duplication of *PAX6* has also been associated with various ocular defects, including congenital cataracts in one family [57].

The molecular mechanisms of PAX6 in lens development, lens epithelial homeostasis, and cataractogenesis are not well understood. Pax6 is an essential gene for lens formation; homozygous mutations in Pax6 prevent the formation of the lens placode [18]. Conditional inactivation of Pax6 in the prospective lens ectoderm further supports its critical function. However, conditional deletion of *Pax6* in the differentiating secondary lens fibers disrupted their cell cycle exit regulation, arrested their differentiation, and induced apoptosis [58]. Several hundreds of Pax6-regulated genes have been identified in the lens placode [59–61], E14.5 embryonic [62], and in newborn Pax6 heterozygous [63] lens. These Pax6 target genes include both regulatory (e.g., c-Maf, Prox1, Sfrp2, and Mab21l1) and structural genes (e.g., crystallins). Thus, it is likely that disrupted profile of crystallins in lenses of heterozygous carriers of *PAX6* mutations is responsible for the cataract formation in young human patients. Another possibility is that reduced levels of PAX6 proteins in the lens epithelium are not compatible with the epithelial homeostasis required to support the lens fiber cell compartment that triggers lens opacification. Finally, it is also possible that PAX6 haploinsufficiency accelerates "natural" ageing of the lens epithelium and in lens fibers.

6.3.2 OTX2, SOX2, SIX3, and EYA1

These four proteins are linked together by their early expression in the prospective lens ectoderm, and this group includes two homeodomain-containing proteins OTX2 and SIX3, HMG-box factor SOX2, and transcriptional coactivator/phosphatase EYA1, an obligatory partner of SIX3 (Table 6.1).

OTX2 represents a homeobox-containing TF related to the *Drosophila orthodenticle* [64]. Complete knockout of mouse *Otx2* results in early embryonic lethality due to abnormal primitive streak formation while heterozygous embryos show numerous abnormalities of head structures including anophthalmia/microphthalmia, micrognathia, holoprosencephaly, and even acephaly [65, 66]. Human *OTX2* mutations were first identified in probands affected with anophthalmia/microphthalmia along with brain malformations and other features in some individuals [38]; since then, multiple additional cases were reported [67]. Cataracts were documented in several *OTX2*-positive cases in association with microphthalmia [38, 68].

SOX2 represents a TF that interacts with DNA through the high-mobility group (HMG) domain. SOX2 and other HMG proteins were shown to have low DNA-binding affinity and therefore require additional proteins to facilitate the interaction [69]. *Sox2* deficiency in mice results in variable phenotypes [70–72]. Taranova and coauthors [72] demonstrated a gene dosage-dependent variation in ocular phenotypes in *Sox2* mutant mice. The variable expressivity of ocular defects was also supported by Kelberman and coauthors (2006) who reported pituitary and male fertility defects but normal eye development in mice carrying heterozygous *Sox2* deletion allele. In humans, *SOX2* mutations represent the most common cause of anophthalmia/microphthalmia, accounting for 10–20 % of cases [38, 73]. Cataracts were reported in some cases affected with *SOX2* mutations, in association with microphthalmia and/or coloboma [38, 73, 74]. The SOX2 and OTX2 proteins were found to co-regulate the expression of the *RAX* gene, thus connecting these factors in the same pathway during retina development [75]. Sox2 and Pax6 were shown to act independently during lens fiber differentiation in mice [58]; at the same time, some studies demonstrated interaction between Sox2 and Pax6 in the regulation of chicken δ1-crystallin gene expression in lens [76].

SIX3 represents a homeobox-containing TF related to *Drosophila sine oculis*, which was shown to be required for eye development. In various vertebrate species, ectopic Six3 expression induces the formation of ectopic lens and retina tissue in specific areas of head territory [77–79]. Complete inactivation of *Six3* locus in mice results in a severe phenotype characterized by telencephalic and opto-preopto-hypothalamic truncations and partial caudalization of the homozygous mutant head, while no obvious phenotype is observed in heterozygous embryos [80]. *Six3* expression in the developing lens placodal ectoderm was shown to be regulated by Pax6 [79]. Human mutations in *SIX3* lead to holoprosencephaly, which is a relatively common severe brain malformation [81]. The reported ocular phenotypes include proptosis of the eyes, microphthalmia, colobomas, deformed lens with attachment to retina, and presenile cortical cataracts [42, 81].

The *EYA1* encodes a homologue of the *Drosophila eyes absent* (*eya*) and belongs to a family of proteins that possess tyrosine phosphatase and threonine phosphatase activities and, in cooperation with a DNA-binding factor, transactivation ability [82, 83]. Inactivation of mouse *Eya1* results in severe anomalies of craniofacial, skeletal, kidney, and ear development [84]. The human *EYA1* mutations are associated with branchio-oto-renal (BOR) and branchio-oto (BO) syndromes, which are characterized by craniofacial and hearing defects with or without kidney anomalies; ocular phenotype is normal in most cases [85, 86]. *EYA1* role in isolated cataract/anterior segment dysgenesis is currently supported by two reports that described three different missense changes [34, 87]. Functional analysis of *EYA1* mutations demonstrated that changes associated with BOR syndrome result in a loss of phosphatase activity, while mutations linked to cataracts and anterior segment defects demonstrate normal or higher levels of catalytic activity [88].

6.3.3 TFAP2A (AP-2α) and FOXE3

Although AP-2α and Foxe3 are structurally different TFs, their mutations cause similar defects, including the incomplete separation of the lens from the surface ectoderm.

Transcription factor AP2A or AP-2α (activating enhancer binding protein 2 alpha) belongs to a family of retinoic acid-inducible transcriptional activators which are involved in a wide range of developmental processes (Table 6.1). The *Tfap2a* homozygous knockout mice display neural tube defects as well as severe craniofacial and body wall abnormalities consistent with *Tfap2a* expression in tissues of ectodermal origin [89]. In respect to vertebrate eye development, *Tfap2a* deletions lead to a range of ocular phenotypes including anophthalmia, lenticulo-corneal adhesions, retinal coloboma, optic stalk abnormalities, and other anomalies [90, 91]. Human mutations in *TFAP2A* result in branchio-oculo-facial syndrome (BOFS), which is characterized by distinctive facial features, malformations of the eyes and ears, cutaneous anomalies, and other variable features [41]. Ocular phenotypes most commonly include microphthalmia, coloboma, and nasolacrimal duct stenosis; cataracts, aphakia, or other lens defects have been observed in some cases [92–94].

FOXE3 belongs to the family of forkhead box (FOX) containing TFs (Table 6.1). The forkhead box encodes 100 highly conserved amino acids domain that mediate interaction of FOX proteins with a consensus DNA response element. The *Foxe3* gene was first connected to lens development through its association with the recessive mouse phenotype *dysgenetic lens* (*dyl*) that involves microphthalmia, small lenses, lenticular-corneal adhesions, and corneal opacity; the phenotype is caused by homozygous mutations in the forkhead domain of *Foxe3* [95, 96]. Heterozygous *dyl* embryos were originally reported as unaffected, but further studies detected lenticular-corneal adhesions, cataract, and other ocular defects in ~40 % of heterozygous animals [97]; Ormestad and coauthors proposed a complete loss of function or haploinsufficiency as mechanisms behind *dyl* homozygous or heterozygous phenotypes, respectively, because of an apparent loss of DNA binding ability by the mutant protein. Recently, the Rinshoken cataract (rct) mouse mutant, characterized by recessive congenital cataracts, was shown to be caused by a 22-bp deletion located ~3.2-kb upstream of *Foxe3* start site and associated with reduced expression of *Foxe3* during lens development [98]. In situ hybridization studies in animal models including mouse and zebrafish demonstrated that *Foxe3* expression is limited to the developing lens and presumptive midbrain regions [95, 96, 99]. Expression studies of human *FOXE3* detected transcripts in the developing lens only, with the strongest expression in the anterior lens epithelium, in embryonic coronal head sections (Carnegie stages 16 and 17) [100]. Human *FOXE3* mutations are associated with both recessive and dominant lens phenotypes; dominant mutations in *FOXE3* have been seen in patients with isolated congenital cataract and/or anterior segment dysgenesis [35, 100] while recessive mutations have been identified in families with nonsyndromic microphthalmia, aphakia, and sclerocornea [101–103]. In fact, *FOXE3* mutations make a significant contribution to bilateral microphthalmia explaining up to 15 % of cases in some studies [101]. *TFAP2A* and *FOXE3* appear to act independently as studies of a conditional deletion of *Tfap2a* in the developing mouse lens demonstrated that *Foxe3* (as well as *Pitx3* and *Pax6*) is not directly affected by the absence of this transcription factor [104].

6.3.4 PITX3

PITX3 (Table 6.1) represents a member of the PITX family of homeobox-containing TFs that is essential to normal embryonic development in vertebrates; all three members of this family are associated with developmental human disorders. The PITX homeodomain, the DNA-binding [105] domain of this transcription factor, belongs to the *paired-like* and the *bicoid-like* subfamilies based on sequence similarity [40, 106, 107] and was shown to interact with 5′-TAATCC-3′ and to other *bicoid* sites [106, 108]. Another conserved domain, a 14 amino acid motif, is located in the C-terminal region and is shared by many *paired-like* homeodomain proteins [105, 107]; function of this motif remains to be defined.

The link between *Pitx3* and lens development was first established through studies of murine *Pitx3* gene that was localized to chromosome 19 to the recessive *aphakia* (*ak*) locus [109]; subsequently, two deletions, removing a noncoding exon 1and an additional upstream sequence of *Pitx3*, were identified in the *ak* genomic DNA [110, 111]. *Pitx3* is normally expressed in the developing lens starting from lens pit stage and can still be detected in adult lens. Rieger and associates showed that the expression of the *Pitx3* is diminished to 5 % of normal level in *ak/ak* embryos [110]. The *ak* mice demonstrate abnormal lens development resulting in the formation of a small and disorganized eye [112]. In *ak* mice, the lens vesicle fails to detach from the overlying ectoderm and the cells at the posterior region do not elongate/differentiate into lens fibers; likely secondary to this, no signs of anterior segment development can be observed in these mutants. A complete knockout of *Pitx3* was also generated and demonstrated similar abnormalities in lens development [113]. The *Pitx3/pitx3* genes have also been identified in various other vertebrates and displayed conserved expression and function during lens development [99].

The first human *PITX3* mutation, a 17-bp insertion C-terminal of the homeodomain was found in a dominant six-generation pedigree that was diagnosed with anterior segment mesenchymal dysgenesis [40]. All the affected members carried a mutant allele and demonstrated cortical cataracts and corneal opacities with or without iris adhesions. Later studies identified the same mutation in five additional dominant pedigrees affected with congenital posterior polar cataract, with anterior segment dysgenesis in some patients [114–118]. Other *PITX3* mutations reported to date include an N-terminal missense change in a family with congenital total cataract [40], and three additional C-terminal frameshift mutations, including deletion of the same 17-bp region duplicated in the recurrent insertion mutation [114, 119, 120]. All insertions result in a frameshift in the 3′ end of the coding region so that the resultant proteins are expected to contain intact N-terminal and homeodomain but abnormal C-terminal regions of varying length. Homozygous *PITX3* mutations result in severe microphthalmia and corneal opacification and were reported in two consanguineous pedigrees [119].

Functional studies of *PITX3* mutations so far included only two disease-associated alleles and uncovered defects in DNA-binding and/or transactivation activities for both mutants, with 17-bp recurrent insertion mutant being more severely affected [108]. In addition, the PITX3 protein associated with the 17-bp insertion demonstrated varied residual activity between lens (~50 % of wild type) and corneal (~80 % of wild type) ocular cell lines, thus suggesting contribution from secondary factors to functional and phenotypic effect(s) of this mutation; these findings seem to be consistent with the observed higher penetrance of cataract phenotype in comparison to corneal/anterior segment features in carriers of this mutant allele [108].

Studies of PITX3 pathway identified links with some other TFs involved in lens development/cataract phenotype. Pax6 was proposed as an upstream regulator because of the decreased *Pitx3* expression in *Pax6* heterozygous mutant mouse lenses [121] and in Pax6 homozygous prospective lens ectoderm [122]; at the same time, downregulation of *Pax6* expression was detected in *Pitx3*-deficient *Xenopus laevis* embryos suggesting that some aspects of *Pax6* expression may be dependent on Pitx3 [123]. *Pitx3* has been positioned upstream of *Foxe3* during lens development by several studies [124–127]. Also, human and zebrafish *PITX3/pitx3* were demonstrated to directly regulate lens expression of the major intrinsic protein of lens fiber, *MIP/aquaporin 0*, associated with human congenital cataracts [128], and finally, the expression of another cataract gene, *Crybb1*, was found to be downregulated in *Pitx3*-deficient *Xenopus laevis* embryos [123].

6.3.5 HSF4 and MAF

HSF4 and MAF (mouse c-Maf) are essential regulatory genes responsible for lens fiber cell differentiation. They function as important regulatory genes of crystallin gene expression. c-Maf appears to regulate all crystallins [129]. In contrast, Hsf4 regulates γ-crystallins but not α- and β-crystallins [130]. Hsf4 is truly a lens-specific factor while c-Maf is more ubiquitously expressed. Several mutations in these genes were identified in human cataract patients.

HSF4/Hsf4 (the major splice variants are 463 and 493 amino acid proteins, Hsf4a and Hsf4b, respectively) is a member of heat shock family of TFs that recognize trimeric heat shock elements, 5′-GAANNTTCNNGAA-3′, including its shorter or longer derivatives [131]. Mouse Hsf4 is expressed from E12.5 in differentiating lens fibers and its expression culminates prior the lens fiber cells denucleation (E16.5–E18.5). Deletion of Hsf4 causes lens cataract mainly due to the retention of nuclei that would be otherwise degraded to form the organelle free zone [130]. Hsf4 regulates the expression of DNase IIβ, an enzyme essential for the degradation of nuclear DNA [26, 132].

HSF4 mutations were found in autosomal dominant lamellar and Marner cataract [36] and recessive cataract [133]. The dominant mutations are located in the N-terminal DNA-binding domain. In contrast, the recessive mutation was found near the C-terminus (Table 6.1).

MAF is a member of a family of four large Maf genes, including MafA, MafB, c-Maf, and NRL, all of them expressed in the eye [134]. However, only mutations in c-Maf cause lens developmental abnormalities in mouse [135]. MAF is a basic leucine zipper TF comprised of 386 amino acids. It recognizes a 13 bp consensus site, 5′-TGCTGANYCNGCA-3′ that is found in all crystallin promoters. It can bind to DNA as homodimer or heterodimer with other Maf and AP-1/CREB proteins [136]. The expression of c-Maf is initiated in the lens placode, continues in the lens vesicle, and is upregulated in the posterior part of the lens vesicle from which the primary lens fibers are formed. Its expression continues in the differentiating secondary lens fibers [129]. Although c-Maf was first identified as an oncogene, it works as a differentiation factor in lens [134].

Several cataract-causing mutations were identified both in human MAF [37, 137] and mouse [138, 139]. Heterozygous mutations in MAF were found in patients with pulverulent and cerulean cataracts accompanied with defects in the cornea and iris [137, 140, 141]. Disruption of distal regulatory region of the MAF locus (16q23) also resulted in human cataract [37, 137].

Two histone lysine acetyltransferases, including CBP/KAT3A and p300/EP300/KAT3B, are obligatory coactivators of AP-2α, ATF4/CREB2, c-Maf, Pax6, Prox1, Sox1, Sox2, and likely other DNA-binding factors that regulate lens development [16]. They are expressed throughout the lens development [142]. Mutations on these genes can cause human cataracts [143]. Recent studies in mouse have shown that loss of three alleles of these genes produces lens cataract [61]. In contrast, deletion of all four alleles is not compatible with the process of lens placode formation from the specified prospective lens ectoderm [61].

6.3.6 Novel TFs: iSyTE Database

Additional DNA-binding TFs that regulate crystallin gene expression, including GATA3, PROX1, and SOX1, are reasonable candidates for cataract genes. In the post-genomic era, novel gene discoveries are mostly driven by a combination of bioinformatics approaches and massively parallel ("next generation") DNA and RNA sequencing. There are several resources that can assist in the analysis/identification of novel cataract factors by providing information about gene expression or structural variation and possible phenotypic associations. The gene expression resources include iSyTe (Integrated Systems Tool for Eye gene discovery; http://bioinformatics.udel.edu/Research/iSyTE) that provides data about embryonic lens expression of mouse orthologs of human genes [144] and BioGPS (Biology Gene Portal System; http://biogps.org) that contains microarray-analysis-based expression information from various species including mouse lens and six other ocular tissues. The DECIPHER (Database of Chromosomal Imbalance and Phenotype in

Humans using Ensemble Resources; http://decipher.sanger.ac.uk/) allows access to human copy number variation data and associated phenotypes, including cataracts and other eye anomalies, and therefore may provide additional useful information to ocular researchers. The OMIM (Online Mendelian Inheritance in Man; www.omim.org) database contains comprehensive reports on human disease-associated loci including multiple regions linked to various cataract phenotypes still awaiting causative gene(s) identification.

6.3.7 Gene Regulatory Networks

Gene regulatory networks (GRNs) provide system-level description of embryonic development and cellular differentiation. GRNs are comprised from a repertoire of subcircuits that govern individual cellular processes. The key regulatory genes of these processes include DNA-binding TFs, including a special class of signal-regulated factors [145]. This design allows formation of new cell types in response to extracellular signaling in a 3D space of the developing embryo. The formation of lens progenitor cells that comprise the lens placode is regulated by Pax6, Six3, and Sox2 with multiple autoregulatory loops [Six3 ⇄ Pax6 ⇄ Sox2] and is reminiscent of the "core" embryonic stem (ES) network of [Oct3/4 ⇄ Nanog ⇄ Sox2] [146, 147]. The expression of Pax6 and Sox2 is regulated by BMP signaling through the activity of Lhx2 in the optic vesicle [148]. Two neuropeptides, somatostatin and nociceptin, control Pax6 expression in lens and olfactory progenitors [149]. Recent studies have also shown that lysine acetyltransferases CBP and p300 are required to complete the process of lens induction from the prospective lens ectoderm [61].

Lens fiber cell differentiation and crystallin gene expression is regulated via a feedforward loop between Pax6, c-Maf, and crystallins that includes Pax6 and c-Maf autoregulation [25]. Hsf4, Prox1, and Sox1 also regulate specific γ-crystallins. CREB and ArH regulate αA- and αB-crystallins, respectively. In addition, evidence exist that CBP and p300, through the interaction with c-Maf, also regulate lens fiber cell differentiation via activation of crystallin genes [150].

6.4 Evaluation and Treatment of Congenital Cataract

6.4.1 Cataract and Amblyopia

Amblyopia is the poor development of vision secondary to abnormal visual stimulation during the critical period of visual development. The ability to reverse amblyopia depends on the development of the visual system at the time of the insult, the duration of the deprivation, and the point at which therapy is instituted. The most critical period is when patients are younger than 2 months. If the amblyopia is not managed properly, it can lead to severe and permanent visual loss as well as nystagmus. After 2–3 months of age the amblyopia can be reversed, but to varying extents. This risk of amblyopia continues to decrease as the patient gets older. Aside from the timing of the visual insult, such as cataract, the size and location of the lens opacity also plays a role in the severity of amblyopia. Larger and denser lesions are at higher risk of causing severe visual loss than those that are smaller. If a cataract is small, and the physician chooses to follow it with serial examination, this must be done at regular intervals to prevent amblyopia. In addition, those children with unilateral cataracts must engage in occlusion therapy. Children with bilateral cataracts may have developmental delay as well as impaired vision, while those with monocular cataracts present with strabismus and are often diagnosed later than their bilateral counterparts. If manifest nystagmus is present at presentation, this is an indication of poor visual prognosis.

6.4.2 Cataract Examination and Surgery in Children

The examination of children with cataracts often requires sedation or general anesthesia. It is imperative to examine both eyes when they are dilated, as often the fellow eye in monocular cataracts may exhibit anatomic abnormalities. Measurement of the corneal diameter, as well as intraocular pressure, is performed. In addition, indirect ophthalmoscopy should be done to detect persistent fetal vasculature or other posterior segment anomalies [151]. Surgery for visually significant cataract should be performed as soon as possible to increase the chances for a useful visual outcome. These cases are performed under general anesthesia. In bilateral cases, the surgeries are usually performed within a few days of each other, to minimize the development of a strong ocular preference by allowing one eye to become dominant.

The lens can be extracted by either a limbal or pars plana approach. The lens is generally fairly soft and can be removed using irrigation and aspiration and traditional adult phacoemulsification is not employed in these cases. An intraocular lens may be inserted depending on the size of the eye, as well as whether the cataract is unilateral or bilateral. Care is taken to perform a posterior capsulorhexis at the time of surgery, as the incidence of secondary membrane formation is approximately 80 %. If formation of the after-cataract occurs and obscures vision, a second surgery is required to remove it, as traditional laser treatment that is performed in adults cannot be employed in the case of children. An anterior vitrectomy is often performed at the time of the posterior capsulorhexis and it is important to check the anterior chamber for the presence of vitreous [152]. Prophylactic antibiotics in the form of a subconjunctival injection are often used to reduce the risk of endophthalmitis, the dreaded complication of cataract surgery. Patients are treated vigorously with antiinflammatory and antibiotics in the postoperative period. Unlike adults, where steroid therapy is instituted about 4–5 times a day and tapered over 1 month, pediatric patients require much more intensive treatment starting with drops 8–10 times per day and tapering slowly over 2–3 month [153].

Occlusion therapy in children with unilateral cataract should be instituted as soon as possible after aphakia is corrected. The decision about postoperative correction of aphakia varies with the age of the child at surgery and whether the cataract is unilateral. Today, most older children receive intraocular lenses, particularly in unilateral cases, and the power calculation is set to achieve emmetropia. These patients will require bifocals for reading, as well as spectacles for any residual refractive error that remains after about 1 month postoperatively.

In younger infants, the decision for refractive correction can be more complicated. The question of whether to implant an intraocular lens to aid in visual rehabilitation is one that has been debated extensively in the ophthalmic community. Prior to the use of IOLs, patients with unilateral cataract used contact lenses for correction of visual acuity. These contact lenses can be difficult to insert, remove, and clean in infants, and the question arose as to whether the visual outcomes using an intraocular lens would be superior to those with contact lens use. In both cases, occlusion therapy for amblyopia must be instituted. Additional questions arose, as to whether there was a difference in the complication rate and the need for secondary surgery using one modality over the other, as well as, what lens power to implant in young children, whose eyes would continue to grow, with a shifting refractive error. Would it be preferable to leave them undercorrected, anticipating that the eyes would lengthen and eventually they would "grow into" their implanted lenses? Or would it be preferable to aim for emmetropia at the time of surgery to better allow for immediate clear vision without the use of spectacles? This could potentially have the advantage of more rapid improvement and tolerance of amblyopia therapy. To that end, the Infant Aphakia Treatment Study (IATS) was developed as a multicenter randomized control trial to evaluate the question of whether there was any difference in the visual outcome at 1 and 4.5 years of age using IOL vs. contact lens in unilateral cataract in children less than 6 months of age. In addition, it looked at other secondary outcomes such as the incidence of secondary glaucoma and secondary surgery [154].

To date, the group has published the 1-year results, showing that there was no significant difference in visual acuity between the two groups at 1 year. However, they did find a statistically significant difference in the need for secondary procedures, with the IOL group having a higher incidence of these procedures [155]. The IATS study performed IOL calculations and implanted lenses that were undercorrected by between 6 and 8 diopters with over-refraction for amblyopia therapy.

6.4.3 Postoperative Complications

Secondary opacification of the visual axis is the most common complication of cataract surgery. The younger the patient, the higher the risk of opacification. This can occur even when posterior capsulotomy has been previously performed. If opacification occurs, a laser capsulotomy can be attempted, but in young children and infants, surgical intervention is necessary and must be done promptly to avoid amblyopia. Topical steroid therapy must be used after surgery to prevent recurrence.

Secondary glaucoma is very common and the most sight-threatening complication of cataract surgery. A recent study that included analysis of 101 patients with bilateral and unilateral congenital/infantile cataracts reported incidence of glaucoma in ~30 % of cases, with this number increasing to ~40 % for patients with bilateral cataracts; the incidence rate was also higher when cataracts were removed early in life [156]. It is generally a form of open-angle glaucoma and can occur within months to years after surgery. The earlier in life the surgery is performed, the higher the incidence. Glaucoma in the infant eye can be detected by edema of the cornea, photophobia, and buphthalmos. These patients must be treated promptly and often this requires surgical intervention. In the IATS, the incidence of glaucoma-related events at 1 year was 12 % overall in both groups, with neither group showing a significant difference. However, there seemed to be a trend toward those patients with persistent fetal vasculature and microphthalmia [157, 158].

The ultimate goal of cataract extraction is to improve visual outcome in the affected eye(s). In Birch et al., a prospective study looking at visual acuity outcomes in IOL vs. contact lens implantation found that there was no significant difference in visual acuity in the two groups at 4 years. This was when IOL was compared to good-to excellent compliance with contact lenses. In the moderate-to-poor compliance contact lens group the visual acuity was worse when compared to those patients that had implanted IOLs. In addition, at 4 years, the individuals who received IOLs after unilateral cataract extraction prior to 6 months of age had the same visual outcomes as those using contact lenses. Those individuals who had IOL implantation with cataract extraction after 1 year of age enjoyed better visual outcome than those with contact lenses [159].

6.5 Conclusions and Perspectives

Since the initial cloning of human *aniridia* and mouse *Sey* genes as *Pax6* in 1991, over two decades of research resulted in many novel insights into the lens and cataract formation. In the next decade, it is very likely that novel genes and novel mutations in known genes will be identified. These studies will fill current gaps between predicted cataract loci and actual gene identification. Through the studies of lens epigenome, novel information will be added to understand the process of DNA methylation in normal and cataractous lenses and during lens ageing. These data will be connected with dysregulated gene expression. Since cataract is a multifactorial disease with a significant environmental component, large-scale epidemiological studies will be required to assess the risks in different populations followed by the dietary and other environmental factors that contribute both to congenital and age-onset cataract.

Studies in model organisms play major role to understand molecular mechanisms of individual genes and their interactions and to elucidate processes and gene regulatory networks that control lens fiber cell differentiation, including degradation of their organelles. Although the focus of this review is cataract, studies of presbyopia and its genetic and environmental factors should be accelerated in the upcoming years.

Suppression of nonsense-mediated mRNA decay is a potential therapy for those eye and lens diseases related to the premature termination codons, not just for PAX6 [51] as described in Sect. 6.3.1. A number of compounds have been identified, including non-aminoglycoside Ataluren [160] that suppress translational termination via increasing the ability of near-cognate aminoacyl tRNAs to block the action of two release proteins, eRF1 and eRF2 [161]. This in turn allows a "readthrough" until the wild-type stop codon is encountered. Eye diseases caused by this type of mutations can be treated by topical formulas that deliver the drug(s) to the tissues of interest [51].

The induced pluripotent stem (iPS) cell technology now allows to produce lens cells that carry the original mutation and study these materials [162, 163] to understand cataract formation cause by known mutations in genes summarized here. To achieve this goal, it is necessary to improve the current methods to generate lentoid bodies and convert them into more advanced lens-like structures.

Another novel area in cataract formation includes studies of noncoding RNAs that regulate early stages of lens development [62] in combination with studies of age-onset cataract, posterior subcapsular opacification, and dysregulation of miRNAs [164, 165].

The significance of pediatric cataract on the clinical practice is going to increase due to the increasing number of families where parents carry on mutations. Ongoing research is aimed to target the nonsense RNA-mediated decay to increase the expression of protein from the nonsense allele.

Acknowledgements This work was supported by NIH grants R01 EY012200 (AC), EY014237 (AC), and EY015518 (EVS) and an unrestricted grant from Research to Prevent Blindness to the Department of Ophthalmology and Visual Sciences.

References

1. Bassnett S, Shi Y, Vrensen GF (2011) Biological glass: structural determinants of eye lens transparency. Philos Trans R Soc Lond B Biol Sci 366:1250–1264
2. Bassnett S (2009) On the mechanism of organelle degradation in the vertebrate lens. Exp Eye Res 88:133–139
3. Shiels A, Hejtmancik JF (2013) Genetics of human cataract. Clin Genet 84:120–127
4. Francis PJ, Berry V, Moore AT, Bhattacharya S (1999) Lens biology: development and human cataractogenesis. Trends Genet 15:191–196
5. Foster A, Gilbert C, Rahi J (1997) Epidemiology of cataract in childhood: a global perspective. J Cataract Refract Surg 23(Suppl 1):601–604
6. Abrahamsson M, Magnusson G, Sjostrom A, Popovic Z, Sjostrand J (1999) The occurrence of congenital cataract in western Sweden. Acta Ophthalmol Scand 77:578–580
7. Wirth MG, Russell-Eggitt IM, Craig JE, Elder JE, Mackey DA (2002) Aetiology of congenital and paediatric cataract in an Australian population. Br J Ophthalmol 86:782–786
8. Grainger RM (1992) Embryonic lens induction: shedding light on vertebrate tissue determination. Trends Genet 8:349–355

9. Cvekl A, Piatigorsky J (1996) Lens development and crystallin gene expression: many roles for Pax-6. Bioessays 18:621–630
10. Ashery-Padan R, Gruss P (2001) Pax6 lights-up the way for eye development. Curr Opin Cell Biol 13:706–714
11. Lovicu FJ, McAvoy JW (2005) Growth factor regulation of lens development. Dev Biol 280:1–14
12. Donner AL, Lachke SA, Maas RL (2006) Lens induction in vertebrates: variations on a conserved theme of signaling events. Semin Cell Dev Biol 17:676–685
13. Griep AE (2006) Cell cycle regulation in the developing lens. Semin Cell Dev Biol 17:686–697
14. Medina-Martinez O, Jamrich M (2007) Foxe view of lens development and disease. Development 134:1455–1463
15. Cvekl A, Duncan MK (2007) Genetic and epigenetic mechanisms of gene regulation during lens development. Prog Retin Eye Res 26:555–597
16. Cvekl A, Mitton KP (2010) Epigenetic regulatory mechanisms in vertebrate eye development and disease. Heredity (Edinb) 105:135–151
17. Smith AN, Radice G, Lang RA (2010) Which FGF ligands are involved in lens induction? Dev Biol 337:195–198
18. Shaham O, Menuchin Y, Farhy C, Ashery-Padan R (2012) Pax6: a multi-level regulator of ocular development. Prog Retin Eye Res 31:351–376
19. Liu W, Lagutin OV, Mende M, Streit A, Oliver G (2006) Six3 activation of Pax6 expression is essential for mammalian lens induction and specification. EMBO J 25:5383–5395
20. Yamada R, Mizutani-Koseki Y, Hasegawa T, Osumi N, Koseki H et al (2003) Cell-autonomous involvement of Mab21l1 is essential for lens placode development. Development 130:1759–1770
21. Wurm A, Sock E, Fuchshofer R, Wegner M, Tamm ER (2008) Anterior segment dysgenesis in the eyes of mice deficient for the high-mobility-group transcription factor Sox11. Exp Eye Res 86:895–907
22. Saravanamuthu SS, Le TT, Gao CY, Cojocaru RI, Pandiyan P et al (2012) Conditional ablation of the Notch2 receptor in the ocular lens. Dev Biol 362:219–229
23. Rowan S, Conley KW, Le TT, Donner AL, Maas RL et al (2008) Notch signaling regulates growth and differentiation in the mammalian lens. Dev Biol 321:111–122
24. Kuszak JR, Zoltoski RK, Tiedemann CE (2004) Development of lens sutures. Int J Dev Biol 48:889–902
25. Xie Q, Cvekl A (2011) The orchestration of mammalian tissue morphogenesis through a series of coherent feed-forward loops. J Biol Chem 286:43259–43271
26. He S, Pirity MK, Wang WL, Wolf L, Chauhan BK et al (2010) Chromatin remodeling enzyme Brg1 is required for mouse lens fiber cell terminal differentiation and its denucleation. Epigenetics Chromatin 3:21
27. Nishimoto S, Kawane K, Watanabe-Fukunaga R, Fukuyama H, Ohsawa Y et al (2003) Nuclear cataract caused by a lack of DNA degradation in the mouse eye lens. Nature 424:1071–1074
28. Wang WL, Li Q, Xu J, Cvekl A (2010) Lens fiber cell differentiation and denucleation are disrupted through expression of the N-terminal nuclear receptor box of NCOA6 and result in p53-dependent and p53-independent apoptosis. Mol Biol Cell 21:2453–2468
29. Tanaka T, Tsujimura T, Takeda K, Sugihara A, Maekawa A et al (1998) Targeted disruption of ATF4 discloses its essential role in the formation of eye lens fibres. Genes Cells 3:801–810
30. Churchill A, Graw J (2011) Clinical and experimental advances in congenital and paediatric cataracts. Philos Trans R Soc Lond B Biol Sci 366:1234–1249
31. Graw J (2009) Mouse models of cataract. J Genet 88:469–486
32. Graw J (2009) Genetics of crystallins: cataract and beyond. Exp Eye Res 88:173–189
33. Reddy MA, Francis PJ, Berry V, Bhattacharya SS, Moore AT (2004) Molecular genetic basis of inherited cataract and associated phenotypes. Surv Ophthalmol 49:300–315

34. Azuma N, Hirakiyama A, Inoue T, Asaka A, Yamada M (2000) Mutations of a human homologue of the Drosophila eyes absent gene (EYA1) detected in patients with congenital cataracts and ocular anterior segment anomalies. Hum Mol Genet 9:363–366
35. Semina EV, Brownell I, Mintz-Hittner HA, Murray JC, Jamrich M (2001) Mutations in the human forkhead transcription factor FOXE3 associated with anterior segment ocular dysgenesis and cataracts. Hum Mol Genet 10:231–236
36. Bu L, Jin Y, Shi Y, Chu R, Ban A et al (2002) Mutant DNA-binding domain of HSF4 is associated with autosomal dominant lamellar and Marner cataract. Nat Genet 31:276–278
37. Jamieson RV, Perveen R, Kerr B, Carette M, Yardley J et al (2002) Domain disruption and mutation of the bZIP transcription factor, MAF, associated with cataract, ocular anterior segment dysgenesis and coloboma. Hum Mol Genet 11:33–42
38. Ragge NK, Brown AG, Poloschek CM, Lorenz B, Henderson RA et al (2005) Heterozygous mutations of OTX2 cause severe ocular malformations. Am J Hum Genet 76:1008–1022
39. Glaser T, Jepeal L, Edwards JG, Young SR, Favor J et al (1994) PAX6 gene dosage effect in a family with congenital cataracts, aniridia, anophthalmia and central nervous system defects. Nat Genet 7:463–471
40. Semina EV, Ferrell RE, Mintz-Hittner HA, Bitoun P, Alward WL et al (1998) A novel homeobox gene PITX3 is mutated in families with autosomal-dominant cataracts and ASMD. Nat Genet 19:167–170
41. Milunsky JM, Maher TA, Zhao G, Roberts AE, Stalker HJ et al (2008) TFAP2A mutations result in branchio-oculo-facial syndrome. Am J Hum Genet 82:1171–1177
42. Pineda-Alvarez DE, Solomon BD, Roessler E, Balog JZ, Hadley DW et al (2011) A broad range of ophthalmologic anomalies is part of the holoprosencephaly spectrum. Am J Med Genet A 155A:2713–2720
43. Ragge NK, Lorenz B, Schneider A, Bushby K, de Sanctis L et al (2005) SOX2 anophthalmia syndrome. Am J Med Genet A 135:1–7, discussion 8
44. Slavotinek AM (2011) Eye development genes and known syndromes. Mol Genet Metab 104:448–456
45. Hever AM, Williamson KA, van Heyningen V (2006) Developmental malformations of the eye: the role of PAX6, SOX2 and OTX2. Clin Genet 69:459–470
46. Epstein J, Cai J, Glaser T, Jepeal L, Maas R (1994) Identification of a Pax paired domain recognition sequence and evidence for DNA-dependent conformational changes. J Biol Chem 269:8355–8361
47. Epstein JA, Glaser T, Cai J, Jepeal L, Walton DS et al (1994) Two independent and interactive DNA-binding subdomains of the Pax6 paired domain are regulated by alternative splicing. Genes Dev 8:2022–2034
48. Hogan BL, Horsburgh G, Cohen J, Hetherington CM, Fisher G et al (1986) Small eyes (Sey): a homozygous lethal mutation on chromosome 2 which affects the differentiation of both lens and nasal placodes in the mouse. J Embryol Exp Morphol 97:95–110
49. van Heyningen V, Williamson KA (2002) PAX6 in sensory development. Hum Mol Genet 11:1161–1167
50. Hanson I, Van Heyningen V (1995) Pax6: more than meets the eye. Trends Genet 11:268–272
51. Gregory-Evans CY, Wang X, Wasan K, Zhao J, Metcalfe AL, Gregory-Evans K (2014) Postnatal manipulation of Pax6 dosage reverses congenital tissue malformation defects. J Clin Invest 124:111–116
52. Chauhan BK, Yang Y, Cveklova K, Cvekl A (2004) Functional interactions between alternatively spliced forms of Pax6 in crystallin gene regulation and in haploinsufficiency. Nucleic Acids Res 32:1696–1709
53. Chauhan BK, Yang Y, Cveklova K, Cvekl A (2004) Functional properties of natural human PAX6 and PAX6(5a) mutants. Invest Ophthalmol Vis Sci 45:385–392
54. Hingorani M, Hanson I, van Heyningen V (2012) Aniridia. Eur J Hum Genet 20:1011–1017

55. Kleinjan DA, Seawright A, Schedl A, Quinlan RA, Danes S et al (2001) Aniridia-associated translocations, DNase hypersensitivity, sequence comparison and transgenic analysis redefine the functional domain of PAX6. Hum Mol Genet 10:2049–2059
56. Lauderdale JD, Wilensky JS, Oliver ER, Walton DS, Glaser T (2000) 3′ deletions cause aniridia by preventing PAX6 gene expression. Proc Natl Acad Sci U S A 97:13755–13759
57. Schilter K, Reis L, Schneider A, Bardakjian T, Abdul-Rahman O et al (2013) Whole-genome copy number variation analysis in anophthalmia and microphthalmia. Clin Genet 84:473–481
58. Shaham O, Smith AN, Robinson ML, Taketo MM, Lang RA et al (2009) Pax6 is essential for lens fiber cell differentiation. Development 136:2567–2578
59. Huang J, Rajagopal R, Liu Y, Dattilo LK, Shaham O et al (2011) The mechanism of lens placode formation: a case of matrix-mediated morphogenesis. Dev Biol 355:32–42
60. Xie Q, Yang Y, Huang J, Ninkovic J, Walcher T et al (2013) Pax6 interactions with chromatin and identification of its novel direct target genes in lens and forebrain. PLoS One 8:e54507
61. Wolf L, Harrison W, Huang J, Xie Q, Xiao N et al (2013) Histone posttranslational modifications and cell fate determination: lens induction requires the lysine acetyltransferases CBP and p300. Nucleic Acids Res 41:10199–10214
62. Shaham O, Gueta K, Mor E, Oren-Giladi P, Grinberg D et al (2013) Pax6 regulates gene expression in the vertebrate lens through miR-204. PLoS Genet 9:e1003357
63. Wolf LV, Yang Y, Wang J, Xie Q, Braunger B et al (2009) Identification of pax6-dependent gene regulatory networks in the mouse lens. PLoS One 4:e4159
64. Simeone A, Acampora D, Mallamaci A, Stornaiuolo A, D'Apice MR et al (1993) A vertebrate gene related to orthodenticle contains a homeodomain of the bicoid class and demarcates anterior neuroectoderm in the gastrulating mouse embryo. EMBO J 12:2735–2747
65. Matsuo I, Kuratani S, Kimura C, Takeda N, Aizawa S (1995) Mouse Otx2 functions in the formation and patterning of rostral head. Genes Dev 9:2646–2658
66. Montalta-He H, Leemans R, Loop T, Strahm M, Certa U et al (2002) Evolutionary conservation of otd/Otx2 transcription factor action: a genome-wide microarray analysis in Drosophila. Genome Biol 3:RESEARCH0015
67. Beby F, Lamonerie T (2013) The homeobox gene Otx2 in development and disease. Exp Eye Res 111:9–16
68. Schilter KF, Schneider A, Bardakjian T, Soucy JF, Tyler RC et al (2011) OTX2 microphthalmia syndrome: four novel mutations and delineation of a phenotype. Clin Genet 79:158–168
69. Scaffidi P, Bianchi ME (2001) Spatially precise DNA bending is an essential activity of the sox2 transcription factor. J Biol Chem 276:47296–47302
70. Kelberman D, Rizzoti K, Avilion A, Bitner-Glindzicz M, Cianfarani S et al (2006) Mutations within Sox2/SOX2 are associated with abnormalities in the hypothalamo-pituitary-gonadal axis in mice and humans. J Clin Invest 116:2442–2455
71. Kiernan AE, Pelling AL, Leung KK, Tang AS, Bell DM et al (2005) Sox2 is required for sensory organ development in the mammalian inner ear. Nature 434:1031–1035
72. Taranova OV, Magness ST, Fagan BM, Wu Y, Surzenko N et al (2006) SOX2 is a dose-dependent regulator of retinal neural progenitor competence. Genes Dev 20:1187–1202
73. Schneider A, Bardakjian T, Reis LM, Tyler RC, Semina EV (2009) Novel SOX2 mutations and genotype-phenotype correlation in anophthalmia and microphthalmia. Am J Med Genet A 149A:2706–2715
74. Wang P, Liang X, Yi J, Zhang Q (2008) Novel SOX2 mutation associated with ocular coloboma in a Chinese family. Arch Ophthalmol 126:709–713
75. Danno H, Michiue T, Hitachi K, Yukita A, Ishiura S et al (2008) Molecular links among the causative genes for ocular malformation: Otx2 and Sox2 coregulate Rax expression. Proc Natl Acad Sci U S A 105:5408–5413

76. Kamachi Y, Uchikawa M, Tanouchi A, Sekido R, Kondoh H (2001) Pax6 and SOX2 form a co-DNA-binding partner complex that regulates initiation of lens development. Genes Dev 15:1272–1286
77. Loosli F, Winkler S, Wittbrodt J (1999) Six3 overexpression initiates the formation of ectopic retina. Genes Dev 13:649–654
78. Oliver G, Loosli F, Koster R, Wittbrodt J, Gruss P (1996) Ectopic lens induction in fish in response to the murine homeobox gene Six3. Mech Dev 60:233–239
79. Lagutin O, Zhu CC, Furuta Y, Rowitch DH, McMahon AP et al (2001) Six3 promotes the formation of ectopic optic vesicle-like structures in mouse embryos. Dev Dyn 221:342–349
80. Lagutin OV, Zhu CC, Kobayashi D, Topczewski J, Shimamura K et al (2003) Six3 repression of Wnt signaling in the anterior neuroectoderm is essential for vertebrate forebrain development. Genes Dev 17:368–379
81. Wallis DE, Roessler E, Hehr U, Nanni L, Wiltshire T et al (1999) Mutations in the homeodomain of the human SIX3 gene cause holoprosencephaly. Nat Genet 22:196–198
82. Li X, Oghi KA, Zhang J, Krones A, Bush KT et al (2003) Eya protein phosphatase activity regulates Six1-Dach-Eya transcriptional effects in mammalian organogenesis. Nature 426:247–254
83. Rayapureddi JP, Kattamuri C, Steinmetz BD, Frankfort BJ, Ostrin EJ et al (2003) Eyes absent represents a class of protein tyrosine phosphatases. Nature 426:295–298
84. Xu PX, Adams J, Peters H, Brown MC, Heaney S et al (1999) Eya1-deficient mice lack ears and kidneys and show abnormal apoptosis of organ primordia. Nat Genet 23:113–117
85. Orten DJ, Fischer SM, Sorensen JL, Radhakrishna U, Cremers CW et al (2008) Branchio-oto-renal syndrome (BOR): novel mutations in the EYA1 gene, and a review of the mutational genetics of BOR. Hum Mutat 29:537–544
86. Vincent C, Kalatzis V, Abdelhak S, Chaib H, Compain S et al (1997) BOR and BO syndromes are allelic defects of EYA1. Eur J Hum Genet 5:242–246
87. Reis LM, Tyler RC, Muheisen S, Raggio V, Salviati L et al (2013) Whole exome sequencing in dominant cataract identifies a new causative factor, CRYBA2, and a variety of novel alleles in known genes. Hum Genet 132:761–770
88. Rayapureddi JP, Hegde RS (2006) Branchio-oto-renal syndrome associated mutations in eyes absent 1 result in loss of phosphatase activity. FEBS Lett 580:3853–3859
89. Zhang J, Hagopian-Donaldson S, Serbedzija G, Elsemore J, Plehn-Dujowich D et al (1996) Neural tube, skeletal and body wall defects in mice lacking transcription factor AP-2. Nature 381:238–241
90. Bassett EA, Williams T, Zacharias AL, Gage PJ, Fuhrmann S et al (2010) AP-2alpha knockout mice exhibit optic cup patterning defects and failure of optic stalk morphogenesis. Hum Mol Genet 19:1791–1804
91. West-Mays JA, Zhang J, Nottoli T, Hagopian-Donaldson S, Libby D et al (1999) AP-2alpha transcription factor is required for early morphogenesis of the lens vesicle. Dev Biol 206:46–62
92. Gestri G, Osborne RJ, Wyatt AW, Gerrelli D, Gribble S et al (2009) Reduced TFAP2A function causes variable optic fissure closure and retinal defects and sensitizes eye development to mutations in other morphogenetic regulators. Hum Genet 126:791–803
93. Al-Dosari MS, Almazyad M, Al-Ebdi L, Mohamed JY, Al-Dahmash S et al (2010) Ocular manifestations of branchio-oculo-facial syndrome: report of a novel mutation and review of the literature. Mol Vis 16:813–818
94. Dumitrescu AV, Milunsky JM, Longmuir SQ, Drack AV (2012) A family with branchio-oculo-facial syndrome with primarily ocular involvement associated with mutation of the TFAP2A gene. Ophthalmic Genet 33:100–106
95. Blixt A, Mahlapuu M, Aitola M, Pelto-Huikko M, Enerback S et al (2000) A forkhead gene, FoxE3, is essential for lens epithelial proliferation and closure of the lens vesicle. Genes Dev 14:245–254

96. Brownell I, Dirksen M, Jamrich M (2000) Forkhead Foxe3 maps to the dysgenetic lens locus and is critical in lens development and differentiation. Genesis 27:81–93
97. Ormestad M, Blixt A, Churchill A, Martinsson T, Enerback S et al (2002) Foxe3 haploinsufficiency in mice: a model for Peters' anomaly. Invest Ophthalmol Vis Sci 43:1350–1357
98. Wada K, Maeda YY, Watanabe K, Oshio T, Ueda T et al (2011) A deletion in a cis element of Foxe3 causes cataracts and microphthalmia in rct mice. Mamm Genome 22:693–702
99. Vihtelic TS (2008) Teleost lens development and degeneration. Int Rev Cell Mol Biol 269:341–373
100. Iseri SU, Osborne RJ, Farrall M, Wyatt AW, Mirza G et al (2009) Seeing clearly: the dominant and recessive nature of FOXE3 in eye developmental anomalies. Hum Mutat 30:1378–1386
101. Reis LM, Tyler RC, Schneider A, Bardakjian T, Stoler JM et al (2010) FOXE3 plays a significant role in autosomal recessive microphthalmia. Am J Med Genet A 152A:582–590
102. Valleix S, Niel F, Nedelec B, Algros MP, Schwartz C et al (2006) Homozygous nonsense mutation in the FOXE3 gene as a cause of congenital primary aphakia in humans. Am J Hum Genet 79:358–364
103. Ali M, Buentello-Volante B, McKibbin M, Rocha-Medina JA, Fernandez-Fuentes N et al (2010) Homozygous FOXE3 mutations cause non-syndromic, bilateral, total sclerocornea, aphakia, microphthalmia and optic disc coloboma. Mol Vis 16:1162–1168
104. Pontoriero GF, Deschamps P, Ashery-Padan R, Wong R, Yang Y et al (2008) Cell autonomous roles for AP-2alpha in lens vesicle separation and maintenance of the lens epithelial cell phenotype. Dev Dyn 237:602–617
105. Mathers PH, Grinberg A, Mahon KA, Jamrich M (1997) The Rx homeobox gene is essential for vertebrate eye development. Nature 387:603–607
106. Amendt BA, Sutherland LB, Semina EV, Russo AF (1998) The molecular basis of Rieger syndrome. Analysis of Pitx2 homeodomain protein activities. J Biol Chem 273:20066–20072
107. Semina EV, Reiter R, Leysens NJ, Alward WL, Small KW et al (1996) Cloning and characterization of a novel bicoid-related homeobox transcription factor gene, RIEG, involved in Rieger syndrome. Nat Genet 14:392–399
108. Sakazume S, Sorokina E, Iwamoto Y, Semina EV (2007) Functional analysis of human mutations in homeodomain transcription factor PITX3. BMC Mol Biol 8:84
109. Semina EV, Reiter RS, Murray JC (1997) Isolation of a new homeobox gene belonging to the Pitx/Rieg family: expression during lens development and mapping to the aphakia region on mouse chromosome 19. Hum Mol Genet 6:2109–2116
110. Rieger DK, Reichenberger E, McLean W, Sidow A, Olsen BR (2001) A double-deletion mutation in the Pitx3 gene causes arrested lens development in aphakia mice. Genomics 72:61–72
111. Semina EV, Murray JC, Reiter R, Hrstka RF, Graw J (2000) Deletion in the promoter region and altered expression of Pitx3 homeobox gene in aphakia mice. Hum Mol Genet 9:1575–1585
112. Varnum DS, Stevens LC (1968) Aphakia, a new mutation in the mouse. J Hered 59:147–150
113. Zhao S, Maxwell S, Jimenez-Beristain A, Vives J, Kuehner E et al (2004) Generation of embryonic stem cells and transgenic mice expressing green fluorescence protein in midbrain dopaminergic neurons. Eur J Neurosci 19:1133–1140
114. Berry V, Yang Z, Addison PK, Francis PJ, Ionides A et al (2004) Recurrent 17 bp duplication in PITX3 is primarily associated with posterior polar cataract (CPP4). J Med Genet 41:e109
115. Addison PK, Berry V, Ionides AC, Francis PJ, Bhattacharya SS et al (2005) Posterior polar cataract is the predominant consequence of a recurrent mutation in the PITX3 gene. Br J Ophthalmol 89:138–141
116. Finzi S, Li Y, Mitchell TN, Farr A, Maumenee IH et al (2005) Posterior polar cataract: genetic analysis of a large family. Ophthalmic Genet 26:125–130
117. Burdon KP, McKay JD, Wirth MG, Russell-Eggit IM, Bhatti S et al (2006) The PITX3 gene in posterior polar congenital cataract in Australia. Mol Vis 12:367–371

118. Summers KM, Withers SJ, Gole GA, Piras S, Taylor PJ (2008) Anterior segment mesenchymal dysgenesis in a large Australian family is associated with the recurrent 17 bp duplication in PITX3. Mol Vis 14:2010–2015
119. Bidinost C, Matsumoto M, Chung D, Salem N, Zhang K et al (2006) Heterozygous and homozygous mutations in PITX3 in a large Lebanese family with posterior polar cataracts and neurodevelopmental abnormalities. Invest Ophthalmol Vis Sci 47:1274–1280
120. Aldahmesh MA, Khan AO, Mohamed J, Alkuraya FS (2011) Novel recessive BFSP2 and PITX3 mutations: insights into mutational mechanisms from consanguineous populations. Genet Med 13:978–981
121. Chauhan BK, Reed NA, Zhang W, Duncan MK, Kilimann MW et al (2002) Identification of genes downstream of Pax6 in the mouse lens using cDNA microarrays. J Biol Chem 277:11539–11548
122. Wolf L, Harrison W, Huang J, Xie Q, Xiao N, Sun J, Kong L, Lachke SA, Kuracha MR, Govindarajan V, Brindle PK, Ashery-Padan R, Beebe DC, Overbeek PA, Cvekl A (2013) HIstone posttranslational modifications and cell fate determination: lens induction requires the lysine acetyltransferases CBP and p300. Nucleic Acids Res 41
123. Hooker L, Smoczer C, KhosrowShahian F, Wolanski M, Crawford MJ (2012) Microarray-based identification of Pitx3 targets during Xenopus embryogenesis. Dev Dyn 241:1487–1505
124. Shi X, Luo Y, Howley S, Dzialo A, Foley S et al (2006) Zebrafish foxe3: roles in ocular lens morphogenesis through interaction with pitx3. Mech Dev 123:761–782
125. Medina-Martinez O, Shah R, Jamrich M (2009) Pitx3 controls multiple aspects of lens development. Dev Dyn 238:2193–2201
126. Ho HY, Chang KH, Nichols J, Li M (2009) Homeodomain protein Pitx3 maintains the mitotic activity of lens epithelial cells. Mech Dev 126:18–29
127. Ahmad N, Aslam M, Muenster D, Horsch M, Khan MA, Carlsson P, BeckersJ GJ (2013) Pitx3 directly regulates Foxe3 during early lens development. Int J Dev Biol 57:741–751
128. Sorokina EA, Muheisen S, Mlodik N, Semina EV (2011) MIP/Aquaporin 0 represents a direct transcriptional target of PITX3 in the developing lens. PLoS One 6:e21122
129. Ring BZ, Cordes SP, Overbeek PA, Barsh GS (2000) Regulation of mouse lens fiber cell development and differentiation by the Maf gene. Development 127:307–317
130. Fujimoto M, Izu H, Seki K, Fukuda K, Nishida T et al (2004) HSF4 is required for normal cell growth and differentiation during mouse lens development. EMBO J 23:4297–4306
131. Fujimoto M, Oshima K, Shinkawa T, Wang BB, Inouye S et al (2008) Analysis of HSF4 binding regions reveals its necessity for gene regulation during development and heat shock response in mouse lenses. J Biol Chem 283:29961–29970
132. Cui X, Wang L, Zhang J, Du R, Liao S et al (2013) HSF4 regulates DLAD expression and promotes lens de-nucleation. Biochim Biophys Acta 1832:1167–1172
133. Smaoui N, Beltaief O, BenHamed S, M'Rad R, Maazoul F et al (2004) A homozygous splice mutation in the HSF4 gene is associated with an autosomal recessive congenital cataract. Invest Ophthalmol Vis Sci 45:2716–2721
134. Yang Y, Cvekl A (2007) Large Maf transcription factors: cousins of AP-1 proteins and important regulators of cellular differentiation. Einstein J Biol Med 23:2–11
135. Takeuchi T, Kudo T, Ogata K, Hamada M, Nakamura M et al (2009) Neither MafA/L-Maf nor MafB is essential for lens development in mice. Genes Cells 14:941–947
136. Vinson C, Myakishev M, Acharya A, Mir AA, Moll JR et al (2002) Classification of human B-ZIP proteins based on dimerization properties. Mol Cell Biol 22:6321–6335
137. Jamieson RV, Munier F, Balmer A, Farrar N, Perveen R et al (2003) Pulverulent cataract with variably associated microcornea and iris coloboma in a MAF mutation family. Br J Ophthalmol 87:411–412
138. Lyon MF, Jamieson RV, Perveen R, Glenister PH, Griffiths R et al (2003) A dominant mutation within the DNA-binding domain of the bZIP transcription factor Maf causes murine cataract and results in selective alteration in DNA binding. Hum Mol Genet 12:585–594

139. Perveen R, Favor J, Jamieson RV, Ray DW, Black GC (2007) A heterozygous c-Maf transactivation domain mutation causes congenital cataract and enhances target gene activation. Hum Mol Genet 16:1030–1038
140. Hansen L, Eiberg H, Rosenberg T (2007) Novel MAF mutation in a family with congenital cataract-microcornea syndrome. Mol Vis 13:2019–2022
141. Vanita V, Singh D, Robinson PN, Sperling K, Singh JR (2006) A novel mutation in the DNA-binding domain of MAF at 16q23.1 associated with autosomal dominant "cerulean cataract" in an Indian family. Am J Med Genet A 140:558–566
142. Yang Y, Wolf LV, Cvekl A (2007) Distinct embryonic expression and localization of CBP and p300 histone acetyltransferases at the mouse alphaA-crystallin locus in lens. J Mol Biol 369:917–926
143. van Genderen MM, Kinds GF, Riemslag FC, Hennekam RC (2000) Ocular features in Rubinstein-Taybi syndrome: investigation of 24 patients and review of the literature. Br J Ophthalmol 84:1177–1184
144. Lachke SA, Ho JW, Kryukov GV, O'Connell DJ, Aboukhalil A et al (2012) iSyTE: integrated systems tool for eye gene discovery. Invest Ophthalmol Vis Sci 53:1617–1627
145. Barolo S, Posakony JW (2002) Three habits of highly effective signaling pathways: principles of transcriptional control by developmental cell signaling. Genes Dev 16:1167–1181
146. Boyer LA, Lee TI, Cole MF, Johnstone SE, Levine SS et al (2005) Core transcriptional regulatory circuitry in human embryonic stem cells. Cell 122:947–956
147. Lee TI, Jenner RG, Boyer LA, Guenther MG, Levine SS et al (2006) Control of developmental regulators by Polycomb in human embryonic stem cells. Cell 125:301–313
148. Yun S, Saijoh Y, Hirokawa KE, Kopinke D, Murtaugh LC et al (2009) Lhx2 links the intrinsic and extrinsic factors that control optic cup formation. Development 136:3895–3906
149. Lleras-Forero L, Tambalo M, Christophorou N, Chambers D, Houart C et al (2013) Neuropeptides: developmental signals in placode progenitor formation. Dev Cell 26:195–203
150. Chen Q, Dowhan DH, Liang D, Moore DD, Overbeek PA (2002) CREB-binding protein/p300 co-activation of crystallin gene expression. J Biol Chem 277:24081–24089
151. Parks MM, Johnson DA, Reed GW (1993) Long-term visual results and complications in children with aphakia. A function of cataract type. Ophthalmology 100:826–840, discussion 840-821
152. Hosal BM, Biglan AW (2002) Risk factors for secondary membrane formation after removal of pediatric cataract. J Cataract Refract Surg 28:302–309
153. Zetterstrom C, Lundvall A, Kugelberg M (2005) Cataracts in children. J Cataract Refract Surg 31:824–840
154. Lambert SR, Buckley EG, Drews-Botsch C, DuBois L, Hartmann E et al (2010) The infant aphakia treatment study: design and clinical measures at enrollment. Arch Ophthalmol 128:21–27
155. Lambert SR, Buckley EG, Drews-Botsch C, DuBois L, Hartmann EE et al (2010) A randomized clinical trial comparing contact lens with intraocular lens correction of monocular aphakia during infancy: grating acuity and adverse events at age 1 year. Arch Ophthalmol 128:810–818
156. Ruddle JB, Staffieri SE, Crowston JG, Sherwin JC, Mackey DA (2013) Incidence and predictors of glaucoma following surgery for congenital cataract in the first year of life in Victoria, Australia. Clin Experiment Ophthalmol 41:653–661
157. Beck AD, Freedman SF, Lynn MJ, Bothun E, Neely DE et al (2012) Glaucoma-related adverse events in the infant aphakia treatment study: 1-year results. Arch Ophthalmol 130:300–305
158. Sharma N, Pushker N, Dada T, Vajpayee RB, Dada VK (1999) Complications of pediatric cataract surgery and intraocular lens implantation. J Cataract Refract Surg 25:1585–1588

159. Birch EE, Cheng C, Stager DR Jr, Felius J (2005) Visual acuity development after the implantation of unilateral intraocular lenses in infants and young children. J AAPOS 9:527–532
160. Peltz SW, Morsy M, Welch EM, Jacobson A (2013) Ataluren as an agent for therapeutic nonsense suppression. Annu Rev Med 64:407–425
161. Keeling KM, Bedwell DM (2011) Suppression of nonsense mutations as a therapeutic approach to treat genetic diseases. Wiley Interdiscip Rev RNA 2:837–852
162. Yang C, Yang Y, Brennan L, Bouhassira EE, Kantorow M et al (2010) Efficient generation of lens progenitor cells and lentoid bodies from human embryonic stem cells in chemically defined conditions. FASEB J 24:3274–3283
163. Qiu X, Yang J, Liu T, Jiang Y, Le Q et al (2012) Efficient generation of lens progenitor cells from cataract patient-specific induced pluripotent stem cells. PLoS One 7:e32612
164. Kubo E, Hasanova N, Sasaki H, Singh DP (2013) Dynamic and differential regulation in the microRNA expression in the developing and mature cataractous rat lens. J Cell Mol Med 17:1146–1159
165. Wang Y, Li W, Zang X, Chen N, Liu T et al (2013) MicroRNA-204-5p regulates epithelial-to-mesenchymal transition during human posterior capsule opacification by targeting SMAD4. Invest Ophthalmol Vis Sci 54:323–332

Lens Regeneration

7

Konstantinos Sousounis, Kenta Nakamura,
and Panagiotis A. Tsonis

Abstract

Lens regeneration after complete lens removal (lentectomy) has mainly been studied in amphibians. Adult newts or axolotl larvae regenerate the lens from the iris while frog tadpoles from the cornea. Transdifferentiation is the term used to describe the cellular and molecular events that underlie lens regeneration, which involves a terminally differentiated tissue (iris) to change and give rise to a different differentiated tissue (lens). This chapter starts with general comparisons among the three amphibians competent for lens regeneration focusing on how these animals have been used as models to study the process. We focus on newt lens regeneration as it can provide a more broad understanding of the process. We concentrate on current knowledge about the mechanism of lens regeneration by presenting the classical histological stages together with molecular signatures that accompany each stage. We address questions regarding the nature of the contributing cells and their relationship to stem cells. We conclude by presenting high-throughput data including the first newt transcriptome and its application during newt lens regeneration. All this information obtained from lower vertebrates is essential for translation research in higher vertebrates where lens regeneration does not occur.

Keywords

Amphibians • Lens regeneration • Transdifferentiation

K. Sousounis • K. Nakamura • P.A. Tsonis (✉)
Department of Biology and Center for Tissue Regeneration and Engineering,
University of Dayton, Dayton, OH 45469-2320, USA
e-mail: sousounisk1@udayton.edu; knakamura0804@gmail.com; ptsonis1@udayton.edu

S. Saika et al. (eds.), *Lens Epithelium and Posterior Capsular Opacification*,
DOI 10.1007/978-4-431-54300-8_7, © Springer Japan 2014

7.1 Introduction to Lens Regeneration

Regeneration is the process where a lost tissue or organ is remade. Partial damage to an organ, like removing the differentiated fibers from the lens, is usually repaired by stem/progenitor cells, cells that reside in the area near the damage [1]. In the case of the lens, repair is conducted by epithelial cells which retain the ability to proliferate and differentiate to lens fibers throughout their life. Lens repair by lens epithelial cells, after removing the lens fibers, is a common model studied in higher vertebrates and is discussed in Chap. 9. In this chapter we will focus on the research that has been conducted in amphibians where the lens is fully removed, a surgical procedure called lentectomy. This procedure involves opening of the cornea using a scalpel and removing the whole lens with forceps. It is critical to note that the lens is fully removed including the lens capsule without rupture, so no trace of lens tissue is left behind which might contribute to lens regeneration as happens in some higher vertebrates.

7.2 Models for Lens Regeneration

The most common animal models for lens regeneration after lentectomy are the frog *Xenopus laevis*, the newts *Notophthalmus viridescens* and *Cynops pyrrhogaster*, and more recently the axolotl *Ambystoma mexicanum*. Gene gain and loss of function experiments, including transgenesis and use of morpholinos, signaling manipulation through drug treatments, and transplantation experiments, are mostly used in these animal models to elucidate the mechanism of lens regeneration [2, 3].

Frogs have a small window, when they are tadpoles, where they can regenerate the lens. They lose this ability later in development. After lentectomy, the lens is regenerated from the cornea [4], a process that, with some differences in crystallins, is recapitulating normal lens development [5–8]. (Note that during development, the lens is induced by the surface ectoderm which later forms the cornea.) Frogs have been used extensively for studies of organism development and organogenesis, so there is a lot of genomic information as well as molecular and genetic tools available, making this organism more accessible in research.

Newts have the ability to regenerate the lens throughout life. In contrast to frogs, lens regeneration in newts is conducted by the iris, specifically the dorsal side of the iris and never from the ventral [9]. (Note that developmentally, the iris is part of the neural ectoderm.) This discrepancy has been utilized in induction experiments. Signaling pathways and other genes including transcriptional factors are tested for their role in lens regeneration by manipulating them in the ventral iris, in an attempt to induce lens regeneration from this noncompetent-for-regeneration side of the iris [10, 11]. This is a gold standard for proving whether a gene is able to promote lens regeneration. The newt regenerative ability of the lens from the iris does not seem to be affected by the development or metamorphosis from larval to adult or during aging [12, 13]. This makes them suitable to be studied in the stage that is more convenient and suitable for the researcher. For example, a gain of function experiment of a certain gene can be studied by transgenesis, while the effects on lens regeneration can be seen during the larval stages [2]. This provides an advantage since the life cycle of *N. viridescens* and *C. pyrrhogaster*, the most studied newts, exceeds 3 years. Furthermore, recent reports add valuable genomic information for newts that was not available previously, making common molecular techniques easier to apply [14]. In addition, since lens regeneration is not affected by aging, newts are suitable for studies on the relationship between aging, regeneration, and tissue homeostasis [12].

A recent study shows that axolotls have also a small window of 2 weeks post hatching where they can regenerate their lens. The lens seems to be regenerated from the iris as shown by BrdU incorporation of dividing cells [15].

In the remainder of this chapter, we will focus on the mechanism of lens regeneration in newts. We will explore the stages of lens regeneration and characterize the cells that are responsible for it and the signaling and transcriptional factors that drive the process. Similarities and differences between newts and frogs can be found in a recent review [16]; however, additional comments will be made when appropriate.

7.3 Classical Morphological Stages of Lens Regeneration

Following lentectomy in newts, 13 morphological stages (I–XIII) of lens regeneration have been determined by Sato [17] as shown in cross sections of the eyeball (Fig. 7.1). Four days post-lentectomy, the iris thickens (I), and 1 day later the dorsal iris thickens further, creating a space between the inner and outer lamina (II). Six days post-lentectomy, the first depigmented cells appear in the dorsal iris (III) that 1 day later will form a hollow vesicle (IV). The cells are still cuboidal in shape. Columnar cells appear from the inner lamina 9 days post-lentectomy (V), which will differentiate into lens fibers (VI). These primary lens fibers will change the morphology and topology of their nucleus (VII) and will fill the hollow vesicle 11 days post-lentectomy (VIII). Secondary fibers are starting to develop from the lens equator 16 days post-lentectomy (IX), approximately when a lens suture appears (X). Eighteen days post-lentectomy, the regenerated lens is round in shape, with the secondary lens fibers surrounding symmetrically the primary lens fibers (XI). Twenty days post-lentectomy, the primary lens fibers are fully differentiated with loss of their nuclei (XII) and the lens appears fully developed 10 days later when cell nuclei are present only in anterior epithelia and in the newly formed secondary fibers at the lens equator (XIII). The lens may take up to 3 months to reach the size of the original lens in the adults. Note that between newt species, the timing of the events might be slightly different. These stages are marked by expression of genes that will be mentioned in the next section.

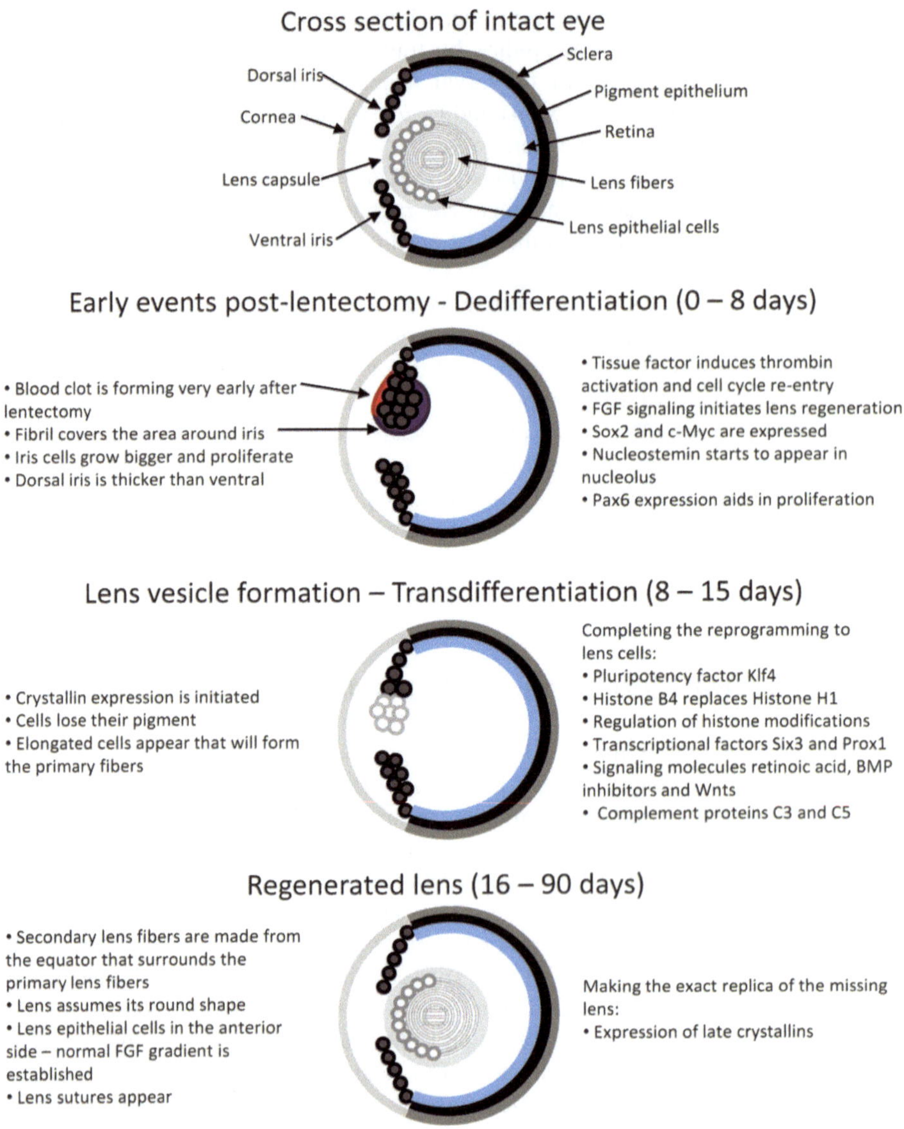

Fig. 7.1 Morphological and molecular stages of newt lens regeneration. Images are schematically representing cross sections of a newt eye in three time periods, the dedifferentiation, the transdifferentiation, and the lens regeneration. *Arrows* indicate the different eye structures. The morphological and molecular changes are presented on the left and right of each cross section, respectively

7.4 Stages of Lens Regeneration Marked by Molecular Events

7.4.1 Early Events

One of the first observations following lens removal is the formation of a blood clot, which is more prominent in the dorsal iris. This event has been attributed to the expression of tissue factor in the dorsal iris, a protein that also mediates thrombin activation. The sequential deposition of fibril allows macrophages and growth factors, like FGF and Wnt, to start the transdifferentiation process in the dorsal margin. Thrombin activation has also been shown to be responsible for cell cycle reentry [18, 19]. FGFs and FGFRs are expressed throughout the process of lens regeneration mediating several stages. FGF signaling acts in the initial stages of lens regeneration since a single FGF2 treatment can initiate the process, even without lentectomy. Experiments where the retina was removed along with the lens resulted in a reduced rate of lens regeneration, which indicates that this is the source of FGF as in normal development, though macrophages cannot be excluded as an additional source of FGF [20–24]. Furthermore, 1 day post-lentectomy, the expression of matrix metalloproteases is highly upregulated, an event that modifies the extracellular matrix, thus changing cell–cell and cell–environment interactions that are essential for transdifferentiation (see Fig. 7.1) [25].

7.4.2 Dedifferentiation

Nucleostemin, a protein associated with stem cells and proliferation, and the pluripotency factors Sox2, c-Myc, and Klf4 are starting to be expressed 2 days post-lentectomy. These proteins are important for the dedifferentiation process since gene expression needs to change from that of a terminally differentiated pigmented iris cell type [26, 27]. In addition, histone modifications are dynamically changing during dedifferentiation in favor of activating marks that can aid the expression of important genes [28]. These events are also accommodated with increased nucleus size, with increasing RNA content starting around 5 days post-lentectomy (see Fig. 7.1) [29, 30].

7.4.3 Transdifferentiation

During the time of transdifferentiation, when the iris cells are not pigmented and are more lenslike, lens-related genes including Six3 and Pax6, with known roles in ectopic lens formation during embryogenesis, are found to be expressed and are involved in the process. Pax6, the master eye gene, has been found to be expressed during lens regeneration in the iris, starting 5 days post-lentectomy and later in the regenerating lens. The role of Pax6 was attributed to proliferation and the expression of crystallin genes [31, 32]. Six3, along with retinoic acid (RA), plays a role in the transdifferentiation. The ventral iris was induced to regenerate a lens, an effect that was related to Six3 overexpression and concomitant treatment with retinoic acid. Similar results were obtained with BMP inhibitors that are known determinants of the dorsal–ventral axis during embryogenesis [10, 31, 32]. The role of retinoic acid in the transdifferentiation process has also been determined by the expression of retinoic acid receptor alpha and delta in the regenerating lens. Blocking this signal led to abnormal regeneration, including reduced ability for lens formation and fiber differentiation [33, 34]. Moreover, another protein with known roles in fiber differentiation during lens development, Prox1, is expressed in the dorsal iris throughout regeneration [35]. Coinciding with the transdifferentiation process, around 8–10 days post-lentectomy, histone B4 is highly expressed and replaces histone H1 in the nucleus. This event is also found during reprogramming of oocytes, and as such, it is important for the transition of gene expression during transdifferentiation [36]. At the same time, Wnt-related genes, Wnt2b and Frizzled4, are expressed in the dorsal iris. This expression has been correlated with the ability of the dorsal iris to transdifferentiate to lens, as shown in vitro using ventral iris aggregates treated with Wnts; however, it might be correlated to the lenslike cell type since normal lens epithelial cells express Wnts [37, 38]. Ten days post-lentectomy, expression of C3 and C5 complement proteins is found in the iris and lens vesicle, respectively [39]. At later stages when the lens vesicle has formed and fibers start to differentiate, expression of alpha, beta, and gamma crystallins appears [40]. Crystallin expression has been used as a marker to detect whether lens regeneration was successful. Lens regeneration is complete when a fully differentiated lens is made, an exact replica of the original (Fig. 7.1).

7.5 Cell Source of Lens Regeneration

An important aspect of regeneration is to identify the source of it. This task is accomplished by tracking experiments using transgenesis and BrdU incorporation, in combination with immunohistochemistry, transplantation, or in vitro cultures. In newts, histological examination and transplantation of different cell types from the eye in newts without a lens showed very early that the pigmented epithelium was responsible for lens regeneration. Interestingly, the entire dorsal iris-pigmented epithelium can transdifferentiate to lens cells in vitro. The same happens for the ventral iris-pigmented epithelium in prolonged in vitro cultures. This indicates that all pigmented iris cells have the ability for transdifferentiation. Gene and signaling regulation must restrict this process only in the dorsal margin. BrdU and ^{3}H-thymidine incorporation reveal that cells from all around the iris reenter the cell cycle early after lentectomy (see Fig. 7.2a) [42]. However, only the dorsal iris shows positive labeling for phosphoH3, a marker that indicates dividing cells (Fig. 7.2b). This shows that cells are responding to a signal, reenter the cell cycle, even in the ventral iris, but progressively dividing cells are restricted only in the dorsal tip where regeneration will occur (Fig. 7.2c) [41].

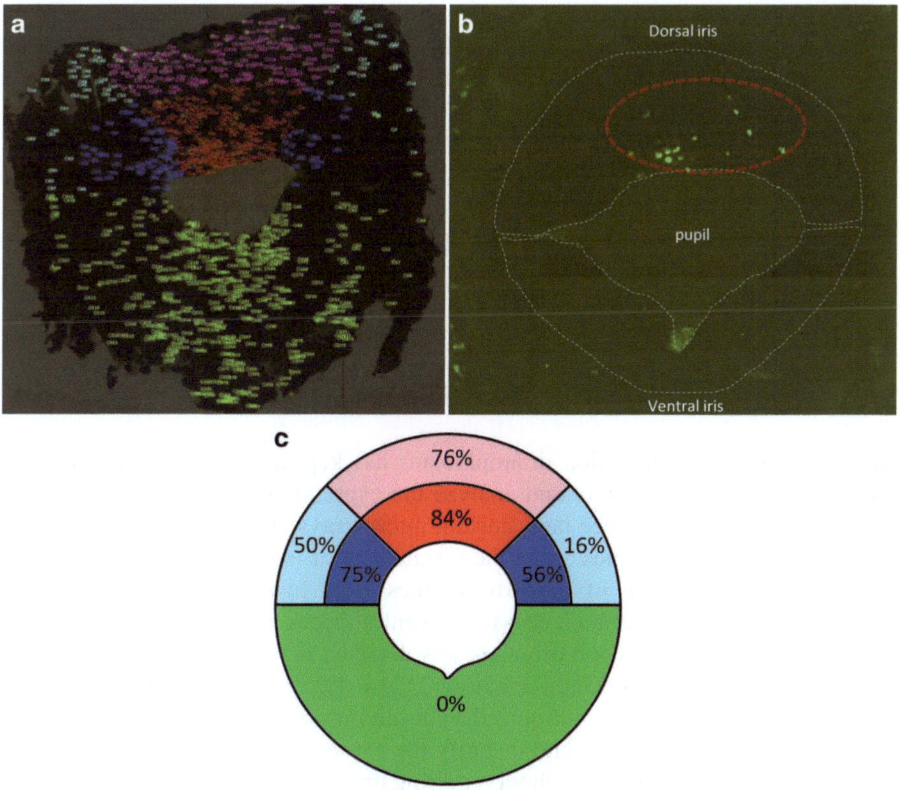

Fig. 7.2 Reentry to the cell cycle and ability of the iris to regenerate a lens. (**a**) Whole-mount BrdU-incorporation detection in the iris cup 8 days post-lentectomy. Many cells from the whole iris are reentering the cell cycle. *Colors* are used to indicate the different sections of the iris. (**b**) PhosphoH3 staining in the iris cup 8 days post-lentectomy. Few cells concentrated in the dorsal iris margin are positive for phosphoH3, which indicates dividing cells. *White-dotted lines* mark dorsal and ventral iris margins. *Red-dotted oval circles* indicate the dorsal iris margin. (**c**) An iris cup divided in colored sections and the potential (percentage) for lens regeneration is shown therein (modified from [41]). Note that while all areas from the dorsal iris have the potential to regenerate, the highest is at the tip of the dorsal iris. *Colors* in (**a**) and (**c**) are indicating the same areas

7.6 Cell Potency During Lens Regeneration

Other questions to address are: What is the potency of these cells? Do they have the ability to differentiate to cell types other than lens? How many times can a lens be regenerated from the same newt? Iris cells have some characteristics of stem cells: they express some of the pluripotency factors, and they remodel their extracellular matrix to a more cancer stem cell niche-like environment that aids cell migration, proliferation, dedifferentiation, and finally the transdifferentiation to lens cells. From this aspect, these cells should become more potent. However, aggregation and transplantation of these cells to other parts of the body, for example, in a limb, results in them regenerating a lens [43]. This indicates that these cells might only be able to transdifferentiate to lens cells, being unipotent. This also shows that the cells are strictly reprogrammed to serve lens regeneration in a very straightforward way. In addition, a 16-year experiment where lens was removed 18 times showed that newts do not decline in their ability to regenerate a normal lens comparable to young ones. This ability might be retained due to expression of DNA repair genes like Rad1 and ROS-related genes such as glutathione peroxidase 1 that can alleviate mutations and stress from the cells, leading to their prolonged ability for regeneration [12, 25].

7.7 Omics Data During Lens Regeneration

Research on lens regeneration in amphibians has kept up with new methods and technologies that could study and analyze the expression of many genes simultaneously. Expressed sequence tags, custom microarrays, and proteomics have been used in order to obtain clues about the mechanism of lens regeneration. Although the scale and the depth of these techniques were low, certain concepts were discovered, including that both dorsal and ventral irides can initiate the process of lens regeneration. Expression of genes was detected in both sides of the iris, also validated with polymerase chain reaction [25, 44–46]. In the last decade, the study of gene expression was revolutionized with the introduction of RNA sequencing. This technique can sequence and quantify the expressed RNAs of a sample. Most importantly it has this ability without requiring prior knowledge on gene sequences, something that in newts is very important and beneficial [47]. The newt transcriptome and its analysis during lens regeneration revealed that dorsal–ventral differences are indeed mostly quantitative. However, by grouping the genes based on function, it revealed patterns of expression. Specifically, cytoskeleton, gene expression, cell cycle, and immune response genes were upregulated in the dorsal iris, 4 and 8 days post-lentectomy, while transposon-related genes were upregulated in the ventral iris at the same time points. These differences are pinpointing how transdifferentiation occurs since iris cells need to change their morphology and gene expression and proliferate in order for the lens to be regenerated [14, 48].

7.8 Lens "Regeneration" in Higher Vertebrates

Vertebrates other than newts do not have the ability to regenerate the lens following complete removal during adulthood. As discussed in Chap. 9, there are animal model systems that study wound healing responses after damaging the lens. In humans, the most common disease of the lens is cataract which affects mostly the elderly. However, when human iris is removed and cultured in vitro, it was found to have the ability to transdifferentiate to lens cells positive for crystallins [49]. This indicates that human iris retains the ability for lens regeneration, but the process is potentially inhibited by an as yet unknown cause [50]. By studying the dorsal–ventral iris differences in newts, this can reveal factors that play a role in the mechanism of lens regeneration and may ultimately unlock the process also in humans.

7.9 Concluding Remarks

Lens regeneration is the remarkable process of rebuilding a lost lens. Amphibians have this ability and are subject to extensive research for finding the mechanism by which it is accomplished. Lens regeneration can ultimately be used in humans to treat cataract, without the side effects and secondary complications that patients face with today's procedures. Advances in technology can aid towards finding the mechanism of lens regeneration, with important potential therapeutic applications.

References

1. Baddour JA, Sousounis K, Tsonis PA (2012) Organ repair and regeneration: an overview. Birth Defects Res C Embryo Today 96:1–29
2. Casco-Robles MM, Yamada S, Miura T, Nakamura K, Haynes T, Maki N, Del Rio-Tsonis K, Tsonis PA, Chiba C (2011) Expressing exogenous genes in newts by transgenesis. Nat Protoc 6:600–608
3. Tsonis PA, Haynes T, Maki N, Nakamura K, Casco-Robles MM, Yamada S, Miura T, Chiba C, Del Rio-Tsonis K (2011) Controlling gene loss of function in newts with emphasis on lens regeneration. Nat Protoc 6:593–599
4. Freeman G (1963) Lens regeneration from the cornea in *Xenopus laevis*. J Exp Zool 154:39–65
5. Brahma SK, McDevitt DS (1974) Ontogeny and localization of the lens crystallins in *Xenopus laevis* lens regeneration. J Embryol Exp Morphol 32:783–794
6. Mizuno N, Mochii M, Takahashi TC, Eguchi G, Okada TS (1999) Lens regeneration in Xenopus is not a mere repeat of lens development, with respect to crystallin gene expression. Differentiation 64:143–149
7. Schaefer JJ, Oliver G, Henry JJ (1999) Conservation of gene expression during embryonic lens formation and cornea-lens transdifferentiation in *Xenopus laevis*. Dev Dyn 215:308–318
8. Mizuno N, Ueda Y, Kondoh H (2005) Requirement for betaB1-crystallin promoter of *Xenopus laevis* in embryonic lens development and lens regeneration. Dev Growth Differ 47:131–140
9. Wolff G (1895) Entwickelungsphysiologische Studien. I. Die Regeneration der Urodelenlinse. Arch EntwMech Org 1:380–390
10. Grogg MW, Call MK, Okamoto M, Vergara MN, Del Rio-Tsonis K, Tsonis PA (2005) BMP inhibition-driven regulation of six-3 underlies induction of newt lens regeneration. Nature 438:858–862
11. Hayashi T, Mizuno N, Kondoh H (2008) Determinative roles of FGF and Wnt signals in iris-derived lens regeneration in newt eye. Dev Growth Differ 50:279–287
12. Eguchi G, Eguchi Y, Nakamura K, Yadav MC, Millan JL, Tsonis PA (2011) Regenerative capacity in newts is not altered by repeated regeneration and ageing. Nat Commun 2:384
13. Inoue T, Inoue R, Tsutsumi R, Tada K, Urata Y, Michibayashi C, Takemura S, Agata K (2012) Lens regenerates by means of similar processes and timeline in adults and larvae of the newt *Cynops pyrrhogaster*. Dev Dyn 241:1575–1583
14. Looso M, Preussner J, Sousounis K, Bruckskotten M, Michel CS, Lignelli E, Reinhardt R, Hoeffner S, Krueger M, Tsonis PA, Borchardt T, Braun T (2013) A de novo assembly of the

newt transcriptome combined with proteomic validation identifies new protein families expressed during tissue regeneration. Genome Biol 14:R16
15. Suetsugu-Maki R, Maki N, Nakamura K, Sumanas S, Zhu J, Del Rio-Tsonis K, Tsonis PA (2012) Lens regeneration in axolotl: new evidence of developmental plasticity. BMC Biol 10:103
16. Henry JJ, Tsonis PA (2010) Molecular and cellular aspects of amphibian lens regeneration. Prog Retin Eye Res 29:543–555
17. Sato T (1940) Vergleichende Studien über die Geschwindigkeit der Wolffschen Linsenregeneration bei *Triton taeniatus* und bei *Diemyctylus pyrrhogaster*. Arch EntwMech Org 140:570–613
18. Imokawa Y, Brockes JP (2003) Selective activation of thrombin is a critical determinant for vertebrate lens regeneration. Curr Biol 13:877–881
19. Godwin JW, Liem KF Jr, Brockes JP (2010) Tissue factor expression in newt iris coincides with thrombin activation and lens regeneration. Mech Dev 127:321–328
20. Reyer RW (1966) The influence of neural retina and lens on lens regeneration from dorsal iris implants in Triturus viridescens larvae. Dev Biol 14:214–245
21. Del Rio-Tsonis K, Jung JC, Chiu IM, Tsonis PA (1997) Conservation of fibroblast growth factor function in lens regeneration. Proc Natl Acad Sci U S A 94:13701–13706
22. McDevitt DS, Brahma SK, Courtois Y, Jeanny JC (1997) Fibroblast growth factor receptors and regeneration of the eye lens. Dev Dyn 208:220–226
23. Del Rio-Tsonis K, Trombley MT, McMahon G, Tsonis PA (1998) Regulation of lens regeneration by fibroblast growth factor receptor 1. Dev Dyn 213:140–146
24. Hayashi T, Mizuno N, Ueda Y, Okamoto M, Kondoh H (2004) FGF2 triggers iris-derived lens regeneration in newt eye. Mech Dev 121:519–526
25. Sousounis K, Michel CS, Bruckskotten M, Maki N, Borchardt T, Braun T, Looso M, Tsonis PA (2013) A microarray analysis of gene expression patterns during early phases of newt lens regeneration. Mol Vis 19:135–145
26. Maki N, Takechi K, Sano S, Tarui H, Sasai Y, Agata K (2007) Rapid accumulation of nucleostemin in nucleolus during newt regeneration. Dev Dyn 236:941–950
27. Maki N, Suetsugu-Maki R, Tarui H, Agata K, Del Rio-Tsonis K, Tsonis PA (2009) Expression of stem cell pluripotency factors during regeneration in newts. Dev Dyn 238:1613–1616
28. Maki N, Tsonis PA, Agata K (2010) Changes in global histone modifications during dedifferentiation in newt lens regeneration. Mol Vis 16:1893–1897
29. Yamada T, Karasaki S (1963) Nuclear RNA synthesis in newt iris cells engaged in regenerative transformation into lens cells. Dev Biol 6:595–604
30. Dumont JN, Yamada T (1972) Dedifferentiation of iris epithelial cells. Dev Biol 29:385–401
31. Del Rio-Tsonis K, Washabaugh CH, Tsonis PA (1995) Expression of pax-6 during urodele eye development and lens regeneration. Proc Natl Acad Sci U S A 92:5092–5096
32. Madhavan M, Haynes TL, Frisch NC, Call MK, Minich CM, Tsonis PA, Del Rio-Tsonis K (2006) The role of Pax-6 in lens regeneration. Proc Natl Acad Sci U S A 103:14848–14853
33. Tsonis PA, Trombley MT, Rowland T, Chandraratna RA, del Rio-Tsonis K (2000) Role of retinoic acid in lens regeneration. Dev Dyn 219:588–593
34. Tsonis PA, Tsavaris M, Call MK, Chandraratna RA, Del Rio-Tsonis K (2002) Expression and role of retinoic acid receptor alpha in lens regeneration. Dev Growth Differ 44:391–394
35. Del Rio-Tsonis K, Tomarev SI, Tsonis PA (1999) Regulation of Prox 1 during lens regeneration. Invest Ophthalmol Vis Sci 40:2039–2045
36. Maki N, Suetsugu-Maki R, Sano S, Nakamura K, Nishimura O, Tarui H, Del Rio-Tsonis K, Ohsumi K, Agata K, Tsonis PA (2010) Oocyte-type linker histone B4 is required for transdifferentiation of somatic cells in vivo. FASEB J 24:3462–3467
37. Stump RJ, Ang S, Chen Y, von Bahr T, Lovicu FJ, Pinson K, de Iongh RU, Yamaguchi TP, Sassoon DA, McAvoy JW (2003) A role for Wnt/beta-catenin signaling in lens epithelial differentiation. Dev Biol 259:48–61

38. Hayashi T, Mizuno N, Takada R, Takada S, Kondoh H (2006) Determinative role of Wnt signals in dorsal iris-derived lens regeneration in newt eye. Mech Dev 123:793–800
39. Kimura Y, Madhavan M, Call MK, Santiago W, Tsonis PA, Lambris JD, Del Rio-Tsonis K (2003) Expression of complement 3 and complement 5 in newt limb and lens regeneration. J Immunol 170:2331–2339
40. Mizuno N, Agata K, Sawada K, Mochii M, Eguchi G (2002) Expression of crystallin genes in embryonic and regenerating newt lenses. Dev Growth Differ 44:251–256
41. Mikami Y (1941) Experimental analysis of the Wolffian lens-regeneration in adult newt, *Triturus pyrrhogaster*. Jpn J Zool 9:269–302
42. Eguchi G, Shingai R (1971) Cellular analysis on localization of lens forming potency in the newt iris epithelium. Dev Growth Differ 13:337–349
43. Ito M, Hayashi T, Kuroiwa A, Okamoto M (1999) Lens formation by pigmented epithelial cell reaggregate from dorsal iris implanted into limb blastema in the adult newt. Dev Growth Differ 41:429–440
44. Makarev E, Call MK, Grogg MW, Atkinson DL, Milash B, Odelberg SJ, Tsonis PA (2007) Gene expression signatures in the newt irises during lens regeneration. FEBS Lett 581:1865–1870
45. Roddy M, Fox TP, McFadden JP, Nakamura K, Del Rio-Tsonis K, Tsonis PA (2008) A comparative proteomic analysis during urodele lens regeneration. Biochem Biophys Res Commun 377:275–279
46. Maki N, Martinson J, Nishimura O, Tarui H, Meller J, Tsonis PA, Agata K (2010) Expression profiles during dedifferentiation in newt lens regeneration revealed by expressed sequence tags. Mol Vis 16:72–78
47. Marioni JC, Mason CE, Mane SM, Stephens M, Gilad Y (2008) RNA-seq: an assessment of technical reproducibility and comparison with gene expression arrays. Genome Res 18:1509–1517
48. Sousounis K, Looso M, Maki N, Ivester CJ, Braun T, Tsonis PA (2013) Transcriptome analysis of newt lens regeneration reveals distinct gradients in gene expression patterns. PLoS One 8: e61445
49. Tsonis PA, Jang W, Del Rio-Tsonis K, Eguchi G (2001) A unique aged human retinal pigmented epithelial cell line useful for studying lens differentiation in vitro. Int J Dev Biol 45:753–758
50. Yasuda K, Okada TS, Eguchi G, Hayashi M (1978) A demonstration of a switch of cell type in human fetal eye tissues in vitro: pigmented cells of the iris or the retina can transdifferentiate into lens. Exp Eye Res 26:591–595

Fibrotic Modifications of the Lens Epithelium

8

I.M. Wormstone, J.A. Eldred, and L.J. Dawes

Abstract

Maintaining lens integrity is a lifetime ambition of the lens. However, if this integrity is breached through an altered ocular environment or physical disruption, such as cataract surgery, then fibrotic modifications to the lens can occur that are ultimately detrimental to vision. Fibrotic changes involve hyperproliferation, transdifferentiation from an epithelial to a myofibroblast phenotype, matrix deposition and matrix contraction. Fibrotic conditions of the lens include anterior subcapsular cataract and posterior capsule opacification, which affect millions. In this chapter we will discuss the regulatory mechanisms that facilitate fibrotic events, which will involve the intimate relationship between growth factors (especially transforming growth factor beta—TGF-β), signal transduction pathways and the extracellular matrix.

Keywords

Epithelium • Fibrosis • Lens • Matrix • Myofibroblast

8.1 Fibrosis and the Lens

The lens epithelium is a critical component in the maintenance of the normal healthy lens; however, fibrotic modifications can occur that are disruptive to visual quality [1–3]. Fibrosis affects many tissues and has been described as "a pathological condition in which tissue structure is disrupted by production of excessive

I.M. Wormstone (✉) • J.A. Eldred
School of Biological Sciences, University of East Anglia, Norwich NR4 7TJ, USA
e-mail: i.m.wormstone@uea.ac.uk

L.J. Dawes
Save Sight Institute, University of Sydney, Sydney 2001, Australia

S. Saika et al. (eds.), *Lens Epithelium and Posterior Capsular Opacification*,
DOI 10.1007/978-4-431-54300-8_8, © Springer Japan 2014

extracellular matrix (ECM)" [4, 5]. Fibrotic disorders are linked to a number of pathological processes, which include hyperproliferation, matrix deposition, matrix contraction and expression of myofibroblast cells (transdifferentiation). It has been assumed for many years that myofibroblast formation is a critical process in the early stages of tissue fibrosis and through the associated generation of a pro-contractile apparatus is required for matrix contraction. Studies carried out on lens cells have begun to question the established dogma and suggest that rather than be profibrogenic, myofibroblasts may in fact play a protective role to regulate the degree of fibrotic response [6, 7]. Further clarification of the role myofibroblasts play in fibrosis and the signalling pathways regulating different facets of fibrotic disease is vital to gain a greater understanding of lens fibrotic disorders, of which the most common are anterior subcapsular cataract (ASC) and posterior capsule opacification (PCO) (Fig. 8.1). ASC is characterised by the development of fibrotic regions underneath the anterior capsule. ASC is often linked to trauma of the eye, which includes impact injury, inflammation or irritation of the eye [8–10]. PCO progressively forms following cataract surgery and is the most prevalent lens fibrotic disorder [11]. PCO results when lens epithelial cells that remain following cataract surgery grow onto the previously cell-free posterior capsule. These cells exhibit classical fibrotic modifications including cell transdifferentiation, matrix deposition and contraction of the lens capsule, which give rise to light scatter and secondary visual loss. Currently, surgical intervention is the only treatment for cataract, initially restoring visual quality. Unfortunately, PCO develops in a significant proportion of patients to such an extent that a secondary loss of vision occurs.

8.2 Regulatory Mechanisms of Fibrosis

A number of experimental tools are available to study and evaluate fibrotic events in the lens [3]. These include cell lines [12], transgenic animals [13, 14], lens explants, whole lens cultures [15], in vitro capsular bag models (generated by simulated cataract surgery) [16–19] and analysis of post-mortem specimens [16, 20]. As with all experimental systems there are limitations with each, but through their collective use, we have gleaned valuable understanding of how lens fibrotic events are regulated.

8.2.1 Inflammation

There is a well-established link between inflammation and fibrotic changes; however, the fundamental regulatory systems that mediate this link are poorly understood. Tissue damage or injury including surgery results in an inflammatory response with the aim of repairing and protecting surrounding tissues. Damaged tissues release chemo-attractants, inducing cells such as the lymphocyte T helper cells that assist in the ensuing immune response by releasing cytokines such as

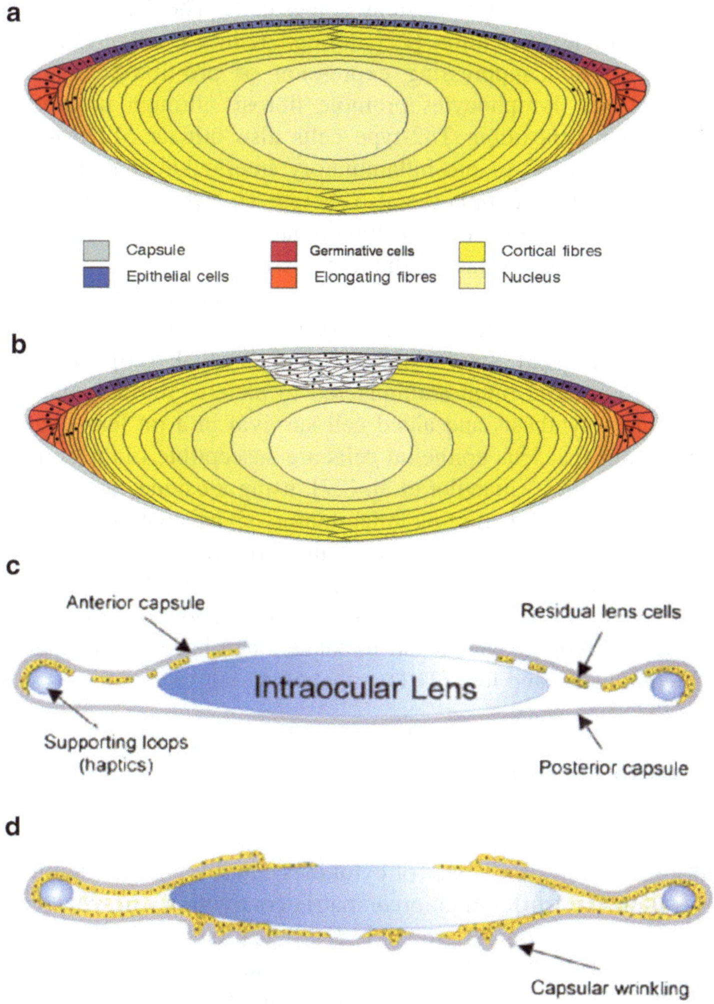

Fig. 8.1 Schematic illustrations showing the cellular arrangement in the normal human lens (**a**) and a lens presenting anterior subcapsular cataract (**b**). Treatment of ASC, along with all other cataracts, is surgery, which produce a lens capsular bag that can house an artificial intraocular lens (**c**). At the point of surgery a clear uninterrupted light path is observed within the visual axis. Due to fibrotic modifications to the lens epithelial cells and consequent matrix alterations, the posterior capsule exhibits light scattering structures that reduce visual quality (**d**); this condition is known as posterior capsule opacification. First published in Eldred et al. [3] with permission from the Royal Society and Wormstone [2] with permission from Experimental Eye Research

interleukins. These inflammatory signals can activate resident macrophages, which in turn lead to increased production of cytokines and chemokines and recruitment of monocytes [21]; the objective of such responses is to return injured tissue back to its normal state.

There are two types of lymphocyte T helper cells. The Th1-type are typically associated with tissue restoration [22] through reduced cell proliferation rates, collagen synthesis and suppressing expression of pro-fibrotic cytokines [23]. In contrast Th2-type lymphocytes promote fibrosis through promotion of cell growth and matrix deposition. Th2-type cells also initiate M2 macrophages to release PDGF and IL-10 (a pro-inflammatory interleukin that decreases IFN-γ and TNF-α). Class M2 macrophages also secrete TGF-β and inhibitors to MMPs, which can contribute to fibrosis [24]. Inflammation is therefore likely to play an important role in fibrotic changes within lens tissue.

Anti-inflammatory regimes are employed during cataract surgery, but curiously these treatments appear to increase the likelihood of developing fibrotic PCO 4 years post-surgery [25]. Interestingly, Symonds et al. [26] have shown that dexamethasone, a steroid often administered to patients during cataract surgery, increased collagen synthesis and aided cell survival in a rat lens explant model. These studies infer that lens epithelial cells are susceptible to the actions of anti-inflammatory molecules. Chandler et al. [27], using a canine capsular bag model, investigated the effects of disrupting inflammatory signals within lens cells through application of cyclooxygenase 2 (COX-2) inhibitors rofecoxib or celecoxib; both COX-2 inhibitors suppressed fibrotic change. Careful consideration must be taken in the management of inflammation following surgery to maintain homeostasis of specific subsets of lymphocytes as an imbalance can have detrimental effects on ECM production (Th2) and loss (Th1).

8.2.2 Growth Factors

A variety of cytokines are elevated in response to injury (both accidental and prescribed, i.e. surgery). A number of cytokines can drive proliferation or mediate matrix deposition, transdifferentiation or matrix contraction. Growth factors such as FGF, EGF and PDGF promote growth [12, 28, 29], but the foremost regulator of fibrotic changes is TGF-β [4]. As a means of regulating the degree of inflammatory response within the eye, elevated TGF-β is reported to suppress white blood cell activity, which in turn limits the degree of inflammatory response observed [30]. Unfortunately the actions of TGF-β are not limited to white blood cells and thus active TGF-β can stimulate receptive cells that can promote fibrosis in these tissues, including the lens. Improving our understanding and knowledge of TGF-β in lens fibrosis is of great importance.

The cytokine transforming growth factor beta is strongly implicated as a key regulator in the fibrotic response [4] (Fig. 8.2). TGF-β can be found in the aqueous humour of the eye and under normal conditions is largely present in a latent, inactive form [31, 32]. Activation of TGF-β from its latent form occurs through protease cleavage via degradation of pro-segments. In the normal eye the active form is of low abundance, and in fact activity is further regulated by proteins in the ocular humours such as α2-macroglobulin, which have a high affinity for free active TGF-β [33]. Following trauma to the lens, e.g. by surgical injury, active levels of TGF-β can be dramatically elevated. This occurs because the trauma of surgery

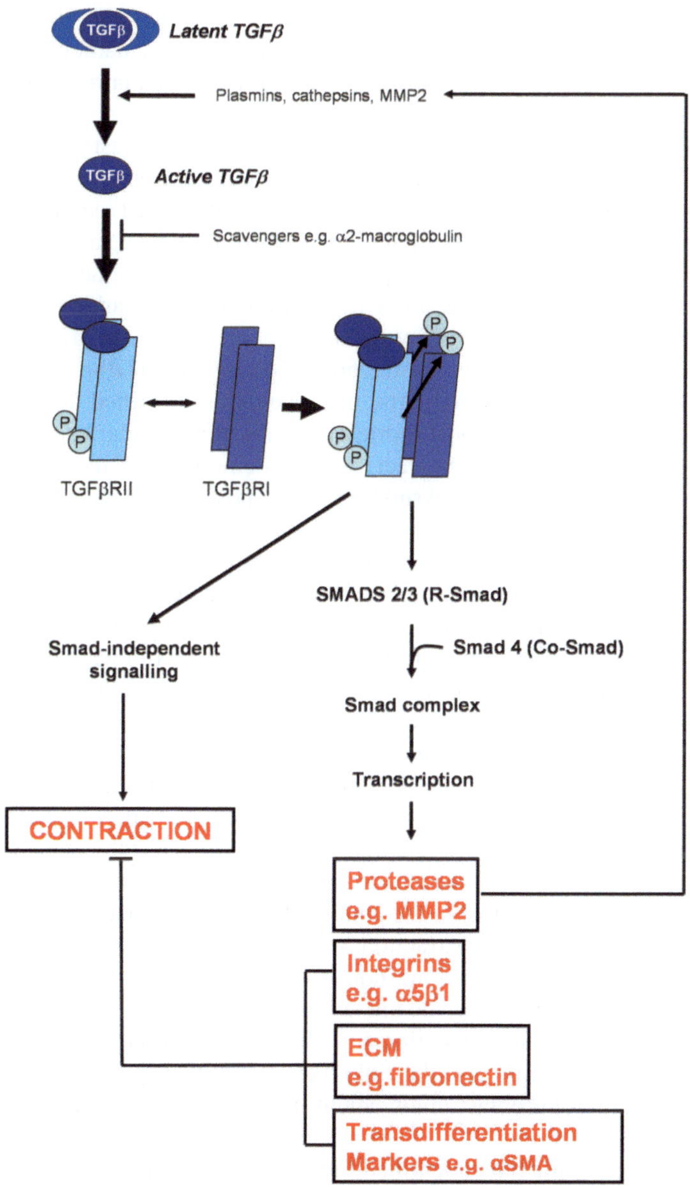

Fig. 8.2 A schematic overview of TGF-β signalling and the control of fibrotic events in the lens

can induce expression of many TGF-β activators, such as plasmins, cathepsins, MMPs [34] and thrombospondin-1 [35] which cleave latent TGF-β precursor protein. Furthermore, TGF-β itself is capable of increasing expression of MMPs, including MMP2 which could further accelerate conversion of latent TGF-β to its active form [16, 36, 37]. In addition, elevated numbers of reactive oxygen species are also reported to promote TGF-β activity [38, 39].

A large body of evidence is reported that demonstrates the ability of active TGF-β to promote cellular responses that define fibrosis, namely, epithelial to myofibroblast transdifferentiation; increased extracellular matrix production and deposition and matrix contraction [4, 11, 40].

Transdifferentiation of lens epithelial cells to a myofibroblast phenotype can be promoted by TGF-β. Analysis of human tissue specimens from anterior subcapsular cataract patients revealed increased levels of the myofibroblast markers α-SMA and fibronectin [41]. In addition, TGF-β has been shown to induce anterior subcapsular cataract in a rat lens culture model [42]; these appear as plaques on the anterior surface of the lens and also express transdifferentiation markers. Interestingly, sensitivity to TGF-β is influenced by age and gender. In the case of sex differences, female rat lenses demonstrated better protection against TGF-β-induced fibrotic changes than their male counterparts; oestrogen was identified as the protective component [43]. In addition, post-mortem analysis of a capsular bag received from a donor 1 month following cataract surgery demonstrated increased levels of α-SMA and matrix contraction/wrinkling of the posterior capsule [16]. Exposure of in vitro human capsular bags to TGF-β gave rise to increased expression of α-SMA and marked wrinkling of the posterior capsule. It has been assumed that TGF-β-induced transdifferentiation is critical for inducing the extracellular matrix synthesis and contraction [44]. Contraction of the extracellular matrix of the lens capsule causes a deformation, which is seen as "wrinkles" that cause light scatter and with increasing severity can lead to visual loss [45, 46]. Human lens cell lines have been utilised to allow detailed investigations of the signalling pathways regulating fibrotic events in the lens. Using this experimental tool, assays have been developed to determine matrix production, transdifferentiation and matrix contraction [7, 12, 47]. Using inhibition studies the pathways regulating TGF-β-induced fibrosis have become clearer. Dawes et al. [7] performed a study using human lens epithelial cells to investigate the relationship between the transdifferentiation and matrix contraction. In contrast to the commonly held view that transdifferentiation is a prerequisite for matrix contraction, it was found that α-SMA expression and fibronectin/fibronectin receptors are not essential for TGF-β-induced matrix contraction to occur. These findings do not however sit in isolation as data presented in lung cells also suggests a reduction in α-SMA expression can promote matrix deposition [48]. Consequently, the links between matrix production, deposition, contraction and myofibroblast involvement needs additional study in a variety of experimental systems.

A number of reports have identified the classical TGF-β/Smad signalling pathway in the pathogenesis of lens fibrotic disorders [6, 13, 14, 20]. Saika et al. [20] have identified Smad3 and Smad4 in cell nuclei of postoperative human lenses, which indicates an active signalling pathway [49]. In addition, murine lenses subjected to injury demonstrated increased myofibroblast expression [49]. Pretreatment of murine lenses with TGF-β-neutralising antibodies in advance of a puncture wound injury prevented Smad3 and Smad4 translocation to the nucleus [49]. Smad3 in preference to Smad2-dependent signalling is generally implicated in fibrotic changes throughout the body [50]. For example, in model systems assessing bleomycin-induced fibrosis, increased detection of activated Smad3 has been found during hepatic stellate cell activation and at the leading edge of scleroderma lesions [51–53]. Consequently, Smad3 knockout transgenic models have been used to establish the relevance of TGF-β/Smad3-dependent pathway in regulating fibrotic events in the lens [14, 13]. Saika et al. [13] utilised the Smad3 knockout transgenic mouse and observed that both TGF-β2-induced and injury-induced α-SMA expressions were inhibited in the lens epithelium of Smad3 knockout mice relative to wild type. Interestingly, Banh et al. [14] still identified some α-SMA expression in a TGF-β1/Smad3$^{-/-}$ mouse; however, the degree of α-SMA expression was relatively low compared to Smad3 expressing TGF-β1/Smad3$^{+/-}$ lenses. Loss of Smad3 in mice also reduces the expression of trauma-induced type I collagen in the lens [13]. In a Smad4 knockdown human lens epithelial cell line, Dawes et al. [6] observed a significant inhibition of TGF-β-induced fibrotic markers, α-SMA and fibronectin, which suggests that TGF-β Smad3-Smad4 signalling is likely to control transdifferentiation of human lens epithelial cells. Curiously, this study found TGF-β-induced matrix contraction and Smad7 expression to be independent of Smad4 expression, suggesting that TGF-β-Smad-independent pathways may also regulate lens fibrosis.

It is now well established that TGF-β can signal independently of Smad function [54]. For instance, TGF-β type I receptor can phosphorylate both serine and tyrosine residues in the SHCA adaptor, recruiting adaptor protein GRB2 and the Ras guanine exchange factor (GEF) son of sevenless (SOS) to activate Ras-Raf-MEK-ERK mitogen-activated protein kinase (MAPK) in mammalian cells [55]. The tyrosine kinase Src can phosphorylate the cytoplasmic domain of the TGF-βRII leading to GRB2 and SHC recruitment allowing the activation of the p38 MAPK pathway [56]. Moreover, TGF-β can also activate the JNK MAP kinase and Rho kinase signalling pathways [57]. Smad-independent ERK and p38 MAP kinase signalling pathways are activated by TGF-β in human lens epithelial cells [6]. The concept that TGF-β-induced matrix contraction by human lens epithelial cells is regulated by Smad-independent signalling [6] is given weight by findings in non-ocular systems where Rho/Rho kinase [58], ERK [59] and p38 MAP kinase [60] promote matrix contraction. In addition, the ERK signalling pathway can increase matrix contraction through activation of myosin light chain kinase (MLCK), a key enzymatic regulator of contractile force [59]. The role of Smad-independent signalling in TGF-β-mediated fibrotic events and matrix contraction in particular requires continued investigation.

The Wnt signalling pathway is reported to play a key role in TGF-β-induced transdifferentiation and has also been implicated in fibrotic disorders [61–64]. The canonical Wnt signalling pathway regulates β-catenin activity. Wnt ligand binding to receptors (Frizzled) and co-receptor low-density lipoprotein receptor-related protein (LRP) leads to the phosphorylation of LRP6, by glycogen synthase kinase-3β (GSK-3β) and casein kinase γ, in its cytoplasmic region. This leads to the recruitment of cytosolic proteins, dishevelled and axin (93). As a result of the association of GSK-3β with Frizzled receptors and LRP, β-catenin phosphorylation is reduced. In this form, β-catenin is no longer targeted for degradation and consequently accumulates in the cytoplasm before translocating to the nucleus where it associates with DNA binding factors T cell factor (TCF) and lymphoid enhancer factor (LEF) to regulate gene expression [62]. It is believed that TGF-β regulates the canonical Wnt signalling pathway through Smad-dependent and Smad-independent mechanisms. It has been reported that TGF-β/Smad signalling can induce ILK, which suppresses E-cadherin at the transcriptional level, potentially through the transcriptional repressor SNAI-1 [64]. Furthermore, ILK can phosphorylate GSK directly or by active Akt. Phosphorylation of GSK leads to its inactivation, which again promotes β-catenin translocation [64]. In both cases the result is to promote transdifferentiation. An investigation using transgenic mice overexpressing TGF-β1 and whole rat lens cultures exposed to TGF-β2 demonstrated elevated levels of Wnt ligands and Frizzled receptors compared to controls; this was associated with increased nuclear β-catenin in cataract forming cells, expression of α-SMA and plaque formation [65].

Future lines of research investigating fibrotic changes to the lens will need to consider both Smad-independent and Smad-dependent pathways and the interactions between them.

8.2.3 Extracellular Matrix

The extracellular matrix (ECM) was once thought of as a static structure that merely served as a scaffold to maintain tissue integrity. This opinion has changed dramatically over the years and we can now appreciate that the ECM regulates many aspects of cell function, which in the lens include proliferation, migration and differentiation. Alterations of the ECM by increased synthesis and deposition of ECM proteins are a key aspect in the pathogenesis of fibrotic disorders [4]. Lens epithelial cells and lens fibres are associated with a basement membrane termed the lens capsule, which is largely composed of collagen IV, along with laminin, heparin sulphate proteoglycans and tenascin [66]. These ECM components provide a physical framework to ensure structural integrity of the lens capsule and permit lens cell attachment and migration. In lens fibrosis, cells over-synthesise and deposit new ECM components, which include fibronectin, vitronectin and collagen types I and III; these deposits contribute to the fibrotic plaque formation and light scattering regions.

Cells from explant cultures are reported to adhere and migrate on vitronectin and fibronectin matrices. These cells, however, demonstrated a more elongated/fibroblast-like cell appearance, increased α-SMA staining and nuclear Smad expression [67]. ED-A domain of fibronectin is believed to play a crucial role in the induction of α-SMA expression [68]; treatment of corneal fibroblasts with an RGD peptide prevents the interaction of fibronectin and its integrin receptor and reduced expression of α-SMA. It is therefore likely that fibronectin expression will play a pivotal role in transdifferentiation events that take place in lens fibrosis. The proteoglycan lumican has also been identified in post-mortem capsular bag specimens [69]. It has also been shown in lumican knockout mice that following injury to the anterior lens capsule by a needle puncture, there was a delay in expression of α-SMA and the appearance of transdifferentiated cells [69]. These studies provide firm evidence that modifications in ECM constituents will influence lens cell characteristics that permit the development of fibrotic lens pathologies.

Secreted protein acidic and rich in cysteine (SPARC) may play a role in lens fibrotic events by altering the expression of ECM proteins [70]. Reports indicate that SPARC null mice develop cataracts by 3–4 months of age because of reduced lens cell growth [71]. SPARC knockout mice exhibit decreased laminin deposition relative to wild-type lenses [72]. Additional studies investigated the effects of TGF-β in lens epithelial cells from wild-type and SPARC null mice [73]. TGF-β was found to promote expression of both fibronectin and α-SMA in both groups, but this was greatest in the SPARC null group, which implies SPARC, to some extent, suppress TGF-β-induced transdifferentiation. In addition, dexamethasone was found to increase SPARC expression in lens cells, which was associated with a decline in fibronectin and collagen type IV. This again suggests SPARC modulation of the ECM can suppress transdifferentiation events.

The actions of a number of growth factors are regulated by matrix interactions. For example, FGF and HGF require heparin binding to facilitate interaction with their associated receptor; as a result, the distribution of heparin sulphate proteoglycans and potentially other ECM components within the lens capsule could play important roles in FGF- and HGF-regulated fibrotic events [74–76]. FGF is reported to exacerbate the effects of TGF-β in a rat whole lens culture model. FGF alone was not capable of inducing anterior capsule plaques, but the severity of response observed with TGF-β was increased when FGF was also added [77]. TGF-β itself can also bind to several matrix components including decorin and collagen type IV [78]. As the level of many growth factors within the eye is elevated following surgery, matrix components within the lens capsule have the capacity to adsorb these proteins and thus provide an enriched environment that is likely to promote fibrotic events [79, 80]. Direct presentation of ligands to receptors or through slow release resulting from proteolytic cleavage of this reservoir is therefore likely to provide long-term contributions to fibrotic change.

Two-way communications between the cell and its underlying matrix are ongoing and can have a marked influence on the ultimate behaviour of the cell. This process is generally mediated by integrins, which are a group of distinct cell surface receptors, composed of α and β subunits. The interaction of specific integrins with their corresponding ECM ligands allows cells to adhere to and migrate across the ECM [81]. In addition, integrins function as cell signalling centres allowing signal transduction to and from the microenvironment [81]; therefore, dys-regulation of integrin expression or function can give rise to fibrotic changes.

Microarray analysis of the human lens epithelial cell line FHL 124 has been performed to identify which integrins are expressed in human lens epithelial cells and identify those of potential importance in lens fibrosis [47]. It was found that human lens epithelial cells express integrin β1, β2, β3, β5 and β6 subunits. The β1 integrin was the most abundant and is the most promiscuous integrin subunit; it is known to associate with 12 different α integrin chains [82] and is the most widely expressed integrin throughout the body [82]. With regard to the alpha subunits, α1, α2, α3, α4, α5, α6, α7, α10, α11, αE, αM and αV were detected in human lens epithelial cells [47]. Exposure of cells to TGF-β resulted in an increased expression of α5, α11, αV and β5 integrin subunits.

Of particular interest to fibrotic changes in the lens epithelium is the expression of α5β1 integrin, which together with its matrix ligand fibronectin demonstrates enhanced expression in response to TGF-β exposure in both the human capsular bag and lens epithelial cell line [7, 47]. Fibronectin interaction with α5β1 integrin is also reported to regulate the expression of α-SMA and transdifferentiation of corneal fibroblast cells to myofibroblasts [83]. As suggested earlier, it is a commonly held view that α5β1 integrin and fibronectin form a putative contractile apparatus with α-SMA [83, 84]. This theory is questioned through experiments using lens cells. Application of an RGDS peptide to block the RGD binding site of α5 integrin revealed that the fibronectin/fibronectin receptor interaction was not required to promote matrix contraction by human lens epithelial cells in response to TGF-β [7]. In addition, Marcantonio et al. [85] demonstrated that following TGF-β exposure, expression of α5β1 integrin was altered in FHL 124 cells, such that a diffuse pattern across the cell was observed, with membrane-bound α5β1 integrin having no association with actin filaments. These findings infer that in the human lens the α5β1 integrin distribution in response to TGF-β could provide multiple sites of attachment to the underlying matrix to counter matrix contraction.

Integrin αVβ5 is reported to play an important role in fibrotic pathologies. This is mainly because it permits mechanotransduction in response to extracellular microenvironments [86] and is implicated in the transdifferentiation of cells to myofibroblasts [87]. Exposure of TGF-β to lens epithelial cells gives rise to an increase in αVβ5 integrin expression [47] in association with increased levels of transdifferentiation markers [7, 16]. On the basis of these findings it has been proposed that αVβ5 integrin can play a bidirectional role in lens fibrosis [88]. Firstly, it could mediate TGF-β-induced transdifferentiation following trauma to the lens and secondly act by facilitating signals from myofibroblasts back to the ECM, which results in the activation of matrix-associated TGF-β. Disruption to lens integrity through injury can alter integrin expression in lens epithelial cells. Sponer et al. [89] showed that expression of αVβ6 integrin was up-regulated in cells residing in vitro human capsular bags cultured in protein-free medium compared with cultured intact lenses. αVβ6 integrin can activate TGF-β through its association with an RGD peptide in the latency-associated peptide [90]. Integrin αVβ6 and αVβ5 are likely to play important roles in regulating the levels of active TGF-β and the formation of microenvironments capable of driving fibrotic modifications of lens epithelial cells.

Integrin-linked kinase (ILK), a serine threonine kinase that binds to the cytoplasmic tail of β1, β2 and β3 integrin subunits, has been detected in lens cells of both murine and human origins [91]. ILK is a multidomain focal adhesion protein that regulates ECM adhesion and signal transduction. Evidence supports the notion that ILK is a regulator in myofibroblast formation [91–93]. For example, expression of ILK in cultured mouse and human lens cells is associated with elevated expression of transdifferentiation markers, fibronectin and α-SMA. In addition, the introduction of ILK-expressing constructs resulted in an altered morphology that was fibroblast-like in nature [92]. Moreover, it has been observed that ILK co-localises with α5β1 integrin; the presence of fibronectin further promoted this association [91]. ILK appears to be a potentially important protein in the regulation of transdifferentiation and is likely to involve association with integrins and matrix components classically linked with fibrosis. Elucidating the role of ILK in the grand scheme of lens fibrotic changes would be an interesting line of investigation.

8.3 Summary

Fibrotic modification of the lens epithelium is likely to result from a disruption to lens integrity. This could result from a change in the ocular environment as a consequence of diseases, external injury or surgical trauma, such as cataract removal. Once lens integrity is breached alterations to lens cells occur, which renders them susceptible to elevated proteins introduced to the aqueous humour and to autocrine factors, which through matrix association provide rich microenvironments capable of driving fibrotic changes. Fibrotic changes involve hyperproliferation, transdifferentiation from an epithelial to a myofibroblast phenotype, matrix deposition and matrix contraction. Growth factors, and in particular transforming growth factor β, drive many of these events. Changes to the matrix composition and integrin expression of lens cells as a consequence of injury and growth factor-induced responses alter the characteristics of the cells, such that they exhibit a contractile phenotype and exhibit myofibroblast properties. It would however appear that the expression of myofibroblast markers is not a prerequisite for matrix contraction. Over the past couple of decades our knowledge of fibrotic modifications of the lens epithelium has grown dramatically and due to its importance in lens pathologies will remain a key area of research for many years to come.

References

1. Wormstone IM, Wang L, Liu CS (2009) Posterior capsule opacification. Exp Eye Res 88 (2):257–269. doi:10.1016/j.exer.2008.10.016
2. Wormstone IM (2002) Posterior capsule opacification: a cell biological perspective. Exp Eye Res 74(3):337–347. doi:10.1006/exer.2001.1153
3. Eldred JA, Dawes LJ, Wormstone IM (2011) The lens as a model for fibrotic disease. Philos Trans R Soc Lond B Biol Sci 366(1568):1301–1319. doi:10.1098/rstb.2010.0341
4. Leask A, Abraham DJ (2004) TGF-beta signaling and the fibrotic response. FASEB J 18 (7):816–827. doi:10.1096/fj.03-1273rev
5. Radisky DC, Przybylo JA (2008) Matrix metalloproteinase-induced fibrosis and malignancy in breast and lung. Proc Am Thorac Soc 5(3):316–322. doi:10.1513/pats.200711-166DR
6. Dawes LJ, Sleeman MA, Anderson IK, Reddan JR, Wormstone IM (2009) TGFbeta/Smad4-dependent and -independent regulation of human lens epithelial cells. Invest Ophthalmol Vis Sci 50(11):5318–5327. doi:10.1167/iovs.08-3223

7. Dawes LJ, Eldred JA, Anderson IK, Sleeman M, Reddan JR, Duncan G, Wormstone IM (2008) TGF beta-induced contraction is not promoted by fibronectin-fibronectin receptor interaction, or alpha SMA expression. Invest Ophthalmol Vis Sci 49(2):650–661. doi:10.1167/iovs.07-0586
8. Fisher RF (1981) The lens in uveitis. Trans Ophthalmol Soc U K 101((Pt 3)(3)):317–320
9. Marcantonio JM, Syam PP, Liu CS, Duncan G (2003) Epithelial transdifferentiation and cataract in the human lens. Exp Eye Res 77(3):339–346
10. Sasaki K, Kojima M, Nakaizumi H, Kitagawa K, Yamada Y, Ishizaki H (1998) Early lens changes seen in patients with atopic dermatitis applying image analysis processing of Scheimpflug and specular microscopic images. Ophthalmologica 212(2):88–94
11. Rhodes JD, Russell SL, Illingworth CD, Duncan G, Wormstone IM (2009) Regional differences in store-operated Ca2+ entry in the epithelium of the intact human lens. Invest Ophthalmol Vis Sci 50(9):4330–4336. doi:10.1167/iovs.08-3222
12. Wormstone IM, Tamiya S, Eldred JA, Lazaridis K, Chantry A, Reddan JR, Anderson I, Duncan G (2004) Characterisation of TGF-beta2 signalling and function in a human lens cell line. Exp Eye Res 78(3):705–714
13. Saika S, Kono-Saika S, Ohnishi Y, Sato M, Muragaki Y, Ooshima A, Flanders KC, Yoo J, Anzano M, Liu CY, Kao WW, Roberts AB (2004) Smad3 signaling is required for epithelial-mesenchymal transition of lens epithelium after injury. Am J Pathol 164(2):651–663
14. Banh A, Deschamps PA, Gauldie J, Overbeek PA, Sivak JG, West-Mays JA (2006) Lens-specific expression of TGF-beta induces anterior subcapsular cataract formation in the absence of Smad3. Invest Ophthalmol Vis Sci 47(8):3450–3460. doi:10.1167/iovs.05-1208
15. Hales AM, Chamberlain CG, McAvoy JW (1995) Cataract induction in lenses cultured with transforming growth factor-beta. Invest Ophthalmol Vis Sci 36(8):1709–1713
16. Wormstone IM, Tamiya S, Anderson I, Duncan G (2002) TGF-beta2-induced matrix modification and cell transdifferentiation in the human lens capsular bag. Invest Ophthalmol Vis Sci 43(7):2301–2308
17. Saxby L, Rosen E, Boulton M (1998) Lens epithelial cell proliferation, migration, and metaplasia following capsulorhexis. Br J Ophthalmol 82(8):945–952
18. Dawes LJ, Illingworth CD, Wormstone IM (2012) A fully human in vitro capsular bag model to permit intraocular lens evaluation. Invest Ophthalmol Vis Sci 53(1):23–29. doi:10.1167/iovs.11-8851
19. Cleary G, Spalton DJ, Koch DD (2007) Effect of square-edged intraocular lenses on neodymium:YAG laser capsulotomy rates in the United States. J Cataract Refract Surg 33 (11):1899–1906. doi:10.1016/j.jcrs.2007.06.056
20. Saika S, Miyamoto T, Ishida I, Shirai K, Ohnishi Y, Ooshima A, McAvoy JW (2002) TGFbeta-Smad signalling in postoperative human lens epithelial cells. Br J Ophthalmol 86 (12):1428–1433
21. Martinez FO, Gordon S, Locati M, Mantovani A (2006) Transcriptional profiling of the human monocyte-to-macrophage differentiation and polarization: new molecules and patterns of gene expression. J Immunol 177(10):7303–7311
22. Keane MP (2008) The role of chemokines and cytokines in lung fibrosis. Eur Respir Rev 17:136–151
23. Bajwa EK, Ayas NT, Schulzer M, Mak E, Ryu JH, Malhotra A (2005) Interferon-gamma1b therapy in idiopathic pulmonary fibrosis: a metaanalysis. Chest 128(1):203–206. doi:10.1378/chest.128.1.203
24. Strieter RM (2008) What differentiates normal lung repair and fibrosis? Inflammation, resolution of repair, and fibrosis. Proc Am Thorac Soc 5(3):305–310. doi:10.1513/pats.200710-160DR
25. Laurell CG, Zetterstrom C (2002) Effects of dexamethasone, diclofenac, or placebo on the inflammatory response after cataract surgery. Br J Ophthalmol 86(12):1380–1384
26. Symonds JG, Lovicu FJ, Chamberlain CG (2006) Differing effects of dexamethasone and diclofenac on posterior capsule opacification-like changes in a rat lens explant model. Exp Eye Res 83(4):771–782. doi:10.1016/j.exer.2006.03.017

27. Chandler HL, Barden CA, Lu P, Kusewitt DF, Colitz CM (2007) Prevention of posterior capsular opacification through cyclooxygenase-2 inhibition. Mol Vis 13:677–691
28. Wormstone IM, Del Rio-Tsonis K, McMahon G, Tamiya S, Davies PD, Marcantonio JM, Duncan G (2001) FGF: an autocrine regulator of human lens cell growth independent of added stimuli. Invest Ophthalmol Vis Sci 42(6):1305–1311
29. Wormstone IM, Tamiya S, Marcantonio JM, Reddan JR (2000) Hepatocyte growth factor function and c-Met expression in human lens epithelial cells. Invest Ophthalmol Vis Sci 41 (13):4216–4222
30. Streilein JW, Ohta K, Mo JS, Taylor AW (2002) Ocular immune privilege and the impact of intraocular inflammation. DNA Cell Biol 21(5–6):453–459. doi:10.1089/10445490260099746
31. Ohta K, Yamagami S, Taylor AW, Streilein JW (2000) IL-6 antagonizes TGF-beta and abolishes immune privilege in eyes with endotoxin-induced uveitis. Invest Ophthalmol Vis Sci 41(9):2591–2599
32. Schlotzer-Schrehardt U, Zenkel M, Kuchle M, Sakai LY, Naumann GO (2001) Role of transforming growth factor-beta1 and its latent form binding protein in pseudoexfoliation syndrome. Exp Eye Res 73(6):765–780. doi:10.1006/exer.2001.1084
33. Schulz MW, Chamberlain CG, McAvoy JW (1996) Inhibition of transforming growth factor-beta-induced cataractous changes in lens explants by ocular media and alpha 2-macroglobulin. Invest Ophthalmol Vis Sci 37(8):1509–1519
34. Yu Q, Stamenkovic I (2000) Cell surface-localized matrix metalloproteinase-9 proteolytically activates TGF-beta and promotes tumor invasion and angiogenesis. Genes Dev 14(2):163–176
35. Frazier WA (1991) Thrombospondins. Curr Opin Cell Biol 3(5):792–799
36. Eldred JA, Hodgkinson LM, Dawes LJ, Reddan JR, Edwards DR, Wormstone IM (2012) MMP2 activity is critical for TGFbeta2-induced matrix contraction–implications for fibrosis. Invest Ophthalmol Vis Sci 53(7):4085–4098. doi:10.1167/iovs.12-9457
37. Dwivedi DJ, Pino G, Banh A, Nathu Z, Howchin D, Margetts P, Sivak JG, West-Mays JA (2006) Matrix metalloproteinase inhibitors suppress transforming growth factor-beta-induced subcapsular cataract formation. Am J Pathol 168(1):69–79
38. Chamberlain CG, Mansfield KJ, Cerra A (2009) Glutathione and catalase suppress TGFbeta-induced cataract-related changes in cultured rat lenses and lens epithelial explants. Mol Vis 15:895–905
39. Fatma N, Kubo E, Sharma P, Beier DR, Singh DP (2005) Impaired homeostasis and phenotypic abnormalities in Prdx6−/−mice lens epithelial cells by reactive oxygen species: increased expression and activation of TGFbeta. Cell Death Differ 12(7):734–750. doi:10.1038/sj.cdd.4401597
40. Saika S, Yamanaka O, Okada Y, Tanaka S, Miyamoto T, Sumioka T, Kitano A, Shirai K, Ikeda K (2009) TGF beta in fibroproliferative diseases in the eye. Front Biosci (Schol Ed) 1:376–390
41. Lee EH, Joo CK (1999) Role of transforming growth factor-beta in transdifferentiation and fibrosis of lens epithelial cells. Invest Ophthalmol Vis Sci 40(9):2025–2032
42. Hales AM, Schulz MW, Chamberlain CG, McAvoy JW (1994) TGF-beta 1 induces lens cells to accumulate alpha-smooth muscle actin, a marker for subcapsular cataracts. Curr Eye Res 13 (12):885–890
43. Hales AM, Chamberlain CG, Murphy CR, McAvoy JW (1997) Estrogen protects lenses against cataract induced by transforming growth factor-beta (TGFbeta). J Exp Med 185 (2):273–280
44. Grinnell F (1994) Fibroblasts, myofibroblasts, and wound contraction. J Cell Biol 124 (4):401–404
45. van Bree MC, van der Meulen IJ, Franssen L, Coppens JE, Zijlmans BL, van den Berg TJ (2012) In-vitro recording of forward light-scatter by human lens capsules and different types of posterior capsule opacification. Exp Eye Res 96(1):138–146. doi:10.1016/j.exer.2011.12.008
46. van Bree MC, van den Berg TJ, Zijlmans BL (2013) Posterior capsule opacification severity, assessed with straylight measurement, as main indicator of early visual function deterioration. Ophthalmology 120(1):20–33. doi:10.1016/j.ophtha.2012.07.050

47. Grocott T, Frost V, Maillard M, Johansen T, Wheeler GN, Dawes LJ, Wormstone IM, Chantry A (2007) The MH1 domain of Smad3 interacts with Pax6 and represses autoregulation of the Pax6 P1 promoter. Nucleic Acids Res 35(3):890–901. doi:10.1093/nar/gkl1105
48. Liu T, Warburton RR, Guevara OE, Hill NS, Fanburg BL, Gaestel M, Kayyali US (2007) Lack of MK2 inhibits myofibroblast formation and exacerbates pulmonary fibrosis. Am J Respir Cell Mol Biol 37(5):507–517. doi:10.1165/rcmb.2007-0077OC
49. Saika S, Okada Y, Miyamoto T, Ohnishi Y, Ooshima A, McAvoy JW (2001) Smad translocation and growth suppression in lens epithelial cells by endogenous TGFbeta2 during wound repair. Exp Eye Res 72(6):679–686. doi:10.1006/exer.2001.1002
50. Flanders KC, Major CD, Arabshahi A, Aburime EE, Okada MH, Fujii M, Blalock TD, Schultz GS, Sowers A, Anzano MA, Mitchell JB, Russo A, Roberts AB (2003) Interference with transforming growth factor-beta/Smad3 signaling results in accelerated healing of wounds in previously irradiated skin. Am J Pathol 163(6):2247–2257
51. Liu C, Gaca MD, Swenson ES, Vellucci VF, Reiss M, Wells RG (2003) Smads 2 and 3 are differentially activated by transforming growth factor-beta (TGF-beta) in quiescent and activated hepatic stellate cells. Constitutive nuclear localization of Smads in activated cells is TGF-beta-independent. J Biol Chem 278(13):11721–11728. doi:10.1074/jbc.M207728200
52. Mori Y, Chen SJ, Varga J (2003) Expression and regulation of intracellular SMAD signaling in scleroderma skin fibroblasts. Arthritis Rheum 48(7):1964–1978. doi:10.1002/art.11157
53. Takagawa S, Lakos G, Mori Y, Yamamoto T, Nishioka K, Varga J (2003) Sustained activation of fibroblast transforming growth factor-beta/Smad signaling in a murine model of scleroderma. J Invest Dermatol 121(1):41–50. doi:10.1046/j.1523-1747.2003.12308.x
54. Derynck R, Zhang YE (2003) Smad-dependent and Smad-independent pathways in TGF-beta family signalling. Nature 425(6958):577–584. doi:10.1038/nature02006
55. Lee MK, Pardoux C, Hall MC, Lee PS, Warburton D, Qing J, Smith SM, Derynck R (2007) TGF-beta activates Erk MAP kinase signalling through direct phosphorylation of ShcA. EMBO J 26(17):3957–3967. doi:10.1038/sj.emboj.7601818
56. Galliher AJ, Schiemann WP (2007) Src phosphorylates Tyr284 in TGF-beta type II receptor and regulates TGF-beta stimulation of p38 MAPK during breast cancer cell proliferation and invasion. Cancer Res 67(8):3752–3758. doi:10.1158/0008-5472.CAN-06-3851
57. Moustakas A, Heldin CH (2009) The regulation of TGFbeta signal transduction. Development 136(22):3699–3714. doi:10.1242/dev.030338
58. Parizi M, Howard EW, Tomasek JJ (2000) Regulation of LPA-promoted myofibroblast contraction: role of Rho, myosin light chain kinase, and myosin light chain phosphatase. Exp Cell Res 254(2):210–220. doi:10.1006/excr.1999.4754
59. Klemke RL, Cai S, Giannini AL, Gallagher PJ, de Lanerolle P, Cheresh DA (1997) Regulation of cell motility by mitogen-activated protein kinase. J Cell Biol 137(2):481–492
60. Borbiev T, Birukova A, Liu F, Nurmukhambetova S, Gerthoffer WT, Garcia JG, Verin AD (2004) p38 MAP kinase-dependent regulation of endothelial cell permeability. Am J Physiol Lung Cell Mol Physiol 287(5):L911–L918. doi:10.1152/ajplung.00372.2003
61. Crosby LM, Waters CM (2010) Epithelial repair mechanisms in the lung. Am J Physiol Lung Cell Mol Physiol 298(6):L715–L731. doi:10.1152/ajplung.00361.2009
62. Konigshoff M, Eickelberg O (2010) WNT signaling in lung disease: a failure or a regeneration signal? Am J Respir Cell Mol Biol 42(1):21–31. doi:10.1165/rcmb.2008-0485TR
63. Liu Y (2010) New insights into epithelial-mesenchymal transition in kidney fibrosis. J Am Soc Nephrol 21(2):212–222. doi:10.1681/ASN.2008121226
64. Willis BC, Borok Z (2007) TGF-beta-induced EMT: mechanisms and implications for fibrotic lung disease. Am J Physiol Lung Cell Mol Physiol 293(3):L525–L534. doi:10.1152/ajplung.00163.2007
65. Chong CC, Stump RJ, Lovicu FJ, McAvoy JW (2009) TGFbeta promotes Wnt expression during cataract development. Exp Eye Res 88(2):307–313. doi:10.1016/j.exer.2008.07.018
66. Cammarata PR, Cantu-Crouch D, Oakford L, Morrill A (1986) Macromolecular organization of bovine lens capsule. Tissue Cell 18(1):83–97

67. Taliana L, Evans MD, Ang S, McAvoy JW (2006) Vitronectin is present in epithelial cells of the intact lens and promotes epithelial mesenchymal transition in lens epithelial explants. Mol Vis 12:1233–1242
68. Serini G, Bochaton-Piallat ML, Ropraz P, Geinoz A, Borsi L, Zardi L, Gabbiani G (1998) The fibronectin domain ED-A is crucial for myofibroblastic phenotype induction by transforming growth factor-beta1. J Cell Biol 142(3):873–881
69. Saika S, Miyamoto T, Tanaka S, Tanaka T, Ishida I, Ohnishi Y, Ooshima A, Ishiwata T, Asano G, Chikama T, Shiraishi A, Liu CY, Kao CW, Kao WW (2003) Response of lens epithelial cells to injury: role of lumican in epithelial-mesenchymal transition. Invest Ophthalmol Vis Sci 44(5):2094–2102
70. Yan Q, Sage EH (1999) SPARC, a matricellular glycoprotein with important biological functions. J Histochem Cytochem 47(12):1495–1506
71. Bassuk JA, Birkebak T, Rothmier JD, Clark JM, Bradshaw A, Muchowski PJ, Howe CC, Clark JI, Sage EH (1999) Disruption of the Sparc locus in mice alters the differentiation of lenticular epithelial cells and leads to cataract formation. Exp Eye Res 68(3):321–331. doi:10.1006/exer.1998.0608
72. Weaver MS, Sage EH, Yan Q (2006) Absence of SPARC in lens epithelial cells results in altered adhesion and extracellular matrix production in vitro. J Cell Biochem 97(2):423–432. doi:10.1002/jcb.20654
73. Gotoh N, Perdue NR, Matsushima H, Sage EH, Yan Q, Clark JI (2007) An in vitro model of posterior capsular opacity: SPARC and TGF-beta2 minimize epithelial-to-mesenchymal transition in lens epithelium. Invest Ophthalmol Vis Sci 48(10):4679–4687. doi:10.1167/iovs.07-0091
74. Aviezer D, Hecht D, Safran M, Eisinger M, David G, Yayon A (1994) Perlecan, basal lamina proteoglycan, promotes basic fibroblast growth factor-receptor binding, mitogenesis, and angiogenesis. Cell 79(6):1005–1013
75. Yayon A, Klagsbrun M, Esko JD, Leder P, Ornitz DM (1991) Cell surface, heparin-like molecules are required for binding of basic fibroblast growth factor to its high affinity receptor. Cell 64(4):841–848
76. Zarnegar R, Michalopoulos GK (1995) The many faces of hepatocyte growth factor: from hepatopoiesis to hematopoiesis. J Cell Biol 129(5):1177–1180
77. Cerra A, Mansfield KJ, Chamberlain CG (2003) Exacerbation of TGF-beta-induced cataract by FGF-2 in cultured rat lenses. Mol Vis 9:689–700
78. Paralkar VM, Vukicevic S, Reddi AH (1991) Transforming growth factor beta type 1 binds to collagen IV of basement membrane matrix: implications for development. Dev Biol 143 (2):303–308
79. Ishida I, Saika S, Okada Y, Ohnishi Y (2005) Growth factor deposition in anterior subcapsular cataract. J Cataract Refract Surg 31(6):1219–1225. doi:10.1016/j.jcrs.2004.11.039
80. Wormstone IM, Anderson IK, Eldred JA, Dawes LJ, Duncan G (2006) Short-term exposure to transforming growth factor beta induces long-term fibrotic responses. Exp Eye Res 83 (5):1238–1245. doi:10.1016/j.exer.2006.06.013
81. Schwartz MA (2001) Integrin signaling revisited. Trends Cell Biol 11(12):466–470
82. Elner SG, Elner VM (1996) The integrin superfamily and the eye. Invest Ophthalmol Vis Sci 37(5):696–701
83. Jester JV, Huang J, Barry-Lane PA, Kao WW, Petroll WM, Cavanagh HD (1999) Transforming growth factor(beta)-mediated corneal myofibroblast differentiation requires actin and fibronectin assembly. Invest Ophthalmol Vis Sci 40(9):1959–1967
84. Jester JV, Petroll WM, Barry PA, Cavanagh HD (1995) Expression of alpha-smooth muscle (alpha-SM) actin during corneal stromal wound healing. Invest Ophthalmol Vis Sci 36 (5):809–819
85. Marcantonio JM, Reddan JR (2004) TGFbeta2 influences alpha5-beta1 integrin distribution in human lens cells. Exp Eye Res 79(3):437–442. doi:10.1016/j.exer.2004.06.014

86. Kass L, Erler JT, Dembo M, Weaver VM (2007) Mammary epithelial cell: influence of extracellular matrix composition and organization during development and tumorigenesis. Int J Biochem Cell Biol 39(11):1987–1994. doi:10.1016/j.biocel.2007.06.025
87. Lygoe KA, Norman JT, Marshall JF, Lewis MP (2004) AlphaV integrins play an important role in myofibroblast differentiation. Wound Repair Regen 12(4):461–470. doi:10.1111/j.1067-1927.2004.12402.x
88. Walker J, Menko AS (2009) Integrins in lens development and disease. Exp Eye Res 88 (2):216–225. doi:10.1016/j.exer.2008.06.020
89. Sponer U, Pieh S, Soleiman A, Skorpik C (2005) Upregulation of alphavbeta6 integrin, a potent TGF-beta1 activator, and posterior capsule opacification. J Cataract Refract Surg 31 (3):595–606. doi:10.1016/j.jcrs.2004.05.058
90. Sheppard D (2004) Roles of alphav integrins in vascular biology and pulmonary pathology. Curr Opin Cell Biol 16(5):552–557. doi:10.1016/j.ceb.2004.06.017
91. Weaver MS, Toida N, Sage EH (2007) Expression of integrin-linked kinase in the murine lens is consistent with its role in epithelial-mesenchymal transition of lens epithelial cells in vitro. Mol Vis 13:707–718
92. de Iongh RU, Wederell E, Lovicu FJ, McAvoy JW (2005) Transforming growth factor-beta-induced epithelial-mesenchymal transition in the lens: a model for cataract formation. Cells Tissues Organs 179(1–2):43–55. doi:10.1159/000084508
93. Li Y, Yang J, Dai C, Wu C, Liu Y (2003) Role for integrin-linked kinase in mediating tubular epithelial to mesenchymal transition and renal interstitial fibrogenesis. J Clin Invest 112 (4):503–516. doi:10.1172/JCI17913

Wound Healing and Epithelial–Mesenchymal Transition in the Lens Epithelium: Roles of Growth Factors and Extracellular Matrix

9

Kumi Shirai, Ai Kitano-Izutani, Takeshi Miyamoto, Sai-ichi Tanaka, and Shizuya Saika

Abstract

The process of tissue fibrosis is characterized by the appearance of myofibroblasts, the key cell type involved in the fibrogenic reaction, and by the excess and disorganized accumulation of extracellular matrix, with resultant tissue contraction and impaired tissue function. Local inflammation is involved in this process with the supply of profibrogenic factors. This is also the case in posterior capsular opacification (PCO). In PCO, tissue myofibroblasts are generated from lens epithelial cells, the only cell lineage in the crystalline lens, through epithelial–mesenchymal transition (EMT), a process through which an epithelial cell changes its phenotype to become more like a mesenchymal cell, with the exception of fibroblast-derived myofibroblasts. Transforming growth factor β (TGFβ) is one of the major growth factors/cytokines involved in the process of EMT, although various other factors expressed by injured tissues orchestrate the EMT process. Among TGFβ signaling cascades, Smad signaling is considered to play a critical role, although other classical mitogen-activated protein kinases also have important roles in modulating lens EMT. The lens epithelium also has the ability to form other types of PCO, regenerated lentoid structures of Soemmering's rings and Elschnig's pearls, both containing crystalline lens-like components. PCO tissue is also characterized by the accumulation of EMT-lens cell-derived matricellular components, i.e., tenascin C, osteopontin, and lumican, that further modulate Smad signaling and EMT of these cells. Fibroblast growth factor reportedly also plays an important role in lens tissue regeneration. PCO-related signal transduction cascades, such as Smad signaling, could be a putative target for the prevention or treatment of unfavorable PCO.

K. Shirai, M.D., Ph.D. • A. Kitano-Izutani, M.D., Ph.D. • T. Miyamoto, M.D., Ph.D.
S.-i. Tanaka, M.D., Ph.D. • S. Saika, M.D., Ph.D. (✉)
Department of Ophthalmology, Wakayama Medical University, 811-1 Kimiidera, Wakayama 641-0012, Japan
e-mail: shizuya@wakayama-med.ac.jp

S. Saika et al. (eds.), *Lens Epithelium and Posterior Capsular Opacification*,
DOI 10.1007/978-4-431-54300-8_9,

Keywords
Epithelial–mesenchymal transition • Myofibroblast • Signal transduction • Smad • Tissue fibrosis • Transforming growth factor β

9.1 Tissue Fibrosis and Epithelial–Mesenchymal Transition (EMT)

Following injury or surgery, the damaged tissue undergoes primary wound healing in order to recover tissue integrity [1]. A complex interplay of cells and extracellular matrix (ECM) takes place for restoration of normal structure and function (tissue remodeling). During this process of tissue repair, growth factors and cytokines orchestrate the behavior of cell types [2, 3]. The main source of such factors is inflammatory cells, although tissue-resident cells also express them. However, failure to restore the normal remodeling process results in the development of tissue fibrosis/scarring and organ dysfunction, characterized by excess and disorganized accumulation and contraction of extracellular matrix [4, 5]. A fibrotic lesion is characterized by persistence of inflammation and presence of myofibroblasts [4, 5], both of which must decrease for the healing process to complete and for the restoration of normal tissue structure and function.

In mesenchymal tissues, myofibroblasts are derived from a local mesenchymal cell, i.e., a fibroblast, through fibroblast–myofibroblast conversion [6–8]. Myofibroblasts are characterized by the expression of α-smooth muscle actin (αSMA), which produces contractile force in a scarred tissue [6–8]. Besides this process, the myofibroblast is also produced through the process of transdifferentiation, in which an epithelial cell changes its phenotype to a (myo)fibroblast, better known as EMT [9–11]. Various ocular and non-ocular tissues/organs are susceptible to diseases related to EMT. In the development of fibrotic lesions in renal or pulmonary fibrosis, myofibroblasts are derived from local fibroblasts and organ-specific epithelial cells. Circulating bone marrow-derived cells (so-called fibrocytes) are also considered to be a part source of tissue myofibroblasts [12] (Fig. 9.1). The crystalline lens is an ectodermal-derived tissue and thus myofibroblasts found in the intact lens must be derived from lens epithelial cells through EMT, as discussed in detail below.

9.2 Roles of EMT in Posterior Capsular Opacification (PCO)

Ocular lens tissue is susceptible to EMT-based fibrotic diseases. Post-cataract surgery fibrosis in lens capsule (PCO) and anterior subcapsular cataract are both caused by EMT of the lens epithelium [13, 14]. In vitro experiments reportedly show evidence that TGFβ is one of the most critical growth factors involved in the EMT of cultured epithelial cell types. The blockage of TGFβ signaling prevents fibrotic behaviors of EMT cells in cell culture [15], although the fibrogenic process

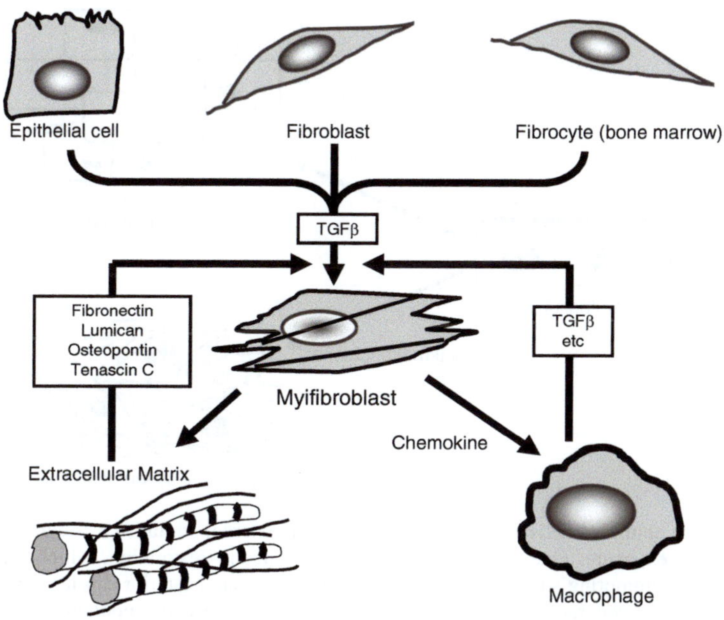

Fig. 9.1 Myofibroblast is derived from multiple origins. The myofibroblast, the main player in the process of tissue fibrosis, is derived from either a local fibroblast or an epithelial cell. Circulating bone marrow-derived cells (fibrocytes) are also presumed to be one of its origins. The myofibroblast exerts a critical role in the process of tissue fibrosis and extracellular matrix reconstruction. Accumulated matrix components and cytokines/chemokines expressed by infiltrated inflammatory cells further modulate myofibroblast generation and tissue fibrosis (reproduced from Saika et al. [70])

of EMT is regulated by other growth factors and cytokines. For example, connective tissue growth factor or platelet-derived growth factors are involved in the generation of myofibroblasts and tissue fibrosis [16]. In cell culture studies that utilize a uniform cell type or a cell line, it is easier to identify the specific signaling cascade(s) downstream of growth factor receptors that are involved in the EMT process [17]. However, such in vitro studies may not accurately reflect the in vivo situation given the heterogeneity of cellular components in one specific organ; an organ may be composed of epithelial cell lineages and mesenchymal cell lineages.

The crystalline lens is solely of ectodermal origin, has a relatively simple structure, and is wrapped by a specific basement membrane, the lens capsule (see Chap. 3) [18]. During embryonic development, the lens is formed by the invagination of surface head ectoderm overlying the optic cup. The lens basement membrane, the lens capsule, envelopes its outer surface and the lens epithelial cells line the inner surface of the anterior hemisphere of the capsule. Any damage or break to basement membranes could induce transdifferentiation of epithelium towards a mesenchymal-like phenotype and/or cell motility in various tissues, leading to malignant neoplasm [19]. This is also true for the crystalline lens. Once injured (including through surgical trauma), a rupture of the anterior lens capsule breaks

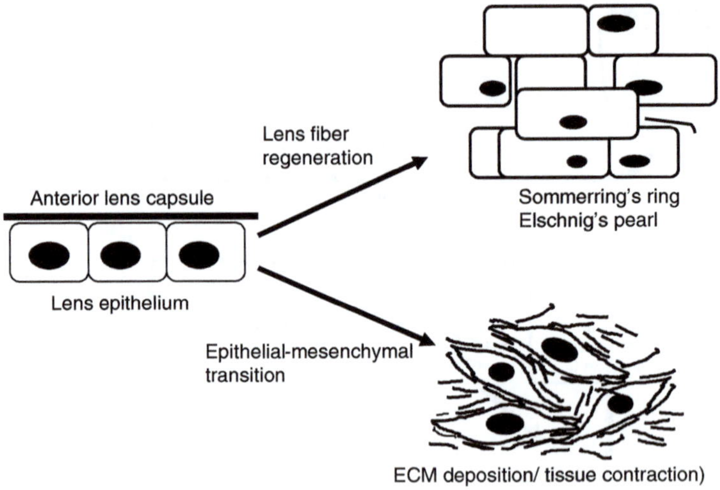

Fig. 9.2 Lens epithelial cells transform two ways upon external stimuli, i.e., trauma and excess growth factor exposure. One is the process of lens structure restoration, which leads to the formation of Soemmering's ring and Elschnig's pearls in human posterior capsular opacification. Another way is the generation of myofibroblasts through a process called epithelial–mesenchymal transition. Myofibroblasts derived from the lens epithelium express fibrous extracellular matrix components and also exert a contractile characteristic

cell–cell contact among the lens epithelium and subsequently results in the formation of fibrous tissue that contains significant numbers of myofibroblasts [20]. The dissociation of cell–cell contact evokes intracellular signaling in response to growth factors [21]. The myofibroblast in an injured lens or in the human capsular bag following cataract extraction is generated through the process of EMT of lens epithelial cells (Fig. 9.2). During the developmental process of this PCO, fibrous extracellular matrix components, including collagen types I and III, proteoglycans, laminin, etc., are accumulated by EMT-derived myofibroblasts on the inner surface of the residual lens capsule [22]. This newly formed tissue in the fibrous PCO is very similar to that observed in the anterior subcapsular cataract.

9.3 Quick Review of TGFβ Signal Transduction

Although EMT in epithelia and subsequent tissue fibrosis are orchestrated by growth factors and cytokines, it is widely believed that, especially, the fibrogenic cytokine transforming growth factor β (TGFβ) is largely involved in various organs [23]. TGFβ family members, i.e., TGFβ1–β3, utilize the Smad (Smad2, Smad3, and Smad4) signaling pathway that is specific to these members of TGFβ superfamily, as well as classical signaling cascades of the mitogen-activated protein kinase (MAP kinase), c-Jun N-terminal kinase (JNK), and p38 MAPK [24, 25]. Other TGFβ superfamily members, such as bone morphogenetic proteins (BMPs), also

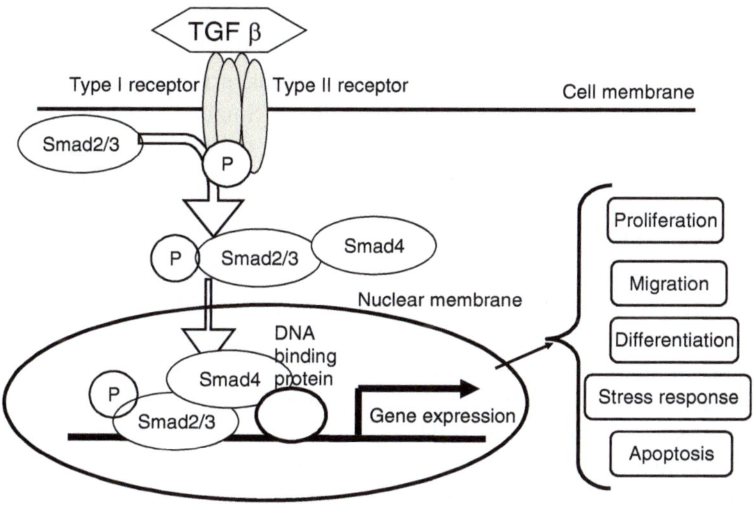

Fig. 9.3 TGFβ activates Smad2/Smad3 signaling cascade besides classical mitogen-activated protein kinase pathways. Smad2 or Smad3 is phosphorylated at its C-terminal region upon TGFβ binding to its specific receptor. Phosphorylated Smad2 or Smad3 binds Smad4, the common Smad, and forms a Smad complex that then translocates to the nucleus to bind to gene promoters. Various cofactors, i.e., AP-1 and β-catenin, modulate Smad-dependent gene expression (reproduced from Saika et al. [71] with a minor modification)

utilize different Smad members (Smad1, Smad5, and Smad8) for transmitting their signaling [26].

TGFβ family members exist in tissues or in aqueous humor as latent forms and are rapidly activated by various stimuli. The activation of TGFβ2 that predominates in the aqueous humor [27] seems to be different from that of TGFβ1 and TGFβ3, as latent TGFβ2 lacks the RGD sequence that is required for its activation on the cell surface. Upon ligand (TGFβ) binding to its receptor, a pair of transmembrane receptor serine–threonine kinases are activated. Receptor-activated Smad proteins, Smad2 and Smad3, are phosphorylated (at their C-terminals) by the TGFβ receptor type I kinase (ALK5). Smad2 and Smad3 then partner with the common mediator, Smad4, and translocate to the nucleus where they modulate the expression of TGFβ-/Smad-dependent gene targets (Fig. 9.3). The roles of Smad2 and Smad3 differ given the loss of Smad2 is required for embryonic mouse viability, whereas mice lacking Smad3 survive [28]. In vitro cell culture studies using fibroblasts derived from mouse embryos lacking either Smad2 or Smad3 provided detailed data concerning the difference of gene expression modulation by each of Smad2 or Smad3 signaling [29]. For example, TGFβ1-mediated induction of matrix metalloproteinase-2 was selectively dependent on Smad2, whereas induction of c-fos, Smad7, and TGFβ1 auto-induction relied on the expression of Smad3. Smad6 and Smad7 are known to be the inhibitory Smads that block the phosphorylation of Smad2/Smad3 [24]. Besides the phosphorylation of the C-terminal of Smad2 or Smad3, the middle-linker region of each Smad is also susceptible to the phosphorylation by MAP kinase. However, the roles of the phosphorylation of this Smad2/Smad3 middle-linker region in the process of EMT are not fully understood. In general, the phosphorylation in the middle-linker region of Smads inhibits signaling mediated by the phosphorylation of the Smad C-terminal, i.e., leading to EMT and apoptosis. However, the roles of this middle-linker Smad region in EMT are still controversial [30, 31]. EMT in lung epithelial cells is mediated by both the C-terminal and the middle-linker phosphorylation of Smad3, leading to c-Jun N-terminal kinase activity [32].

As mentioned earlier, the BMPs, members of the TGFβ superfamily, bind to their own receptors and phosphorylate Smad1, Smad5, and Smad8, which then bind to Smad4 for translocation to the nucleus. BMP signaling upregulates Id2 or Id3, both of which attenuate Smad2/Smad3 signals [33] (Fig. 9.6). In vitro experiments showed that various signal transmitters including Smad2 or Smad3 are critical in the EMT process. However, no reliable in vivo data on the roles of these signaling molecules is available. This results from the fact that it is not easy to define the origin of local myofibroblasts in fibrosis/scarring lesions.

Wnt/β-catenin signaling is also involved in TGFβ-mediated EMT of cultured renal epithelial cells via cross talk with the TGFβ/Smad pathway and is involved in the profibrotic process [21, 34]. The molecular mechanism of this cross talk between TGFβ/Smad3 and Wnt/β-catenin signaling pathways involves the C-terminal Smad3 interacting with both the N-terminal region and the middle region of β-catenin protein in a TGFβ-dependent manner. The interaction of β-catenin with Smad3 and Smad4 protects β-catenin from an ubiquitin–proteasome-dependent degradation, promoting signaling [35].

9.4 Modulation of Lens Epithelium EMT by TGFβ Signaling

In vitro experiments reportedly show evidence that TGFβ is one of the most critical growth factors involved in the EMT of cultured lens epithelial cells, and the blockage of TGFβ signaling prevents fibrotic behavior of cells [36]. The role of TGFβ signaling in the lens epithelium in situ was not fully investigated until its role in wound healing in mouse lines was reported [37]. Postoperative EMT in the lens epithelium and formation of fibrotic tissue inside the lens capsule can be modeled by a puncture injury in the mouse lens. Following this puncture injury in the mouse lens, Smad4, the common Smad in TGFβ/Smad signaling, translocates to the nucleus of the lens epithelium within 12 h post-injury [20] (Fig. 9.4), indicating that TGFβ signaling is activated during this period. The Smad4 nuclear translocation is blocked by intraocular administration of anti-TGFβ2 neutralizing antibody, indicating that TGFβ2 is the TGFβ family member that activates the lens epithelium post-injury also in mice [20]. Given TGFβ2 predominates in the aqueous humor in humans and that it does not possess the RGD sequence-dependent activation of its latent form, this indicates that TGFβ2 is activated in the aqueous humor in a different mode to TGFβ1 or TGFβ3, although the exact mechanism of its activation upon puncture injury is still not clear. The lens epithelial cells adjacent to the injury

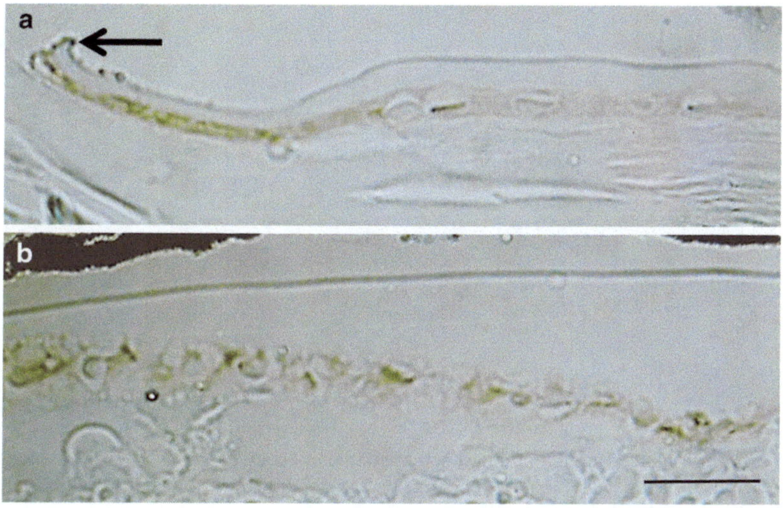

Fig. 9.4 Immediate activation of Smad4 signal upon breaking injury in the anterior capsule in a mouse crystalline lens. In an uninjured mouse lens, Smad4 locates to the cytoplasm of in vivo mouse lens epithelial cells (not shown). Once the center of the anterior capsule is broken using a needle, Smad4 translocates to the nuclei within 12 h in the area adjacent to the capsular break (**a**) and then Smad4 nuclear translocation is observed in the mid-peripheral area of the lens epithelial cells by 24 h post-injury (**b**). *Arrow* indicates a break in the anterior lens capsule. *Bar*: 50 μm

undergo EMT and form fibrotic tissue at this site. The lens is unique in that it contains an epithelial cell lineage, with the lens epithelium and lens fibers derived from the epithelial cells. Therefore, myofibroblasts observed in an injured lens are most likely derived from EMT cells. This puncture injury-induced EMT and fibrotic tissue formation is not observed in Smad3-null mice (Fig. 9.5), indicating that such an EMT is mediated by Smad3 signaling [38]. However, high level of active TGFβ1 overexpression under the αA-crystallin promoter in a Smad3-null lens can still induce EMT of the epithelium, leading to the formation of subcapsular fibrotic plaque lesions [39]. The detailed signaling mechanism that allows the cells to bypass Smad3 signaling to the induction of an EMT is still unknown. However, forced overexpression of TGFβ1 might utilize Smad2 or other signaling cascades. TGFβ/non-Smad signaling is also involved in steps during the EMT process. Signals such as Rho kinase, PI3 kinase, or Src are reported to modulate the process of EMT in lens cells in culture [40–42]. However, the in vivo involvement has not been fully uncovered.

Not only Smad but also other signaling cascades are involved in EMT and PCO formation as reviewed by Martinez and de Iongh [14]. TGFβ also activates MAP kinases, the classical growth factor signals, upon ligand binding to the receptor. Like other growth factor signaling TGFβ-signaling cascades via MAP kinases include three subfamilies, i.e., p42/44 ERK (external regulated kinase), c-Jun N-terminal kinase (JNK), and p38 MAP kinase. These signaling pathways mediate different biological responses. The p42/44 MAP kinase/Erk signal is

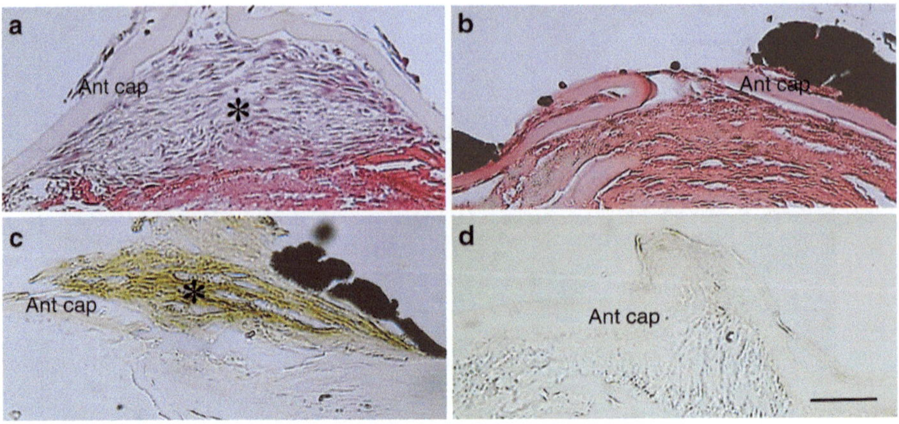

Fig. 9.5 Injury-induced epithelial–mesenchymal transition as examined by the expression of α-smooth muscle actin was blocked by Smad3 gene ablation in a mouse crystalline lens. The Smad3-null mouse develops a crystalline lens of a normal structure (not shown). Frame (**a**) shows the central anterior lens capsule. At week four post-puncture/post-injury, the cells form a multilayer (*asterisk*) of elongated fibroblast-like cells that are generated through epithelial–mesenchymal transition in wild-type mice (**a**), whereas such tissue is not observed in Smad3-null injured lens (**b**). Immunohistochemical examination shows that the elongated cells in the multilayer (*asterisk*) are labeled for α-smooth muscle actin, the marker for myofibroblast (**c**). No such accumulation of myofibroblasts is observed in a Smad3-null injured anterior lens (**d**). *Bar*: 50 μm

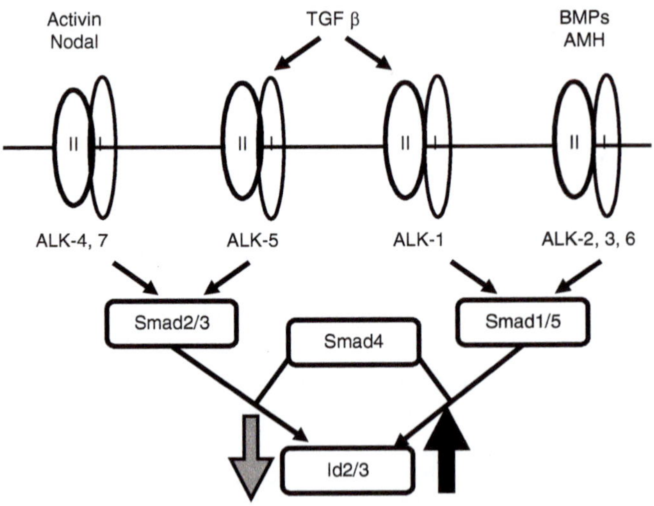

Fig. 9.6 Smad and Id (inhibitory factor of differentiation)2/Id3 expression. Smad signaling network and expression of Id2 and Id3. Signaling cascades derived from TGFβ or activin suppress the expression of Id2 and Id3. On the other hand, Smad1 upregulates the expression of these two genes, Id2 and Id3. Id2 and Id3 block TGFβ/Smad signaling and thus exhibit an inhibitory effect on epithelial–mesenchymal transition. In certain cell types, TGFβ transduces Smad1/Smad5 activation though ALK1 (reproduced from Saika [72])

involved in the regulation of cell proliferation, whereas JNK and p38 are activated upon cellular stresses and modulate cell survival/cell death or control expression of stress-response genes [43]. MAP kinases modulate Smad signaling via cross talk. Besides the canonical phosphorylation of Smad molecules in the C-terminal region, the middle-linker region of Smad2 or Smad3 can be phosphorylated by JNK or p38 MAP kinase following the activation of various ligands and/or external stimuli. It is quite possible that such phosphorylation in the Smad middle-linker region by such signals might further modulate Smad function during the processes of wound healing or tissue fibrosis in vivo, as observed in vitro [30]. Smad signals can also be modulated by cross talks between non-MAP kinase signals. For example, the upregulation of Smad7, nuclear factor-κB, and interferon-γ inhibit the TGFβ/Smad signal [44–47]. In vivo lens epithelium expresses Sprouty 1 and Sprouty 2, family members of endogenous receptor tyrosine kinase inhibitors. Sprouty counteracts TGFβ activation of Smad and Erk and subsequently blocks EMT of the lens epithelium in in vivo that was shown by using conditional mutant mice. On the other hand, the deletion of Sprouty 1 and Sprouty 2 leads to spontaneous EMT–fibrosis in a mouse lens [48]. EMT-related signaling cascades finally modulate the expression of EMT-related transcription factors, i.e., ZEB (*Sip1*/*δEF1*), bHLH (*E47*/*Twist*), and *Snail1*/*Snail2* [49].

Smad utilizes β-catenin as a cofactor during gene expression regulation. Wnt5a, Wnt5b, Wnt7b, Wnt8a, and Wnt8b and their frizzled receptors are upregulated in association with TGFβ-induced EMT and cataract development in animals [50]. EMT of lens epithelial cells is also observed in the process of PCO formation. In vitro studies showed that the overexpression of Wnt3a resulted in the upregulated expression of β-catenin, c-Myc, and cyclin D1 [51]. Wnt signal is also involved in EMT of lens epithelial cells, like in TGFβ-induced EMT in other systems. The expression of the epithelial marker E-cadherin was downregulated in Wnt3a-overexpressing HLE B-3 cells, whereas that of the mesenchymal marker fibronectin was upregulated [14, 51]. Wnt3a promotes epithelial–mesenchymal transition, migration, and proliferation of lens epithelial cells [51]. Together, it is quite possible that Wnt/β-catenin signaling is involved in the EMT of PCO development by the lens epithelium in humans.

9.5 Modulation of Lens Cell EMT by Extracellular Matrix Components

During the process of development of PCO, EMT myofibroblasts express various fibrogenic extracellular matrix components, i.e., fibrillar or basement membrane-type collagens, proteoglycans, laminin, fibronectin, hyaluronan, etc. [13, 52]. Cell culture studies report that such extracellular matrix macromolecules affect behaviors of cells, i.e., adhesion, migration, proliferation, and cell death (apoptosis) [53]. Similar effects of extracellular matrix components on lens epithelial cell behavior, e.g., EMT, were reported. Overall non-collagenous matricellular proteins modulate EMT and fibrogenic response of lens epithelial cells via affecting

TGFβ/Smad signaling and/or other signaling cascades. For example, osteopontin or tenascin C is expressed in tissues as a wound healing-related component [54]. The former protein is originally found as an immune modulator and the latter is known to be an anti-adhesive wound healing-related molecule. The lens epithelium begins to express osteopontin or tenascin C following injury or cataract surgery in patients, as well as in mice. In mice, the loss of osteopontin or tenascin C delayed the activation of Smad2/Smad3 and subsequent injury-induced EMT in the lens epithelium [55, 56]. Lumican is a core protein component of keratan sulfate proteoglycan. Although lumican (keratan sulfate proteoglycan) is detected in corneal stroma and required for the maintenance of corneal transparency, lumican glycoprotein is expressed in other tissues as a wound healing-related component, like tenascin C or osteopontin [57]. Lumican is also expressed in human PCO and mouse lens epithelium post-injury. The loss of lumican impairs injured EMT of lens epithelial cells in mice [58]. Cell culture studies showed that vitronectin, fibronectin, and collagen type I can also promote lens epithelial cell EMT through enhancement of Smad signaling [59, 60]. However, the exact mechanism by which such ECM molecules affect Smad signaling and the process of EMT is unknown. An explanation may include that integrin-mediated cytoplasmic signaling might affect Smad signaling or other signal(s) that is (are) involved in the process of EMT or that extracellular matrix components modulate TGFβ binding to its cell surface receptor. The expression pattern of matrix receptors might also be altered during tissue repair process in the lens. For example, CD44, a receptor for hyaluronan, is upregulated in human PCO samples or in an injured mouse lens epithelium [61, 62]. Integrin expression patterns are also affected by a puncture injury in animal lens, suggesting this may be the case in human lens epithelium post-cataract surgery.

Signaling mechanisms that promote lens fiber-type PCO are yet to be uncovered. However, the lens fiber differentiation system might be related to the development of this type of PCO, namely, Soemmering's ring and Elschnig's pearls. Both fibroblast growth factor signaling and Wnt–frizzled signaling are reportedly required for lens fiber differentiation in the physiological condition [63]. It is to be investigated if Wnt–frizzled system is also involved in postoperative lens fiber differentiation that forms Elschnig's pearls and Soemmering's rings. Recently, it was reported that a novel protein, Equarin, which is a secreted molecule expressed at the equatorial region of the lens, also plays an important role in chick lens fiber differentiation through fibroblast growth factor signaling [64].

9.6 Inhibiting EMT-Related PCO by Targeting TGFβ Signals: Experimental Approaches

The inhibition of proliferation and EMT of the lens epithelium migt be included in future strategies for the prevention of fibrous post-cataract surgery PCO. Various experimental trials have been tested to block the cell proliferation on the residual lens capsule post-cataract extraction. Antiproliferative agents, e.g., mitomycin C,

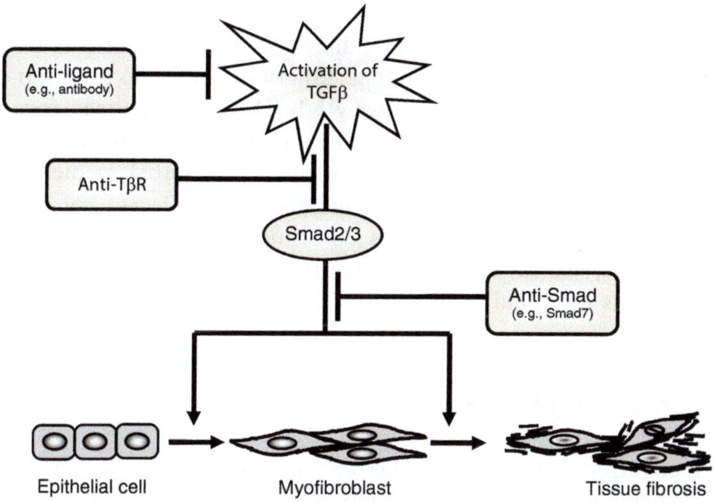

Fig. 9.7 Strategies of inhibition of epithelial–mesenchymal transition (EMT) by targeting TGFβ actions. Lens epithelial cells acquire a myofibroblastic phenotype in association with the expression of α-smooth muscle actin and extracellular matrix components upon injury, surgery, or exposure to excess growth factors. This process is regulated mainly by TGFβ/Smad signaling. Therapeutic strategies against EMT include deletion of a ligand by administration of neutralizing antibodies, for example, blocking ligand binding to its specific receptor using anti-TGFβ receptor antibodies or by signaling inhibition using anti-Smad gene transfer techniques (reproduced from Saika et al. [70])

reportedly exhibited anti-PCO effects in animals, but have not yet been applied to human patients [65]. Cell culture studies suggest that blocking growth factor receptors or signaling cascades may also be effective [38]. As for the strategies that focus on anti-EMT effects, the inhibition of Smad signaling might be a possible concept for the prevention of PCO formation. Blocking antibodies against specific growth factors/cytokines is widely applied in the treatment of inflammatory diseases or pathological neovascularization. For example, neutralizing antibodies against vascular endothelial growth factor (VEGF) are clinically used in the treatment of malignancy or in VEGF-involved eye diseases, i.e., diabetic retinopathy or age-related macular degeneration. Antibodies against tumor necrosis factor α (TNFα) have anti-rheumatoid arthritis effects and therapeutic effects against persistent uveitis of Behcet's disease. Fibrogenic effects, including induction of EMT, can be suitable targets for the prevention of PCO, although they are not yet clinically applied.

The loss of Smad3 markedly impairs injury-induced EMT in the lens epithelium in mice, suggesting that blocking the Smad2/Smad3 signal might be a reasonable strategy to inhibit EMT in the lens epithelium post-injury [38]. Adenoviral gene transfer of anti-Smad components, i.e., Smad7, bone morphogenetic protein-7 (BMP-7), or inhibitors of differentiation 2/3, also blocks injury-induced lens epithelium EMT in mice [33, 66] (Fig. 9.7). These findings suggest that we will be able to use natural compounds that possess anti-Smad effects, for the purpose of prevention of PCO. Natural compounds that have anti-Smad effects may also be candidates for anti-fibrotics. Halofuginone, otherwise known as 7-bromo-6-chloro-3-[3-(3-hydroxy-2-piperidinyl)-2-oxopropyl]-4(3H)-quinazolinone (one of the quinazolinone derivatives), reportedly blocks tissue fibrosis by the upregulation of Smad7, the inhibitory Smad [67]. Constituents of herbal medicines are also potential candidates for anti-fibrogenic/inflammatory effects. For example, an herbal medicine, Inchin-ko-to, and its constituents, genipin and emodin, have anti-TGFβ effects, as revealed by experiments using cultured fibroblasts or α-TN4 mouse lens epithelial cell line [68–70].

9.7 Concluding Remarks

It is well recognized that inhibiting TGFβ/Smad2/Smad3 signals is one of the most promising therapeutics in the prevention/treatment of EMT-related fibrotic diseases. Although PCO is currently surgically treated by Nd:YAG laser, the procedure potentially increases the risk of rhegmatogenous retinal detachment. PCO in children and infants has to be blocked to escape the development of amblyopia. Gene therapy or strategies of non-gene therapy are to be clinically established to inhibit the development of PCO [71, 73].

Acknowledgment The authors express deep thanks to Dr. John W. McAvoy and Dr. Frank J Lovicu, Save Sight Institute, University of Sydney, Sydney, New South Wales, Australia, for their critical review of the manuscript.

References

1. Gurtner GC, Werner S, Barrandon Y, Longaker MT (2008) Wound repair and regeneration. Nature 453:314–321
2. Kiwanuka E, Junker J, Eriksson E (2012) Harnessing growth factors to influence wound healing. Clin Plast Surg 39:239–248
3. Douglas HE (2010) TGF-ß in wound healing: a review. J Wound Care 19:403–406
4. Wynn TA (2008) Cellular and molecular mechanisms of fibrosis. J Pathol 214:199–210
5. Tomasek JJ, Gabbiani G, Hinz B, Chaponnier C, Brown RA (2002) Myofibroblasts and mechano-regulation of connective tissue remodeling. Nat Rev Mol Cell Biol 3:349–363
6. Gabbiani G (2003) The myofibroblast in wound healing and fibrocontractive diseases. J Pathol 200:500–503
7. Hinz B, Gabbiani G (2003) Cell-matrix and cell-cell contacts of myofibroblasts: role in connective tissue remodeling. Thromb Haemost 90:993–1002
8. Micallef L, Vedrenne N, Billet F, Coulomb B, Darby IA, Desmouliere A (2012) The myofibroblast, multiple origins for major roles in normal and pathological tissue repair. Fibrogenesis Tissue Repair 5(Suppl 1):S5
9. Willis BC, duBois RM, Borok Z (2006) Epithelial origin of myofibroblasts during fibrosis in the lung. Proc Am Thorac Soc 3:377–382
10. Kalluri R, Neilson EG (2003) Epithelial-mesenchymal transition and its implications for fibrosis. J Clin Invest 112:1776–1784

11. Zeisberg M, Kalluri R (2004) The role of epithelial-to-mesenchymal transition in renal fibrosis. J Mol Med 82:175–181
12. Quan TE, Cowper SE, Bucala R (2006) The role of circulating fibrocytes in fibrosis. Curr Rheumatol Rep 8:145–150
13. de Iongh RU, Wederell E, Lovicu FJ, McAvoy JW (2005) Transforming growth factor-β-induced epithelial-mesenchymal transition in the lens: a model for cataract formation. Cells Tissues Organs 179:43–55
14. Martinez G, de Iongh RU (2010) The lens epithelium in ocular health and disease. Int J Biochem Cell Biol 42:1945–1963
15. Ishida W, Mori Y, Lakos G, Sun L, Shan F, Bowes S, Josiah S, Lee WC, Singh J, Ling LE, Varga J (2006) Intracellular TGF-β receptor blockade abrogates Smad-dependent fibroblast activation in vitro and in vivo. J Invest Dermatol 126:1733–1744
16. Xiao H, Wei H, Yang GB, Peng HL, Zhang C (2011) Effects of paeoniflorin on expressions of CTGF, PDGF and TNF-α in mice with hepatic fibrosis due to *Schistosoma japonicum* infection. Zhongguo Xue Xi Chong Bing Fang Zhi Za Zhi 23:288–291
17. Zavadil J, Böttinger EP (2005) TGF-β and epithelial-to-mesenchymal transitions. Oncogene 24:5764–5774
18. Menko S, Philp N, Veneziale B, Walker J (1998) Integrins and development: how might these receptors regulate differentiation of the lens. Ann N Y Acad Sci 842:36–41
19. Paterson FL, Kazenwadel J, Bert AG, Khew-Goodall Y, Ruszkiewicz A, Goodall GJ (2013) Extracellular matrix remodeling and cellular differentiation. Neoplasia 15:180–191
20. Saika S, Okada Y, Miyamoto T, Ohnishi Y, Ooshima A, McAvoy JW (2001) Smad translocation and growth suppression in lens epithelial cells by endogenous TGFβ2 during wound repair. Exp Eye Res 72:679–686
21. Masszi A, Fan L, Rosivall L, McCulloch CA, Rotstein OD, Mucsi I, Kapus A (2004) Integrity of cell-cell contacts is a critical regulator of TGF-β 1-induced epithelial-to-myofibroblast transition: role for β-catenin. Am J Pathol 165:1955–1967
22. Saika S (2004) Intraocular lens: biocompatibility related to lens epithelial cells. Prog Retinal Eye Res 23:283–305
23. Biernacka A, Dobaczewski M, Frangogiannis NG (2011) TGF-β signaling in fibrosis. Growth Factor 29:196–202
24. Massague J (2012) TGFβ signaling in context 13:616–630
25. Mu Y, Gudey SK, Landstrom M (2012) Non-Smad signaling pathways. Cell Tissue Res 347:11–20
26. Itoh S, Itoh F, Goumans MJ, Ten Dijke P (2000) Signaling of transforming growth factor-β family members through Smad proteins. Eur J Biochem 267:6954–6967
27. Jampel HD, Roche N, Stark WJ, Roberts AB (1990) Transforming growth factor-β in human aqueous humor. Curr Eye Res 9:963–969
28. Weinstein M, Yang X, Deng C (2000) Functions of mammalian Smad genes as revealed by targeted gene disruption in mice. Cytokine Growth Factor Rev 11:49–58
29. Piek E, Ju WJ, Heyer J, Escalante-Alcalde D, Stewart CL, Weinstein M, Deng C, Kucherlapati R, Bottinger EP, Roberts AB (2001) Functional characterization of transforming growth factor β signaling in Smad2- and Smad3-deficient fibroblasts. J Biol Chem 276:19945–19953
30. Kamato D, Burch ML, Piva TJ, Rezaei HB, Rostam MA, Xu S, Zheng W, Little PJ, Osman N (2013) Transforming growth factor-β signaling: role and consequences of Smad linker region phosphorylation. Cell Signal 25:2017–2024
31. Matsuzaki K (2011) Smad phosphoisoform signaling specificity: the right place at the right time. Carcinogenesis 32:1578–1588
32. Velden JL, Alcorn JF, Guala AS, Badura EC, Janssen-Heininger YM (2011) c-Jun N-terminal kinase 1 promotes transforming growth factor-β1-induced epithelial-to-mesenchymal transition via control of linker phosphorylation and transcriptional activity of Smad3. Am J Respir Cell Mol Biol 44:571–581

33. Saika S, Ikeda K, Yamanaka O, Flanders KC, Ohnishi Y, Nakajima Y, Muragaki Y, Ooshima A (2006) Adenoviral gene transfer of BMP-7, Id2, or Id3 suppresses injury-induced epithelial-to-mesenchymal transition of lens epithelium in mice. Am J Physiol Cell Physiol 290:282–289
34. Tian X, Zhang J, Tan TK, Lyons JG, Zhao H, Niu B, Lee SR, Tsatralis T, Zhao Y, Wang Y, Cao Q, Wang C, Wang Y, Lee VW, Kahn M, Zheng G, Harris DC (2013) Association of β-catenin with P-Smad3 but not LEF-1 dissociates in vitro profibrotic from anti-inflammatory effects of TGF-β1. J Cell Sci 126:67–76
35. Zhang M, Wang M, Tan X, Li TF, Zhang YE, Chen D (2010) Smad3 prevents β-catenin degradation and facilitates β-catenin nuclear translocation in chondrocytes. J Biol Chem 285:8703–8710
36. Xu J, Lamouille S, Derynck R (2009) TGF-β-induced epithelial to mesenchymal transition. Cell Res 19:156–172
37. Ashcroft GS, Yang X, Glick AB, Weinstein M, Letterio JL, Mizel DE, Anzano M, Greenwell-Wild T, Wahl SM, Deng C, Roberts AB (1999) Mice lacking Smad3 show accelerated wound healing and an impaired local inflammatory response. Nat Cell Biol 1:260–266
38. Saika S, Kono-Saika S, Ohnishi Y, Sato M, Muragaki Y, Ooshima A, Flanders KC, Yoo J, Anzono M, Liu C-Y, Kao WW-Y, Roberts AB (2004) Smad3 signaling is required for epithelial-mesenchymal transition of lens epithelium after injury. Am J Pathol 164:651–663
39. Banh A, Deschamps PA, Gauldie J, Overbeek PA, Sivak JG, West-Mays JA (2006) Lens-specific expression of TGF-β induces anterior subcapsular cataract formation in the absence of Smad3. Invest Ophthalmol Vis Sci 47:3450–3460
40. Cho HJ, Yoo J (2007) Rho activation is required for transforming growth factor-β-induced epithelial-mesenchymal transition in lens epithelial cells. Cell Biol Int 31:1225–1230
41. Cho HJ, Baek KE, Saika S, Jeong MJ, Yoo J (2007) Snail is required for transforming growth factor-β-induced epithelial-mesenchymal transition by activating PI3 kinase/Akt signal pathway. Biochem Biophys Res Commun 353:337–343
42. Walker JL, Wolff IM, Zhang L, Menko AS (2007) Activation of SRC kinases signals induction of posterior capsule opacification. Invest Ophthalmol Vis Sci 48:2214–2223
43. Ferrer I, Blanco R, Carmona M, Puig B, Dominguez I, Vinals F (2002) Active, phosphorylation-dependent MAP kinases, MAPK/ERK, SAPK/JNK and p38, and specific transcription factor substrates are differentially expressed following systemic administration of kainic acid to the adult rat. Acta Neuropathol 103:391–407
44. Saika S, Ikeda K, Yamanaka O, Miyamoto T, Ohnishi Y, Sato M, Muragaki Y, Ooshima A, Nakajima Y, Kao WW-Y, Flanders KC, Roberts AB (2005) Expression of Smad7 in mouse eyes accelerates healing of corneal tissue following exposure to alkali. Am J Pathol 166:1405–1418
45. Bitzer M, von Gersdorff G, Liang D, Dominguez-Rosales A, Beg AA, Rojkind M, Bottinger EP (2000) A mechanism of suppression of TGF-β/SMAD signaling by NF-κB/RelA. Genes Dev 14:187–197
46. Weng H, Mertens PR, Gressner AM, Dooley S (2007) IFN-gamma abrogates profibrogenic TGF-β signaling in liver by targeting expression of inhibitory and receptor Smads. J Hepatol 46:295–303
47. Ulloa L, Doody J, Massague J (1999) Inhibition of transforming growth factor-β/SMAD signaling by the interferon-γ/STAT pathway. Nature 397:710–713
48. Shin EH, Basson MA, Robinson ML, McAvoy JW, Lovicu FJ (2012) Sprouty is a negative regulator of transforming growth factor β-induced epithelial-to-mesenchymal transition and cataract. Mol Med 18:861–873
49. Zheng H, Kang Y (2014) Multilayer control of the EMT master regulators. Oncogene 33:1755–1763
50. Chong CC, Stump RJ, Lovicu FJ, McAvoy JW (2009) TGFbeta promotes Wnt expression during cataract development. Exp Eye Res 88:307–313
51. Bao XL, Song H, Chen Z, Tang X (2012) Wnt3a promotes epithelial-mesenchymal transition, migration, and proliferation of lens epithelial cells. Mol Vis 18:1983–1990

52. Wormstone IM (2002) Posterior capsule opacification: a cell biological perspective. Exp Eye Res 74:337–347
53. Streuli C (1999) Extracellular matrix remodeling and cellular differentiation. Curr Opin Cell Biol 11:634–640
54. Saika S, Sumioka T, Okada Y, Yamanaka O, Kitano A, Miyamoto T, Shirai K, Kokado H (2013) Wakayama symposium: modulation of wound healing response in the corneal stroma by osteopontin and tenascin-C. Ocul Surf 11:12–15
55. Saika S, Shirai K, Yamanaka O, Miyazaki K, Okada Y, Kitano A, Flanders KC, Kon S, Uede T, Kao WW-Y, Rittling S, Denhardt D (2007) Loss of osteopontin perturbs the epithelial-mesenchymal transition in an injured mouse lens epithelium. Lab Invest 87:130–138
56. Tanaka S, Sumioka T, Fujita N, Kitano A, Okada Y, Yamanaka O, Flanders KC, Miyajima M, Saika S (2010) Suppression of injury-induced epithelial-mesenchymal transition in a mouse lens epithelium lacking tenascin-C. Mol Vis 16:1194–1205
57. Saika S, Shiraishi A, Saika S, Liu C-Y, Funderburgh JL, Kao CW-C, Converse RL, Kao WW-Y (2000) Role of lumican in the corneal epithelium during wound healing. J Biol Chem 275:2607–2612
58. Saika S, Miyamoto T, Tanaka S, Tanaka T, Ishida I, Ohnishi Y, Ooshima A, Ishiwata T, Assano G, Chikama T-I, Shiraishi A, Liu C-Y, Kao CW-C, Kao WW-Y (2003) Response of lens epithelial cells to injury: role of lumican in epithelial-mesenchymal transition. Invest Ophthalmol Vis Sci 44:2094–2102
59. Taliana L, Evans MD, Ang S, McAvoy JW (2006) Vitronectin is present in epithelial cells of the intact lens and promotes epithelial mesenchymal transition in lens epithelial explants. Mol Vis 12:1233–1242
60. Shintani Y, Maeda M, Chaika N, Johnson KR, Wheelock MJ (2008) Collagen I promotes EMT in lung cancer cells via TGF-β3 signaling. Am J Respir Cell Mol Biol 38:95–104
61. Saika S, Kawashima Y, Miyamoto T, Okada Y, Tanaka S, Yamanaka O, Ohnishi Y, Ooshima A, Yamanaka A (1998) Immunolocalization of hyaluronan and CD44 in quiescent and proliferating human lens epithelial cells. J Cataract Refract Surg 24:1266–1270
62. Desai VD, Wang Y, Simirskii VN, Duncan MK (2010) CD44 expression is developmentally regulated in the mouse lens and increases in the lens epithelium after injury. Differentiation 79:111–119
63. Dawes LJ, Sugiyama Y, Tanedo AS, Lovicu FJ, McAvoy JW (2013) Wnt-frizzled signaling is part of an FGF-induced cascade that promotes lens fiber differentiation. Invest Ophthalmol Vis Sci 54:1582–1590
64. Song X, Sato Y, Felemban A, Ito A, Hossain M, Ochiai H, Yamamoto T, Sekiguchi K, Tanaka H, Ohta K (2012) Equarin is involved as an FGF signaling modulator in chick lens differentiation. Dev Biol 368:109–117
65. Miyamoto T, Saika S, Okada Y, Ishida-Nishikawa I, Sumioka T, Fujita N, Ohnishi Y (2004) Topical exposure of mitomycin C reduces opacification of the residual anterior lens capsule and lenticular regeneration after vitrectomy and lensectomy in rabbits. Graefes Arch Clin Exp Ophthalmol 242:327–331
66. Saika S, Ikeda K, Yamanaka O, Sato M, Muragaki Y, Ohnishi Y, Ooshima A, Nakajima Y, Namikawa K, Kiyama H, Flanders KC, Roberts AB (2004) Transient adenoviral gene transfer of Smad7 prevents injury-induced epithelial-mesenchymal transition of lens epithelium in mice. Lab Invest 84:1245–1258
67. Xavier S, Piek E, Fujii M, Javelaud D, Mauviel A, Flanders KC, Samuni AM, Felici A, Reiss M, Yarkoni S, Sowers A, Mitchell JB, Roberts AB, Russo A (2004) Amelioration of radiation-induced fibrosis: inhibition of transforming growth factor-β signaling by halofuginone. J Biol Chem 279:15167–15176
68. Kitano A, Saika S, Yamanaka O, Ikeda K, Reinach PS, Nakajima Y, Okada Y, Shirai K, Ohnishi Y (2006) Effects of genipin on the behaviors of human subconjunctival fibroblasts in vitro. Ophthalmic Res 38:355–360

69. Kitano A, Saika S, Yamanaka O, Reinach PS, Ikeda K, Okada Y, Shirai K, Ohnishi Y (2006) Genipin suppression of fibrogenic behaviors of the α-TN4 lens epithelial cell line. J Cataract Refract Surg 32:1727–1735
70. Kitano A, Saika S, Yamanaka O, Ikeda K, Okada Y, Shirai K, Reinach PS (2007) Emodin suppression of ocular surface inflammatory reaction. Invest Ophthalmol Vis Sci 48:5013–5022
71. Saika S, Yamanaka O, Flanders KC, Okada Y, Miyamoto T, Sumioka T, Shirai K, Kitano A, Miyazaki K, Tanaka S, Ikeda K (2008) Epithelial-mesenchymal transition as a therapeutic target for prevention of ocular tissue fibrosis. Endocr Metab Immune Disord Drug Targets 8 (1):69–76
72. Saika S (2005) J Jpn Soc Cataract Res 17:17–21 (in Japanese)
73. Saika S, Yamanaka O, Okada Y, Tanaka S, Miyamoto T, Sumioka T, Kitano A, Shirai K, Ikeda K (2009) TGF β in fibroproliferative diseases in the eye. Front Biosci (Schol Ed) 1:376–390

Part II

Clinical Science: Pathology

Histology of Posterior Capsular Opacification

10

Takeshi Miyamoto, Nobuyuki Ishikawa, Kumi Shirai,
Ai Kitano-Izutani, Sai-ichi Tanaka, and Shizuya Saika

Abstract

Lens epithelial cells (LECs) undergo tissue repair reaction following cataract-intraocular lens (IOL) surgery. However, the reaction in turn produces opacification of the lens capsule that potentially impairs vision. Histology and immunohistochemistry are employed for characterization of the tissues of capsular opacification. LECs transform into regenerated lens fiber structures and myofibroblasts. The former behavior develops Soemmering's ring in the peripheral capsular bag and Elschnig's pearls on the inner surface of the posterior capsule. Histology shows lens-like cellular arrangement in these structures. The latter response is called epithelial-mesenchymal transition (EMT) and generates fibrous tissue accumulation on the capsule. EMT-derived cells no longer exhibit an epithelial feature, but accumulate fibrous extracellular matrix around themselves. Contraction of EMT myofibroblasts shrinks the postoperative lens capsule around an IOL.

Keywords

Epithelial-mesenchymal transition • Extracellular matrix • Histology • Immunohistochemistry • Intraocular lens • Myofibroblast • Posterior capsular opacification

T. Miyamoto, M.D., Ph.D. • N. Ishikawa, M.D., Ph.D. • K. Shirai, M.D., Ph.D., Ph.D.
A. Kitano-Izutani, M.D., Ph.D. • S.-i. Tanaka, M.D., Ph.D. • S. Saika, M.D., Ph.D. (✉)
Department of Ophthalmology, Wakayama Medical University, 811-1 Kimiidera,
Wakayama 641-0012, Japan
e-mail: shizuya@wakayama-med.ac.jp

S. Saika et al. (eds.), *Lens Epithelium and Posterior Capsular Opacification*,
DOI 10.1007/978-4-431-54300-8_10,

10.1 Overview of Tissue Repair Reaction

When a tissue is damaged, local tissue repair (or wound healing) reaction takes place. Tissue-specific resident cells and influx of inflammatory cells are involved in the reaction. Following initial inflammation and subsequent primary tissue repair, tissue resident cells proliferate and differentiate for the purpose of restoration of normal tissue architecture and its function (tissue remodeling process) [1, 2]. For example, the wound is first closed by blood clot in cutaneous wound healing following an incision injury. Then polymorphonuclear leukocytes and macrophages invade the tissue adjacent to the wound. Inflammatory cells secrete growth factors and cytokines that modulate tissue resident cell behaviors and activate further invasion of inflammatory cells. Epidermal keratinocytes migrate for the purpose of recovering the wound surface and fibroblasts transform into myofibroblasts that possess contractile characteristic and express various extracellular matrix involved in tissue repair and fibrosis. Neovascularization also occurs in a newly formed granulation tissue, promoted by angiogenic growth factors, e. g., vascular endothelial growth factor (VEGF). Suppression of neovascularization impairs wound healing, which further supports the critical roles of newly formed vessels there. Once tissue is primary healed with granulation tissue, inflammatory cells and myofibroblasts undergo presumably apoptosis and the inflammation and fibrotic process decline, followed by tissue remodeling and restoration of normal cutaneous structure. Neovascularization also gradually regressed. Complex machinery of growth factor system has to mediate the regression of inflammation and the granulation tissue [3, 4]. For example, the loss of tumor necrosis factor blocks the regression on tissue inflammation caused by wound healing reaction.

This is also the case in a crystalline lens tissue, although different from skin as the lens tissue lacks fibroblasts and blood vessels and is structured only by lens epithelium and lens fibers. Following cataract extraction the residual LECs proliferate and regenerate lenticular tissues [5–8]. However, such behavior fails to restore whole functional crystalline lens, but forms immature lens structure of Soemmering's ring and Elschnig's pearls (discussed below). LECs also produce mesenchymal-type cells and fibrous tissue of capsular opacification (Fig. 10.1) (discussed in detail below).

10.2 Behaviors of Postoperative Lens Epithelial Cells

Lens tissue is developed from invaginated surface ectoderm above the optic cup of the neural tissue during embryonic development. Therefore, the basement membrane locates at the surface of the crystalline lens, which finally forms the lens capsule (Fig. 10.2a, b). The components of the lens capsule are quite similar to those of the epidermal basement membrane; type IV collagen, laminin, and different kinds of proteoglycans are detected [12]. The embryonic ectodermal cells on the inner surface of the anterior capsule differentiate into LECs and those on the

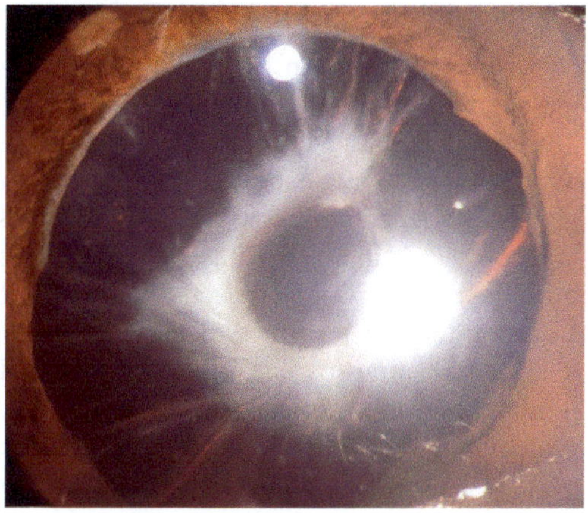

Fig. 10.1 Excess fibrosis at the edge of the anterior capsulotomy and posterior capsular opacification in a patient with pseudo-exfoliation syndrome

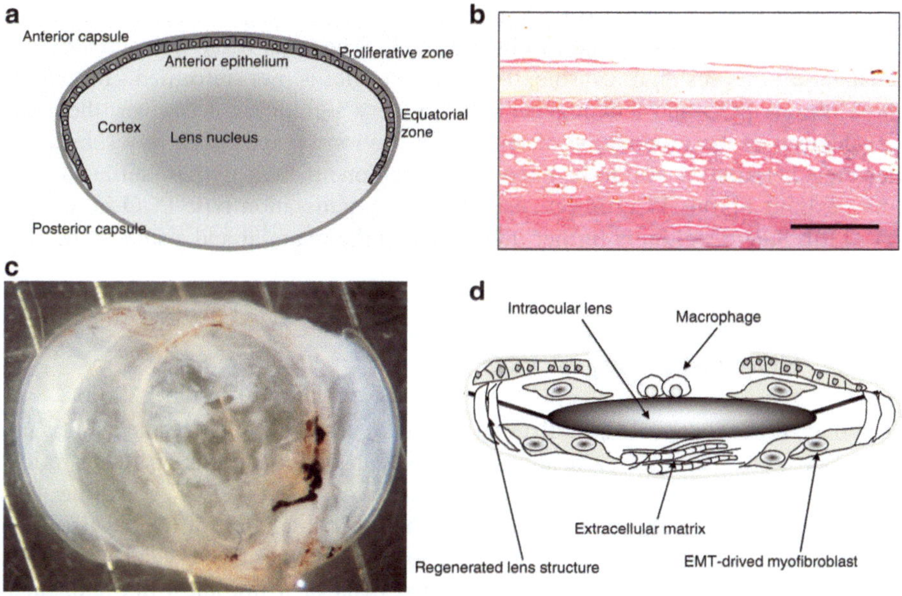

Fig. 10.2 (**a**) Outline of posterior capsular opacification following cataract-intraocular lens (IOL) surgery. In normal human crystalline lenses, a monolayer of lens epithelium locates beneath the anterior capsule, being packed by anterior lens cortical fibers. Posterior capsule lacks inner epithelium unlike anterior capsule (reproduced from Saika et al. [9] with a minor modification). (**b**) Histological section stained by hematoxylin and eosin shows a flattened monolayer of the epithelium beneath the capsule. *Bar*: 50 μm (Paraffin section) (reproduced from Saika [10] with a minor modification). (**c**) An intraocular lens is observed inside an opaque capsular bag. The edge of the anterior capsulotomy is more opaque as compared with the anterior capsule attaching to the anterior surface of the lens optic part (reproduced from Saika [11] with a minor modification). (**d**) A scheme of lens capsule post-cataract-IOL surgery. Fibroblastic lens cells generated through epithelial-mesenchymal transition and fibrous extracellular matrix are observed on the inner surface of the capsule. Peripheral capsular bag is occupied with regenerated lenticular fibers of Soemmering's ring. Lentoid structures of Elschnig's pearls are also formed on the posterior capsule, although not included in the scheme (reproduced from Saika et al. [9] with a minor modification)

posterior capsule form lens fibers that are well organized for the purpose of the maintenance of the tissue transparency.

Once lens tissue is injured by cataract-IOL surgery, the residual lens capsule may become opaque during the long-term healing interval, which potentially impairs patients' vision (Fig. 10.2c). LECs take charge of the wound healing reaction. The main finding of the developing process of posterior capsular opacification (PCO) is an accumulation of abnormal tissue formed by the residual LECs on the inner surface of the lens capsule [13–17], although the lens capsule itself remains transparent. The lens capsule does not regenerate even though LECs produce matrix components of the capsule, i.e., laminin, type IV collagen, and proteoglycans. The LECs on the inner surface of the residual anterior lens capsule migrate onto the inner surface of the posterior capsule and transform their phenotype into others much different from the original ones. As summarized above, the LECs (de)differentiate in two ways; regeneration of lenticular structure and fibroblastic transformation (or EMT) (Fig. 10.2d) [18, 19]. The former reaction yields Soemmering's ring and Elschnig's pearls. The latter behavior causes fibrous change of the residual capsule, but the capsule is not reformed. In infants both fibrous PCO and lens regeneration are very significant. However, the exact mechanism of the modulation of lens cell behavior toward either way (fibrous or lens regeneration) is still to be uncovered; even one patient with infantile cataract may exhibit fibrous PCO in one eye and lens regeneration-type PCO in another eye [20]. Such reaction is further influenced by the material and shape of the IOL implanted in the capsule [21, 22].

10.3 Histopathology of PCO

Although the lens capsule is quite thin (around 20 μm in the anterior and about 5 μm in the posterior capsules), a healing-opacified capsule containing an IOL is thick. The central portion of the residual posterior capsule of the crystalline lens containing an IOL may become opaque with an accumulation of fibrous tissue on the inner surface of the capsule during the long-term healing interval.

Scanning electron microscopy shows that cells of a fibroblast-like elongated morphology are observed on the outer (anterior) and inner surface of the residual capsule (Fig. 10.3a–c) [23, 24]. Such fibrotic contraction is also formed at the edge of the anterior capsulotomy. Higher-magnification observation of the inner surface of the posterior capsule behind the optic part of the IOL shows accumulation of fibroblast-like cells and fibrous components in fibrous PCO tissue and spherical lentoid structures of Elschnig's pearls (Fig. 10.3d, e). Posterior lens capsule is originally free from cells and therefore both fibroblast-like cells and Elschnig's pearl-type cells are considered to be derived from LECs that had located to the inner surface of the anterior lens capsule (Fig. 10.4) [23, 24]. Regenerated lens-like tissue may grow in the peripheral bag of the capsule, which is called as Soemmering's ring (Fig. 10.5). Fibrous opacification produced by EMT-lens cells is more marked on the inner surface of the anterior capsule that covers the optic portion of an IOL

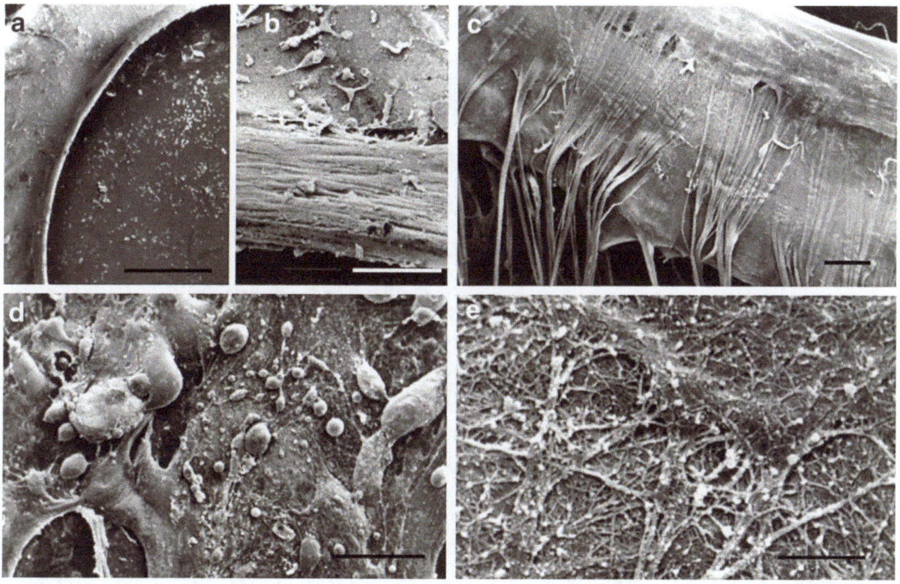

Fig. 10.3 Scanning electron micrographs of a human lens capsule following cataract-intraocular lens (IOL) surgery. (**a**) Removing the IOL allows observation of the inner surface of the posterior capsule. *Bar*: 500 μm. (**b**) Observation at a higher magnification shows fibrous change at the edge of the anterior capsulotomy and the presence of fibroblast-like-shaped cells on the outer surface of the anterior capsule. *Bar*: 50 μm. (**c**) Anterior portion of zonular bundles is seen on the outer surface of the peripheral capsule. *Bar*: 100 μm. (**d**) Lentoid structures of Elschnig's pearls are detected on the inner surface of the posterior capsule. *Bar*: 50 μm. (**e**) Accumulation of collagenous fibers and presumed presence of a cell beneath the fibrosis are also observed on the inner surface of the posterior capsule. *Bar*: 5 μm

implanted in the capsular bag as compared with the posterior capsule behind the IOL optic (Fig. 10.6).

Transmission electron microscopy shows that fibrous capsular opacification contains elongated cells and abundant extracellular matrix (Fig. 10.6e) [28, 29]. The matrix consists of classical collagenous fibers and basement membrane-like structures surrounding cells. On the other hand, non-collagenous extracellular matrix is observed in the area of lenticular structure regeneration. The cells there are much more similar to the original LECs in terms of intracellular structure and the packed arrangement of each cell.

10.4 Immunohistochemical Analysis of the Nature of PCO Tissue

Histology has a limitation in the characterization of tissues. Immunohistochemistry is therefore utilized in tissue specimens to gain further knowledge on the nature of the PCO tissue and cells therein [12, 30–33]. Fibrous tissue accumulated between the

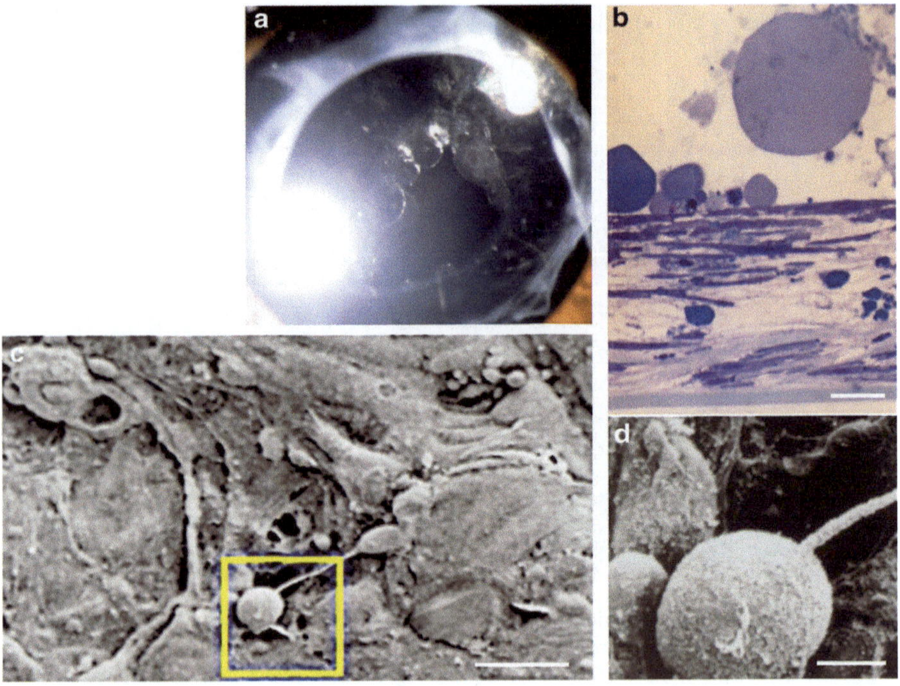

Fig. 10.4 Histology of Elschnig's pearls. (**a**) Elschnig's pearls are observed at the edge of a posterior capsulotomy performed by Nd:YAG laser (reproduced from Saika [25] with a minor modification). (**b**) Light microscopic histology of Elschnig's pearls and elongated lens cells accumulated on the inner surface of the posterior capsule in a rabbit eye following healing after an experimental cataract extraction and implantation of an intraocular lens. Elongated fibroblast-like cells presumed generated through epithelial-mesenchymal transition of lens epithelial cells and homogenous Elschnig's pearls are observed. *Bar*: 10 μm (Epon section with toluidine blue staining) (reproduced from Saika [26] with a minor modification). (**c**) Scanning electron microscopy shows accumulation of regenerated lenticular structures on the inner surface of the posterior capsule in a human case of posterior capsular opacification. *Bar*: 50 μm. (**d**) High magnification observation in the frame C indicates velutinous surface of the structure. *Bar*: 10 μm

IOL optic and the inner surface of the capsule has been stained for various types of fibrillar collagen types, fibronectin, laminin, hyaluronan, etc. (Figs. 10.6 and 10.7). These extracellular matrix components were the ones detected in the tissue under wound healing process, i.e., in granulation tissue of the skin wound. The lens cells after an injury or surgery alter the expression pattern of integrins, the receptor family of extracellular matrix ligands. The fibroblast-like cells in the fibrous tissue in the capsule stained for α-smooth muscle actin, the marker of myofibroblasts [34]. The presence of myofibroblasts explains the gradual contraction of the residual capsule during healing following cataract extraction because of the contractile characteristic of the cells. Although in tissues a myofibroblast is in general derived from a fibroblast upon stimulation, i.e., specific cytokine binding to the receptor, such cell in the lens capsule is produced from a LEC through the process of EMT [9, 35–37].

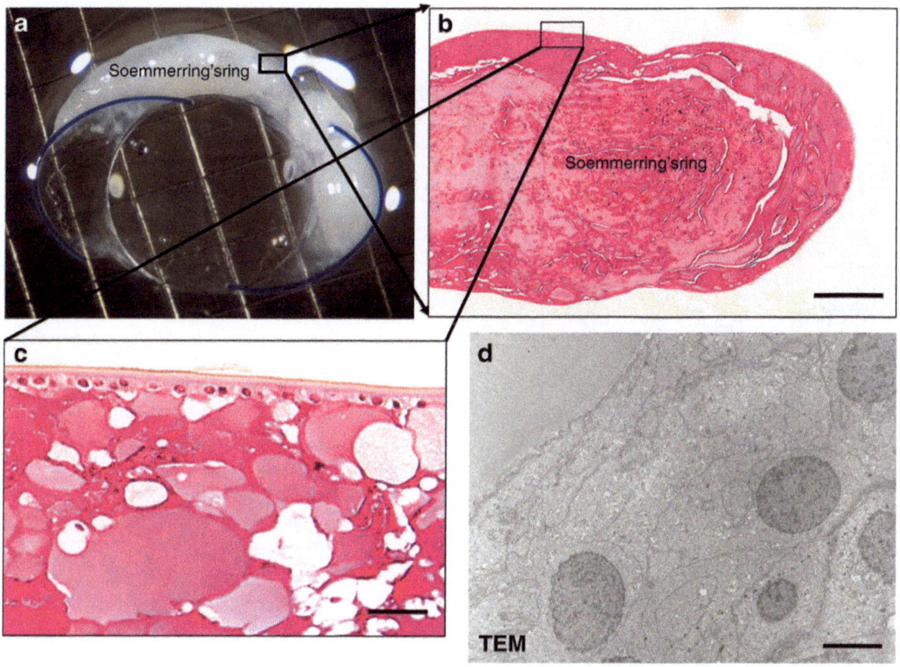

Fig. 10.5 (**a**) Observation of Soemmering's ring formed between the peripheral capsular bag and an intraocular lens (IOL). (**b**) Hematoxylin and eosin staining shows presence of lens fiber-like cells in peripheral lens capsular bag. *Bar*: 1 mm. (**c**) Higher-magnification examination showed that regenerated lenticular fibers are packed beneath a line of lens epithelium attaching to the inner surface of the lens capsule. *Bar*: 100 μm (Frames B, C, paraffin section). (**d**) Transmission electron microscopy demonstrates each lenticular fiber cell connected to each other. Cytoplasmic organelles seem to be poorly developed. *Bar*: 5 μm (**a–c**, reproduced from Saika [10] with a minor modification)

Detailed examination also detected the expression of transforming growth factor β (TGF-β) subtypes on the cell surface receptors in healing LECs in human PCO specimens. Aqueous humor also contains abundant TGF-β2. TGF-β activates signal transduction cascades, i.e., Smad and non-Smad signals, and are considered to modulate behaviors, i.e., EMT of LECs (discussed in Chap. 9). Behaviors of LECs around an implanted IOL are further affected by the nature of the material of the IOL [6, 13] (discussed in detail below).

10.5 Cell Adhesion to the Anterior Surface of an Implanted IOL

In usual cases the peripheral part of the anterior surface of the implanted IOL is covered by the edge of the anterior lens capsule. Macrophages and foreign-body giant cells are usually observed on the IOL surface through slit-lamp biomicroscopy

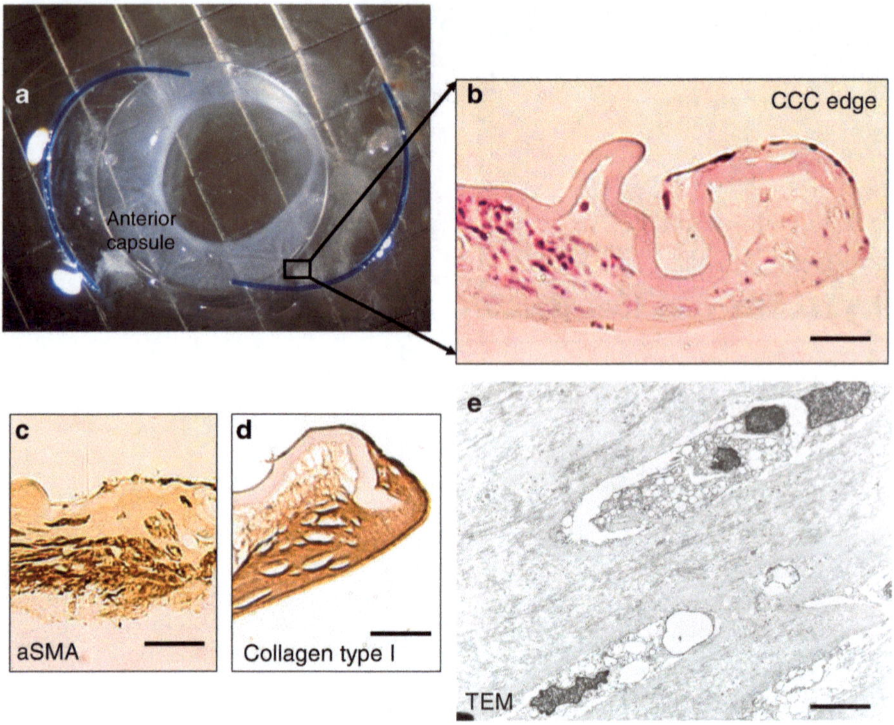

Fig. 10.6 (**a**) Observation of fibrous opacification of the anterior capsule that covers the periphery of the optic of an intraocular lens (IOL). (**b**) Hematoxylin and eosin staining shows presence of elongated cells in the accumulation of eosinophilic matrix substance beneath the anterior capsule. Immunohistochemical examination of the specimen reveals the cells are myofibroblasts positive for α-smooth muscle actin (**c**) and the matrix is labeled with anti-collagen type I antibody.(**d**) (frames **b–d**, paraffin section) *Bar*: 50 μm in frames **b–d**. (**e**) Transmission electron microscopy showed the presence of elongated cells in the matrix. Each cell is surrounded by basement membrane-like relatively homogenous matrix layer. *Bar*: 5 μm (**a–d**, reproduced from Saika [27] with a minor modification)

as well as under histological examination of an extracted specimen [38, 39]. However, the foreign-body reaction by macrophages and macrophage-derived giant cells fails to rap an implanted IOL and the IOL maintains optical function. LECs also migrate onto the anterior IOL surface (outgrowth) besides moving toward the equator of the lens capsular bag and posterior capsule (Fig. 10.8). The degree of the outgrowth of the LECs is affected by the characteristics of the IOL material [13]. Lens epithelium outgrowth is less common with a silicone IOL as compared with an IOL made of PMMA or hydrophobic acrylic material. On the other hand, fibrosis and subsequent tissue contraction is more marked with a silicone IOL as compared with other IOL materials as discussed below.

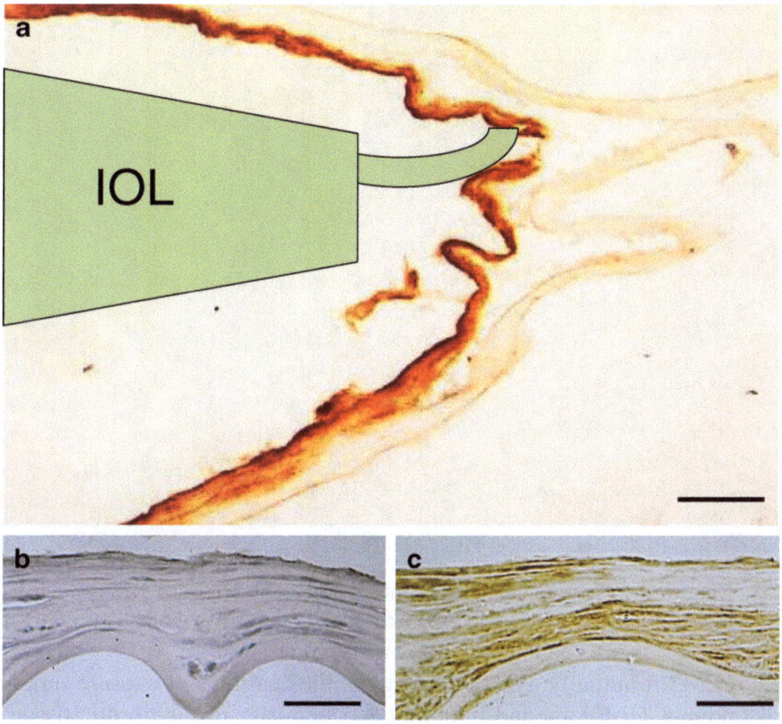

Fig. 10.7 (**a**) Immuno-detection of collagen type I on the inner surface of the capsular bag. IOL, intraocular lens. *Bar*: 25 μm (cryosection) (reproduced from Saika S et al. [12] with a minor modification). (**b**) During postoperative healing, lens epithelial cells transform into fibroblastic-shaped cells and migrate to the inner surface of the posterior capsule behind the optic of an intraocular lens in association with accumulation of extracellular matrix. *Bar*: 50 μm (paraffin section). (**c**) Immunohistochemistry detected the expression of α-smooth muscle actin in the cells, indicating that the cells are myofibroblasts generated through epithelial-mesenchymal transition. *Bar*: 50 μm (paraffin section)

10.6 IOL Materials and PCO

Surface characteristics of artificial materials affect cell behaviors, i.e., cell migration, proliferation, and cell death as revealed by in vitro cell culture studies [13]. This is considered to be the case in in vivo conditions when a foreign-body material is implanted in tissue [40–42]. Clinical observation indicates that the degree and nature of PCO or fibrosis of the anterior lens capsule at the capsulotomy edge following cataract extraction and implantation of an IOL is affected by the characteristics of the IOL material surface, i.e., hydrophilic or hydrophobic. A hydrophobic material, i.e., silicone, reportedly more markedly activates fibroblastic reaction, i.e., EMT of LECs as compared with other materials. On the other hand, a hydrophilic material of hydrogel accelerates proliferation of LECs, but affects the

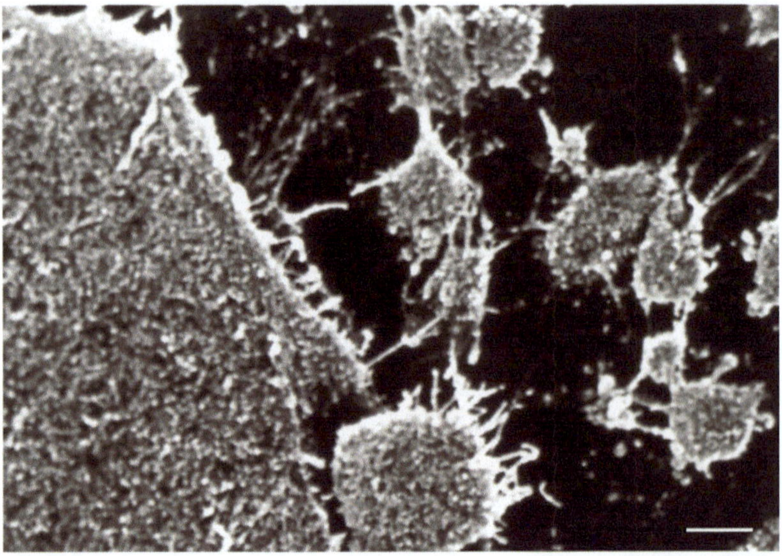

Fig. 10.8 Outgrowth of lens epithelial cells onto the anterior surface of the optic portion of an intraocular lens (IOL) during post-cataract surgery healing interval in a rabbit eye. *Bar*: 10 μm

process of EMT minimally. Histopathology of the healing lens capsule supports the clinical findings. Histology and immunohistochemistry of the fibrotic tissue at the capsulotomy edge of the anterior capsule with a silicone IOL show that the cells are more fibroblastic in morphology as compared with the cells with a hydrophilic IOL and labeled for α-smooth muscle actin, the hallmark of an EMT-derived myofibroblast [43].

References

1. Martin P (1997) Wound healing–aiming for perfect skin regeneration. Science 276:75–81
2. Flanders KC, Major CD, Arabshahi A, Aburime EE, Okada MH, Fujii M, Blalock TD, Schultz GS, Sowers A, Anzano MA, Mitchell JB, Russo A, Roberts AB (2003) Interference with transforming growth factor-b/Smad3 signaling results in accelerated healing of wounds in previously irradiated skin. Am J Pathol 163:2247–2257
3. Saika S, Ikeda K, Yamanaka O, Flanders KC, Okada Y, Miyamoto T, Kitano A, Ooshima A, Nakajima Y, Ohnishi Y, Kao WW (2006) Loss of tumor necrosis factor alpha potentiates transforming growth factor beta-mediated pathogenic tissue response during wound healing. Am J Pathol 168(6):1848–1860
4. Saika S (2007) Yin and yang in cytokine regulation of corneal wound healing: roles of TNFα. Cornea 26(9 Suppl 1):S70–S74
5. Saika S, Yamanaka O, Sumioka T, Miyamoto T, Miyazaki K, Okada Y, Kitano A, Shirai K, Tanaka S, Ikeda K (2008) Fibrotic disorders in the eye: targets of gene therapy. Prog Retin Eye Res 27(2):177–196. doi:10.1016/j.preteyeres.2007.12.002
6. Saika S (2004) Relationship between posterior capsule opacification and intraocular lens biocompatibility. Prog Retin Eye Res 23(3):283–305

7. Blomstedt G, Fagerholm P, Gallo J, Philipson B (1987) After-cataract in the rabbit eye following extracapsular cataract extraction - a wound healing reaction. Acta Ophthalmol Suppl 182:93–99
8. Saika S, Miyamoto T, Ishida I, Okada Y, Ohnishi Y, Ooshima A (2001) Lens epithelial cell regeneration of a capsule-like structure during postoperative healing in rabbits. J Cataract Refract Surg 27(7):1076–1078
9. Saika S, Yamanaka O, Flanders KC, Okada Y, Miyamoto T, Sumioka T, Shirai K, Kitano A, Miyazaki K, Tanaka S, Ikeda K (2008) Epithelial-mesenchymal transition as a therapeutic target for prevention of ocular tissue fibrosis. Endocr Metab Immune Disord Drug Targets 8 (1):69–76
10. Saika S (2006) Practical ophthalmology 8. Imasugu-yakudatsu-ganbyouri, Bunkoudou, Tokyo, Japan, p 138 (in Japanese)
11. Saika S (2006) Practical ophthalmology 8. Imasugu-yakudatsu-ganbyouri, Bunkoudou, Tokyo, Japan, p 137 (in Japanese)
12. Saika S, Kawashima Y, Miyamoto T, Okada Y, Tanaka SI, Ohmi S, Minamide A, Yamanaka O, Ohnishi Y, Ooshima A, Yamanaka A (1998) Immunolocalization of prolyl 4-hydroxylase subunits, alpha-smooth muscle actin, and extracellular matrix components in human lens capsules with lens implants. Exp Eye Res 66(3):283–294
13. Werner L (2008) Biocompatibility of intraocular lens materials. Curr Opin Ophthalmol 19(1):41–49
14. Apple DJ, Werner L (2001) Complications of cataract and refractive surgery: a clinicopathological documentation. Trans Am Ophthalmol Soc 99:95–107
15. Dewey S (2006) Posterior capsule opacification. Curr Opin Ophthalmol 17(1):45–53
16. Wormstone IM (2002) Posterior capsule opacification: a cell biological perspective. Exp Eye Res 74(3):337–347
17. Wormstone IM, Wang L, Liu CS (2009) Posterior capsule opacification. Exp Eye Res 88 (2):257–269
18. de Iongh RU, Wederell E, Lovicu FJ, McAvoy JW (2005) Transforming growth factor-beta-induced epithelial-mesenchymal transition in the lens: a model for cataract formation. Cells Tissues Organs 179(1–2):43–55
19. Saika S, Okada Y, Miyamoto T, Ohnishi Y, Ooshima A, McAvoy JW (2001) Smad translocation and growth suppression in lens epithelial cells by endogenous TGFbeta2 during wound repair. Exp Eye Res 72(6):679–686
20. Shirai K, Saika S, Okada Y, Oda S, Ohnishi Y (2004) Histology and immunohistochemistry of fibrous posterior capsule opacification in an infant. J Cataract Refract Surg 30(2):523–526
21. Buehl W, Findl O (2008) Effect of intraocular lens design on posterior capsule opacification. J Cataract Refract Surg 34(11):1976–1985
22. Saeed MU, Jafree AJ, Saeed MS, Zia R, Sheikh IM, Heravi M (2012) Intraocular lens and capsule opacification with hydrophilic and hydrophobic acrylic materials. Semin Ophthalmol 27(1–2):15–18
23. Kappelhof JP, Pameyer JH, De Jong PT, Jongkind JF, Vrensen GF (1986) The proteinaceous coating and cytology of implant lenses in rabbits. Am J Ophthalmol 102(6):750–758
24. Saika S, Ohmi S, Tanaka S, Ohnishi Y, Yamanaka A, Ooshima A (1997) Light and scanning electron microscopy of rabbit lens capsules with intraocular lenses. J Cataract Refract Surg 23 (5):787–794
25. Saika S (2006) Practical ophthalmology 8. Imasugu-yakudatsu-ganbyouri, Bunkoudou, Tokyo, Japan, p 141 (in Japanese)
26. Saika S (2001) Practical new mook of ophthalmology 1. Ganshikkann-to-shousyoutiyu, Kanehara-syuppann, Tokyo, Japan, p 29 (in Japanese)
27. Saika S (2006) Practical Ophthalmology 8. Imasugu-yakudatsu-ganbyouri, Bunkoudou, Tokyo, p 139 (in Japanese)
28. Ishibashi T, Sugai S, Kubota T, Ohnishi Y, Inomata H (1990) Cellular adhesiveness on implanted lenses in monkeys. Graefes Arch Clin Exp Ophthalmol 228(4):356–362

29. Saika S, Tanaka S, Ohmi S, Minamide A, Ohnishi Y, Yamanaka A, Ooshima A, Kimura M (1997) Deposition of extracellular matrix on intraocular lenses in rabbits: an immunohistochemical and transmission electron microscopic study. Graefes Arch Clin Exp Ophthalmol 235(4):241–247
30. Saika S, Kawashima Y, Miyamoto T, Okada Y, Tanaka S, Yamanaka O, Ohnishi Y, Ooshima A, Yamanaka A (1998) Immunohistochemical identification of proteoglycan types in fibrotic human capsules with intraocular lens implants. Jpn J Ophthalmol 42(5):368–372
31. Ishibashi T, Hatae T, Inomata H (1994) Collagen types in human posterior capsule opacification. J Cataract Refract Surg 20(6):643–646
32. Wederell ED, de Iongh RU (2006) Extracellular matrix and integrin signaling in lens development and cataract. Semin Cell Dev Biol 17(6):759–776
33. Kappelhof JP, Vrensen GF (1992) The pathology of after-cataract. A minireview. Acta Ophthalmol Suppl (205):13–24
34. Gabbiani G (2003) The myofibroblast in wound healing and fibrocontractive diseases. J Pathol 200(4):500–503
35. Saika S, Miyamoto T, Kawashima Y, Okada Y, Yamanaka O, Ohnishi Y, Ooshima A (2000) Immunolocalization of TGF-β1, -β2, and -β3, and TGF-β receptors in human lens capsules with lens implants. Graefes Arch Clin Exp Ophthalmol 238(3):283–293
36. Saika S (2006) TGFβ pathobiology in the eye. Lab Invest 86(2):106–115
37. Roberts AB, Tian F, Byfield SD, Stuelten C, Ooshima A, Saika S, Flanders KC (2006) Smad3 is key to TGF-beta-mediated epithelial-to-mesenchymal transition, fibrosis, tumor suppression and metastasis. Cytokine Growth Factor Rev 17(1–2):19–27
38. Ishikawa N, Miyamoto T, Okada Y, Saika S (2011) Cell adhesion on explanted intraocular lenses: part 1: analysis of explanted IOLs. J Cataract Refract Surg 37(7):1333–1338
39. Ishikawa N, Miyamoto T, Okada Y, Saika S (2011) Cell adhesion on explanted intraocular lenses part 2: experimental study of a surface-modified IOL in rabbits. J Cataract Refract Surg 37(7):1339–1342. doi:10.1016/j.jcrs.2011.04.002
40. Richards RG, Moriarty TF, Miclau T, McClellan RT, Grainger DW (2012) Advances in biomaterials and surface technologies. J Orthop Trauma 26(12):703–707. doi:10.1097/BOT.0b013e31826e37a2
41. Anderson JM, Rodriguez A, Chang DT (2008) Foreign body reaction to biomaterials. Semin Immunol 20(2):86–100
42. Ziats NP, Miller KM, Anderson JM (1988) In vitro and in vivo interactions of cells with biomaterials. Biomaterials 9(1):5–13
43. Saika S, Miyamoto T, Ohnishi Y (2003) Histology of anterior capsule opacification with a polyHEMA/HOHEXMA hydrophilic hydrogel intraocular lens compared to poly(methyl methacrylate), silicone, and acrylic lenses. J Cataract Refract Surg 29(6):1198–1203

PCO Rates in a Large Series of Human Eyes Obtained Postmortem 11

Shannon Stallings and Liliana Werner

Abstract

A significant amount of information can be obtained from pathological analyses of pseudophakic human eyes obtained postmortem in terms of interactions between an intraocular lens material/design and surrounding ocular tissues. The ability to perform direct analyses of an entire organ containing an artificial implant, obtained at different intervals after implantation and in large numbers, is unique to Ophthalmology. In our laboratory located at the Moran Eye Center, University of Utah, we have analyzed more than 600 donor eyes implanted with different lenses. In this chapter we describe some of our major studies using these important specimens, which focused on capsular bag opacification. Relatively large series of eyes were used to compare 1- and 3-piece hydrophobic acrylic lenses in terms of posterior capsule opacification. The outcome of this complication was also compared between 3-piece silicone lenses with round or square posterior optic edges. Furthermore, we had the opportunity to evaluate a small number of eyes implanted with the bag-in-the-lens, which is not available in the United States. We expect to gain better knowledge on preventative measures for posterior capsule opacification and other complications as our series increases and new intraocular lens designs become available in the market.

Keywords

Bag-in-the-lens • Hydrophobic acrylic • Intraocular lens • Posterior capsule opacification • Silicone • Square edge

S. Stallings, M.D. • L. Werner, M.D., Ph.D. (✉)
John A. Moran Eye Center, University of Utah, Salt Lake City, UT, USA
e-mail: liliana.werner@hsc.utah.edu

S. Saika et al. (eds.), *Lens Epithelium and Posterior Capsular Opacification*, DOI 10.1007/978-4-431-54300-8_11, © Springer Japan 2014

11.1 Introduction and Background

Posterior capsule opacification (PCO) remains the most frequent complication affecting cataract surgery with intraocular lens (IOL) implantation [1, 2]. While better understanding of the mechanisms involved in lens epithelial cell (LEC) proliferation and migration across the posterior capsule has led to development of preventive measures, a clear, definitive answer to this problem has yet to emerge. Therefore, it is important to continually evaluate the effectiveness of such measures. Although animal studies have provided significant data, especially with regard to evaluation of new IOL designs and their effect on PCO prevention, and are the preferred method when an appropriate nonanimal model is unavailable, no model exists that duplicates a human eye unerringly. Furthermore, it is often difficult to study long-term effects of a new surgical technique or new implantable device in living patients. They may be lost to follow-up or decide to withdraw from participation, for instance. Complete pathological assessment is also not possible in clinical studies.

Human eyes obtained postmortem from donors who underwent device implantation during their lifespan may provide some of the most accurate insights into the effects of the implant to the eye. The ability to perform direct analysis of an entire organ containing an artificial implant, obtained at different intervals after implantation and in large numbers, is unique to Ophthalmology. Pathologic evaluation of pseudophakic human eyes obtained postmortem is of utmost importance in understanding the interactions between the IOL biomaterial and the IOL design and the intraocular structures.

11.1.1 Moran Eye Center Database of Pseudophakic Human Eyes Obtained Postmortem

At the time of this writing, 661 pseudophakic human eyes mostly obtained postmortem from eye banks within the United States have been evaluated in our laboratory at the Moran Eye Center [3, 4]. The donor eyes had been implanted with different IOLs. In this chapter we describe PCO results from this large eye bank series. To accomplish this, once enucleated, globes were placed into a solution of 10 % neutral buffered formalin for at least 48 h. The staff of the participating eye banks was instructed to attempt to obtain the date of the implantation procedure whenever possible, generally through interaction with the donor family. Gross measurements were obtained using a digital metric ruler; measurements included anterior–posterior length, equatorial diameter, and corneal diameter. Each whole eye then had anterior segment scanning with a very-high-frequency (VHF) ultrasound system with digital enhancement (Artemis, ArcScan). The system uses a broadband 50 MHz transducer (bandwidth 10–60 MHz), which sweeps in a reverse arc high-precision mechanism to acquire B-scans [5]. Information, such as the position of the IOL in relation to intraocular structures, IOL tilt, and IOL

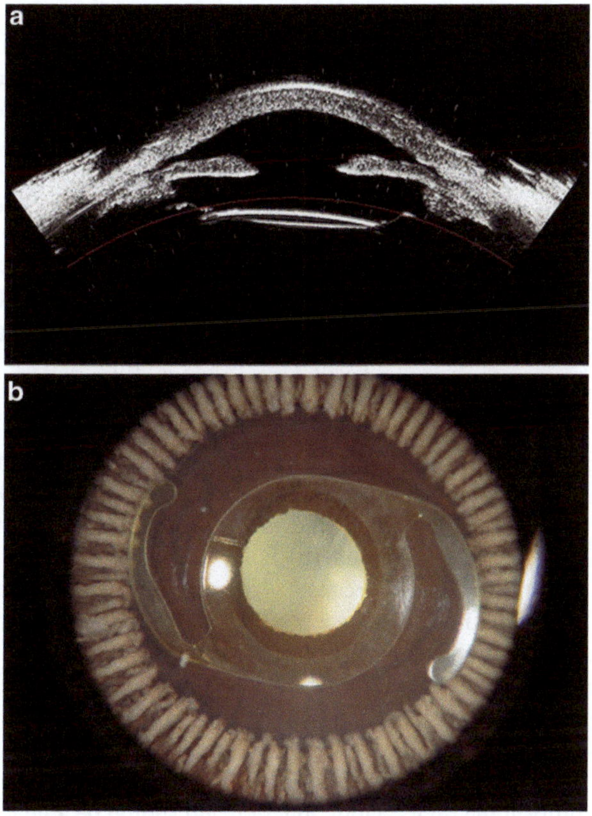

Fig. 11.1 Very-high-frequency ultrasound scan from a human eye obtained postmortem (**a**), obtained before sectioning of the eye, and corresponding gross photograph obtained from the posterior or Miyake–Apple view (**b**). The eye was implanted with a 1-piece hydrophobic acrylic lens, which is overall centered in relation to the capsular bag. The lens is symmetrically fixated within the bag, the capsulorhexis edge covers the optic periphery for 360°, and the capsular bag is overall clear

decentration, was therefore obtained before disruption caused by sectioning (Fig. 11.1a).

Each eye was then sectioned at the equator. Gross examination of the anterior segment was performed from the posterior aspect (Miyake–Apple view) to assess the degree of capsular bag opacification. The intensity of anterior capsule opacification (ACO) was scored from grades zero to four. The intensity of central PCO (behind the central 3.0 mm of the IOL optic), peripheral PCO (behind the periphery of the IOL optic), and Soemmering's ring formation (equatorial region of the capsular bag, outside the IOL optic area) was also scored from grades zero to four, according to previous studies [6–9]. The area of Soemmering's ring formation was noted from zero to four (according to the number of quadrants involved). Other aspects analyzed from the posterior view were coverage of the IOL optic periphery by the anterior capsule (noted from 0° to 360°) and IOL fixation (e.g., bag–bag, bag–sulcus, sulcus–sulcus) (Fig. 11.1b).

Eyes with any degree of PCO were further analyzed for the site of peripheral PCO initiation (initial PCO). After gross examination, the anterior segment of selected eyes was sectioned in the pupil–optic nerve plane, with the cuts oriented parallel to the axis passing through the optic–haptic junctions (in the case of looped single-piece lenses). This secured the entire IOL in the entire capsular bag and ensured that histopathologic cuts would pass through the optic–haptic junctions. Other axis orientations were used in some eyes, according to the IOL model. After dehydration and embedding in paraffin, the eyes were sectioned and stained with different methods such as periodic acid-Schiff (PAS) and Masson's trichrome stains. Multiple 3-micron thick sections of each eye were then examined under a light microscope, and photomicrographs were taken for photodocumentation.

11.2 Evaluation of Hydrophobic Acrylic Lenses in Human Eyes Obtained Postmortem

In vitro [10, 11], animal [7–9], and clinical studies [12–14] have found that an important design feature for PCO prevention is a sharp (square) edge on the posterior optic IOL surface, which creates a more effective barrier against PCO (more details on the role of the square edge on PCO prevention can be found in Chap. 17). Numerous studies have shown that this modification results in lower PCO rates, and modern IOLs generally include this feature. It has been hypothesized that because of this barrier effect from the square edge, a 3-piece IOL would be more successful in preventing PCO than a 1-piece lens. The optic–haptic junction of a 1-piece lens remains vulnerable to PCO formation, because it lacks barrier protection if the lens has a smooth transition between the optic and the haptics. Additionally, rabbit studies performed previously have found that whenever PCO started with such 1-piece IOLs, it had a tendency to start at the level of the optic–haptic junctions [7–9]. On the other hand, a 3-piece lens lacks this vulnerability at the optic–haptic junction and should be able to maintain a 360-degree sharp edge along the posterior optic. However, most available prospective randomized clinical studies report no statistically significant difference in PCO rates between 1- and 3-piece hydrophobic acrylic IOLs [15–21].

In a study done in our laboratory comparing 219 human eyes with hydrophobic acrylic lenses obtained postmortem, no difference was found in terms of peripheral or central rates of PCO among eyes with 1-piece versus 3-piece lenses [4]. The study included 119 eyes with a 1-piece hydrophobic acrylic lens of the AcrySof design (Alcon Laboratories) and 100 with a 3-piece hydrophobic acrylic lens of either the AcrySof or the Sensar AR40e design (Abbott Medical Optics, Inc.). Initial peripheral PCO was found in 84 eyes (71 %) with 1-piece IOLs, which was statistically similar to 61 eyes (61 %) with 3-piece IOLs. There was no statistically significant difference in peripheral or central PCO rates or degrees of continuous curvilinear capsulorhexis (CCC) coverage over the optic between these two groups of eyes, although there were statistically significant higher rates of ACO and Soemmering's ring in eyes with 3-piece IOLs. In 63 (75 %) of the 84 eyes with 1-piece IOLs and peripheral PCO, the site of PCO initiation was at the optic–haptic junction (Figs. 11.2 and 11.3). Furthermore, the authors hypothesized at the beginning of the cadaver eye study that the site of initial PCO in eyes with 3-piece IOLs would be in areas of the optic periphery not covered by the CCC. However, no

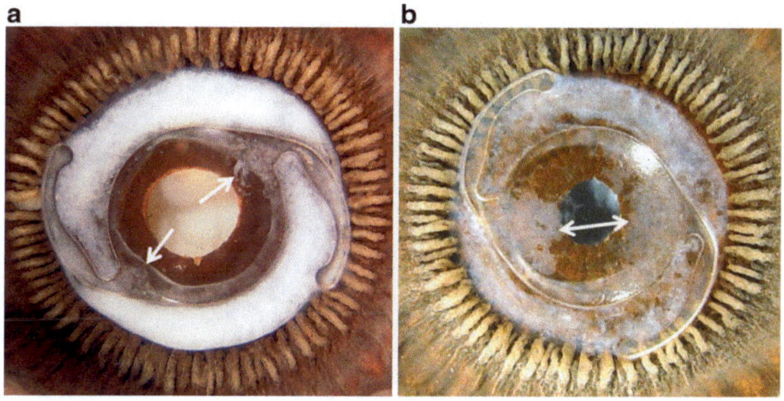

Fig. 11.2 Gross photographs from human eyes obtained postmortem implanted with 1-piece hydrophobic acrylic lenses, obtained from the posterior or Miyake–Apple view (**a** and **b**). The *arrows* show the site of initial posterior capsule opacification. Both eyes have complete capsulorhexis coverage of the optic periphery and posterior capsule opacification started at the optic–haptic junctions. Published in Ness et al. [4]

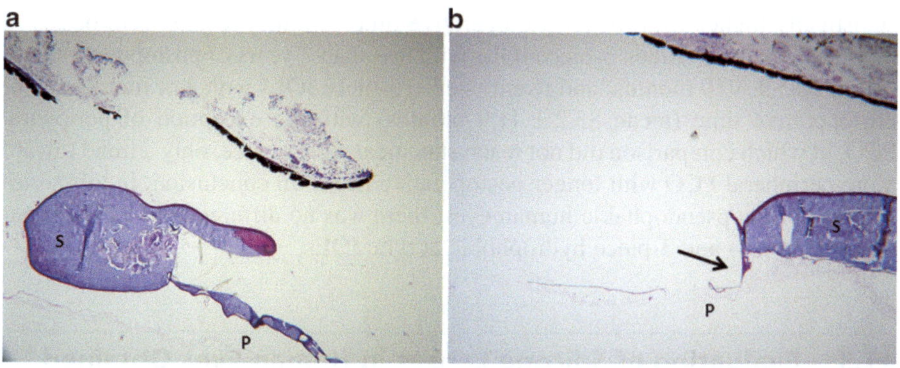

Fig. 11.3 Photomicrographs of histopathologic sections cut from human eyes obtained postmortem implanted with 1-piece (**a**) and 3-piece (**b**) hydrophobic acrylic lenses. The material within the Soemmering's ring (S) started to proliferate onto the posterior capsule (P) in the 1-piece lens, at the level of the optic–haptic junction. The *arrow* indicates the capsular imprint of the square edge of the optic of the 3-piece lens. Published in Ness et al. [4]

statistically significant correlation between the site of initial PCO formation and sites lacking anterior CCC coverage over the optic periphery in these eyes was found (Fig. 11.4).

A comparison of thirty-three 1-piece IOLs with less than 3 years of postoperative time (mean 20.7 ± 8.7 months) and twenty-two 1-piece IOLs with 3 or more years of postoperative time (mean 59.1 ± 21.5 months) showed that the postoperative time had no statistically significant effect on the rate of ACO or central PCO. However, the rate of peripheral PCO and Soemmering's ring formation was

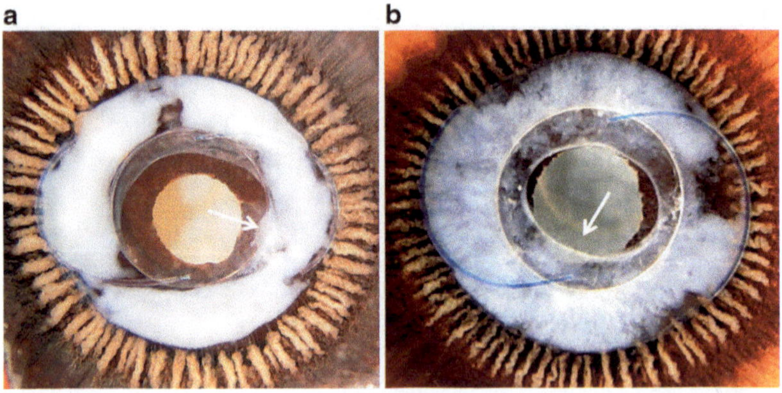

Fig. 11.4 Gross photographs from human eyes obtained postmortem implanted with 3-piece hydrophobic acrylic lenses, obtained from the posterior or Miyake–Apple view. The *arrows* show the site of initial posterior capsule opacification in all eyes. (**a**) The eye has incomplete capsulorhexis coverage of the optic periphery; posterior capsule opacification started at an uncovered optic site. (**b**) Although the eye has complete capsulorhexis coverage of the optic periphery, initial posterior capsule opacification can be observed. Published in Ness et al. [4]

statistically higher with longer postoperative times. Similar results were observed in a comparison of fifteen 3-piece IOLs with less than 3 years of postoperative time (mean 20.5 ± 10.9 months) and twenty-seven 3-piece IOLs with 3 or more years of postoperative time (mean 85.2 ± 41.9 months) with the exception of peripheral PCO, in which comparison did not reach statistical significance, only a trend toward more peripheral PCO with longer postoperative times. In conclusion, in this series of postmortem pseudophakic human eyes, there was no difference in PCO formation between 1- and 3-piece hydrophobic acrylic IOLs.

11.3 Evaluation of Silicone Lenses in Human Eyes Obtained Postmortem

In 2004, Nixon showed the advantage of the square-edged design in vivo in silicone lenses [22]. In general, the edges of silicone IOLs are thought to be sharper than those of hydrophobic acrylic and hydrophilic acrylic IOLs ([23–25]; see Chap. 17). Nixon implanted a square-edged silicone IOL (SoFlex SE, Bausch & Lomb) in 25 patients who had had a conventional round-edged version of the same IOL (SoFlex LI61U, Bausch & Lomb) implanted in the contralateral eye [22]. The IOLs were sequentially photographed between 1 week and 9 months after surgery under high-magnification slit-lamp examination using a high-resolution digital camera. At 1 week, both designs had evidence of LEC migration along the posterior capsule. At 1 month, both IOL designs showed 360° of anterior and posterior capsule adhesion to the edge of the optic. At 1 month, however, migrating LECs encountered a “damming” effect at the square posterior edge, but not at the round edge.

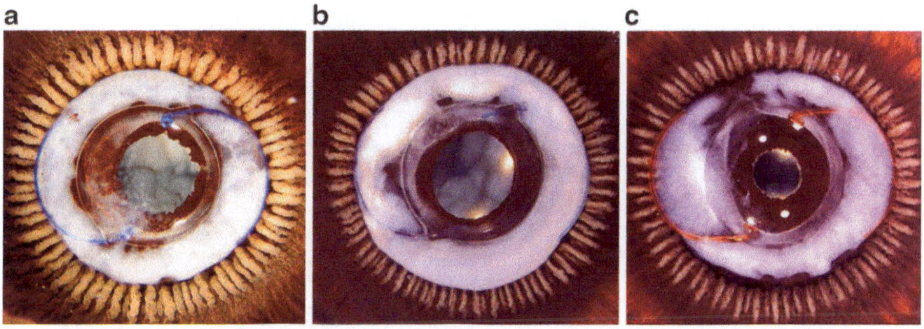

Fig. 11.5 Gross photographs from the posterior or Miyake–Apple view of cadaver eyes implanted with round-edge, 3-piece silicone lenses. All eyes exhibit significant Soemmering's ring formation and variable degrees of peripheral posterior capsule opacification and anterior capsule opacification. (**a**) SI-30 (Abbott Medical Optics, Inc.), with Prolene haptics. (**b**) SI40 (Abbott Medical Optics, Inc.), with PMMA haptics. (**c**) AQ series (Staar Surgical), with polyimide haptics. Published in Maddula et al. [3]

A thin fibrotic ring began to form around the SoFlex SE edge at 2 months and was complete for 360° at 3 months. At 9 months, there was no evidence of LEC migration beyond this ring. With round-edged IOLs, the fibrotic ring never fully formed and LEC migration continued posterior to the optic.

In the following study performed in our laboratory evaluating silicone IOLs, the authors hypothesized that rates of PCO would be lower in cadaver eyes with square-edged IOLs, assuming both groups of silicone IOLs would induce similar amounts of fibrosis and considering the overall superior sharpness of the edge of square-edged silicone IOLs on the market [3]. In this study, the rate of PCO in round versus square-edged silicone IOLs was compared. Donor eyes that had symmetric in the bag fixation of silicone IOLs were included in the analysis. 43 eyes contained a 3-piece silicone square-edged lens, while 87 contained a 3-piece silicone lens with a rounded edge. The 3-piece round-edged IOLs consisted of the SI-30 and SI-40 (both Abbott Medical Optics, Inc.), the LI61U (Bausch & Lomb), and 3-piece silicone models with polyimide haptics (Staar Surgical) (Fig. 11.5). The square-edged group consisted of Clariflex IOLs (Abbott Medical Optics, Inc.) and CeeOn Edge IOLs (Pharmacia) (Fig. 11.6).

The authors found that there was a statistically significant difference in central and peripheral PCO formation between 3-piece IOLs with round edges and square edges (Fig. 11.7). Of the 43 eyes with 3-piece square-edged silicone IOLs, 15 had initial peripheral PCO and 9 of them had incomplete capsulorhexis coverage of the optic periphery. Peripheral PCO started at the site without capsulorhexis coverage in eight of nine eyes. Performance of a centered capsulorhexis with a diameter slightly smaller than the IOL optic, providing coverage of the optic periphery for 360° is apparently beneficial in terms of PCO prevention, and most available clinical studies found that this capsulorhexis configuration decreases the rate of PCO formation [26]. However, a study by Vasavada et al. could not confirm such a benefit [27]. The capsulorhexis coverage of round-edged and square-edged silicone

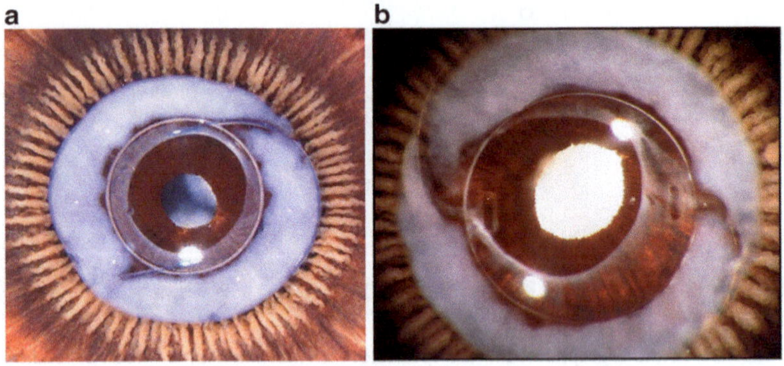

Fig. 11.6 Gross photographs from the posterior or Miyake–Apple view of cadaver eyes implanted with square-edge, 3-piece silicone lenses. Both eyes exhibit significant Soemmering's ring formation, variable degrees of anterior capsule opacification, but no central or peripheral PCO. (**a**) Clariflex (Abbott Medical Optics, Inc.), with PMMA haptics. (**b**) CeeOn Edge (Pharmacia), with PVDF haptics. Published in Maddula et al. [3]

IOLs was essentially similar in the study, but eight of the nine eyes with incomplete capsulorhexis coverage, exhibited peripheral PCO initiation at this site [3].

Eyes with 3-piece square-edged silicone IOLs were further divided into two subgroups according to the postoperative period (<3 or ≥3 years). There was no statistically significant difference in central and peripheral PCO or in Soemmering's ring formation. However, there was a trend toward more Soemmering's ring formation and peripheral PCO with longer postoperative time. Finally, a smaller subset, consisting of 26 silicone plate lenses was received in addition to the 3-piece lenses. All of the eyes generally exhibited significant ACO, central and peripheral PCO, Soemmering's ring formation, and capsular bag contraction, including capsulorhexis phimosis. Out of 26 eyes, 21 had Nd:YAG posterior capsulotomy for central PCO (80.76 %). Mild to moderate IOL decentration was observed in this group, especially in association with asymmetric capsular fibrosis (Fig. 11.8). The results confirm those of previous studies with this lens design [6, 28, 29].

11.4 Evaluation of the Bag-in-the-Lens in Human Eyes Obtained Postmortem

The "bag-in-the-lens" (BIL) concept theoretically prevents PCO by changing the relationship between the IOL and the capsular bag, eliminating contact between the lens and the inner surface of the latter [30–40]. It involves the use of a twin capsulorhexis lens design and performance of anterior and posterior capsulorhexes of the same size. According to this concept, if both capsules are well stretched around the optic of the lens, any remaining LECs will be captured within the remaining space of the capsular bag, and their proliferation will be limited to this

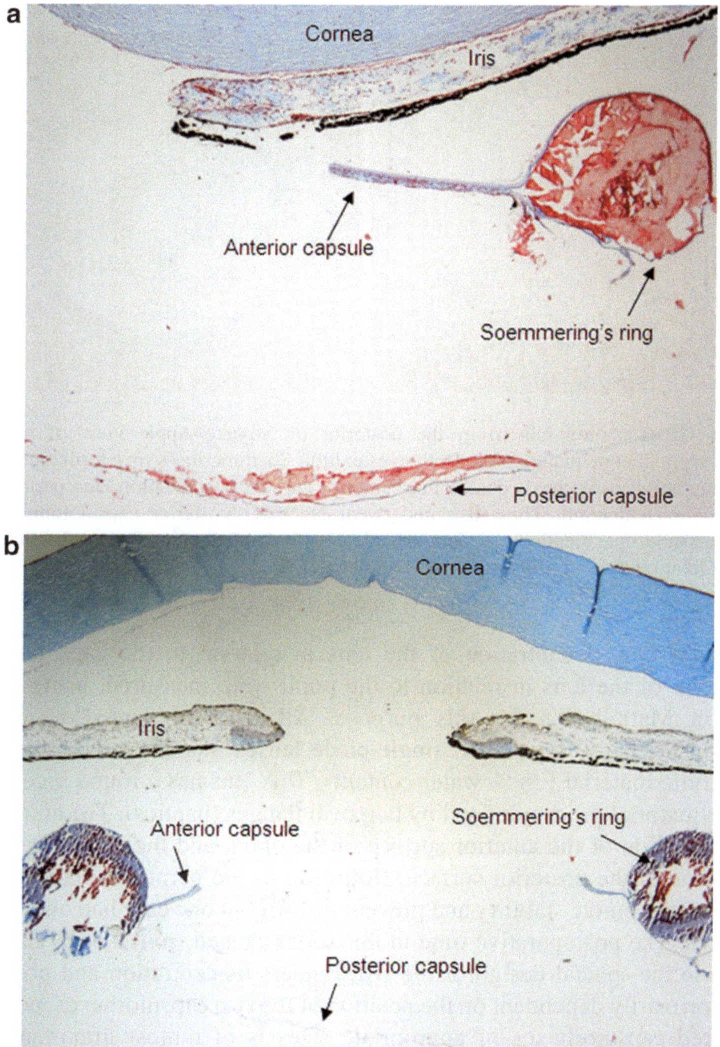

Fig. 11.7 Light photomicrographs obtained from pseudophakic cadaver eyes. (**a**) Eye implanted with an SI-30 (Abbott Medical Optics, Inc.). (**b**) Eye implanted with a Clariflex (Abbott Medical Optics, Inc.) lens. Although both eyes exhibit significant Soemmering's ring formation, there is a lack of material proliferating onto the posterior capsule in (**b**). Published in Maddula et al. [3]

space, so the visual axis will remain clear. More details on the BIL are presented in Chap. 22.

We had the opportunity to receive in our laboratory six eyes implanted with the BIL from patients of Dr. Marie-José Tassignon, inventor of the concept. The patients had terminal cancer and donated their eyes for postmortem research [39, 40]. Capsular bag diameter, pupil diameter, average space in the periphery of

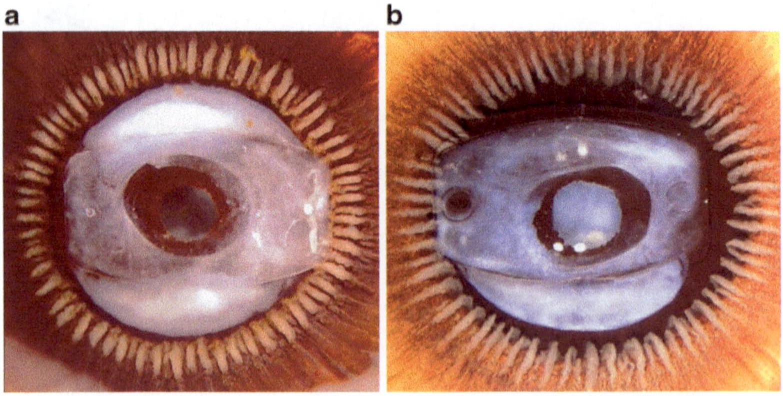

Fig. 11.8 Gross photographs from the posterior or Miyake–Apple view of cadaver eyes implanted with silicone plate lenses. Both eyes exhibit Soemmering's ring formation, significant capsular bag and capsulorhexis contraction, anterior capsule opacification, and peripheral posterior capsule opacification. They also underwent Nd:YAG posterior capsulotomy for central posterior capsule opacification. (**a**) Silicone plate with small fixation holes. (**b**) Silicone plate with large fixation holes. Published in Maddula et al. [3]

the capsular bag, decentration of the lens in relation to the capsular bag, and decentration of the lens in relation to the pupil were measured, using a program created in MatLab 6.5 for this purpose. All BILs implanted were Morcher (Stuttgart, Germany) type 89A, single-piece lenses manufactured from a hydrophilic acrylic material (28 % water content). This lens has a round biconvex optic with an equatorial groove defined by two oval flanges (haptics). The anterior flange is a continuation of the anterior surface of the optic, and the posterior flange is a continuation of the posterior surface. Both flanges are perpendicularly oriented to each other to promote stability and prevent tilt. All but one case had uneventful BIL procedures. The postoperative time in this series ranged from 4 to 39 months.

Because the special design of the BIL renders its centration and postoperative stability primarily dependent on the position of the two capsulorhexes, performance of centered capsulorhexes of appropriate sizes is of utmost importance in BIL implantation. In each case but one cadaver eye, ultrasound examination revealed the presence of a well-fixated, well-centered IOL located at the level of the capsular bag. The maximal BIL decentration observed in the specimens included in this study with uneventful surgery was 0.301 mm (in relation to the capsular bag) and 0.532 mm (in relation to the pupil). In one eye, the anterior capsulorhexis was torn off, and although BIL implantation was still possible, a relatively large degree of decentration was observed postoperatively [40]. The degree of decentration observed was, however, not clinically significant with implantation of a monofocal, spherical lens. Previous studies have estimated the mean decentration in relation to the limbus at 0.304 ± 0.17 mm, and decentration in relation to the pupil at 0.256 ± 0.15 mm by measuring 180 eyes from 125 patients examined at 5 weeks and 6 and 12 months after surgery. The geometric center of the IOL, measured on

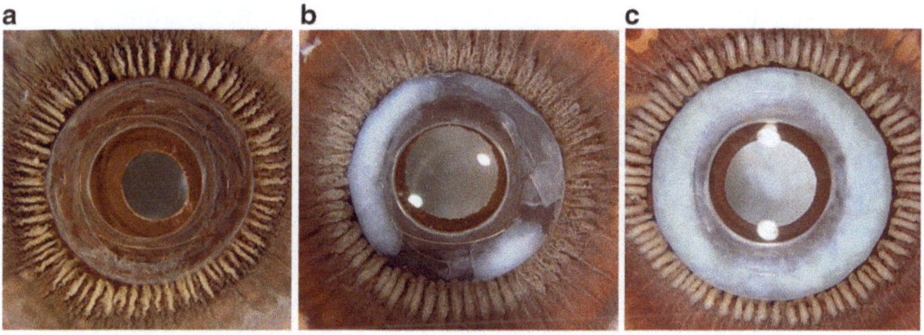

Fig. 11.9 Gross photographs from the posterior or Miyake–Apple view of cadaver eyes implanted with the bag-in-the-lens. From (**a**) to (**c**) the proliferative material within the capsular bag forming the Soemmering's ring increases with increasing postoperative time. The area behind the optic remains perfectly clear. Published in Werner et al. [40]

a red reflex slit-lamp photograph, was compared with the geometric center of the limbus and the pupil by using the same computer program mentioned earlier [36].

In a donor eye with a postoperative time of 4 months, 3 patches of regenerative/proliferative whitish material were observed in the equatorial region of the capsular bag (beginning of Soemmering's ring formation). The capsular bag was otherwise clear. The bag was also mostly clear in two other donor eyes with a postoperative time of 9 months (Fig. 11.9a). More significant Soemmering's ring formation was observed in a donor eye with a postoperative time of 27 months, occupying approximately one half of the periphery of the capsular bag (Fig. 11.9b). Abundant Soemmering's ring formation was observed for 360° in the capsular bag of two donor eyes with postoperative times of 38 and 39 months (Fig. 11.9c). In all instances, the visual axis remained perfectly clear [40].

Whenever the anterior and posterior capsules were properly secured in the peripheral groove of the IOL, BIL implantation has been proven to be highly effective in preventing PCO. In this series, histopathologic sections passing through the center of the capsular bag showed anterior and posterior capsule openings directed to the groove at the periphery of the lens. In this configuration, any regenerative/proliferative material would remain confined to the intercapsular space of the capsular bag remnant outside the optic rim/groove, which was observed in the gross analysis of the eyes and confirmed by histopathologic evaluation. Cortical material and pearls were found within the Soemmering's ring formation. Tissue composed of LECs and fibrosis was present on the inner surface of the anterior capsule, apparently mediating adhesion between anterior and posterior capsules at the rhexis sites (Fig. 11.10).

It appears that a fibrocellular tissue develops during the first postoperative year on the inner surface of the anterior rhexis margin (rhexis fibrosis), which mediates adhesion between anterior and posterior capsules at that site, inside the IOL groove. This most likely helps enhance the postoperative stability of the lens and the confinement of any regenerative/proliferative material to the remaining space of

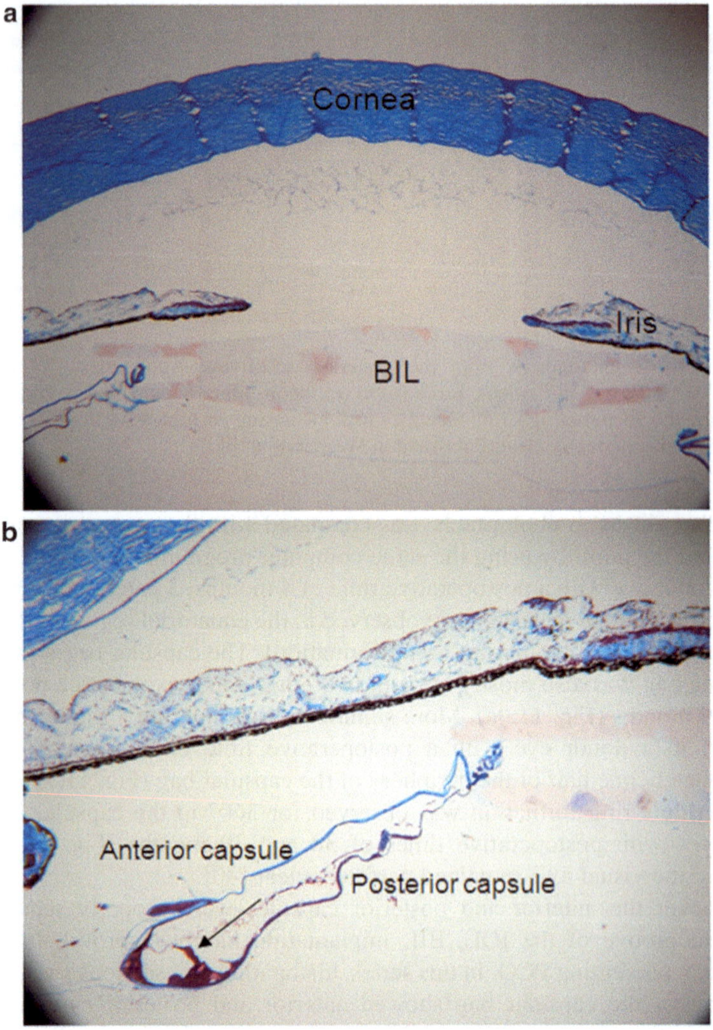

Fig. 11.10 Histopathologic sections cut through the anterior segment of a pseudophakic cadaver eye implanted with the bag-in-the lens (BiL). (**a**) The outline of the lens can be seen behind the iris. Anterior and posterior capsule openings are directed to the peripheral groove of the lens. (**b**) The *arrow* shows the presence of proliferative/regenerative material confined to the equatorial region of the capsular bag. Published in Werner et al. [39]

the capsular bag. Indeed, this was clearly observed during gross and histologic analyses of eyes that exhibited significant amounts of Soemmering's ring formation [40]. These results confirm previous studies on preclinical (in vitro studies using a human capsular bag model and studies in rabbit eyes) and clinical evaluations of the BIL in adult and pediatric eyes [30–38].

11.5 Summary

In conclusion, factors in PCO prevention have been described as IOL related or surgery related [6]. In terms of IOL design, this chapter has examined 1- versus 3-piece designs and square versus round edges. In the first series of postmortem pseudophakic human eyes, there was no difference in PCO formation between 1- and 3-piece hydrophobic acrylic IOLs [4]. When PCO occurred with the 1-piece design, it had a tendency to start at the optic–haptic junctions. It has previously been proven that square edges are more beneficial in preventing PCO than round edges, but this feature has to be present for 360° around the optic edge for maximal efficiency [7–9]. This particularly cadaver eye study did not show any statistically significant correlation between the site of initial PCO formation and sites lacking anterior CCC coverage over the optic periphery in eyes with 3-piece hydrophobic acrylic lenses [4]. In the large series of eyes obtained containing silicone lenses, a statistically significant difference in central and peripheral PCO between round-edged IOLs and square-edged 3-piece silicone IOLs was found [3]. This confirms that the role played by the IOL design in PCO prevention is more preponderant than the role of the IOL material, as lenses manufactured from other than hydrophobic acrylic can also be effective in preventing PCO provided they have a square (and continuous) optic edge. Contrary to the study on hydrophobic acrylic lenses [4], the study on silicone lenses showed a trend toward initiation of PCO in areas without CCC coverage over the optic periphery [3]. We also had the opportunity to analyze the first series of human eyes implanted with the BIL, obtained postmortem at different postoperative times [39, 40]. BIL centration depends on the performance of centered capsulorhexes, of appropriate size. The results confirm the concept of the lens design in that any proliferative/regenerative material remains confined to the intercapsular space of the capsular bag remnant outside the optic rim. We hope to be able to analyze other, more recently commercially available IOLs as our series of pseudophakic human eyes obtained postmortem increases. It would be particularly interesting to evaluate eyes implanted with specialized IOLs, such as accommodating lenses.

References

1. Apple DJ, Solomon KD, Tetz MR et al (1992) Posterior capsular opacification. Major review. Surv Ophthalmol 37(2):73–116
2. Schaumberg DA, Dana MR, Christen WG, Glynn RJ (1998) A systematic overview of the incidence of posterior capsular opacification. Ophthalmology 105:1213–1221
3. Maddula S, Werner L, Ness P et al (2011) Pathology of 157 human cadaver eyes with round-edged or modern square-edged silicone intraocular lenses: analyses of capsule bag opacification. J Cataract Refract Surg 37:740–748
4. Ness P, Werner L, Maddula S et al (2011) Pathology of 219 human cadaver eyes with 1-piece or 3-piece hydrophobic acrylic intraocular lenses: capsular bag opacification and sites of square-edged barrier breach. J Cataract Refract Surg 37:923–930

5. Werner L, Lovisolo C, Chew J, Tetz M, Müller M (2008) Meridional differences in internal dimensions of the anterior segment in human eyes evaluated with 2 imaging systems. J Cataract Refract Surg 34:1125–1132
6. Apple D, Peng Q, Visessook N et al (2001) Eradication of posterior capsule opacification; documentation of a marked decrease in Nd:Yag laser posterior capsulotomy rates noted in an analysis of 5416 pseudophakic human eyes obtained postmortem. Ophthalmology 108:505–518
7. Werner L, Mamalis N, Izak A et al (2005) Posterior capsule opacification in rabbit eyes implanted with 1 piece and 3-piece hydrophobic acrylic intraocular lenses. J Cataract Refract Surg 31:805–811
8. Vargas L, Izak A, Apple D et al (2003) Implantation of a single-piece, hydrophilic acrylic, minus-power foldable posterior chamber intraocular lens in a rabbit model; clinicopathologic study of posterior capsule opacification. J Cataract Refract Surg 29:1613–1620
9. Werner L, Mamalis N, Pandey S et al (2004) Posterior capsule opacification in rabbit eyes implanted with hydrophilic acrylic intraocular lenses with enhanced square edge. J Cataract Refract Surg 30:2403–2409
10. Boyce J, Bhermi G, Spalton D et al (2002) Mathematic modeling of the forces between an intraocular lens and the capsule. J Cataract Refract Surg 28:1853–1859
11. Nishi O, Yamamoto N, Nishi K et al (2007) Contact inhibition of migrating lens epithelial cells at the capsular bend created by a sharp-edged intraocular lens after cataract surgery. J Cataract Refract Surg 33:1065–1070
12. Buel W, Findl O, Menapace R et al (2002) Effect of an acrylic intraocular lens with a sharp posterior optic edge on posterior capsule opacification. J Cataract Refract Surg 28:1105–1111
13. Abela-Formanek C, Amon M, Schild G et al (2002) Uveal and capsular biocompatibility of hydrophobic acrylic, hydrophilic acrylic, and silicone intraocular lenses. J Cataract Refract Surg 28:50–61
14. Hayashi K, Hayashi H (2005) Posterior capsule opacification in the presence of an intraocular lens with a sharp versus rounded optic edge. Ophthalmology 112:1550–1556
15. Sacu S, Finle O, Menapace R et al (2004) Comparison of posterior capsule opacification between the 1-piece and 3-piece AcrySof intraocular lenses; two-year results of a randomized trial. Ophthalmology 111:1840–1846
16. Leydolt C, Davidovic S, Sacu S et al (2007) Long-term effect of 1-piece and 3-piece hydrophobic acrylic intraocular lens on posterior capsule opacification; a randomized trial. Ophthalmology 114:1663–1669
17. Bender L, Nimsgern C, Jose R et al (2004) Effect of 1-piece and 3-piece AcrySof intraocular lenses on the development of posterior capsule opacification after cataract surgery. J Cataract Refract Surg 30:786–789
18. Bilge A, Aykan U, Akin T et al (2004) The effects of three-piece or single-piece acrylic intraocular lens implantation on posterior capsule opacification. Eur J Ophthalmol 14:375–380
19. Nejima R, Miyata K, Honbou M et al (2004) A prospective, randomized comparison of single and three piece acrylic foldable intraocular lenses. Br J Ophthalmol 88:746–749
20. Nejima R, Mmiyai K, Kataoka Y et al (2006) Prospective intrapatient comparison of 6.0 millimeter optic single-piece and 3-piece hydrophobic acrylic foldable intraocular lenses. Ophthalmology 113:585–590
21. Zemaitiene R, Jasinskas V, Auffarth G (2007) Influence of three-piece and single-piece designs of two sharp-edge optic hydrophobic acrylic intraocular lenses on the prevention of posterior capsule opacification: a prospective, randomized, long-term clinical trial. Br J Ophthalmol 91:644–648
22. Nixon D (2004) In vivo digital imaging of the square-edged barrier effect of a silicone intraocular lens. J Cataract Refract Surg 30:2574–2584
23. Werner L, Muller M, Tetz M (2008) Evaluating and defining the sharpness of intraocular lenses: microedge structure of commercially available square-edged hydrophobic lenses. J Cataract Refract Surg 34:310–317

24. Werner L, Tetz M, Feldmann I et al (2009) Evaluating and defining the sharpness of intraocular lenses: microedge structure of commercially available square-edged hydrophilic intraocular lenses. J Cataract Refract Surg 35:556–566
25. Nanavaty M, Spalton D, Boyce J et al (2008) Edge profile of commercially available square-edged intraocular lenses. J Cataract Refract Surg 34:677–686
26. Smith D, Daynes T, Hinckley M et al (2004) The effect of lens edge design versus anterior capsule overlap on posterior capsule opacification. Am J Ophthalmol 138:521–526
27. Vasavada A, Raj S (2004) Anterior capsule relationship of the Acry-Sof intraocular lens optic and posterior capsule opacification: a prospective randomized clinical trial. Ophthalmology 111:886–894
28. Werner L, Pandey S, Escobar-Gomez M et al (2000) Anterior capsule opacification: a histopathological study comparing different IOL styles. Ophthalmology 107:463–471
29. Werner L, Pandey S, Apple D et al (2001) Anterior capsule opacification: correlation of pathological findings with clinical sequelae. Ophthalmology 108:1675–1681
30. Tassignon M, De Groot V, Vrensen G (2002) Bag-in-the-lens implantation of intraocular lenses. J Cataract Refract Surg 28:1182–1188
31. De Groot V, Tassignon M, Vrensen G (2005) Effect of bag-in-the-lens implantation on posterior capsule opacification in human donor eyes and rabbit eyes. J Cataract Refract Surg 31:398–405
32. Tassignon M, De Groot V, Van Tenten Y (2005) Searching the way out for posterior capsule opacification. Verh K Acad Geneeskd Belg 67:277–288
33. De Groot V, Leysn I, Neuhann T et al (2006) One-year follow-up of bag-in-lens intraocular lens implantation in 60 eyes. J Cataract Refract Surg 32:1632–1637
34. Leysen I, Coeckelbergh T, Gobin L et al (2006) Cumulative neodymium:YAG laser rates after bag-in-the-lens and lens-in-the-bag intraocular lens implantation: comparative study. J Cataract Refract Surg 32:2085–2090
35. Tassignon M, De Veuster I, Godts D et al (2007) Bag-in-the-lens intraocular lens implantation in the pediatric eye. J Cataract Refract Surg 33:611–617
36. Verbruggen K, Rozema J, Gobin L et al (2007) Intraocular lens centration and visual outcomes after bag-in-the-lens implantation. J Cataract Refract Surg 33:1267–1272
37. Tassignon M, Rozema J, Gobin L (2006) Ring-shaped caliper for better anterior capsulorhexis sizing and centration. J Cataract Refract Surg 32:1253–1255
38. De Keyzer K, Leysen I, Timmermans J et al (2008) Lens epithelial cells in an in vitro capsular bag model: lens-in-the-bag versus bag-in-the-lens technique. J Cataract Refract Surg 34:687–695
39. Werner L, Tassignon M, Gobin L et al (2008) Bag-in-the-lens: first pathological analysis of a human eye obtained postmortem. J Cataract Refract Surg 34:2163–2165
40. Werner L, Tassignon M, Zaugg B et al (2010) Clinical and histopathologic evaluation of six human eyes implanted with the bag-in-lens. Ophthalmology 117:55–62

Part III

Clinical Outcomes

Natural Course of Elschnig Pearl Formation and Disappearance

12

Nino Hirnschall and Oliver Findl

Abstract

This book chapter summarizes different observational studies to show morphological short-term changes of Elschnig pearls. Firstly, a classification of different types of regeneratory posterior capsule opacification is given. In a second part, the natural course of Elschnig pearls within days and weeks without treatment is explained. Elschnig pearls disappear and appear within days. The degree of progression and regression varies greatly between the eyes. These morphological changes were observed to an even greater extent in Elschnig pearls that survived a Nd:YAG capsulotomy. Even after low-energy Nd:YAG laser treatments with and without opening the posterior capsule, rapid morphological changes of Elschnig pearls were observed. Within the years after Nd:YAG capsulotomy groups of Elschnig pearls at the capsulotomy margin were observed, so-called pearl strings.

Knowledge about Elschnig pearl turnover with and without treatment could be of importance for attempts to modulate lens epithelial cell regeneration or lens regrowth as well as for lens refilling procedures.

The authors have no proprietary or financial interest in any of the materials or equipment mentioned in this study.

Thanks go to Thomas Neumayer, MD, who provided us with some of the retroillumination images.

N. Hirnschall, M.D.
Vienna Institute for Research in Ocular Surgery (VIROS), A Karl Landsteiner Institute, Hanusch Hospital, 1140 Vienna, Austria
e-mail: n.hirnschall@viros.at

O. Findl, M.D., M.B.A. (✉)
Vienna Institute for Research in Ocular Surgery (VIROS), A Karl Landsteiner Institute, Hanusch Hospital, 1140 Vienna, Austria

Moorfields Eye Hospital, NHS Foundation Trust, London, UK
e-mail: oliver@findl.at

S. Saika et al. (eds.), *Lens Epithelium and Posterior Capsular Opacification*,
DOI 10.1007/978-4-431-54300-8_12, © Springer Japan 2014

Keywords
Lens epithelial cell • Life span • Morphological changes • Regeneratory PCO

12.1 Introduction

A better understanding of the natural course of posterior capsule opacification (PCO) development may be of critical importance to tackle PCO, which is still the main long-term complication after cataract surgery and an obstacle for lens refilling or Phaco-Ersatz, the ultimate attempt to treat presbyopia. Since it has not yet been possible to remove all lens epithelial cells (LECs) during cataract surgery and a complete depletion may compromise lens capsule integrity in the long term, modulating LECs is a possible solution. To attempt to do this, however, a better understanding of the dynamics of PCO change and Elschnig pearl turnover are needed.

Clinically, PCO has two different components: a regeneratory and a fibrotic component (Fig. 12.1). Regeneratory PCO has more clinical impact and is the main reason for a decrease in visual function after implantation of an intraocular lens (IOL) [1–3].

12.2 Classification of Regeneratory PCO

12.2.1 Elschnig Pearls

Clinically the most important cells of PCO are the (Hirschberg) Elschnig pearls, which form lentoid bodies. Kappelhof and coauthors [4] stated that these pearls show a nucleus sometimes located within the basal and slender part of the cell. The morphology of the nuclei was found to vary from round to oval to lobulated.

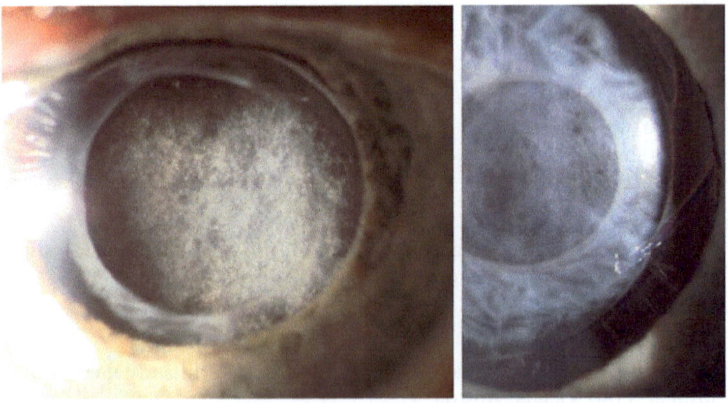

Fig. 12.1 Two different types of posterior capsule opacification (PCO) are shown. The *left image* shows regeneratory (Elschnig pearl) type, and the *right one* shows fibrotic PCO

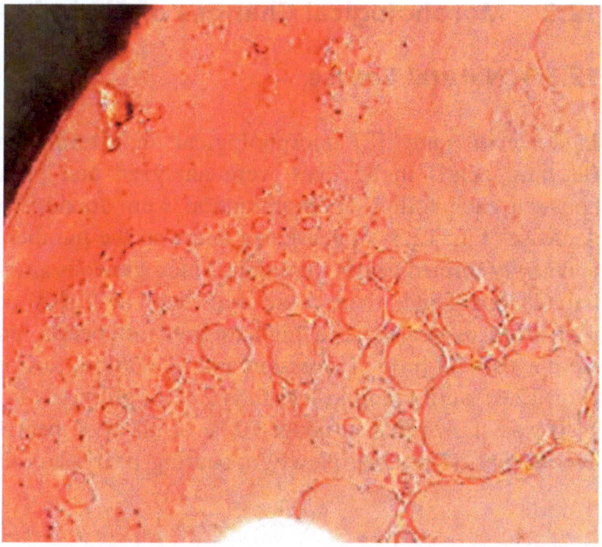

Fig. 12.2 Cheese holes (acellular spaces) surrounded by bridges of lens epithelial cells on the posterior lens capsule

Apart from the nucleus a rare vesicular body was found. The pearls appeared as globular structures with an elongated form of the basal part, and the structures of the pearls varied, from smooth to covered with microvilli. The abnormal globular/spherical shape of the Elschnig pearls is thought to be a result from the absence of the normal internal lenticular pressure. Elschnig pearls are interconnected by gap junction and desmosome-like structures. The membranes of the most superficial Elschnig pearls exhibited a microvillous surface and between the posterior capsule and the Elschnig pearls there was a space measuring 0.5–2 μm wide [5].

12.2.2 Soemmering's Ring

First the Soemmering's ring was described in connection with ocular trauma [6]. Its morphology was described as a dumbbell or donut-shaped lesion of lens remnants, the E-cells, that may occur after rupture of the anterior lens capsule or anterior capsulotomy during extracapsular cataract extraction (ECCE). After the implantation of an IOL, the Soemmering's ring is often localized in the capsule bag fornix, where the haptics of the IOL are located.

In the mid-periphery the opened anterior capsule adheres to the intact posterior capsule and due to the mitosis and proliferation of the remaining epithelial cells the space in between is filled up by lens fibers [6].

12.2.3 Monolayer and Acellular Spaces

Monolayers are fine relatively transparent layers of LECs that sometimes contain acellular spaces, appearing like "cheese holes" (Fig. 12.2) [7].

12.3 Morphological Changes of Elschnig Pearls

12.3.1 Natural Course

In a recent study [8], morphological short-term changes of more than 6,000 Elschnig pearls in 85 eyes were analyzed and pronounced changes of pearls between each follow-up examination were found. The degrees of increase and decrease (Fig. 12.3) in pearl size and also the number of disappearing (Fig. 12.4) and newly formed pearls were quite heterogeneous among eyes and among patients. On the one hand, there were eyes with a highly active progression and also regression of pearls, so a high turnover of pearls, and on the other hand there were eyes with little progression and regression. Fusion (Fig. 12.5) and separation (Fig. 12.6) were found in less than 2 % of all pearls in a time period of 2 weeks.

It was shown that in about 70 % of all eyes, pearls increased in size during a 2-week period [8, 10]. Newly appearing pearls are always small, but pearls that

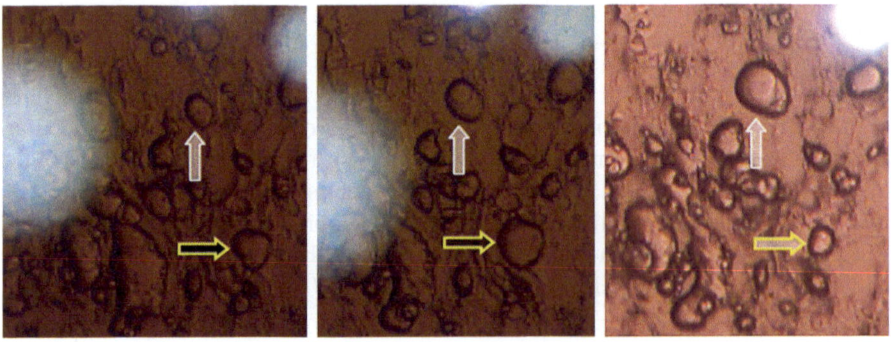

Fig. 12.3 Retroillumination images were taken 2 weeks (*left image*), 1 week (*middle image*), and 10 min (*right image*) before Nd:YAG capsulotomy and demonstrate the increase in pearl size (*white arrow*) and the decrease in pearl size (*yellow arrow*) in the same eye

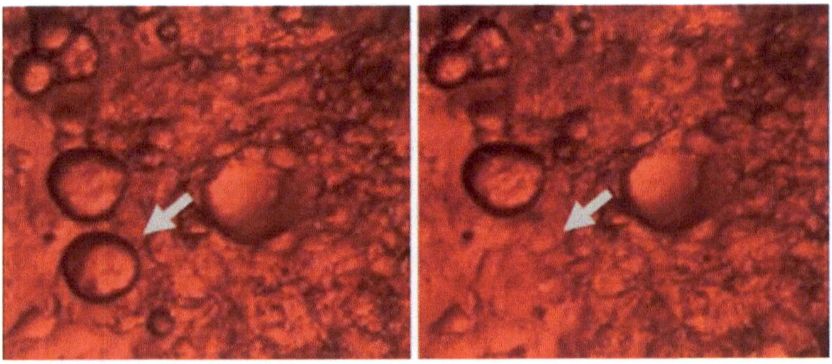

Fig. 12.4 The *arrow* points at a large round Elschnig pearl that disappears within 2 weeks [9]

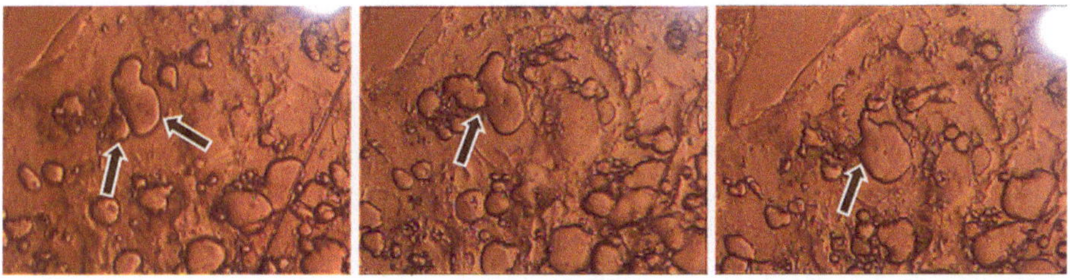

Fig. 12.5 Retroillumination images taken 10 min (*left image*) and 1 (*middle image*) and 2 (*right image*) weeks before the Nd:YAG capsulotomy of the same eye to demonstrate the fusion of pearls

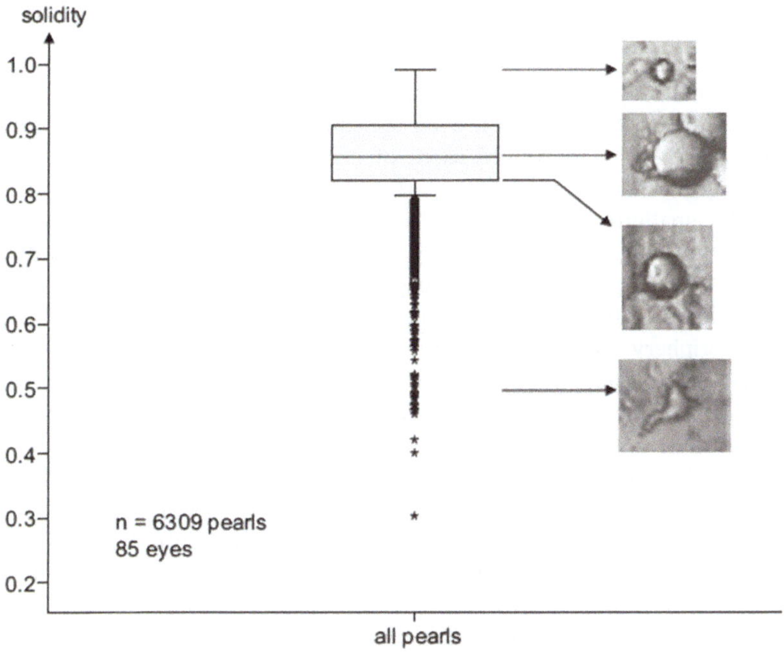

Fig. 12.6 Box plot of the solidity of all pearls. The higher the solidity, the less frayed and the more regular the pearl as also illustrated by the sample photographs. Representative pearls chosen at each quartile are depicted for better understanding. The upper most pearl has a high solidity of 1 (*perfect circle*), whereas the lower most pearl shows a very frayed shape [8]

disappear range from large to small indicating a different mechanism for loss of pearl volume than apoptosis. Osmotic variations seem likely as the mechanism, with rapid change in pearl size instead of pearl death. However, an extreme decrease in pearl volume does not necessarily mean the complete disappearance of the pearl, since it may still be present in a very small size, but not identifiable in retroillumination photography.

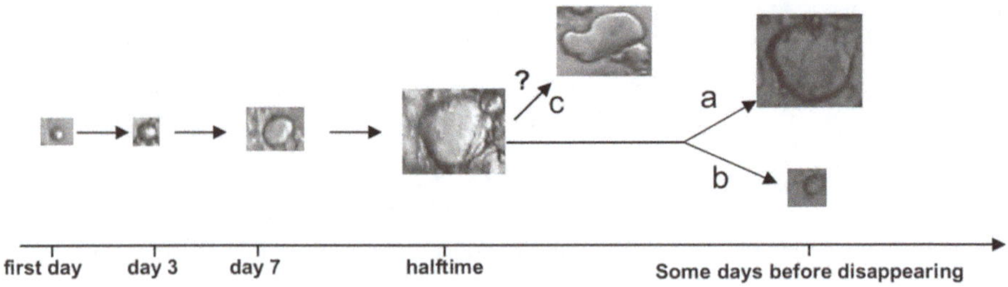

Fig. 12.7 Illustration of the typical morphological changes of a pearl during its life span. Size and shape characteristics taken from different pearls to fit the mean values. On the right side two possible pathways at the end of a pearl's lifetime are depicted: growth to a large pearl and then disappearance (*a*) or alternatively shrinking of the pearl until undetectable (*b*). Also, the questionable fragmentation of a pearl is shown (*c*) [8]

Even within single days, pronounced changes of pearls were found showing increase or decrease in size and even appearance or disappearance of pearls in numerous cases [9, 11]. Single cases of significant spontaneous regression or disappearing Elschnig pearls were reported by Nakashima et al. [12] and Caballero et al. [13]. Caballero et al. assumed that the most likely cause of pearl regression was cell death. They hypothesized, that after several years of proliferation and migration, the cells may lose their capacity to proliferate and begin to die by apoptosis. In long-term studies [14] it was shown that Elschnig pearls migrate from the periphery towards the center of the posterior lens capsule. It was shown that in the second year after IOL implantation about 100 % of all patients have PCO in the peripheral zone, but only 8.3 % in the central zone, and 66.6 % in the mid-periphery.

Even the shape of pearls is quite variable among eyes (Fig. 12.6), but in general newly formed pearls and pearls shortly before disappearing showed a higher solidity and roundness than pearls during the middle part of their life span (Fig. 12.7) [8]. Pearls that disappear are typically quite round with a smooth shape. Similar characteristics were found for newly appearing pearls. This raises the question whether these are different subsets of pearls than those that are more stationary showing little change and have a more frayed shape that also deviates from the round form or this change and transition in shape is actually part of the life cycle of a pearl.

In a previous pharmaceutical study [15] the effect of anti-inflammatory drugs on Elschnig pearls was tested. In this study significant short-term changes were found in the treatment group, but these changes were not significantly different to the placebo-control group. Therefore, testing of pharmaceuticals on Elschnig pearls should always be tested against placebo.

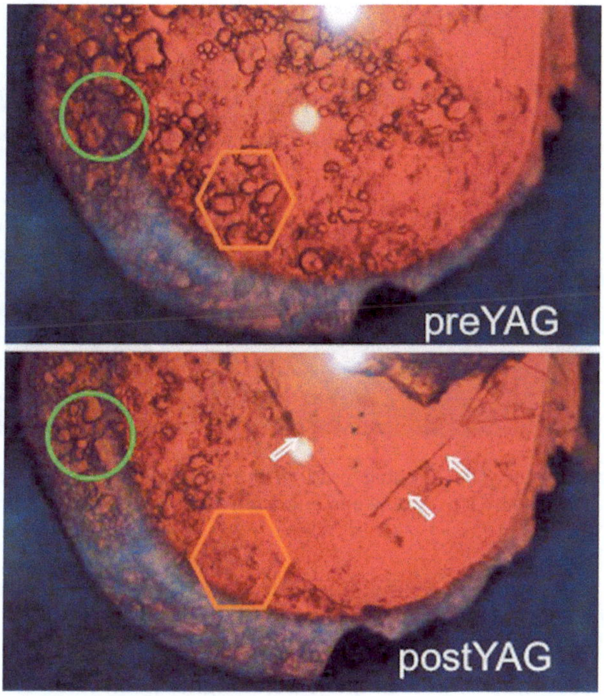

Fig. 12.8 Retroillumination images taken 10 min before and 10 min after Nd:YAG laser capsulotomy. On the remaining posterior capsule pearls disappeared (*hexagon*) or survived (*circle*). Inside the Nd:YAG opening all Elschnig pearls disappeared [17]

12.3.2 Morphological Changes After Nd:YAG Capsulotomy

Posterior capsulotomy seems to influence the formation of the remaining PCO structures outside of the capsulotomy opening, especially the formation of Elschnig pearls. Cases of proliferation and ensuing regression of Elschnig pearls after Nd: YAG capsulotomy were published [15–18]. Hollick and coauthors [7] mentioned a higher regression of LECs in patients with polyacrylic IOLs than in PMMA or silicone IOLs. Georgopoulos and coauthors [16] mentioned a spontaneous regression of Elschnig pearls after Nd:YAG capsulotomy with a loss of pearls in the areas outside the YAG opening in 45 % of the cases.

In a larger and more detailed study [17], observing short-time changes immediately and within weeks after Nd:YAG capsulotomy, a similar percentage of pearls outside the capsulotomy opening disappeared and survived the capsulotomy (Fig. 12.8). The number of disappeared pearls on the remaining capsule was several orders of magnitude higher compared to the normal turnover of pearls, especially taking into account that this change was observed in a short time span of only 20 min. The reason could be a reduced contact pressure between the posterior capsule and the IOL surface that could facilitate posterior migration of the cells towards the margin of the posterior capsulotomy opening and, in consequence, pearls may fall into the vitreous cavity. In single cases an Elschnig pearl survived the Nd:YAG capsulotomy on the posterior surface of the IOL within the Nd:YAG

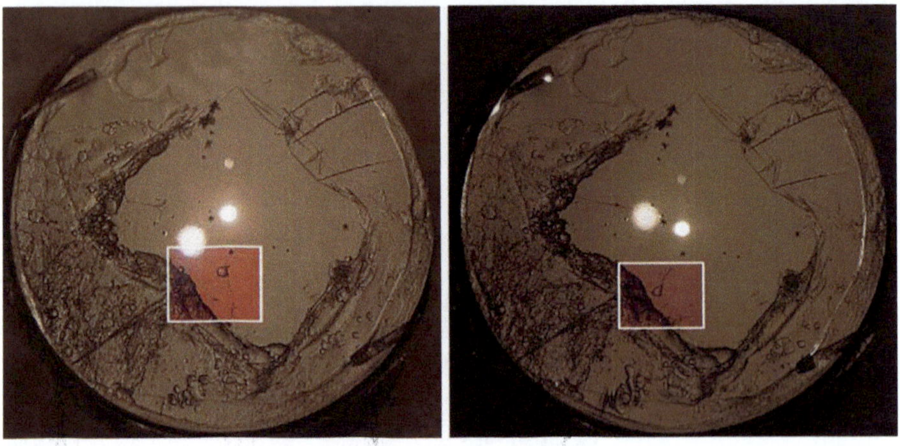

Fig. 12.9 The Elschnig pearl in the rectangle inside the Nd:YAG opening after Nd:YAG capsulotomy is very active in migration

capsulotomy opening, but these cells were shown to disappear soon or to migrate to the margin of the Nd:YAG capsulotomy (Fig. 12.9).

Another possible explanation could be the direct mechanical detonation of the laser shock wave. In this case, a higher number of disappearing pearls around the Nd:YAG opening would be expected compared with pearls more peripheral in location. However, although pearls that were adjacent to the rhexis showed a slightly higher stability, no significant differences were found concerning the distance to the Nd:YAG opening and number of disappearing pearls. No correlation between the used Nd:YAG laser energy and the pearl size changes was observed [17]. Alternatively, apoptosis of the pearls could be induced by cytokines of the aqueous humor, which has gained access to the LECs through the opened posterior capsule, although this would be expected to take longer than just minutes. Also, phagocytosis of pearls by macrophages, probably having migrated from the vitreous, could cause these changes.

Alternatively, osmotic processes could lead to bursting of cells when aqueous from Berger's space rushes into the subcapsular room once the capsule is opened.

It was observed that many new pearls appeared within the first week after Nd:YAG capsulotomy. Probably the increase in space between the posterior capsule and the IOL due to the released tension in the posterior capsule after capsulotomy leads to the higher number of newly formed pearls. Another possible reason for this increase in the number of Elschnig pearls could be an inflammatory process following the capsulotomy. The remaining pearls showed a sudden decrease in size and displayed a significantly increased turnover during the weeks following capsulotomy [18, 19]. This decrease could be due to the fact that the pearls were flattened between the posterior capsule and the surface of the IOL prior to capsulotomy. After Nd:YAG capsulotomy, the space between the posterior capsule and the IOL surface is larger, and the pearls change from a flattened into a more

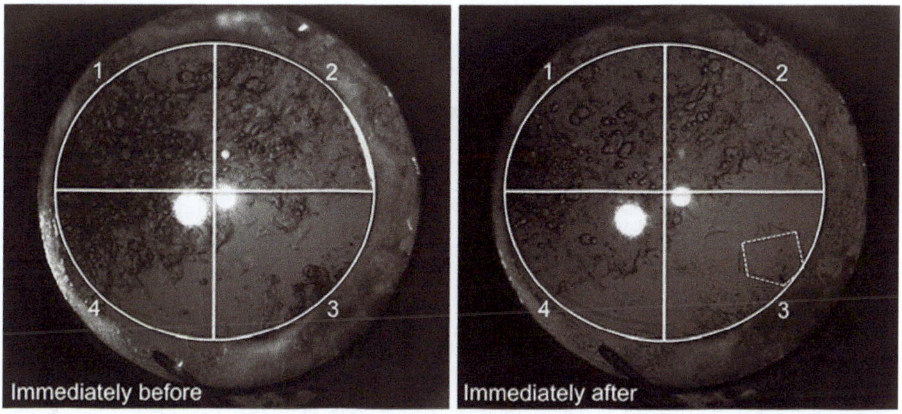

Fig. 12.10 Retroillumination images were taken before (*left*) and immediately after (*right*) capsulotomy. The *dotted line* in the *right image* depicts the mini YAG opening. Note the significant changes in pearl density far from the opening, such as in quadrant 1 [20]

spherical shape. Alternatively, as mentioned above, osmotic changes due to an altered milieu in the subcapsular region around the pearls as a result of aqueous inflow from Berger's space could cause a rapid change in pearl size. This change in milieu would need to be hyperosmolar compared with the usual capsular bag milieu.

The clinical importance of these findings may be to attempt to find new strategies of laser "polishing" of PCO without opening the posterior capsule.

12.3.3 Morphological Changes After Modified Nd:YAG Laser Treatment

In a recent study the effect of a "mini Nd:YAG" laser treatment in the periphery and "polishing" of the posterior lens capsule without opening the posterior lens capsule with a modified low-energy Nd:YAG laser was examined [17–20]. Performing a very small Nd:YAG capsulotomy in the periphery of the posterior lens capsule resulted in a significant reduction of Elschnig pearls on the entire lens capsule (Fig. 12.10). Therefore, mini Nd:YAG appears to alter regeneratory PCO also in an area where the capsule is still intact outside of the small opening. These findings were also observed to a smaller extent, when the posterior lens capsule was "polished" with a modified low-energy Nd:YAG laser without opening the posterior lens capsule (Fig. 12.11).

Different theories may explain this phenomenon: one of these explanations is the direct influence by the mechanical forces of the laser shock wave [18, 19]. The laser creates a plasma that leads to a pressure wave. The break of the capsule is caused by amplified and focused infrared light that rips away the electrons from their nuclei [21]. This would suggest that the plasma itself should not lead to the disappearance

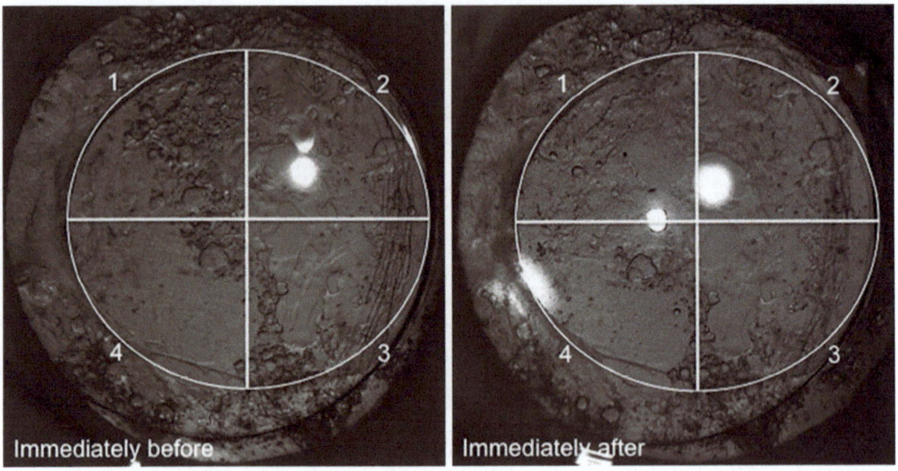

Fig. 12.11 Retroillumination images taken before (*left*) and immediately after (*right*) gentle Nd: YAG. One quadrant was treated and the post-YAG image shows a relevant decrease of Elschnig pearls [20]

of Elschnig pearls located at a larger distance to the YAG opening, since they should be far out of reach, but possibly the laser shock wave could have this effect. The laser shock wave expands into all directions with the same velocity. It is likely that a factor additional to the laser shock wave is responsible for the disappearance of Elschnig pearls immediately after Nd:YAG laser treatment.

Probably, the combination of two factors such as the reduced contact between the posterior capsule and the IOL surface together with the effect of the laser shock wave could be a reason for the reduction of PCO immediately after the Nd:YAG capsulotomy. The energy plasma created by the laser leads to the breakdown first, so that the tension of the capsule is reduced and the pressure wave follows and leads to a kind of "bounce" effect of the posterior capsule. This "bounce" effect could be, in combination with the direct pressure wave of the laser, the reason for the reduction of regeneratory PCO.

If it would be possible to remove PCO and to keep the posterior lens capsule intact, the functional properties of the capsule such as its elasticity and its function as separating posterior from anterior segment, as a result, concepts that allow restoration of accommodation, such as lens refilling, or "Phaco-Ersatz," where PCO appears to be the main limiting factor, may be possible [22]. Furthermore, the combination of very low Nd:YAG energy and maintaining an intact posterior capsule may result in a decreased risk of complications as seen with conventional Nd:YAG capsulotomy, such as cystoid macula edema or retinal detachment. Additionally, remnants of the Elschnig pearls would not be dissipated into the posterior segment of the eye after Nd:YAG laser treatment.

Fig. 12.12 Formations of Elschnig pearls (pearl strings) at the Nd:YAG capsulotomy margin

12.3.4 Morphological Changes of Pearl Strings

Kurosaka and coauthors [19] observed a probability of developing formations of Elschnig pearls at the Nd:YAG capsulotomy margin, so-called pearl strings (Fig. 12.12), in 77 % of all cases within the first 2 years after Nd:YAG capsulotomy.

The implantation of an IOL and continuous curvilinear capsulorhexis increased the development of pearl strings, whereas age, sex, total YAG energy, diabetes mellitus, or high myopia had no effect on the morphology of pearl strings [18].

Caballero et al. [13] reported that Elschnig pearls on the capsulotomy margin were not only located between the posterior capsule and the IOL, but also on the posterior face of the posterior capsule [23]. Elschnig pearls may undergo hyperproliferation at the edge of the YAG capsulotomy, which may even close it in rare instances.

Georgopoulos and coauthors [16] observed the correlation between pearl strings and IOL materials: silicone and PMMA IOLs led to significant pearl formation on the capsulotomy margin, often combined with a reduction of peripheral regeneratory PCO (silicone). On the other hand, hydrogel IOLs led to a higher incidence of reclosure of the YAG-capsulotomy opening.

12.4 Remaining Questions

Several questions remain to be answered. What are the factors influencing the dynamics of change of PCO? There is a great variability between patients, but factors, such as IOL material, age, or gender, did not have an effect on pearl turnover [8, 16, 17].

The question remains, what is an Elschnig pearl morphologically? In the literature, Fagerholm [24] compared lens fiber cells with Elschnig pearls. His analysis showed that Elschnig pearls may be cells that are differentiating to lens fiber cells. However, contrary to lens fiber cells, Elschnig pearls still contained a nucleus and few cell organelles. These so-called lens fiber-like cells may form extracellular vacuoles that appear as transparent spheres. These vacuoles were found to be surrounded by very thin lens fiber-like cells. The theory was based on a model by Sakuragawa [25] that the intact lens may respond, when exposed to extracellular fluid, by increasing the transport of sodium and thereby water into the extracellular space with the resulting formation of a cyst-like cavity. Kappelhof et al. [4, 6] described Elschnig pearls in rabbits as subcapsular epithelial cells originally from the Soemmering's ring that escape and migrate towards the center of the posterior capsule. A contrary theory by Jongebloed et al. [26] is based on transmission electron microscopy (TEM) after collecting Elschnig pearls with a glass cannula onto a Millipore filter. In his study, no cell membrane remnants were found indicating that Elschnig pearls are biophysical products of a slowly developing process of degradation of lens fibers and having no cellular origin. One argument that confirms this theory is that the TEM images showed erythrocytes close to Elschnig pearls. These erythrocytes were deformed because of the osmium staining before using the TEM, whereas Elschnig pearls retained their shape. The explanation was that Elschnig pearls are products of the ballooning of the cell membrane, emerging from the cytoplasm of the lens fibers. Similar formations are found in cataractous lenses. The origin of the Elschnig pearls could be the cell membrane of the degenerating lens fiber that consists of unsaturated lipids. This would explain why the osmotic staining had no effect on the pearl size and shape, whereas real cells were deformed.

Marcantonio et al. [27] describe the phenomenon of epithelial mesenchymal transition of LECs as the cellular substrate of PCO including Elschnig pearls driven by cytokines, such as TGF-beta.

In fact, the literature to date has conflicting theories on what Elschnig pearls actually are. From our clinical findings, there are some characteristics that speak in favor of the pearl being a cyst-like structure. Firstly, the rapid change in size is seen in some cases, especially the disappearance of large pearls within few days. Secondly, the very large size of some of the pearls would make these extremely large cells. Thirdly, the very low rate of fragmentation/division of Elschnig pearls. Also, in these cases of questionable fragmentation/division, it is feasible that a large pearl has decreased in size with a concomitant increase in size of a previously not detectable small neighboring pearl.

On slit lamp examination using indirect illumination and a method to assess the refractive index of round structures, as described by Brown [28], Elschnig pearls appear to be filled with material of higher refractive index than the surrounding tissue. This would speak in favor of a solid and cellular origin of pearls or, alternatively, a cyst-like formation filled with lens fiber material.

References

1. Findl O, Buehl W, Bauer P, Sycha T (2010) Interventions for preventing posterior capsule opacification. Cochrane Database Syst Rev 2, CD003738
2. Buehl W, Sacu S, Findl O (2005) Association between intensity of posterior capsule opacification and visual acuity. J Cataract Refract Surg 31(3):543–547
3. Buehl W, Sacu S, Findl O (2005) Association between intensity of posterior capsule opacification and contrast sensitivity. Am J Ophthalmol 140(5):927–930
4. Kappelhof JP, Vrensen GF, de Jong PT, Pameyer J, Willekens B (1986) An ultrastructural study of Elschnig's pearls in the pseudophakic eye. Am J Ophthalmol 101(1):58–69
5. Sveinsson O (1993) The ultrastructure of Elschnig's pearls in a pseudophakic eye. Acta Ophthalmol (Copenh) 71(1):95–98
6. Kappelhof JP, Vrensen GF, de Jong PT, Pameyer J, Willekens BL (1987) The ring of Soemmerring in man: an ultrastructural study. Graefes Arch Clin Exp Ophthalmol 225 (1):77–83
7. Hollick EJ, Spalton DJ, Ursell PG, Pande MV (1998) Lens epithelial cell regression on the posterior capsule with different intraocular lens materials. Br J Ophthalmol 82(10):1182–1188
8. Findl O, Neumayer T, Hirnschall N, Buehl W (2010) Natural course of Elschnig pearl formation and disappearance. Invest Ophthalmol Vis Sci 51(3):1547–1553
9. Neumayer T, Findl O, Buehl W, Georgopoulos M (2006) Daily changes in the morphology of Elschnig pearls. Am J Ophthalmol 141(3):517–523
10. Neumayer T, Findl O, Buehl W, Sacu S, Menapace R, Georgopoulos M (2005) Long-term changes in the morphology of posterior capsule opacification. J Cataract Refract Surg 31 (11):2120–2128
11. Buehl W, Findl O, Neumayer T, Georgopoulos M, Sacu S (2005) Short-term changes in the morphology of posterior capsule opacification. J Cataract Refract Surg 31(5):962–968
12. Nakashima Y, Yoshitomi F, Oshika T (2002) Regression of Elschnig pearls on the posterior capsule in a pseudophakic eye. Arch Ophthalmol 120(3):397–398
13. Caballero A, Marin JM, Salinas M (2000) Spontaneous regression of Elschnig pearl posterior capsule opacification. J Cataract Refract Surg 26(5):779–780
14. Dholakia SA, Vasavada AR, Singh R (2005) Prospective evaluation of phacoemulsification in adults younger than 50 years. J Cataract Refract Surg 31(7):1327–1333
15. Neumayer T, Buehl W, Findl O (2006) Effect of topical prednisolone and diclofenac on the short-term change in morphology of posterior capsular opacification. Am J Ophthalmol 142 (4):550–556
16. Georgopoulos M, Findl O, Menapace R, Buehl W, Wirtitsch M, Rainer G (2003) Influence of intraocular lens material on regeneratory posterior capsule opacification after neodymium: YAG laser capsulotomy. J Cataract Refract Surg 29(8):1560–1565
17. Hirnschall N, Neumayer T, Georgopoulos M, Findl O (2009) Effect of Nd:YAG capsulotomy on the morphology of surviving Elschnig pearls. Br J Ophthalmol 93(12):1643–1647
18. Kato K, Kurosaka D, Bissen-Miyajima H, Negishi K, Hara E, Nagamoto T (1997) Elschnig pearl formation along the posterior capsulotomy margin after neodymium:YAG capsulotomy. J Cataract Refract Surg 23(10):1556–1560
19. Kurosaka D, Kato K, Kurosaka H, Yoshino M, Nakamura K, Negishi K (2002) Elschnig pearl formation along the neodymium:YAG laser posterior capsulotomy margin. Long-term follow-up. J Cataract Refract Surg 28(10):1809–1813
20. Hirnschall N, Neumayer T, Rossmann M, Georgopoulos M, Chen YA, Findl O (2012) Influence of a modified Nd:YAG laser treatment outside the visual axis on the morphology of Elschnig pearls. Ophthalmic Res 48(3):151–155
21. Aslam TM, Devlin H, Dhillon B (2003) Use of Nd:YAG laser capsulotomy. Surv Ophthalmol 48(6):594–612
22. Nishi Y, Mireskandari K, Khaw P, Findl O (2009) Lens refilling to restore accommodation. J Cataract Refract Surg 35(2):374–382

23. Caballero A, Garcia-Elskamp C, Losada M, Salinas M, Marin JM (2001) Natural evolution of Elschnig pearl posterior capsule opacification after posterior capsulotomy. J Cataract Refract Surg 27(12):1979–1986
24. Fagerholm PP, Philipson BT (1977) Formation of after-cataract by regeneration of human and rabbit lens epithelium in tissue culture. Acta Ophthalmol (Copenh) 55(3):362–371
25. Sakuragawa M, Kuwabara T, Kinoshita JH, Fukui HN (1975) Swelling of the lens fibers. Exp Eye Res 21(4):381–394
26. Jongebloed WL, Kalicharan D, Los LI, van der Veen G, Worst JG (1991) A combined scanning and transmission electronmicroscopic investigation of human (secondary) cataract material. Doc Ophthalmol 78(3–4):325–334
27. Marcantonio JM, Vrensen GF (1999) Cell biology of posterior capsular opacification. Eye (Lond) 13(Pt 3b):484–488
28. Brown N (1971) Visibility of transparent objects in the eye by retroillumination. Br J Ophthalmol 55(8):517–524

Effect of Posterior Capsule Opacification and Anterior Capsule Contraction on Visual Function

13

Ken Hayashi

Abstract

In this chapter, a study done to examine the effect of (1) the size of Nd:YAG laser posterior capsulotomy for posterior capsule opacification (PCO) and (2) anterior capsule contraction (ACC) on visual function is described. (1) Eyes with PCO first underwent posterior capsulotomy smaller than pupillary size, after which the capsulotomy was secondarily enlarged to greater than pupillary size. Visual acuity (VA), contrast VA, and that with glare (glare VA) under photopic and mesopic conditions were measured after the small and large capsulotomies (in the same eyes). After enlargement, the mean capsulotomy area increased from 4.8 to 15.3 mm^2. Mean corrected-distance VA (CDVA) did not improve significantly after enlargement, but photopic contrast VA and glare VA at moderate-to-low-contrasts and mesopic contrast VA and glare VA improved significantly. (2) Eyes with ACC underwent Nd:YAG laser anterior capsulotomy. The anterior capsule opening area was measured before and after capsulotomy and was correlated with VA and contrast sensitivity. The mean anterior capsule opening area increased from 8.2 to 18.0 mm^2 after capsulotomy. The mean contrast sensitivity at most visual angles improved significantly after capsulotomy, although CDVA did not. The anterior capsule opening area before capsulotomy was correlated significantly with contrast sensitivity, but not with CDVA. In conclusion, in eyes with PCO, contrast sensitivity was worse with a small capsulotomy than with a large capsulotomy, suggesting that a capsulotomy larger than the pupil is necessary to restore contrast sensitivity. ACC impairs significantly contrast sensitivity in proportion to the opening area, but does not worsen CDVA markedly.

K. Hayashi, M.D. (✉)
Hayashi Eye Hospital, 4-7-13 Hakataekimae, Hakata-Ku, Fukuoka 812, Japan
e-mail: hayashi-ken@hayashi.or.jp

S. Saika et al. (eds.), *Lens Epithelium and Posterior Capsular Opacification*,
DOI 10.1007/978-4-431-54300-8_13, © Springer Japan 2014

Keywords
Anterior capsule contraction • Contrast sensitivity • Neodymium:yttrium–aluminum–garnet laser capsulotomy • Posterior capsule opacification • Visual acuity

13.1 Introduction

Posterior capsule opacification (PCO) after cataract surgery impairs various visual function parameters, including visual acuity (VA), contrast sensitivity, and glare disability [1–7]. The impairment of visual function parameters is related to the degree of PCO. In eyes with dense PCO that require neodymium:yttrium–aluminum–garnet (Nd:YAG) laser posterior capsulotomy, all visual function parameters are impaired. When PCO is slight, contrast sensitivity or glare disability may deteriorate with no marked loss of VA [6].

Nd:YAG laser posterior capsulotomy improves these visual function parameters. However, surgeons are perplexed about what size of posterior capsulotomy should be made. When a large capsulotomy is made, adverse effects of the posterior capsulotomy such as retinal detachment [8–10], cystoid macular edema [10, 11], or an elevation in intraocular pressure (IOP) [12, 13] tend to occur. However, if the capsulotomy is too small, visual function may not be recovered. Previous studies showed that glare disability in eyes with a small capsulotomy is worse than that in eyes with a large capsulotomy despite no significant difference in VA [14–16]. However, to date, there is no study about the relationship between capsulotomy size and contrast sensitivity.

Furthermore, anterior capsule contraction (ACC) occurs within several months after cataract surgery [17–20]. The degree of ACC varies from patient to patient [21] and depends on the presence of comorbidity such as pseudoexfoliation syndrome [22], retinitis pigmentosa [23], or diabetes mellitus [24]. Previous studies showed that the anterior capsule opening occasionally becomes smaller than the pupillary area or, in some cases, even occludes completely [25–30]. When the area of the anterior capsule opening becomes smaller than the pupillary area, the light entering the eye is limited, which leads to impairment in visual function. However, there is no report showing how ACC is associated with deterioration of visual function.

The aim of the study described in this chapter was twofold: the first was to examine the effect of Nd:YAG laser posterior capsulotomy size on visual function parameters. Because contrast sensitivity and glare disability vary from patient to patient, we made the Nd:YAG laser posterior capsulotomy smaller and then larger than the pupillary size in each eye, and the visual function with small capsulotomy was compared to the visual function with large capsulotomy in the same eye. The second aim was to examine the effect of ACC on visual function. VA and contrast sensitivity in eyes with severe ACC that required Nd:YAG laser anterior capsulotomy were examined, and the degree of ACC was correlated with these visual function parameters.

13.2 Methods

13.2.1 Patients

Pseudophakic patients with monofocal intraocular lens (IOL) who were consecutively scheduled for Nd:YAG laser posterior and anterior capsulotomy were screened for inclusion in this study. An Nd:YAG laser capsulotomy was scheduled when an eye lost two or more decimal lines of VA or when the patient complained of blurred vision. The author verified the presence of PCO or ACC (usually when the anterior capsular rim was seen within the pupillary area) on slit-lamp biomicroscopy and performed patient screening. When both eyes were affected, only the eye that was scheduled for Nd:YAG laser capsulotomy first was included. Pre-laser exclusion criteria were any pathology of the macula or optic nerve, previous history of inflammation, or liquefied aftercataract. Post-laser exclusion criteria were corrected-distance VA (CDVA) of less than 0.5 due to any pathology of unknown cause, media opacities other than PCO, and any difficulties with data collection or analysis.

13.2.2 Procedures of Nd:YAG Laser Posterior Capsulotomy

A physician first made an Nd:YAG laser capsulotomy smaller than the pupillary diameter using a Q-switched Nd:YAG laser (YC-1300; Nidek, Gamagori, Japan). After topical anesthesia, a contact lens was applied to enhance power density at the level of the posterior capsule. The central posterior capsule within the pupillary area was cleared as much as possible without pupil dilation by emitting laser energy on the posterior capsule; energy levels were between 0.5 and 1.5 mJ. Special care was taken to not pit the IOL optic. Approximately 2 weeks after the first capsulotomy, after full pupil dilation, the capsulotomy opening was enlarged to greater than the pupillary area (approximately 5.0 mm in diameter) by the same surgeon who used almost the same technique as used for the smaller capsulotomy.

13.2.3 Procedures of Nd:YAG Laser Anterior Capsulotomy

After full dilation of the pupil, all patients underwent Nd:YAG laser anterior capsulotomy with a Q-switched Nd:YAG laser (YC-1300; Nidek). After topical anesthesia, a contact lens was applied to enhance power density at the level of the anterior capsule. Either four or six relaxing incisions were made from the margin of the anterior capsule opening to the edge of the IOL optic by emitting laser energy on the anterior capsular rim; energy levels were between 1.0 and 2.5 mJ. Special care was taken to not pit the IOL optic.

13.2.4 Outcome Measures

In eyes with PCO, CDVA was measured using decimal charts, while contrast VA and that in the presence of a glare source (glare VA) were measured 2 weeks after the first posterior capsulotomy and 2 weeks after the enlargement. Contrast VA and glare VA were examined using the contrast sensitivity accurate tester (CAT-2000; Menicon, Tokyo, Japan) [6]. This device measures the logarithm of the minimal angle of resolution (logMAR) VA using five contrast visual targets (100, 25, 10, 5, and 2.5 %) under photopic and mesopic conditions. Measurement under photopic conditions was made with chart lighting of 100 cd mm^{-2}, and that under the mesopic conditions was done with chart lighting of 3 cd mm^{-2}. A glare light source of 200 lx was located in the periphery at 20° along the visual axis. In eyes with ACC, contrast sensitivity was measured using the contrast glare tester (CGT-1000; Takagi Seiko) before and approximately 2 weeks after Nd:YAG laser anterior capsulotomy [5]. The CGT-1000 measures 12-step contrast thresholds using concentric ring-shaped visual targets that are equivalent to visual angles of 6.3°, 4.0°, 2.5°, 1.6°, 1.0°, and 0.7° at 0.35 m. Contrast sensitivity is calculated as a logarithm of the inverse value of the contrast threshold.

The area of the posterior capsulotomy opening was measured using the Scheimpflug photography system (EAS-1000; Nidek, Gamagori, Japan) 2 weeks after the small capsulotomy was created and 2 weeks after its enlargement. The area of the anterior capsule opening was also measured using the EAS-1000. The refractive status and keratometric cylinder were measured using an autorefractometer (KR-7100; Topcon, Tokyo, Japan). The manifest spherical equivalent was determined as the spherical power plus half the cylindrical power. The pupillary diameter under photopic and mesopic conditions was examined using an electronic pupillometer (FP-10000; TMI, Saitama, Japan). The central retinal (foveal) thickness was measured using the Stratus optical coherence tomography-3 (OCT-3; Carl Zeiss Meditec, Dublin, California, USA). Intraocular pressure (IOP) was measured using an applanation tonometer before and 2 h after the Nd:YAG laser capsulotomy.

13.2.5 Statistical Analysis

Decimal VA was converted to a logMAR scale for statistical analysis. Contrast VA and glare VA were also converted to the logarithm of inverse values for statistical analysis. CDVA, contrast VA, and glare VA, area of the posterior capsulotomy and anterior capsule opening, and other continuous variables were compared using the Wilcoxon signed-rank test or Mann–Whitney *U* test. Univariate associations between visual functions and the area of the anterior capsule opening were evaluated using Pearson's correlation coefficient. Any differences showing a *P* value of less than 0.05 were considered to be statistically significant.

13.3 Results

In eyes with PCO, the mean area of posterior capsulotomy opening was 4.8 ± 1.5 mm^2 after small capsulotomy and 15.3 ± 4.7 mm^2 after enlargement ($P < 0.0001$). Mean CDVA was 0.49 ± 0.27 before Nd:YAG posterior capsulotomy, 0.95 ± 0.14 after the small capsulotomy, and 0.96 ± 0.14 after enlargement. Mean CDVA improved significantly after the small capsulotomy ($P \leq 0.0001$), and it did not improve significantly more after enlargement ($P = 0.1729$).

Under photopic conditions (Fig. 13.1), contrast VA and glare VA at moderate-to-low-contrast visual targets were significantly better after enlargement than those before enlargement ($P \leq 0.0242$), although no significant difference was found at high-contrast visual targets. Under mesopic conditions (Fig. 13.2), contrast VA and glare VA at high to moderate visual targets improved significantly after enlargement to the large capsulotomy size ($P \leq 0.0136$), but no significant difference was found in the contrast VA and glare VA at low contrasts because it was immeasurable in most of the patients.

The mean foveal thickness as determined with the OCT-3 was almost the same at the baseline (before the small capsulotomy), before the large capsulotomy, and at

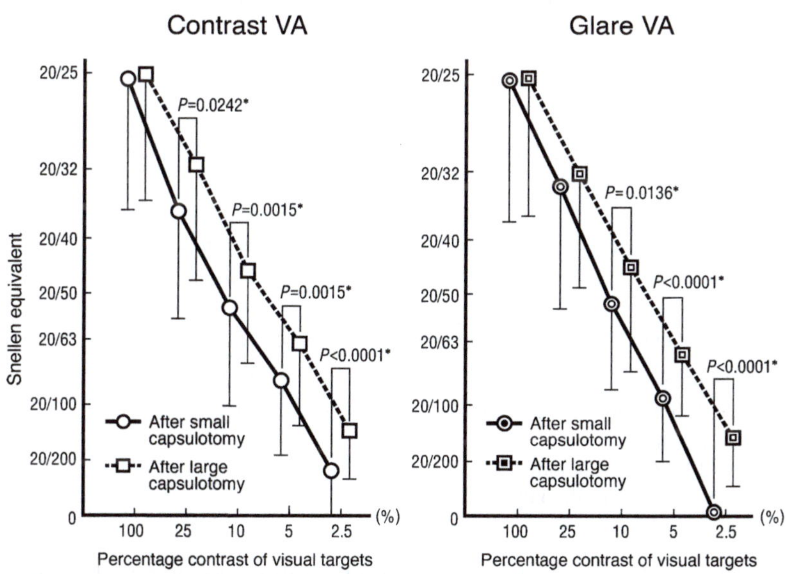

Fig. 13.1 Mean contrast visual acuity (contrast VA) and that in the presence of a glare (glare VA) under photopic conditions after small posterior capsulotomy and large posterior capsulotomy using the Nd:YAG laser. Mean contrast VA and glare VA under photopic conditions at moderate-to-low-contrast visual targets after small capsulotomy were significantly worse than that after large capsulotomy. *Asterisk*, statistically significant difference

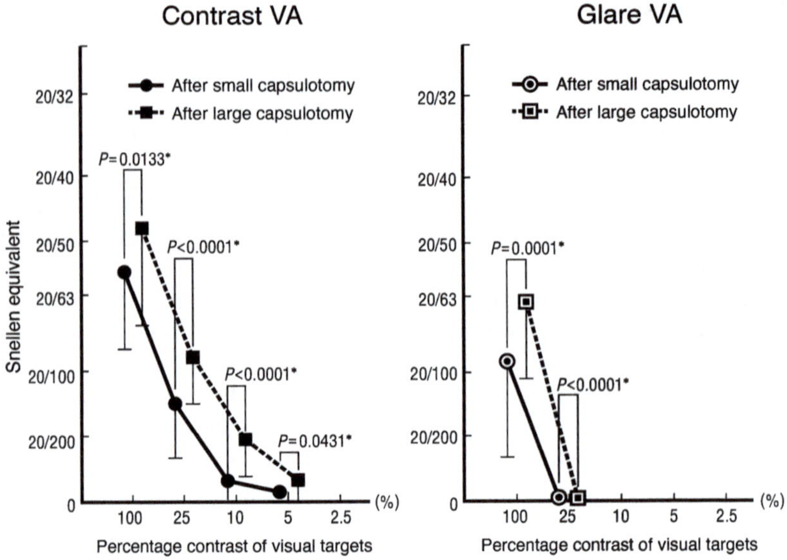

Fig. 13.2 Mean contrast visual acuity (contrast VA) and that in the presence of a glare (glare VA) under mesopic condition after small posterior capsulotomy and large posterior capsulotomy using the Nd:YAG laser. Mean contrast VA and glare VA under mesopic conditions at high to moderate visual targets after small capsulotomy were significantly worse than that after large capsulotomy, and that at low visual targets could not be measured. *Asterisk*, statistically significant difference

Table 13.1 Mean (±SD) foveal thickness, intraocular pressure (IOP), and increase in IOP after small capsulotomy and large capsulotomy

	After small capsulotomy	After large capsulotomy	*P* value
Foveal thickness (μ)	174 ± 23	175 ± 20	0.8590*
IOP (mmHg)			
Before capsulotomy[a]	14.0 ± 3.1	13.1 ± 2.7	0.2189*
After capsulotomy[b]	15.7 ± 3.5	15.1 ± 3.5	0.0926*
Increase in IOP (mmHg)	1.7 ± 2.0	2.0 ± 3.5	0.5349*

*No significant difference
[a]Just before Nd:YAG capsulotomy
[b]At 2 h after Nd:YAG capsulotomy

2 weeks after the large capsulotomy. During the follow-up, no eye showed cystoid macular edema or retinal breaks and retinal detachment due to the Nd:YAG laser capsulotomy. The mean IOP was similar just before the small capsulotomy, at 2 h after the small capsulotomy, immediately before enlargement of the capsulotomy, and at 2 h after enlargement to the large capsulotomy size. The increase in IOP before and after the small capsulotomy and before and after enlargement to a large capsulotomy size was also similar (Table 13.1).

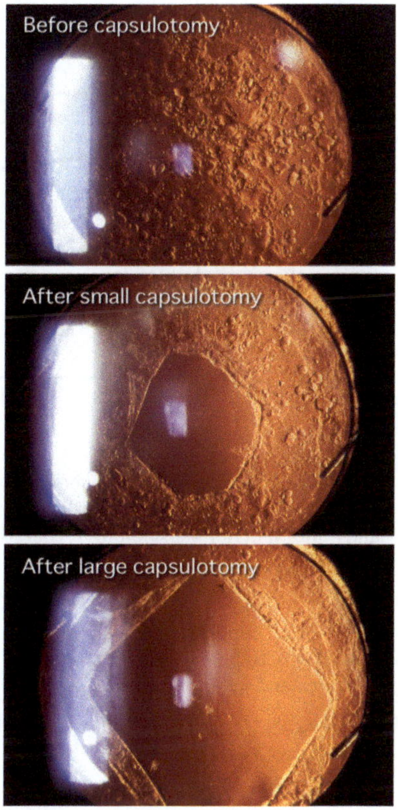

Fig. 13.3 Photographs of a representative eye that underwent a 2-step Nd:YAG laser capsulotomy. The Elschnig pearl type of posterior capsule opacification is seen before Nd:YAG laser capsulotomy (*top*). An Nd:YAG laser posterior capsulotomy smaller than the pupillary size was first made (*middle*). At 2 weeks after this small capsulotomy, the capsulotomy opening was enlarged to greater than pupillary size (*bottom*)

Figure 13.3 shows a representative eye that underwent this 2-step Nd:YAG laser capsulotomy. The Elschnig pearl type of PCO was seen 38 months after cataract surgery. An Nd:YAG laser posterior capsulotomy smaller than the pupillary area was first made; then 2 weeks after the small capsulotomy, the capsulotomy opening was enlarged to greater than the pupillary area.

In eyes with ACC, the mean area of the anterior capsule opening increased significantly from 8.2 ± 3.6 mm^2 to 18.0 ± 7.4 mm^2 after YAG laser anterior capsulotomy ($P < 0.0001$). Mean CDVA did not improve significantly after anterior capsulotomy ($P = 0.1729$). The mean threshold at most visual angles, except for that at 0.7°, improved significantly after anterior capsulotomy (Fig. 13.4).

Both before and after anterior capsulotomy, no significant correlation was found between the opening area and CDVA. There was also no significant correlation between increase in the opening area and improvement in CDVA (Table 13.2). Before anterior capsulotomy, significant correlations were found between the area of opening and contrast threshold at visual angles of 6.3°, 4.0° (Fig. 13.5), 2.5°, 1.6°, or 1.0° (all except 0.7°). However, after anterior capsulotomy, no significant correlation was found between the opening area and contrast sensitivity. There were also significant correlations between the increase in the opening area and improvement in contrast sensitivity.

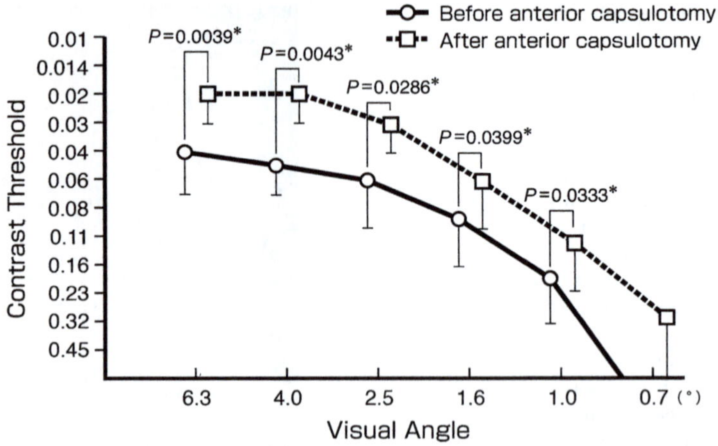

Fig. 13.4 Changes in mean contrast sensitivity before and after Nd:YAG laser anterior capsulotomy. Mean contrast threshold at most visual angles, except for 0.7°, improved significantly after anterior capsulotomy. *Asterisk*, statistically significant difference

Table 13.2 Simple correlation between area of anterior capsule opening (ACO) and contrast sensitivity (CS) before and after Nd:YAG laser anterior capsulotomy and between the increase in ACO area and improvement in CS

Visual angle (°)	Correlation coefficient (*r*)	*P* value
Between ACO area and CS		
Before anterior capsulotomy		
6.3	0.598	0.0003*
4.0	0.725	<0.0001*
2.5	0.609	0.0002*
1.6	0.590	0.0004*
1.0	0.521	0.0023*
0.7	0.193	0.2890
After anterior capsulotomy		
6.3	0.042	0.8190
4.0	0.040	0.8286
2.5	0.005	0.9797
1.6	0.120	0.5121
1.0	0.197	0.2803
0.7	0.216	0.2346
Between the increase in ACO area and improvement in visual acuity		
6.3	0.640	<0.0001*
4.0	0.532	0.0017*
2.5	0.464	0.0075*
1.6	0.455	0.0089*
1.0	0.493	0.0042*
0.7	0.192	0.2932

*Statistically significant correlation

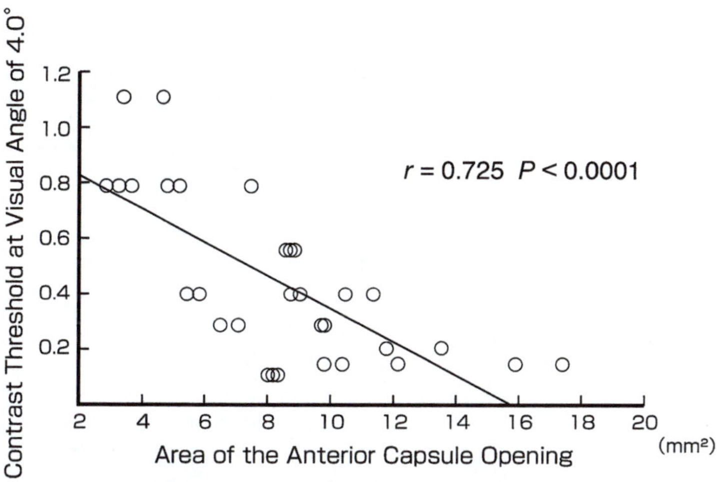

Fig. 13.5 Scatterplots of simple correlation between the area of the anterior capsule opening and contrast threshold at the visual angle of 4.0° before Nd:YAG laser anterior capsulotomy. Prior to capsulotomy, a smaller area of the anterior capsule opening was associated with worse contrast threshold at the visual angle of 4.0°

Figure 13.6 shows photographs of a representative eye that underwent Nd:YAG laser anterior capsulotomy. The anterior capsule opening was severely constricted 6 months after cataract surgery. Six relaxing incisions were made by Nd:YAG laser shots directed to the margin of the anterior capsule opening to the edge of the IOL optic. Two weeks after anterior capsulotomy, the anterior capsule opening was enlarged markedly.

13.4 Discussion

The study described in this chapter has shown that in eyes with PCO, CDVA with Nd:YAG laser posterior capsulotomy smaller than the pupillary area did not differ significantly from that with capsulotomy larger than the pupillary area. However, photopic contrast VA and glare VA at moderate-to-low-contrasts and mesopic contrast VA and glare VA with the large capsulotomy were significantly better than were those with the small capsulotomy. This suggests that Nd:YAG laser posterior capsulotomy of diameter larger than the pupillary area is necessary to recover the contrast sensitivity and to lessen substantially glare disability in eyes with PCO.

When surgeons make a large Nd:YAG laser posterior capsulotomy, they are often concerned about the occurrence of adverse effects, such as retinal detachment [8–10], cystoid macular edema [10, 11], or an increase in IOP [12, 13]. However, in this series, retinal breaks and retinal detachment did not occur during follow-up. The foveal thickness did not increase after enlargement to a large capsulotomy size,

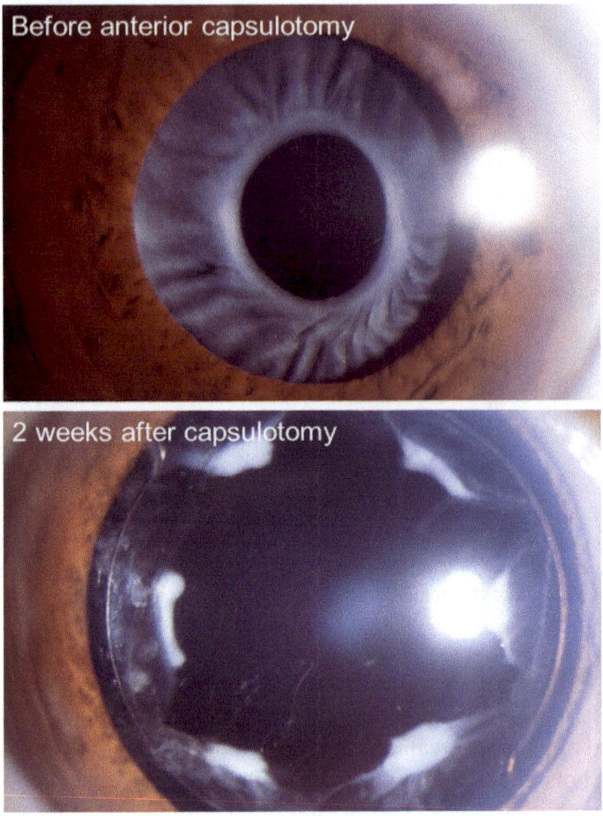

Fig. 13.6 Photographs of a representative eye that underwent Nd:YAG laser anterior capsulotomy. The anterior capsule opening was constricted substantially by 6 months after cataract surgery (*top*). Six relaxing incisions were made from the margin of the anterior capsule opening to the edge of IOL optic. Two weeks after anterior capsulotomy, the anterior capsule opening was enlarged markedly (*bottom*)

and no cystoid macular edema was seen. In addition, the increase in IOP after either the small or large capsulotomy was slight, and there was no significant difference in increase in IOP between the small and large capsulotomies. Furthermore, it has been shown that the incidence of retinal detachment and cystoid macular edema after Nd:YAG laser capsulotomy is uncommon in eyes that have previously undergone phacoemulsification and in-the-bag implantation of an IOL [5, 31–34]. These findings suggest that the incidence of adverse effects does not differ markedly between eyes with a small capsulotomy and those with a large capsulotomy.

In eyes with a marked ACC, the area of the anterior capsule opening before Nd: YAG laser anterior capsulotomy was correlated significantly with contrast sensitivity, but was not correlated with VA. After anterior capsulotomy, contrast sensitivity improved significantly, although VA did not. The opening area after capsulotomy was correlated with neither contrast sensitivity nor VA. In addition, the increase in the opening area was correlated significantly with improvement in contrast sensitivity. These results indicate that extensive ACC after cataract surgery worsens substantially the contrast sensitivity, but does not markedly decrease VA.

The mean area of the anterior capsule opening was increased by Nd:YAG laser anterior capsulotomy. 96.8 % of the eyes that underwent radial anterior capsulotomy showed an increase in the opening area. In this study, either four or six radial incisions were made to the anterior capsule rim by Nd:YAG laser emission, but the increase in the opening area did not differ between four and six incisions, suggesting that four or more radial relaxing incisions using the Nd:YAG laser increase significantly the opening area, leading to improvement in contrast sensitivity.

Many studies showed that PCO impairs various visual function parameters, including VA, contrast sensitivity, and glare disability [1–7]. Our previous studies showed that VA is correlated most strongly with the degree of PCO when it is so dense as to require an Nd:YAG capsulotomy [5], while contrast sensitivity is deteriorated most prominently, without significant loss of VA, when PCO is slight [6]. Thus, it is clear that all of the visual function parameters improved after Nd:YAG capsulotomy, but the degree of improvement may depend on the size of the capsulotomy. Several studies assumed that glare disability was worse with a small capsulotomy than with a large capsulotomy [14, 15]. The study described in this chapter has shown that both contrast sensitivity and glare disability recovered in eyes that underwent capsulotomy of diameter larger than the pupillary size.

The worse contrast sensitivity and glare disability with a small posterior capsulotomy for PCO can be explained as follows. The amount of light entering into the eye is dependent on the pupillary size. When the Nd:YAG laser capsulotomy opening is smaller than the pupillary size, the light entering into the eye, and finally reaching the macula, is intercepted by the unopened part of the opaque posterior capsule, which may lead to the worsening in contrast sensitivity. Furthermore, as the entering light passes through the residual capsule, it will be scattered, which may cause glare disability. However, even when the capsulotomy is small, the optical center of the pupillary area is not occluded, and light entering along the visual axis may thus not be restricted enough to markedly decrease the VA.

The ACC occurs within several months after cataract surgery [17–20]. The degree of ACC varies depending upon the individual and factors inherent to the IOL. Individual risk factors for marked contraction have been reported to be older age [21] and comorbidity including pseudoexfoliation [22], retinitis pigmentosa [23], and diabetes mellitus [24]. Previous studies including ours showed that an IOL factor that may affect anterior capsule contraction is the optic material [35, 36]. When the size of the anterior capsule opening becomes equal to or smaller than the pupillary area, the ACC may impair visual function. However, no study has shown to date how the ACC is related to visual function. The study described in this chapter verified that severe ACC impairs contrast sensitivity.

PCO generally progresses from the periphery to the central pupillary area and, therefore, obstructs the entrance of light into the pupil, which results in impairment of both visual acuity and contrast sensitivity even in its early stages. On the other hand, because ACC stops within several months after surgery, the optical center of the pupillary area is not usually occluded. Accordingly, light entering along the visual axis and ultimately reaching the macula may not be limited enough to markedly decrease the visual acuity. However, portions of the entering light are surely intercepted by the constricted anterior capsule rim, which may lead to a decrease in contrast sensitivity.

13.5 Conclusions

Contrast sensitivity and glare disability with a posterior capsulotomy larger than the pupil size is better than that with a capsulotomy smaller than the pupillary size, although the size of the capsulotomy does not affect significantly the VA. The incidence of adverse effects did not differ between eyes with a small posterior capsulotomy and eyes with a large capsulotomy. To restore contrast sensitivity and to lessen glare disability, an Nd:YAG laser posterior capsulotomy larger than pupil size is recommended. The ACC also worsens significantly contrast sensitivity in proportion to the anterior capsule opening area, but does not affect VA markedly. When patients with a marked ACC complain of blurred vision, performing an Nd: YAG laser anterior capsulotomy is recommended.

References

1. Nadler DJ, Clayman HM, Jaffe MS, Luscombe SM et al (1984) Glare disability in eyes with intraocular lenses. Am J Ophthalmol 97(1):43–47
2. Wilkins M et al (1996) Visual recovery under glare conditions following laser capsulotomy. Eye 10(1):117–120
3. Tan JCH et al (1998) Comparison of methods to assess visual impairment from glare and light scattering with posterior capsule opacification. J Cataract Refract Surg 24(12):1626–1631
4. Magno BV et al (1997) Evaluation of visual function following neodymium:YAG laser posterior capsulotomy. Ophthalmology 104(8):1287–1293
5. Hayashi K et al (2003) Correlation between posterior capsule opacification and visual function before and after neodymium:YAG laser posterior capsulotomy. Am J Ophthalmol 136 (4):720–726
6. Hayashi K, Hayashi H (2005) Posterior capsule opacification in the presence of an intraocular lens with a sharp versus rounded optic edge. Ophthalmology 112(9):1550–1556
7. van Bree MCJ et al (2008) Effect of neodymium:YAG laser capsulotomy on retinal straylight values in patients with posterior capsule opacification. J Cataract Refract Surg 34 (10):1681–1686
8. Leff SR et al (1987) Rhegmatogenous retinal detachment after YAG laser posterior capsulotomy. Ophthalmology 94(10):1222–1225
9. Javitt JC et al (1992) National outcomes of cataract extraction: increased risk of retinal complications associated with Nd:YAG laser capsulotomy. The cataract patient outcomes research team. Ophthalmology 99(10):1487–1497, discussion 1497–1498
10. Steinert RF et al (1991) Cystoid macular edema, retinal detachment, and glaucoma after Nd: YAG laser posterior capsulotomy. Am J Ophthalmol 112(4):373–380
11. Shah GR (1986) Three thousand YAG lasers in posterior capsulotomies: an analysis of complications and comparison to polishing and surgical discission. Ophthalmic Surg 17 (8):473–477
12. Channell MM, Beckman H (1984) Intraocular pressure changes after neodymium-YAG laser posterior capsulotomy. Arch Ophthalmol 102(7):1024–1026

13. Stark WJ et al (1985) Neodymium:YAG lasers: an FDA report. Ophthalmology 92(2):209–212
14. Lohmann CP et al (1994) iBlendungsempfindlichkeit vor und nach Nd:YAG-kapsulotomie: ein vergleich zwischen kleiner und groβer kapsulotome [The role of glare and light scatter in the degradation of visual performance before and after Nd:YAG capsulotomy: a comparison of small and large size capsultomies]. Klin Monatsbl Augenheilkd 205(2):65–69
15. Goble RR et al (1994) The role of light scatter in the degradation of visual performance before and after Nd:YAG-capsulotomy. Eye 8(Pt5):530–534
16. Yilmaz S et al (2006) The effect of Nd:YAG laser capsulotomy size on refraction and visual acuity. J Refract Surg 22(7):719–721
17. Hayashi K et al (1997) Reduction in the area of the anterior capsule opening after polymethylmethacrylate, silicone, and soft acrylic intraocular lens implantation. Am J Ophthalmol 123(4):441–447
18. Joo CK et al (1996) Capsular opening contraction after continuous curvilinear capsulorhexis and intraocular lens implantation. J Cataract Refract Surg 22(5):585–590
19. Ursell PG et al (1997) Anterior capsule stability in eyes with intraocular lenses made of poly (methyl methacrylate), silicone, and AcrySof. J Cataract Refract Surg 23(10):1532–1538
20. Park TK et al (2002) Changes in the area of the anterior capsule opening after intraocular lens implantation. J Cataract Refract Surg 28(9):1613–1617
21. Kato S et al (2002) Risk factors for contraction of the anterior capsule opening after cataract surgery. J Cataract Refract Surg 28(1):109–112
22. Hayashi H et al (1998) Anterior capsule contraction and intraocular lens dislocation in eyes with pseudoexfoliation syndrome. Br J Ophthalmol 82(12):1429–1432
23. Hayashi K et al (1998) Anterior capsule contraction and intraocular lens dislocation after implant surgery in eyes with retinitis pigmentosa. Ophthalmology 105(7):1239–1243
24. Hayashi H et al (1998) Area reduction in the anterior capsule opening in eyes of diabetes mellitus patients. J Cataract Refract Surg 24(8):1105–1110
25. Hansen SO et al (1993) Progressive constriction of the anterior capsular opening following intact capsulorhexis. J Cataract Refract Surg 19(1):77–82
26. Davison JA (1993) Capsule contraction syndrome. J Cataract Refract Surg 19(5):582–589
27. Nishi O, Nishi K (1993) Intraocular lens encapsulation by shrinkage of the capsulorhexis opening. J Cataract Refract Surg 19(4):544–545
28. Scorolli L et al (1996) Capsule contraction after continuous curvilinear capsulorhexis. J Cataract Refract Surg 22(9):1245–1246
29. Tuft SJ, Talks SJ (1998) Delayed dislocation of foldable plate-haptic silicone lenses after Nd: YAG laser anterior capsulotomy. Am J Ophthalmol 126(4):586–588
30. Spang KM et al (1999) Complete occlusion of the anterior capsular opening after intact capsulorhexis: clinicopathologic correlation. Am J Ophthalmol 127(3):343–345
31. Olsen G, Olson RJ (2000) Update on a long-term, prospective study of capsulotomy and retinal detachment rates after cataract surgery. J Cataract Refract Surg 26(7):1017–1021
32. Jahn CE et al (2003) Pseudophakic retinal detachment after uneventful phacoemulsification and subsequent neodymium:YAG capsulotomy for capsule opacification. J Cataract Refract Surg 29(5):925–929
33. Lewis H et al (1987) A prospective study of cystoid macular edema after neodymium:YAG laser posterior capsulotomy. Ophthalmology 94(5):478–482
34. Billotte C, Berdeaux G (2004) Adverse clinical consequences of neodymium:YAG laser treatment of posterior capsule opacification. J Cataract Refract Surg 30(10):2064–2071
35. Hayashi K, Hayashi H (2005) Intraocular lens factors that may affect anterior capsule contraction. Ophthalmology 112(2):286–292
36. Werner L et al (2001) Anterior capsule opacification: correlation of pathologic findings with clinical sequelae. Ophthalmology 108(9):1675–1681

Part IV

Surgical Methods for PCO Prevention

Size of Continuous Curvilinear Capsulorhexis for Prevention of PCO

14

Yong Eun Lee and Choun-ki Joo

Abstract

Several systemic and ocular associations have been cited as influencing the development of posterior capsule opacification (PCO). Surgical methods such as the size of CCC, in-the-bag IOL implantation, and sealed capsule irrigation also can influence the formation of PCO.

Size of the CCC is discussed in this chapter. Between large and small capsulorhexis, significantly greater wrinkling of the posterior capsule and worse posterior capsular opacification are seen in the large capsulorhexis group than with small capsulorhexis group. Visual acuity is worse with large capsulorhexis. Both equatorial and anterior capsule lens epithelial cells are important in the production of posterior capsular opacification, and their relative contribution depends on the relationship of the anterior capsule edge to the lens optic.

A well-centered capsulorhexis smaller than the lens optic is preferable to a large capsulorhexis, so that the edge lies completely on the surface of the implant.

Keywords

Capsulorhexis • PCO • Size

Y.E. Lee • C.-k. Joo (✉)
Department of Ophthalmology and Visual Science, Seoul St. Mary's Hospital, College of Medicine, The Catholic University of Korea, 505 Banpo-dong, Seocho-gu, Seoul, South Korea
e-mail: ckjoo@catholic.ac.kr

S. Saika et al. (eds.), *Lens Epithelium and Posterior Capsular Opacification*, DOI 10.1007/978-4-431-54300-8_14, © Springer Japan 2014

14.1 Background

Continuous curvilinear capsulorhexis (CCC) is a routine part of cataract surgery since its introduction by Gimbel and Neuhann in 1990 [20, 45]. Because CCC makes true capsular bag fixation of the intraocular lens (IOL) possible with better centration [2, 6, 49]. An intact capsulorhexis reduces the severity of the blood–aqueous barrier breakdown and the foreign-body reaction on the intraocular lens surface and prevents contact between the intraocular lens and uveal tissue. It also produces a safer environment during phacoemulsification and results in a more predictable refractive outcome [52, 59, 60].

However, there is much debate as to what the ideal diameter of the capsulorhexis should be. Smaller capsulorhexis is easier to perform but makes removal of the cataract more difficult and has the major disadvantage that the anterior capsular opening can shrink, leading to phimosis, which causes decreased vision [14, 22, 74]. Large capsulorhexis makes the subsequent phacoemulsification and removal of soft lens matter easier and enables a wide fundal view, which is an important consideration when undertaking surgery for patients with diabetes or patients at risk of retinal detachment. The effect of the size of the capsulorhexis on posterior capsular opacification (PCO) is unclear.

PCO is caused by residual lens epithelial cells that remain in the capsular bag after cataract surgery and proliferate and migrate over the capsule to produce Elschnig pearls or transform into myofibroblasts to cause capsular fibrosis [11, 51]. After surgery, these residual cells are found in the equator of the capsular bag and under the anterior capsule. Many surgeons believe that the lens epithelial cells at the equator of the capsular bag are the most important in the pathogenesis of posterior capsular opacification [5, 72]. They postulate that a large capsulorhexis with a diameter greater than the intraocular lens optic will allow fusion of the anterior capsular flap to the posterior capsule, setting up a mechanical barrier to lens epithelial cell (LEC) migration from the equator, leading to less PCO [33, 72]. Mechanical separation of the anterior capsule leaf from the posterior capsule by the intraocular lens as occurs with a capsulorhexis opening smaller than the lens optic might then lead to an increase in lens epithelial cell ingrowth [27].

An alternative view is that lens epithelial cells originating from the anterior capsule make an important contribution to the production of PCO [51]. If this were the case, a large CCC would allow anterior capsular lens epithelial cells access to the acellular posterior capsule with the subsequent production of PCO. Other views are that a larger capsulotomy will reduce PCO by removing more LEC and increasing the distance for the cells to migrate [51]. Another hypothesis is that lens epithelial cells proliferate after capsulotomy because they have been released from contact inhibition and that a smaller anterior CCC will release fewer cells from this inhibition and therefore decrease cellular proliferation [30].

PCO usually occurs in between <5 and 50 % of uncomplicated senile cataracts during the first 2 years following cataract surgery [64, 67].

PCO develops over a clear posterior capsule from a few months to a few years after uncomplicated cataract surgery. PCO is caused by residual equatorial lens

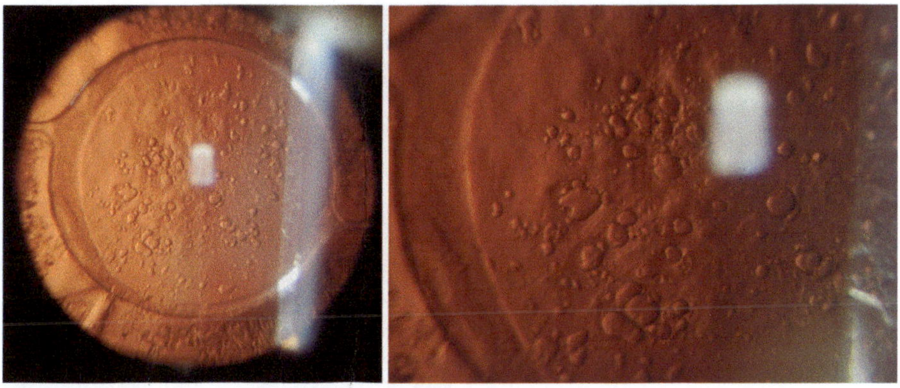

Fig. 14.1 Elschnig-pearl-type PCO (*left*) with high magnification (*right*). Anterior capsule is covered 360° on the optic of IOL

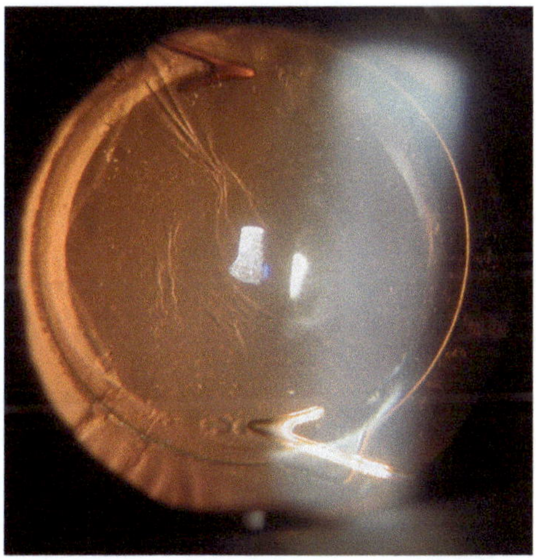

Fig. 14.2 Fibrosis-type PCO

epithelial cells proliferation and migration from the equatorial region of the lens capsule to the posterior capsule surface. In regions of the anterior and posterior capsule junction, Elschnig pearls (Fig. 14.1) can form, located behind the iris or filling the pupil space. The formation of Elschnig pearls causes a decrease in visual acuity and sometimes double vision following the implantation of an intraocular lens.

Capsular fibrosis type is less common and usually appears earlier than Elschnig pearls; it is thought to be caused by LEC metaplasia with myofibroblast development. Clinically, capsular fibrosis type PCO is seen as wrinkling, haze, gray-white streaks and plaques on the surface of the posterior capsule (Fig. 14.2). Mixed type of Elschnig-pearl and capsular fibrosis can also appear (Fig. 14.3). Symptoms reported by patients include distortion of image, glare, and reduction of visual acuity [40, 73].

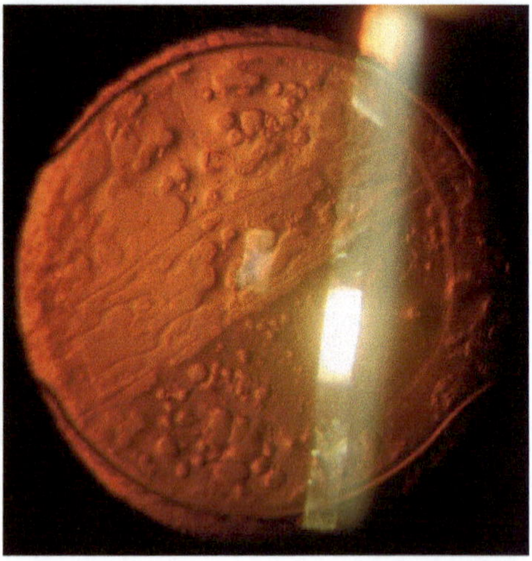

Fig. 14.3 Mixed type of PCO with fibrous and pearl-type areas

Several authors suggest using a primary posterior continuous curvilinear capsulorhexis (PCCC) to prevent this complication [10, 37, 39]. The PCCC technique is frequently used in children to prevent PCO, but is rarely used in adults [8, 18, 19, 38]. The question of the efficacy of PCCC in preventing PCO remains unanswered.

14.2 Risk Factors

Several systemic and ocular associations have been known as influencing the development of PCO. The incidence of PCO is more common in young patients and those with pseudoexfoliation, uveitis, or traumatic cataracts [41]. Myopes have been postulated to increase the risk of PCO; this probably occurs because IOL implantation was deferred in them, but a study of IOL implantation in myopic eyes showed no association between the degree of myopia and the degree of PCO [48]. When compared with nondiabetic patients, diabetic patients had more PCO after cataract surgery, but the stage of diabetic retinopathy and the systemic status of the diabetes do not seem to correlate with the degree of PCO [26]. Patients with retinitis pigmentosa showed a higher incidence and density of PCO [70]. Patients with myotonic dystrophy required multiple capsulotomies following cataract surgery due to PCO and progressive capsulorhexis contracture [21].

Other factors that influence PCO development are the intraocular lens type (material, design, optic size and edge, haptic design), accurate hydrodissection and removal of cortical masses, anterior capsule polishing, in-the-bag fixation of the optic and the haptic part of IOL, and anterior capsulorhexis localization and diameter. Numerous studies have examined the influence of physical properties of the IOL and accurate surgical lens removal technique on the formation of the PCO [1, 12, 16, 17, 24, 71]. Relatively few studies have investigated the influence of anterior capsulorhexis on the development of PCO [7, 28, 65]. So in this chapter, evaluation for the impact of anterior capsulorhexis diameter, localization, and shape on PCO development following cataract extraction with phacoemulsification will be discussed.

14.3 Surgical Techniques for the Prevention of PCO

Current research on the prevention of PCO focuses on surgical techniques, changes in IOL material and designs, and pharmacologic methods. Researchers have examined extensive intraoperative polishing of the anterior and posterior capsules [58], the use of lenses with posterior convexity to ensure closer IOL–capsule contact [9, 68, 69], and application of several antimitotic drugs or anti-LEC immunologic agents [46, 63].

The ideal diameter of the capsulorhexis is particularly important. After surgery, a wound healing response with cellular proliferation and laying down of extracellular matrix is induced and these residual cells proliferate and migrate over the capsule. It has been shown that the anterior epithelial cells in both rabbits and humans initially undergo proliferation and by 4 days after surgery transform into myofibroblasts starting at the cut edge of anterior capsule and by 1 week a ring of fibrosis forms around the rhexis where it touches the posterior capsule in the aphakic eye [50, 51].

14.3.1 About the Size and Shape of the CCC

A study of posterior capsule changes in monkeys after extracapsular cataract extraction showed fibroblasts, which appeared to be transformed lens epithelial cells, in the region where the anterior capsular remnant was in contact with the posterior capsule [11]. Histopathologic findings from 10 cases of fibrosis-type PCO have shown that anterior lens epithelial cells at the capsulotomy edge undergo fibrous metaplasia, causing wrinkling of the posterior capsule [51]. These cells then migrated over the posterior capsule. The ultrastructure of transformed cells found in the posterior capsule folds of postmortem pseudophakic eyes showed that they contain microfilaments (contractile capabilities), reflected in the wrinkling of the capsule. α-smooth muscle actin is a marker for myofibroblastic differentiation.

Kurosaka have shown that LECs remaining in the capsular bag after cataract extraction in rabbits become positive for a smooth muscle actin around the rim of the anterior capsule, whereas the lens epithelial cells at the equatorial region and on the central posterior capsule were only transiently positive for α-smooth muscle actin [42]. These authors showed that these myofibroblastically differentiated cells produced wrinkling of the capsule. These observations support the hypothesis that it is the contact of the anterior capsule edge with the posterior capsule that is important in the production of wrinkling of capsule. When the capsulorhexis is small, most of these anterior capsular lens epithelial cells are held away from the posterior capsule by the lens optic, and wrinkling is much less likely to occur. However, equatorial lens epithelial cells are now able to migrate across the capsule, and consequently lens epithelial cell sheets are seen with this size of capsulotomy. When a large capsulorhexis is eccentric on the intraocular lens surface, a mixed picture is seen, with a fine sheet of lens epithelial cells without wrinkles on the side where the lens optic holds the capsulotomy margin away from the posterior capsule and fibrosis with wrinkles in the area where the anterior capsulotomy edge comes into contact with the posterior capsule off the lens optic. The adhesion of the anterior capsule to the posterior capsule does not produce a complete barrier to the migration of lens epithelial cells, which can be seen extending underneath the lens optic.

The increase in PCO with large capsulorhexes is reflected by a poorer visual performance in this group of patients. Visual acuity was significantly worse with large CCC, and there was a trend toward worse contrast sensitivity in this group compared with patients with small capsulorhexes. Wrinkling of the posterior capsule is important optically because light scattering is caused by the wrinkles [47].

There was a trend for patients with large capsulorhexis to have higher anterior chamber flare and significantly greater anterior chamber cell readings compared with the small capsulorhexis group. Increased postoperative inflammation was not a result of more difficult surgery, as the phacoemulsification times and powers were lower for the large capsulorhexis group because the phacochop technique used is quicker and easier to perform with a large capsulorhexis. We believe that the increased blood–aqueous barrier breakdown with the patients in the large capsulorhexis group is caused by the increased turbulence of the infusion fluid in the anterior chamber and the lens fragments coming into contact with uveal tissue. In a small capsulotomy, the anterior capsule may act as a protective barrier, keeping the fluid and fragments mainly in the capsular bag and away from the iris.

This study demonstrated significantly greater wrinkling of the posterior capsule and worse PCO with large capsulorhexis than with small capsulorhexes. Visual acuity was worse with large capsulorhexes. Both equatorial and anterior capsule lens epithelial cells are important in the production of PCO, and their relative contribution depends on the relationship of the anterior capsule edge to the lens optic.

Here is an interesting result of a study [43]. In Fig. 14.4, 86.79 % of the patients with a small capsulorhexis had no or mild PCO, whereas 68.18 % of the patients with a large capsulorhexis had moderate or severe PCO. Patients with a small capsulorhexis had significantly less PCO than those with a large capsulorhexis ($p < 0.001$).

The amount of PCO in the groups with central (66 patients) and paracentral (33 patients) capsulorhexis localization was similar (Fig. 14.5)—89.4 % of the patients with a central capsulorhexis had no or mild PCO, whereas 75.75 % of the patients with a paracentral capsulorhexis had moderate or severe PCO.

In Fig. 14.6, patients with a regular anterior capsulorhexis rim had significantly less posterior capsular opacification than those with an irregular anterior capsulorhexis rim ($p < 0.001$). In conclusion, small capsulorhexis diameter

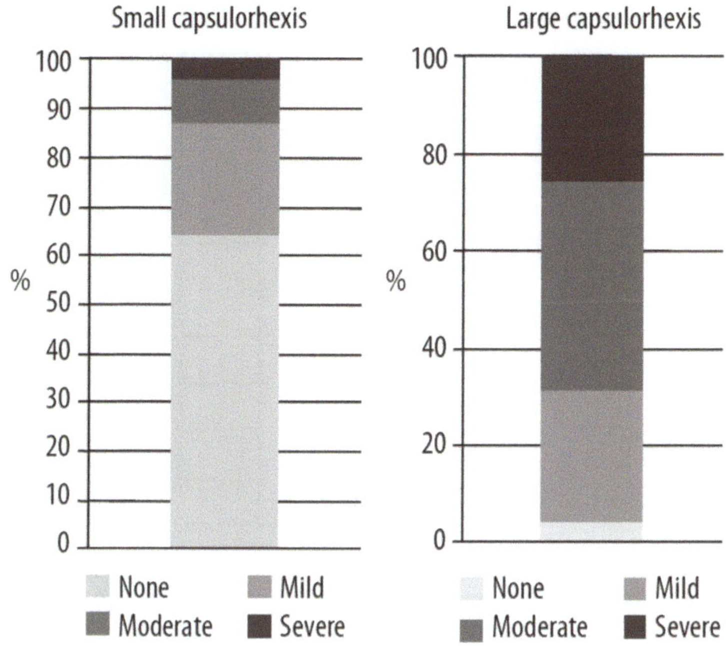

Fig. 14.4 The amount of PCO in the groups with small and large capsulorhexis at 6 months follow-up

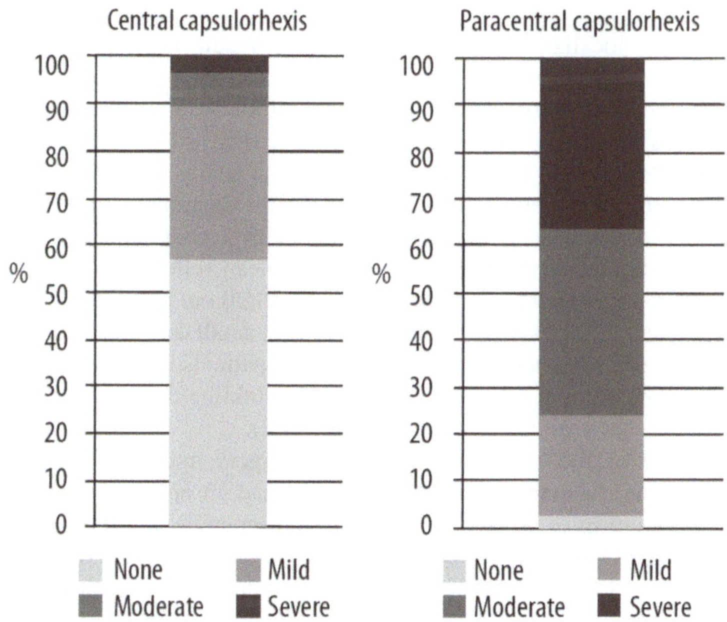

Fig. 14.5 Patients with a central capsulorhexis had significantly less posterior capsular opacification than those with a paracentral anterior capsulorhexis ($p < 0.001$)

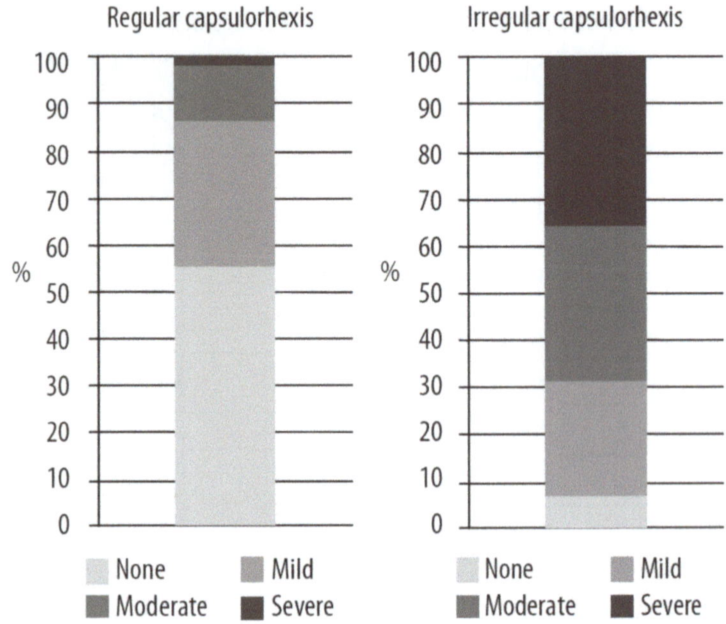

Fig. 14.6 The amount of PCO in the groups with regular and irregular anterior capsulorhexis rim shapes—86.44 % of the patients with a regular rim of the anterior capsulorhexis had no or mild PCO and 13.55 % had moderate or severe PCO, while 69.04 % of the patients with an irregular capsulorhexis rim had moderate or severe PCO

and its central localization and regular shape result in less PCO following phacoemulsification.

Randomized prospective study of Emma J. Hollick shows the mean logMAR visual acuities at each visit for the two groups (Fig. 14.7) [28]. Patients were refracted at 90 days, and the mean refracted visual acuity over the 1-year follow-up was 20.02 for the patients in the small capsulorhexis group, which is worse than 20/15 on the Snellen chart but better than 20/20 (95 % confidence interval, 20.04–20.002). For the large capsulorhexis group the mean refracted visual acuity was 0.02 over the year, which is slightly worse than 20/20 but better than 20/30 (95 % confidence interval, 20.01–0.26). The patients with small capsulorhexis had significantly better mean refracted visual acuity than the patients with large capsulorhexis ($P = 0.025$). The amount of posterior capsular wrinkling at 3 months was significantly different between the two groups (Fig. 14.8).

In the study of Joo et al., 3 months after surgery, most eyes with an initial capsular opening diameter of less than 5.5 mm had an opening diameter smaller than 5.0 mm. In most eyes with an initial capsular opening larger than 5.5 mm, the opening remained larger than 5.0 mm. There results suggest that the ideal CCC size is 5.5–6.0 mm or larger to prevent progressive capsular opening shrinkage [32].

In the prospective study of Aykan et al., 496 eyes of 367 patients underwent standardized phacoemulsification with capsulorhexis and capsular bag foldable

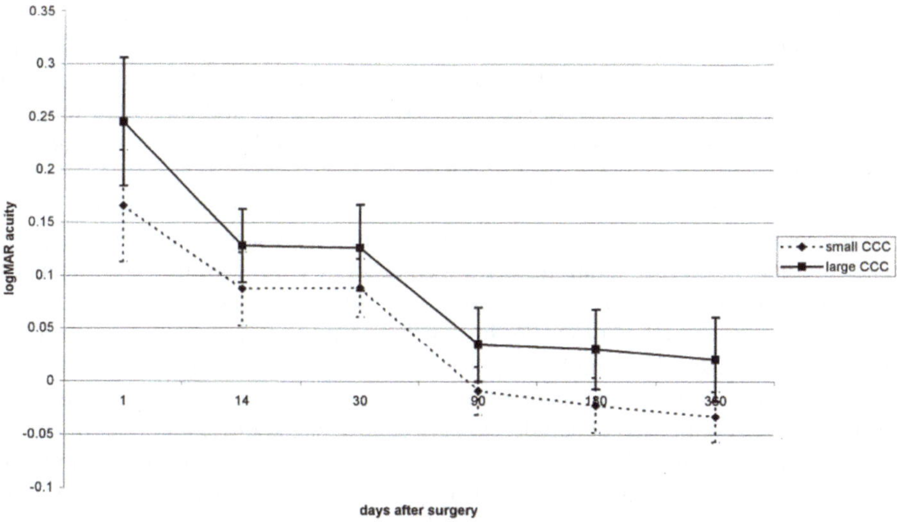

Fig. 14.7 The logMAR visual acuities for small and large capsulorhexis (CCC) groups over the year (error bars represent 95 % confidence intervals) ($P = 0.025$)

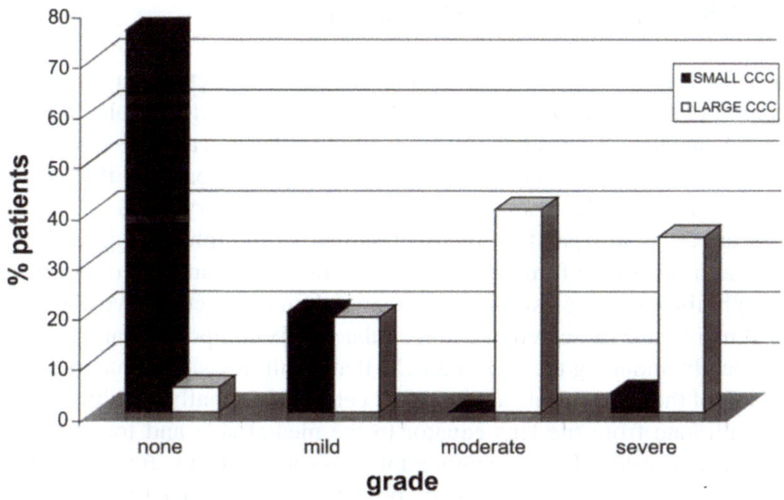

Fig. 14.8 The percentage of patients with each grade of wrinkling at 3 months postoperatively for small and large capsulorhexis groups ($P = 0.0001$)

acrylic IOL implantation [7]. The patients were randomly assigned to receive either a small capsulorhexis of 4.5–5 mm to lie completely on the IOL optic or a large capsulorhexis of 6–7 mm to lie completely off the lens optic. Retroillumination photographs were taken at 6 months and then yearly. Throughout the follow-up, there was less PCO in the small capsulorhexis group than in the large capsulorhexis group. Small capsulorhexis was associated with less wrinkling of the posterior capsule and less PCO than was large capsulorhexis. PCO after IOL implantation has a multifactored pathogenesis. Small (4.5–5.0 mm) capsulorhexis and capsular bag implantation of 5.5 mm acrylic IOL are likely to reduce the PCO incidence when compared with the 6.0–7.0 mm capsulorhexis. This study indicates that the frequency of PCO is much lower in cases with small capsulorhexis where the anterior capsule was 360° on the optic than with large capsulorhexis.

Ravalico et al. reported that capsulorhexis with slightly smaller diameter than the IOL optic decreased PCO incidence when compared to large capsulorhexis [65]. The complete in-the-bag fixation provides an accurate lens centration and enhances the IOL optic barrier effect. In cases where one or both haptics are out of the bag, a potential space exists that allows for ingrowth of cells and PCO formation. Instead, a tight fit of the capsule around the optic may be provided by creating an ideal CCC with edge on the IOL surface. In one study, the tight adhesion of capsulorhexis edge around the lens optic was concluded to help sequester the interior compartment of the capsule, containing the IOL optic from the surrounding aqueous humor and any potentially deleterious factor [31]. IOL implantation in the capsular bag was also suggested to cause less inflammation and PCO formation when compared with sulcus fixation [44].

14.3.2 Effect of the Capsular Tension Ring

The capsular tension ring has been used for the role of structural support when lens subluxation or lens zonulysis occurs, and it has been reported to play a role in the fixation of stable IOL position by preventing anterior capsular contracture. In addition, it has been expected to have a preventive effect on PCO that is currently accepted as an unavoidable side effect after cataract surgery.

Posterior capsular opacification is classified by morphology into two types: fibrosis and pearl. In the former, A cells remaining in the anterior capsule proliferate through fibrous metaplasia, express α-smooth muscle actin, have elasticity, and form cell membrane by secreting extracellular matrix components and basal lamina analog, thereby inducing the fold and PCO that result in vision reduction. E cells in the equator of the lens capsule are germinal cells that normally and throughout their life time migrate from the lens equator to the medial side and form the nucleus, epinucleus, and cortex. They are important cells causing PCO after cataract surgery and are a major cause of the pearl-type posterior capsular opacification. However, since both A cells and E cells can cause the two types of PCO, and thus cause PCO developed after cataract surgery, these two types may be mixed. Based on the assumption that PCO is the phenomenon of the migration of remnants cells to the posterior capsule, and since a capsule tension ring contacts the entire 360° of the lens equator, it has been known to physically block the migration of lens epithelial cells and the posterior invasion of deformed fibroblasts and thus to effectively suppress PCO [55, 56].

In Fig. 14.9, which was studied by Joo et al., three cases (7.3 %) developed PCO in insertion group A and ten (25 %) in control group B [34]. The development

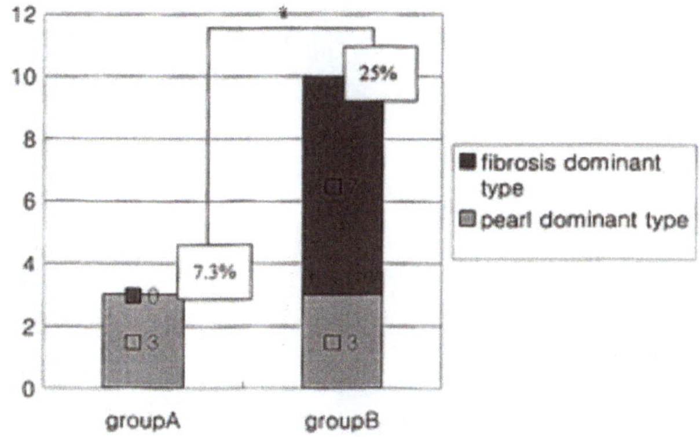

Fig. 14.9 Comparison of the incidence and type of posterior capsular opacity between groups A and B. *Asterisks*: statistically significant difference ($p=0.037$ on Fisher's exact test)

frequency in the insertion group was significantly lower ($p=0.037$). All three eyes in insertion group A were pearl dominant type and among the ten eyes in control group B, 3 were fibrosis dominant type and seven were pearl dominant type.

For prevention of PCO, the factors important to surgery are the removal of the cortex that can be strengthened by hydrodissection, the in-the-bag fixation of IOL and the incision smaller than the optics that barely covers the IOL surface, whereas the factors pertinent to IOL are the biocompatibility that can suppress the proliferation of lens epithelial cells, the maximal IOL optic-posterior capsule contact, and the barrier function of the optic itself and the square truncated edge [3, 4, 61, 62]. This last factor is based on the capsular bend theory that explains the preventive effect of the capsule tension ring on PCO [53, 57]. Conclusively, the capsule tension ring is effective and safe for the reduction of the incidence of PCO. And the ring insertion not only reduced the incidence but also delayed the development time in this study.

14.3.3 Pharmacologic Methods to Prevent PCO

Recently, several reports have recommended that PCO might be prevented by inhibiting the proliferation and migration of the LECs remaining after cataract surgery. Among these studies, Behar-Cohen et al. reported that FGF2-saporin is bound to heparin surface PMMA IOLs and prevents the growth of epithelial cells on the surface of the lens in vitro and in vivo [75]. IOLs coated with thapsigargin, a hydrophobic inhibitor of endoplasmic reticulum (Ca^{2+})-ATPase, were also reported to greatly reduce human lens cell growth in the capsular bag [15].

Nishi et al. reported that the sustained release of ethylenediaminetetraacetic acid (EDTA) and Arg–Gly–Asp (RGD) peptide can inhibit LEC migration because EDTA can disrupt LEC adhesions by chelating Ca^{2+}, which is important for integrin function and that RGD peptide can competitively inhibit for cellular integrin binding to the lens capsule [54]. Joo et al. reported that salmosin was shown to specifically inhibit LEC migration and proliferation in an animal model [35].

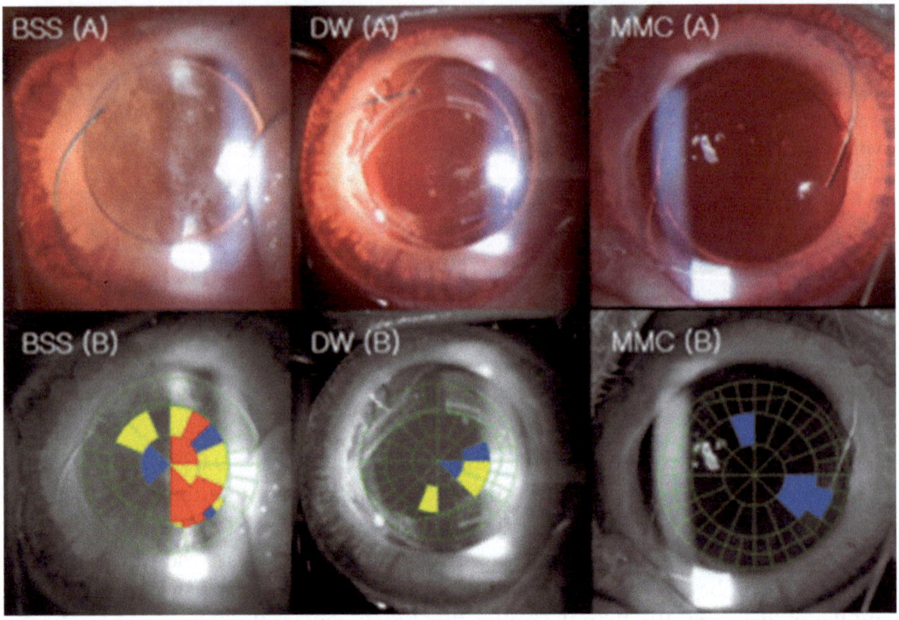

Fig. 14.10 (**a**) Slitlamp photographs 3 months postoperatively in rabbit eyes. (**b**) PCO score with POCOman software. The severity score is calculated according to the formula: (*Blue* × 1 + *Yellow* × 2 + *Red* × 3)/total area. The grade of PCO in the MMC group was less prominent than in the other groups 3 months postoperatively. BSS, balanced salt solution; DW, distilled water

Antimetabolites and other pharmacological agents inhibited the development of LECs and prevent PCO. Several previous studies have found that antimetabolites such as methotrexate, mitomycin-C (MMC), 5-fluorouracil (5-FU), and colchicines inhibit the development of LECs in vivo and in vitro [23, 25, 29, 66]. However, these agents may be harmful and toxic to surrounding intraocular tissues. Another study found that DDW resulted in extensive lysis of LEC when exposed for 120 s in vitro [13]. MMC and DW (distilled water) significantly inhibited the development of PCO 3 months after surgery using the sealed capsule irrigation (SCI) device compared with controls (BSS) in the study of SY Kim et al. [36]. MMC was significantly more effective in preventing PCO compared with the DW (Fig. 14.10). MMC and DW did not cause significant toxicity. These results suggest that the MMC is more effective in reducing PCO than DW and that the SCI device can protect the surrounding ocular tissues from MMC toxicity in rabbit eyes.

14.4 Conclusion

The standard treatment for PCO is Nd:YAG laser capsulotomy, which is expensive and has been associated with retinal detachment, intraocular pressure increase, cystoid macular edema, and IOL damage. Several techniques have been advocated to reduce the incidence of PCO, including intraoperative polishing of capsules, the use of lenses with square edge and posterior convexity to ensure closer IOL–capsule contact, and application of several antimitotic drugs or anti-LEC immunologic agents.

The complete in-the-bag fixation provides an accurate lens centration and enhances the IOL optic barrier effect. IOL implantation in the capsular bag was also suggested to cause less inflammation and PCO formation when compared with sulcus fixation. The capsulorhexis should be continuous and of a diameter that will just overlap the lens edges completely. Small capsulorhexis covers the edges of the lens and increases contact between the IOL and the posterior capsule and reduces the incidence of PCO. Researchers suggest that 360-degree full interaction of anterior capsular flap to the lens has an important role in reducing PCO development. A possible mechanism may result from full capsule-IOL interaction yielding LEC inhibition.

References

1. Abhilakh Missier KA, Nuijts RM, Tjia KF (2003) Posterior capsule opacification: silicone plate-haptic versus AcrySof intraocular lenses. J Cataract Refract Surg 29(8):1569–1574
2. Akkin C, Ozler SA, Mentes J (1994) Tilt and decentration of bag-fixated intraocular lenses: a comparative study between capsulorhexis and envelope techniques. Doc Ophthalmol 87 (3):199–209
3. Apple DJ (2000) Influence of intraocular lens material and design on postoperative intracapsular cellular reactivity. Trans Am Ophthalmol Soc 98:257–283
4. Apple DJ, Peng Q, Visessook N, Werner L, Pandey SK, Escobar-Gomez M, Ram J, Whiteside SB, Schoderbeck R, Ready EL, Guindi A (2000) Surgical prevention of posterior capsule opacification. Part 1: progress in eliminating this complication of cataract surgery. J Cataract Refract Surg 26(2):180–187
5. Apple DJ, Solomon KD, Tetz MR, Assia EI, Holland EY, Legler UF, Tsai JC, Castaneda VE, Hoggatt JP, Kostick AM (1992) Posterior capsule opacification. Surv Ophthalmol 37 (2):73–116
6. Assia EI, Legler UF, Merrill C, Hicklin JC, Castaneda VE, Hoggatt JP, Wasserman D, Apple DJ (1993) Clinicopathologic study of the effect of radial tears and loop fixation on intraocular lens decentration. Ophthalmology 100(2):153–158
7. Aykan U, Bilge AH, Karadayi K, Akin T (2003) The effect of capsulorhexis size on development of posterior capsule opacification: small (4.5 to 5.0 mm) versus large (6.0 to 7.0 mm). Eur J Ophthalmol 13(6):541–545
8. BenEzra D, Cohen E (1997) Posterior capsulectomy in pediatric cataract surgery: the necessity of a choice. Ophthalmology 104(12):2168–2174
9. Born CP, Ryan DK (1990) Effect of intraocular lens optic design on posterior capsular opacification. J Cataract Refract Surg 16(2):188–192
10. Castaneda VE, Legler UF, Tsai JC, Hoggatt JP, Assia EI, Hogan C, Apple DJ (1992) Posterior continuous curvilinear capsulorhexis. An experimental study with clinical applications. Ophthalmology 99(1):45–50
11. Cobo LM, Ohsawa E, Chandler D, Arguello R, George G (1984) Pathogenesis of capsular opacification after extracapsular cataract extraction. An animal model. Ophthalmology 91 (7):857–863
12. Corydon C, Lindholt M, Knudsen EB, Graakjaer J, Corydon TJ, Dam-Johansen M (2007) Capsulorhexis contraction after cataract surgery: comparison of sharp anterior edge and

modified anterior edge acrylic intraocular lenses. J Cataract Refract Surg 33(5):796–799. doi:10.1016/j.jcrs.2007.01.020
13. Crowston JG, Healey PR, Hopley C, Neilson G, Milverton EJ, Maloof A (2004) Water-mediated lysis of lens epithelial cells attached to lens capsule. J Cataract Refract Surg 30(5):1102–1106. doi:10.1016/j.jcrs.2004.01.028
14. Davison JA (1993) Capsule contraction syndrome. J Cataract Refract Surg 19(5):582–589
15. Duncan G, Wormstone IM, Liu CS, Marcantonio JM, Davies PD (1997) Thapsigargin-coated intraocular lenses inhibit human lens cell growth. Nat Med 3(9):1026–1028
16. Gallagher SP, Pavilack MA (1999) Risk factors for anterior capsule contraction syndrome with polypropylene or poly(methyl methacrylate) haptics. J Cataract Refract Surg 25(10):1356–1361
17. Georgopoulos M, Menapace R, Findl O, Petternel V, Kiss B, Rainer G (2003) After-cataract in adults with primary posterior capsulorhexis: comparison of hydrogel and silicone intraocular lenses with round edges after 2 years. J Cataract Refract Surg 29(5):955–960
18. Gimbel HV (1996) Posterior capsulorhexis with optic capture in pediatric cataract and intraocular lens surgery. Ophthalmology 103(11):1871–1875
19. Gimbel HV, DeBroff BM (1994) Posterior capsulorhexis with optic capture: maintaining a clear visual axis after pediatric cataract surgery. J Cataract Refract Surg 20(6):658–664
20. Gimbel HV, Neuhann T (1990) Development, advantages, and methods of the continuous circular capsulorhexis technique. J Cataract Refract Surg 16(1):31–37
21. Gjertsen IK, Sandvig KU, Eide N, Olsen BA (2003) Recurrence of secondary opacification and development of a dense posterior vitreous membrane in patients with myotonic dystrophy. J Cataract Refract Surg 29(1):213–216
22. Hansen SO, Crandall AS, Olson RJ (1993) Progressive constriction of the anterior capsular opening following intact capsulorhexis. J Cataract Refract Surg 19(1):77–82
23. Hansen TJ, Tyndall R, Soll DB (1987) Methotrexate-anticollagen conjugate inhibits in vitro lens cell outgrowth. Invest Ophthalmol Vis Sci 28(7):1206–1209
24. Hanson RJ, Rubinstein A, Sarangapani S, Benjamin L, Patel CK (2006) Effect of lens epithelial cell aspiration on postoperative capsulorhexis contraction with the use of the AcrySof intraocular lens: randomized clinical trial. J Cataract Refract Surg 32 (10):1621–1626. doi:10.1016/j.jcrs.2006.04.035
25. Haus CM, Galand AL (1996) Mitomycin against posterior capsular opacification: an experimental study in rabbits. Br J Ophthalmol 80(12):1087–1091
26. Hayashi Y, Kato S, Fukushima H, Numaga J, Kaiya T, Tamaki Y, Oshika T (2004) Relationship between anterior capsule contraction and posterior capsule opacification after cataract surgery in patients with diabetes mellitus. J Cataract Refract Surg 30(7):1517–1520. doi:10.1016/j.jcrs.2003.11.045
27. Hiles DA, Johnson BL (1980) The role of the crystalline lens epithelium in postpseudophakos membrane formation. J Am Intraocul Implant Soc 6(2):141–147
28. Hollick EJ, Spalton DJ, Meacock WR (1999) The effect of capsulorhexis size on posterior capsular opacification: one-year results of a randomized prospective trial. Am J Ophthalmol 128(3):271–279
29. Ismail MM, Alio JL, Ruiz Moreno JM (1996) Prevention of secondary cataract by antimitotic drugs: experimental study. Ophthalmic Res 28(1):64–69
30. Jacob TJ, Humphry RC, Davies EG, Thompson GM (1987) Cytological factors relating to posterior capsule opacification following cataract surgery. Br J Ophthalmol 71(9):659–663
31. Javdani SM, Huygens MM, Callebaut F (2002) Neodymium: YAG capsulotomy rates after phacoemulsification with hydrophobic and hydrophilic acrylic intraocular lenses. Bull Soc Belge Ophtalmol 283:13–17
32. Joo CK, Shin JA, Kim JH (1996) Capsular opening contraction after continuous curvilinear capsulorhexis and intraocular lens implantation. J Cataract Refract Surg 22(5):585–590
33. Kappelhof JP, Vrensen GF, de Jong PT, Pameyer J, Willekens B (1986) An ultrastructural study of Elschnig's pearls in the pseudophakic eye. Am J Ophthalmol 101(1):58–69

34. Kim JH, Kim H, Joo CK (2005) The effect of capsular tension ring on posterior capsular opacity in cataract surgery. Korean J Ophthalmol 19(1):23–28
35. Kim JT, Lee DH, Chung KH, Kang IC, Kim DS, Joo CK (2002) Inhibitory effects of salmosin, a disintegrin, on posterior capsular opacification in vitro and in vivo. Exp Eye Res 74 (5):585–594. doi:10.1006/exer.2001.1150
36. Kim SY, Kim JH, Choi JS, Joo CK (2007) Comparison of posterior capsule opacification in rabbits receiving either mitomycin-C or distilled water for sealed-capsule irrigation during cataract surgery. Clin Experiment Ophthalmol 35(8):755–758. doi:10.1111/j.1442-9071.2007.01588.x
37. Koch DD, Kohnen T (1997) Retrospective comparison of techniques to prevent secondary cataract formation after posterior chamber intraocular lens implantation in infants and children. J Cataract Refract Surg 23(Suppl 1):657–663
38. Kohnen T, Pena-Cuesta R, Koch DD (1996) Secondary cataract formation following pediatric intraocular lens implantation: 6-month results. Ger J Ophthalmol 5(3):171–175
39. Kraff MC, Sanders DR, Lieberman HL (1979) Total cataract extraction through a 3-mm incision: a report of 650 cases. Ophthalmic Surg 10(2):46–54
40. Kratz RP (1974) Symposium: phacoemulsification. Difficulties, complications, and management. Trans Am Acad Ophthalmol Otolaryngol 78(1):OP18–OP21
41. Kuchle M, Naumann GO (2001) Pseudoexfoliation and posterior capsular opacification. Am J Ophthalmol 131(6):820
42. Kurosaka D, Kato K, Nagamoto T (1996) Presence of alpha smooth muscle actin in lens epithelial cells of aphakic rabbit eyes. Br J Ophthalmol 80(10):906–910
43. Langwinska-Wosko E, Broniek-Kowalik K, Szulborski K (2011) The impact of capsulorhexis diameter, localization and shape on posterior capsule opacification. Med Sci Monit 17(10): CR577–CR582
44. Laurell CG, Zetterstrom C, Lundgren B (1998) Phacoemulsification and lens implantation in rabbit eyes: capsular bag versus ciliary sulcus implantation and 4.0 versus 7.0 mm capsulorhexis. J Cataract Refract Surg 24(2):230–236
45. Leaming DV (1997) Practice styles and preferences of ASCRS members–1996 survey. J Cataract Refract Surg 23(4):527–535
46. Legler UF, Apple DJ, Assia EI, Bluestein EC, Castaneda VE, Mowbray SL (1993) Inhibition of posterior capsule opacification: the effect of colchicine in a sustained drug delivery system. J Cataract Refract Surg 19(4):462–470
47. Liu CS, Wormstone IM, Duncan G, Marcantonio JM, Webb SF, Davies PD (1996) A study of human lens cell growth in vitro. A model for posterior capsule opacification. Invest Ophthalmol Vis Sci 37(5):906–914
48. Luke C, Dietlein TS, Jacobi PC, Konen W, Krieglstein GK (2001) Massive anterior capsule shrinkage after plate-haptic silicone lens implantation in uveitis. J Cataract Refract Surg 27(2):333–336
49. Martin RG, Sanders DR, Souchek J, Raanan MG, DeLuca M (1992) Effect of posterior chamber intraocular lens design and surgical placement on postoperative outcome. J Cataract Refract Surg 18(4):333–341
50. McDonnell PJ, Rowen SL, Glaser BM, Sato M (1985) Posterior capsule opacification. An in vitro model. Arch Ophthalmol 103(9):1378–1381
51. McDonnell PJ, Zarbin MA, Green WR (1983) Posterior capsule opacification in pseudophakic eyes. Ophthalmology 90(12):1548–1553
52. Miyake K, Asakura M, Kobayashi H (1984) Effect of intraocular lens fixation on the blood-aqueous barrier. Am J Ophthalmol 98(4):451–455
53. Nishi O, Nishi K (1999) Preventing posterior capsule opacification by creating a discontinuous sharp bend in the capsule. J Cataract Refract Surg 25(4):521–526
54. Nishi O, Nishi K, Akaishi T, Shirasawa E (1997) Detection of cell adhesion molecules in lens epithelial cells of human cataracts. Invest Ophthalmol Vis Sci 38(3):579–585

55. Nishi O, Nishi K, Mano C, Ichihara M, Honda T (1998) The inhibition of lens epithelial cell migration by a discontinuous capsular bend created by a band-shaped circular loop or a capsule-bending ring. Ophthalmic Surg Lasers 29(2):119–125
56. Nishi O, Nishi K, Menapace R (1998) Capsule-bending ring for the prevention of capsular opacification: a preliminary report. Ophthalmic Surg Lasers 29(9):749–753
57. Nishi O, Nishi K, Sakanishi K (1998) Inhibition of migrating lens epithelial cells at the capsular bend created by the rectangular optic edge of a posterior chamber intraocular lens. Ophthalmic Surg Lasers 29(7):587–594
58. Nishi O, Nishi K, Sakka Y, Sakuraba T, Maeda S (1991) Intercapsular cataract surgery with lens epithelial cell removal. Part IV: capsular fibrosis induced by poly(methyl methacrylate). J Cataract Refract Surg 17(4):471–477
59. Olsen T, Gimbel H (1993) Phacoemulsification, capsulorhexis, and intraocular lens power prediction accuracy. J Cataract Refract Surg 19(6):695–699
60. Pande M, Spalton DJ, Marshall J (1996) Continuous curvilinear capsulorhexis and intraocular lens biocompatibility. J Cataract Refract Surg 22(1):89–97
61. Peng Q, Apple DJ, Visessook N, Werner L, Pandey SK, Escobar-Gomez M, Schoderbek R, Guindi A (2000) Surgical prevention of posterior capsule opacification. Part 2: enhancement of cortical cleanup by focusing on hydrodissection. J Cataract Refract Surg 26(2):188–197
62. Peng Q, Visessook N, Apple DJ, Pandey SK, Werner L, Escobar-Gomez M, Schoderbek R, Solomon KD, Guindi A (2000) Surgical prevention of posterior capsule opacification. Part 3: intraocular lens optic barrier effect as a second line of defense. J Cataract Refract Surg 26 (2):198–213
63. Power WJ, Neylan D, Collum LM (1994) Daunomycin as an inhibitor of human lens epithelial cell proliferation in culture. J Cataract Refract Surg 20(3):287–290
64. Raj SM, Vasavada AR, Johar SR, Vasavada VA, Vasavada VA (2007) Post-operative capsular opacification: a review. Int J Biomed Sci 3(4):237–250
65. Ravalico G, Tognetto D, Palomba M, Busatto P, Baccara F (1996) Capsulorhexis size and posterior capsule opacification. J Cataract Refract Surg 22(1):98–103
66. Ruiz JM, Medrano M, Alio JL (1990) Inhibition of posterior capsule opacification by 5-fluorouracil in rabbits. Ophthalmic Res 22(4):201–208
67. Schmidbauer JM, Vargas LG, Apple DJ, Escobar-Gomez M, Izak A, Arthur SN, Golescu A, Peng Q (2002) Evaluation of neodymium:yttrium-aluminum-garnet capsulotomies in eyes implanted with AcrySof intraocular lenses. Ophthalmology 109(8):1421–1426
68. Sellman TR, Lindstrom RL (1988) Effect of a plano-convex posterior chamber lens on capsular opacification from Elschnig pearl formation. J Cataract Refract Surg 14(1):68–72
69. Sterling S, Wood TO (1986) Effect of intraocular lens convexity on posterior capsule opacification. J Cataract Refract Surg 12(6):655–657
70. Sudhir RR, Rao SK (2001) Capsulorhexis phimosis in retinitis pigmentosa despite capsular tension ring implantation. J Cataract Refract Surg 27(10):1691–1694
71. Tadros A, Bhatt UK, Abdul Karim MN, Zaheer A, Thomas PW (2005) Removal of lens epithelial cells and the effect on capsulorhexis size. J Cataract Refract Surg 31(8):1569–1574. doi:10.1016/j.jcrs.2005.01.027
72. Tetz MR, O'Morchoe DJ, Gwin TD, Wilbrandt TH, Solomon KD, Hansen SO, Apple DJ (1988) Posterior capsular opacification and intraocular lens decentration. Part II: experimental findings on a prototype circular intraocular lens design. J Cataract and Refract Surg 14 (6):614–623
73. Werner L, Pandey SK, Apple DJ, Escobar-Gomez M, McLendon L, Macky TA (2001) Anterior capsule opacification: correlation of pathologic findings with clinical sequelae. Ophthalmology 108(9):1675–1681
74. Young DA, Orlin SE (1994) Capsulorhexis contracture in phacoemulsification surgery. Ophthalmic Surg 25(7):477–478
75. Behar-Cohen FF, David T, D'Hermies F, Pouliquen YM, Buechler Y, Nova MP, Houston LL, Courtois Y (1995) In vivo inhibition of lens regrowth by fibroblast growth factor 2-saporin. Invest Ophthalmol Vis Sci 36(12):2434–2448

Effect of Anterior Capsule Polishing on Capsule Opacification and YAG Laser Capsulotomy

15

Rupert Menapace

Abstract

Lens epithelial cells (LECs) at the rhexis edge and in the area surrounding the optic (A-LECs) are the main cause not only for fibrotic whitening and contraction of the capsule but also for permanent collagenous sealing of the capsule bag following intraocular lens (IOL) implantation.

Anterior capsule polishing (ACP) mainly removes these A-LECs. Methods differ in efficiency of LEC removal and in the trauma thereby exerted on capsule, zonules and ciliary body.

Even after thorough removal, fibrosis does partly occur. Regarding the impact on regeneratory posterior capsule opacification (PCO) formation, no increase in PCO rates was found when evaluating retroilluminated photographs. In spite of this, a considerable increase in YAG laser capsulotomy rate was found by one study group. As the cause, accumulation of amorphous LEC material was detected in the retrolental space. Bending and collagenous bonding of the capsules which provide for a lasting barrier effect at the optic circumference was found to be compromised. Thus, ACP though decreasing capsular whitening and shrinkage apparently interferes with barrier formation. ACP also changes the morphology of LECs accumulating in the retrolental space, causing a decrease of contrast vision while not picked up upon retroillumination.

Thus, ACP is not recommended with standard in-the-bag IOL implantation. ACP is a useful adjunct to the posterior optic buttonholing technique where retrolental LEC invasion is independent from capsule bending and thus cannot be affected.

The impact of ACP on capsular bag closure and cellular reaction as visualized by biomicroscopy elucidates the physiology of after-cataract formation and the mechanism of barrier formation at the IOL optic barrier.

R. Menapace (✉)
Department of Ophthalmology, Medical University Vienna, Vienna General Hospital, Waehringer Guertel 18-20, 1090 Vienna, Austria
e-mail: rupert.menapace@meduniwien.ac.at

S. Saika et al. (eds.), *Lens Epithelium and Posterior Capsular Opacification*, DOI 10.1007/978-4-431-54300-8_15, © Springer Japan 2014

Keywords
Barrier failure • Capsular sealing • Capsule contraction • Capsule polishing • Fibrosis • Lens epithelia cell

15.1 The Lens Epithelial Cell and After-Cataract

Removal of the lens contents by phacoemulsification and cortex aspiration creates an empty capsular bag with residual lens epithelial cells (LECs) left behind in the equatorial region and on the undersurface of the peripheral anterior capsule. These LECs proliferate to form monolayers on the capsular surfaces which continue to line the anterior capsule leaflet many years after surgery. Some LECs, however, differentiate or undergo a transition to another cell type. LECs located *in the capsular fornix* differentiate to fibre-like structures that lead to Soemmering's ring formation and peripheral thickening of the capsular bag. LECs located *closer to the rhexis*, termed as equatorial LECs, or E-LECs in the following, exhibit a strong potential to migrate and swell to form globules (Elschnig pearls). LECs located *at the rhexis edge and in the area surrounding the optic*, in the following addressed as anterior LECs, or A-LECs, show less potential to migrate and swell but instead exhibit an exquisite propensity for epithelial-mesenchymal transition by myofibroblastic transdifferentiation. The resulting cells are fibroblastic in morphology, express the smooth muscle isoform of actin and secrete extracellular matrix containing proteins not normally present in the lens [1]. Contraction of muscle fibres leads to shrinkage, and deposition of collagen-like matrix causes whitening and bonding of the capsules. Whitening and shrinkage of the anterior and posterior capsule caused by A-LECs is also termed *fibrotic* after-cataract, while swollen E-LECs accumulating in the capsular bag are addressed as *regeneratory* after-cataract. Regardless of its type, after-cataract is called anterior and posterior capsule opacification (ACO, and PCO) according to its location.

Posterior capsule opacification (PCO) in eyes with in-the-bag implanted intraocular lenses (IOL) is mainly caused by proliferating equatorial LECs migrating unto the posterior capsule behind the IOL optic. Depending on the width of the interspace between the IOL optic and posterior capsule, these proliferating cells may form flat syncytial or honeycomb structures when narrow or mono- or multilayers of pearls of different sizes sometimes interspersed with large globular ("bladder") cells when wide (Fig. 15.1) [2]. Depending on the structure, these accumulating cells more or less interfere with the patient's vision, globular Elschnig pearls representing the main substrate of vision-disturbing PCO, which may finally prompt YAG laser capsulotomy (YAG-LCT).

Anterior capsule opacification (ACO) occurs when the anterior capsule or rhexis leaf touches down on the anterior IOL surface [3]. The IOL material then initiates and catalyses the process of myofibroblastic transdifferentiation ending up in *anterior* capsule fibrosis (Fig. 15.2). The various IOL materials differ in the "fibrogenetic potential", silicones generally exhibiting the strongest potential to

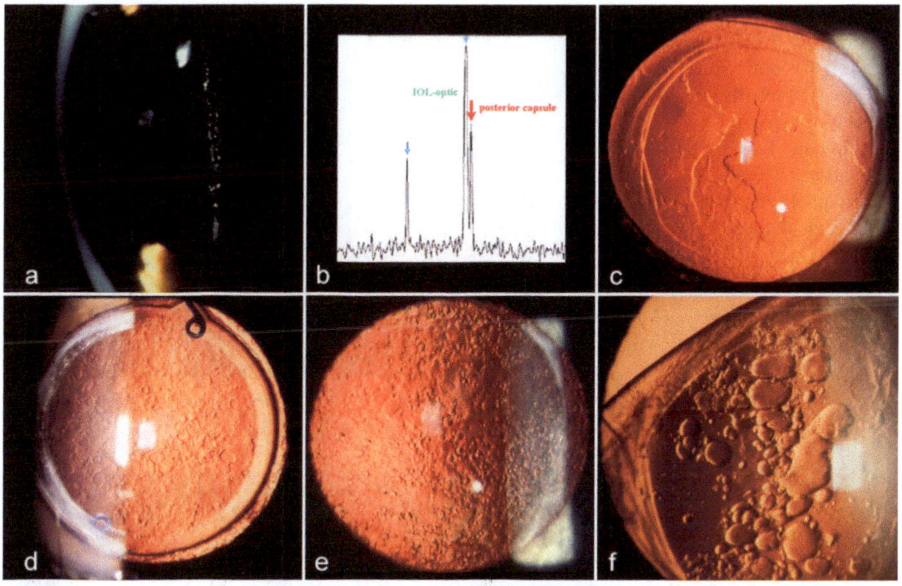

Fig. 15.1 "Regeneratory after-cataract" results from E-LECs invading the retro-optical space (**a** through **f**); depending upon its width, optically homogeneous cell syncytia or inhomogeneous pearl agglomerates may form

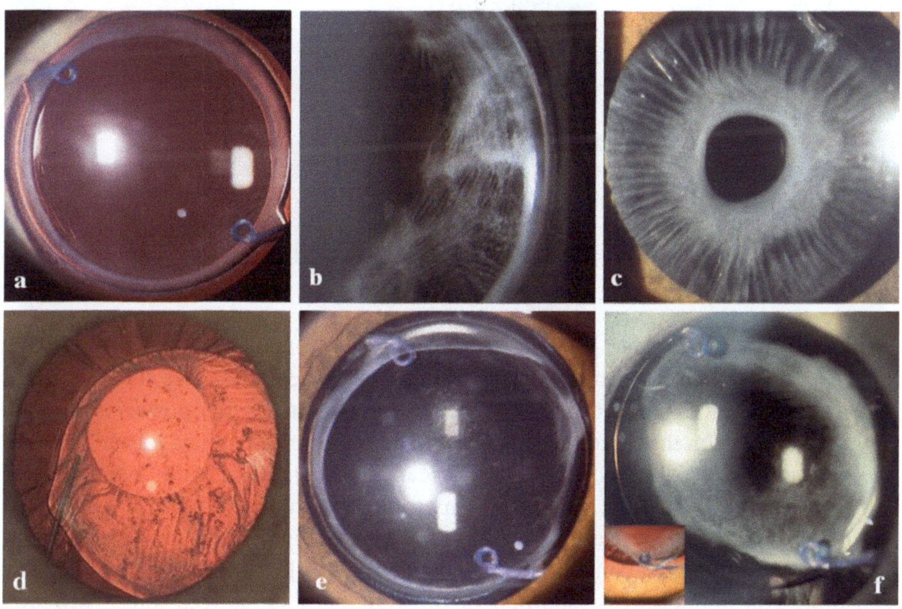

Fig. 15.2 "Fibrotic after-cataract" derives from A-LECs located at and around the rhexis edge (**a** through **f**); upon contact to the IOL, they transdifferentiate into myofibroblast-like cells which contract and deposit collagen

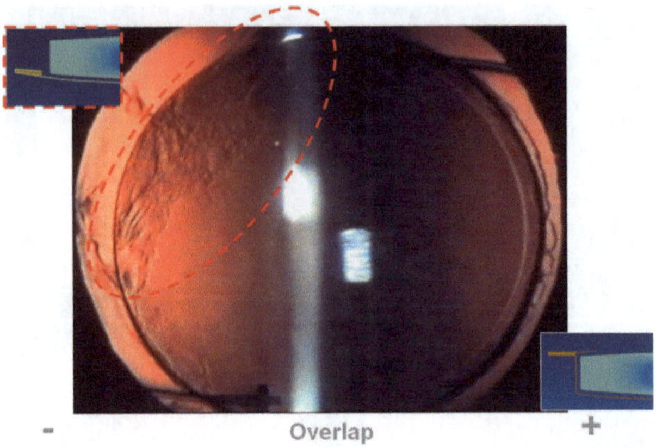

Fig. 15.3 Primary contact of a capsulorhexis edge evading the optic rim (*upper-left*) results in lack of capsule bending and barrier formation; LECs quickly invade retro-optical space

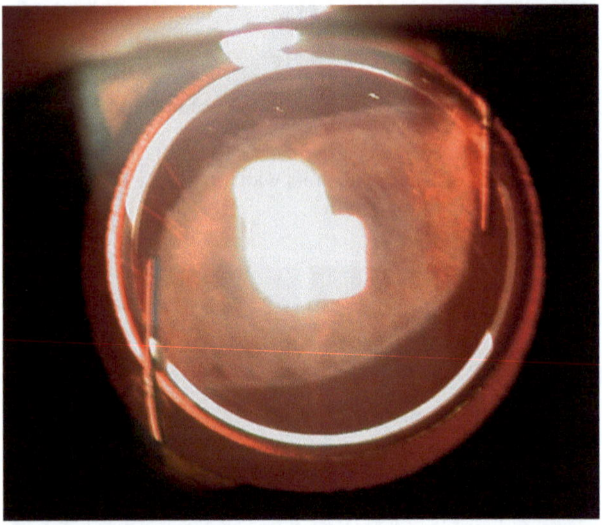

Fig. 15.4 "Primary anterior buttonholing": Chamber flattening makes the IOL optic pop through the capsulorhexis; direct contact of capsulorhexis edge to the posterior capsule and the catalysing effect of silicone cause massive fibrosis of central capsule

induce capsular shrinkage [4–9]. *Posterior* capsule fibrosis occurs when A-LECs migrate around the optic rim unto the retro-optical posterior capsule (Fig. 15.2b) or when the anterior capsulorhexis edge directly contacts the posterior capsule (Fig. 15.2e, f). In the latter case, firm collagenous gluing may ensue between the capsulorhexis edge and the posterior capsule [10]. IOLs with rounded optic edge profiles readily allow for A-LEC migration around the optic rim even when the optic is properly overlapped by the rhexis leaf (Figs. 15.2b and 15.5b-right). Primary capsulorhexis contact to the posterior capsule is established when the capsulorhexis diameter exceeds that of the IOL optic (Fig. 15.3-left). However, such contact may also occur with proper circumferential rhexis-optic overlap: It may happen at the conclusion or shortly after the surgery when *chamber flattening* makes the IOL optic pop through the capsulorhexis (Fig. 15.4) or delayed when

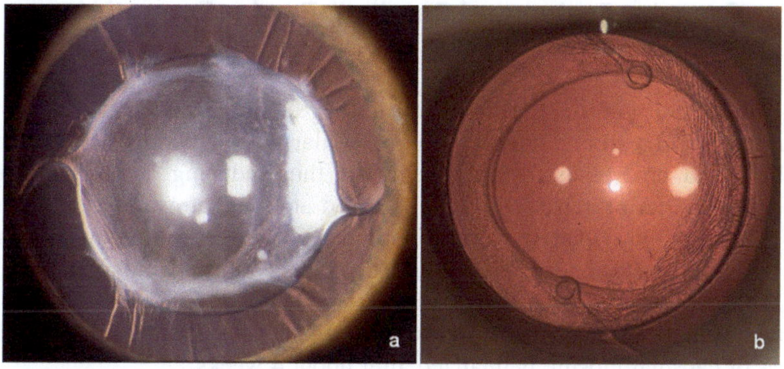

Fig. 15.5 (**a**) Anterior rhexis leaf retraction and "secondary anterior buttonholing": Scarcely overlapping rhexis leaf establishes contact to the posterior capsule when retracted, giving rise to massive retrolental fibrosis emerging from rhexis edge. (**b**) Fibrotic PCO as the cause of retraction only forms along rounded edge (*right*) while blocked by sharp edge (*left* circumference of hybrid optic silicone IOL)

fibrotic retraction of the anterior capsule causes the capsule edge to flip around and settle down on the posterior capsule (primary and secondary *rhexis leaf retraction*) resulting in partial (Fig. 15.2e) or complete "anterior optic buttonholing" (Figs. 15.2f and 15.4) [2]. Delayed retraction occurs when the contraction forces of fibrotic PCO (f-PCO) exceed those of ACO. This is only seen with rounded posterior optic edges (Figs. 15.2e, f, 15.5a, b), since sharp posterior edges withhold LEC migration and thus block f-PCO formation (Fig. 15.5b-left). Migration of A-LECs unto the posterior capsule reduces and mechanically counteracts the contraction forces of anterior capsule fibrosis by decreasing the total number of A-LECs left back on the rhexis leaf and by exerting countertraction when transdifferentiating. This is why acrylic and silicone IOLs with round posterior optic edges allow for significantly more f-PCO, but less ACO than their sharp-edged counterparts, and why rounded-edged silicone IOLs led to significantly less anterior capsulorhexis contraction than sharp-edged silicone IOLs [9]. Thus, the edge profile of IOLs greatly influences fibrosis itself and the configuration of the capsular bend [11, 12].

Since A-LECs migrating from the rhexis leaf overlapping the optic periphery around the optic rim unto the posterior capsule beneath the optic readily undergo myofibroblastic transdifferentiation, f-PCO is usually confined to the very periphery. With the edge of a capsulorhexis larger than the optic and thus contacting the posterior capsule, however, A-LECs migrating from the rhexis edge directly unto the posterior capsule tend to cause more pronounced fibrosis (Fig. 15.3-left). With the capsulorhexis rim lying beneath an anteriorly buttonholed silicone optic, however, massive fibrosis may be induced which may fully encroach upon the centre (Figs. 15.2e, 15.4, and 15.5a).

15.2 Capsular Fusion and Bending, Barrier Formation and Barrier Failure

When implanted with an IOL, the capsule bag starts closing with the posterior capsule wrapping around the implant. Capsular fusion begins at the capsular periphery and progresses towards the optic, thus inducing a posterior capsule bend as the posterior capsule is usually pulled anteriorly towards the rhexis leaf (Fig. 15.6a, b), provided the optic is properly overlapped by the rhexis leaf (Fig. 15.3-right). The posterior capsule bend thus induced along the posterior optic edge constitutes a barrier against centripetal migration of E-LECs. It is unclear whether this is a cell-biological or mechanical effect or both [13–15]. The process of capsular closure is finalized after about 4 weeks.

The process of capsular bending takes place during the first 4 weeks after surgery. Progress and extent of capsular fusion is highlighted by the "capsular fusion line", which delineates the border between fused and non-fused capsule areas (Fig. 15.7). Speed and completeness of capsular fusion depend on various factors: IOL design, specifically width of the optic-haptic junction, haptic bulk as well as haptic diameter and rigidity, but also the IOL material [16]. Large and oversized haptics ovalize and stretch out the capsular bag, causing stress folds and preventing capsule closure and fusion along the IOL axis ("Capsular distension syndrome" (Fig. 15.8)), resulting in barrier failure and retrolental LEC invasion. Independent from the implant characteristics, capsular fusion may be incomplete, or totally absent, resulting in a lack of capsular bending and thus barrier formation. Regardless whether caused by the implant or idiopathic, this is termed "primary barrier failure".

Alongside with capsular fusion, myofibroblastic transdifferentiation of A-LECs starts when the anterior capsule takes up contact with the implant surface. When the process of capsular fusion is finalized, transdifferentiating myofibroblasts contract the capsule and deposit collagen, resulting in "shrink-wrapping" and collagenous "capsular sealing" of the capsular bag. Both are highlighted under oblique slitlamp illumination (Fig. 15.9). In cases where the capsular bag remains partly or fully patent, fibrosis of the anterior capsule will also be incomplete or absent in the areas where capsule-implant contact is lacking.

Later on, capsular fusion and bending may be reversed by Soemmering's ring formation and delayed E-LEC proliferation, which start in the periphery to progress centrally to form an annular mass that mechanically separates the once fused capsules. If collagenous bonding of both capsules along the optic rim is not strong enough, the sealing line may be broken open and posterior capsular bending reversed, and E-LECs allowed to access and fill the retro-optical space ("secondary barrier failure") (Figs. 15.10, 15.11, and 15.12). This typically starts after 3–5 years. Proficiency of Soemmering's ring formation and E-LEC proliferation varies greatly between individuals. While it generally decreases with age, younger patients usually showing strong and older patients weak "proliferative potential", old patients may occasionally also express considerable proliferative forces. If collagenous bonding is strong enough, the barrier effect will be preserved. With capsular sealing along the optic rim strong enough, re-division is halted

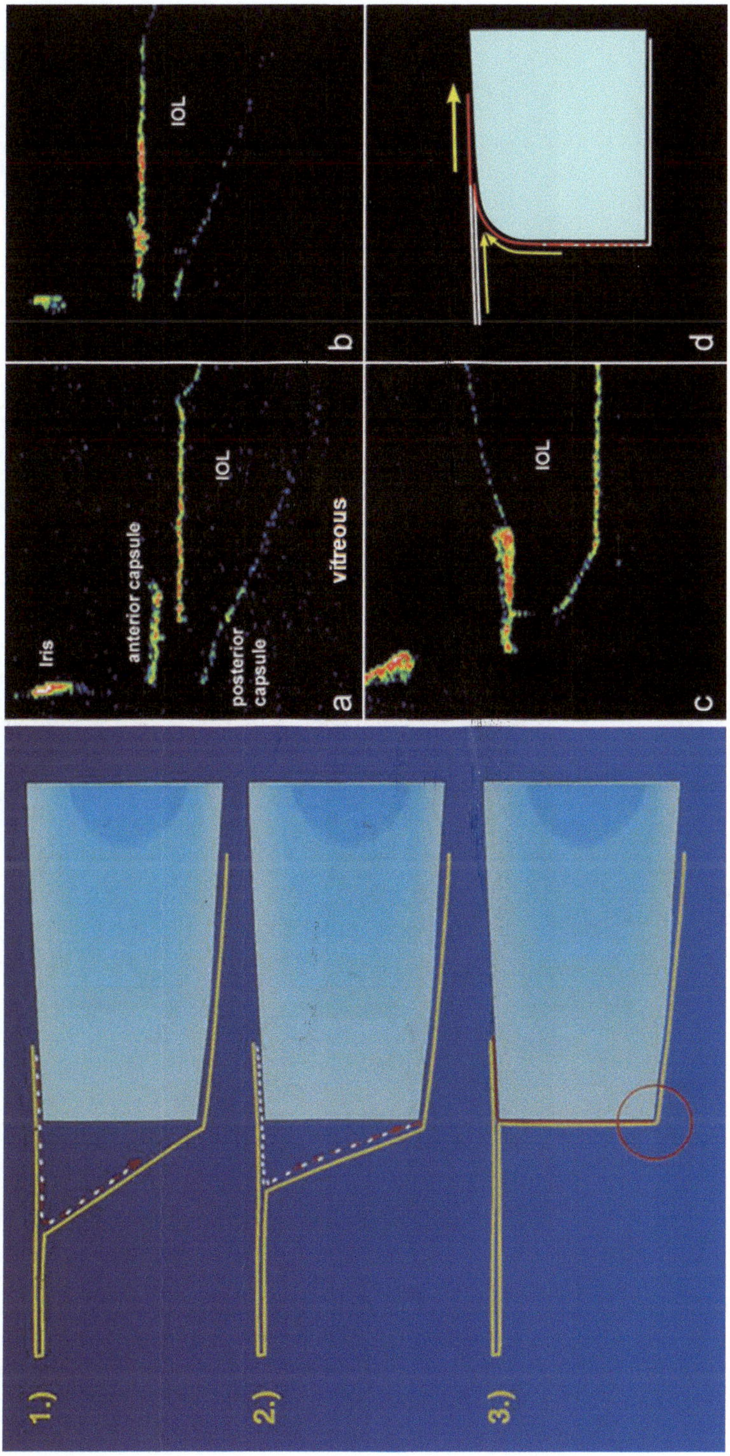

Fig. 15.6 Capsular bag closure and capsular fusion start out in the periphery and induce capsular bending when progressing centrally; note LEC migration unto posterior capsule blocked by sharp posterior optic edge (*1–3*) and "touch down" of rhexis leaf unto optic (**a, b**) and shrinkage causing "shrink-wrapping" of optic (**c, d**)

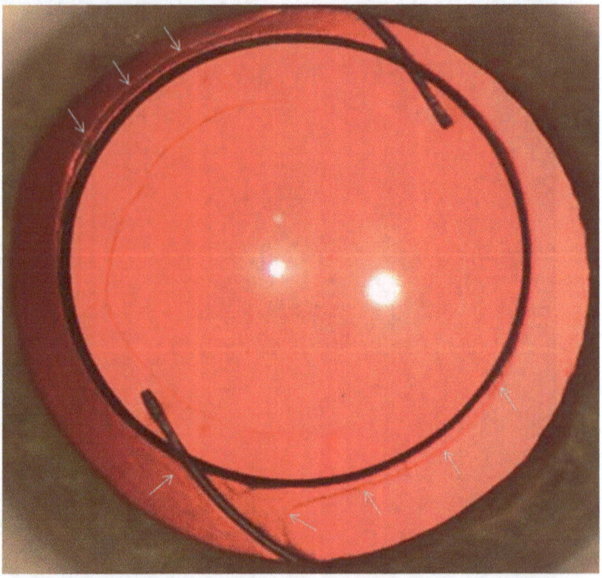

Fig. 15.7 "Capsular fusion line" delineates fused and non-fused areas of the capsule bag

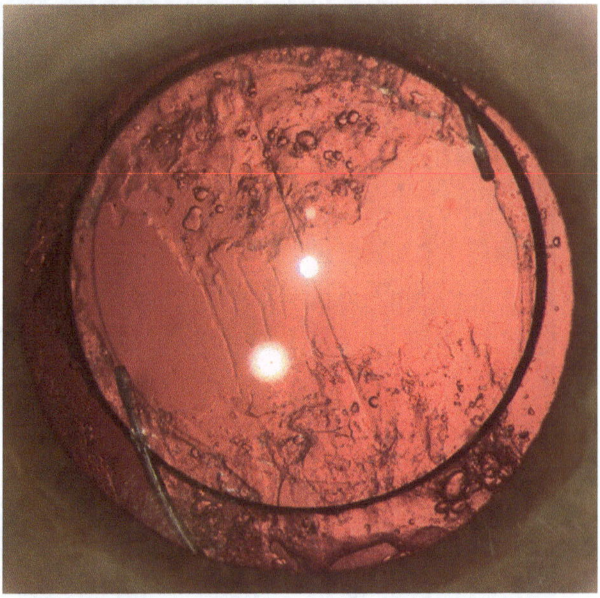

Fig. 15.8 "Capsular distension syndrome": Large and oversized haptics ovalize and stretch out the capsular bag, causing stress folds and preventing capsule closure and fusion along the IOL axis, resulting in barrier failure and retrolental LEC invasion

along the sealing line (Figs. 15.9 and 15.13-right) even with a rounded optic edge. With a sharp posterior edge, the sealing line may be discontinuous and the barrier effect still preserved as long as the posterior capsule between the sealed areas is pulled anteriorly and bent around the posterior optic edge ("capsular suspension effect" (Fig. 15.14)). Delayed-onset regeneratory PCO (r-PCO) formation due to

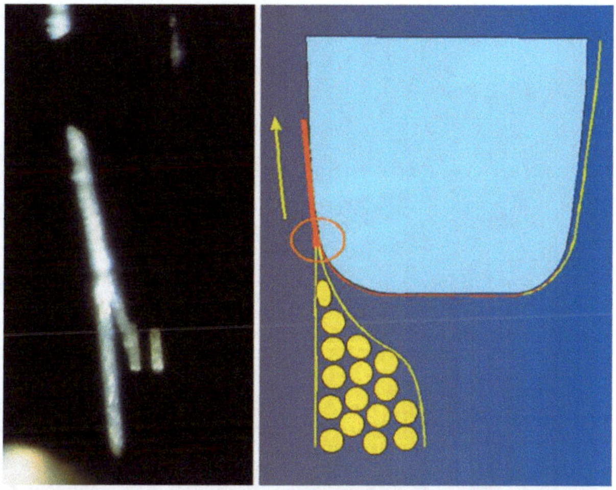

Fig. 15.9 When the process of capsular fusion is finalized, transdifferentiating myofibroblasts contract the capsule and deposit collagen, resulting in "shrink-wrapping" and collagenous "capsular sealing" of the capsular bag. Both are highlighted under oblique slitlamp illumination

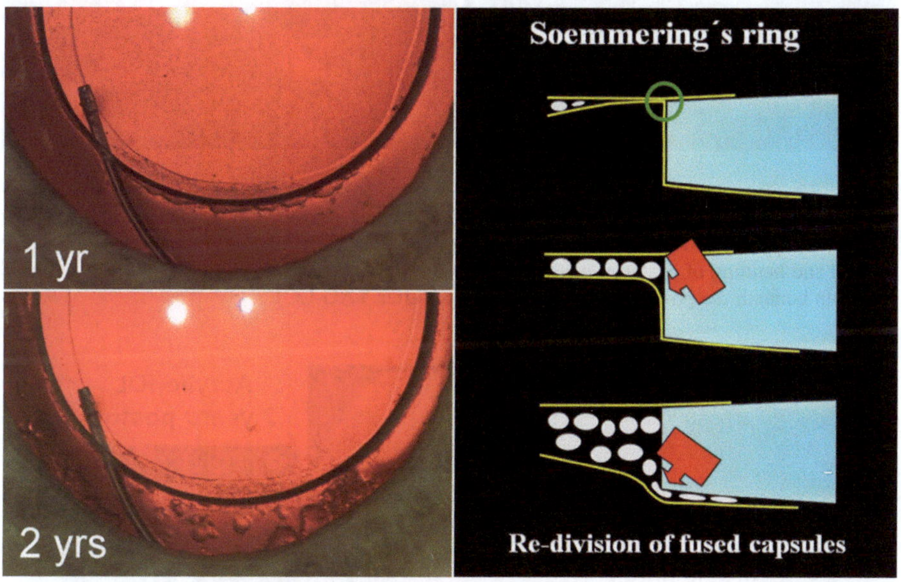

Fig. 15.10 If collagenous bonding of both capsules along the optic rim is not strong enough, the sealing line may be broken open and posterior capsular bending reversed, and E-LECs allowed to access and fill the retro-optical space

delayed secondary barrier failure because of inadequate collagenous capsule sealing is particularly observed with IOLs made from materials with low potential of inducing myofibroblastic transdifferentiation, specifically acrylics, while less with silicone IOLs with high such potential.

This two-stage development of after-cataract formation explains why differences in r-PCO and YAG-LCT rates do not become obvious before 3 years

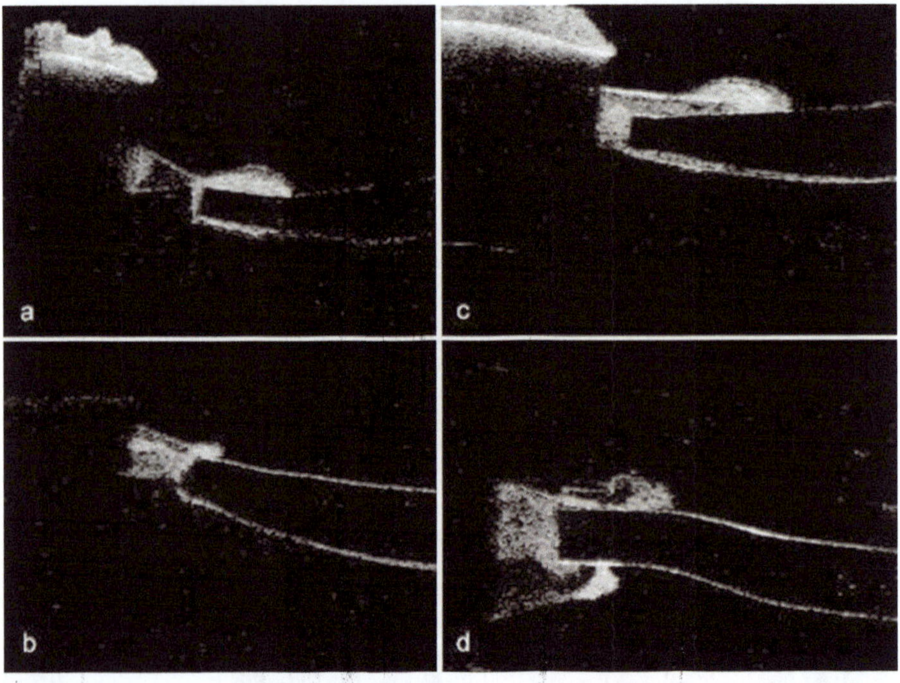

Fig. 15.11 Time course of delayed secondary barrier failure as documented by HR-OCT: (**a**) Soemmering's ring and proliferating LECs having already re-divided the anterior and posterior capsule and pushing towards the fusion line at the optic rim; (**b**) fusion line being progressively opened and bending of posterior capsule at posterior optic edge reversed to (**c**) finally allow LEC ingrowth beneath the posterior IOL optic; (**d**) situs after YAG-LCT

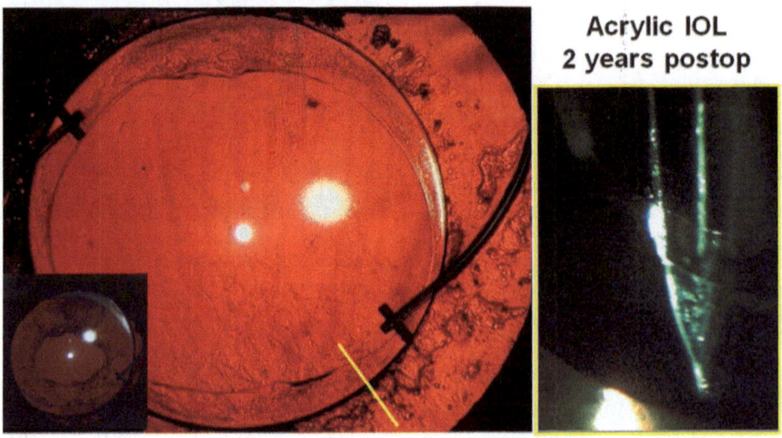

Fig. 15.12 "Secondary barrier failure" with sharp-edged acrylic IOL; capsular re-division highlighted by slitlamp illumination

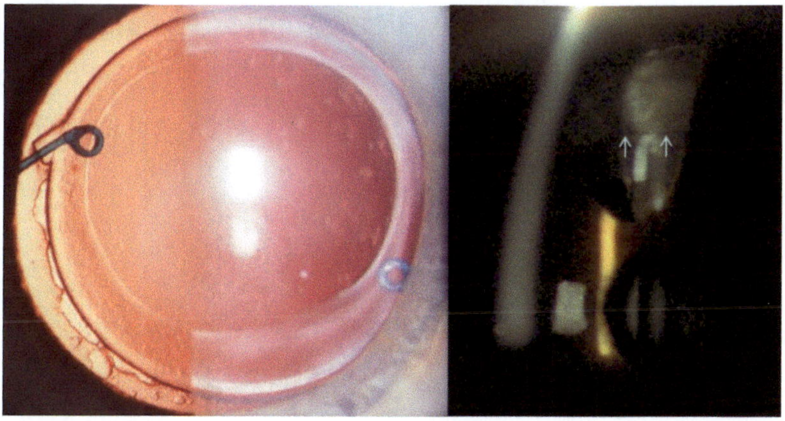

Fig. 15.13 If collagenous sealing along the optic rim is strong enough, however, re-division is halted along the sealing line even with a rounded optic edge. Note annular cell mass encased in peripheral bag and demarcated by sealing line

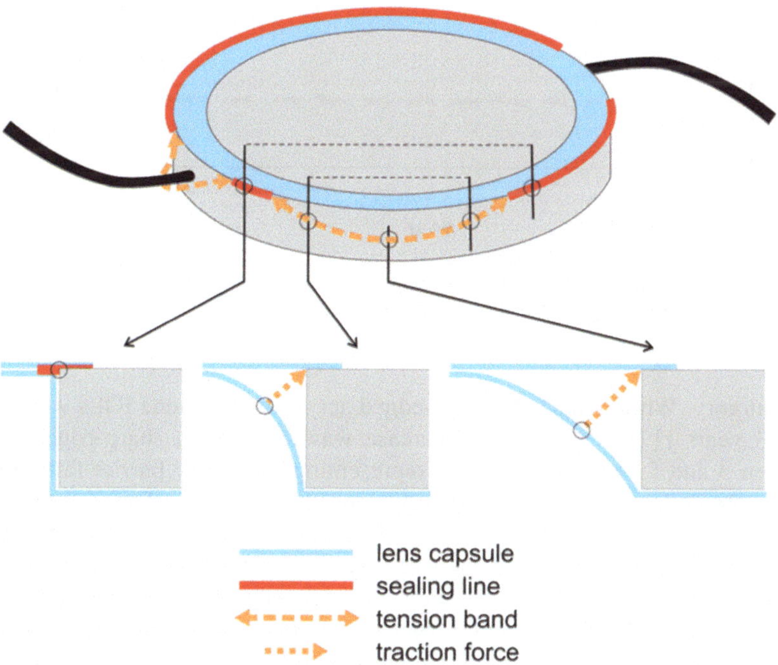

Fig. 15.14 "Capsular suspension effect": With a sharp posterior edge, the sealing line may be discontinuous, and the barrier effect still preserved as long as the posterior capsule between the sealed areas is pulled anteriorly and bent around the posterior optic edge

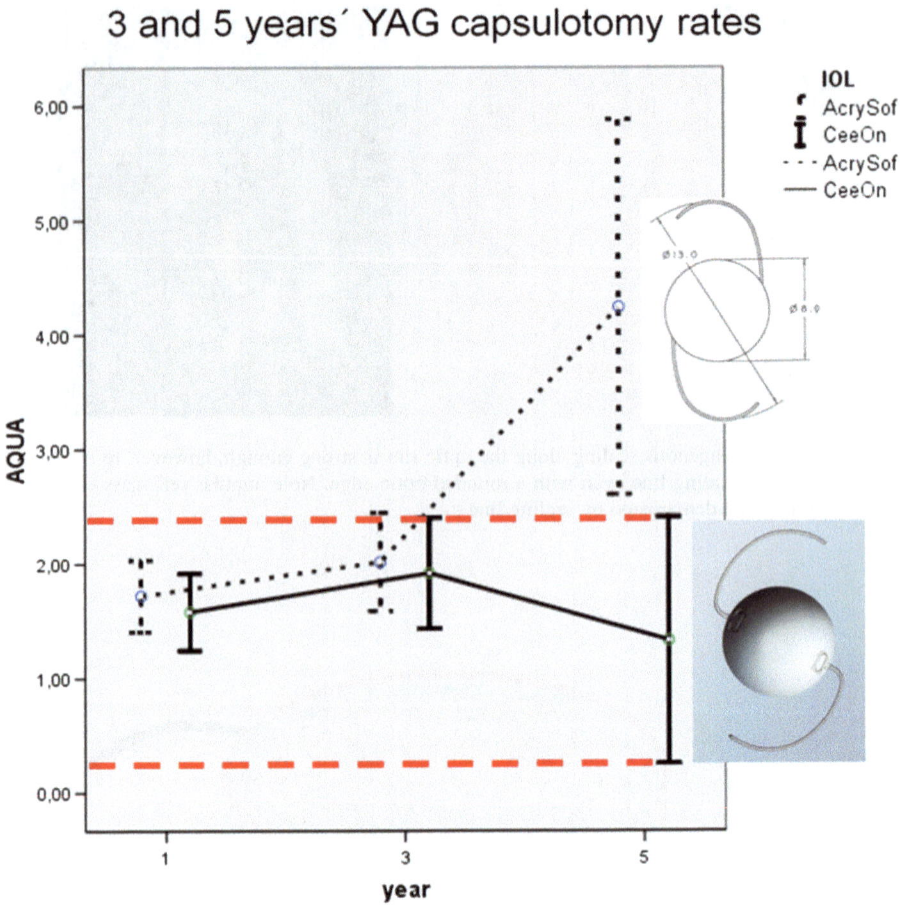

Fig. 15.15 While PCO rates with sharp-edged acrylic and silicone IOLs were similar after 3 years, a continuous increase was found with sharp-edged acrylics between 3 and 5 years

after surgery. While rates with sharp-edged acrylic and silicone IOLs were similar after 3 years [17], a continuous increase was found with sharp-edged acrylics between 3 and 5 years (Fig. 15.15), representing secondary barrier failure due to inadequate collagenous capsule sealing along the optic. The fact that differences were no longer found between one- and three-piece acrylic IOLs after 5 years simply reflects the general propensity of acrylic IOLs for delayed secondary barrier failure even if equipped with a sharp posterior edge [18], while silicone IOLs continue to provide an excellent barrier effect even with round-edged optics due to the pronounced collagenous capsular sealing [19]. It also explains why the significantly better performance of acrylic IOLs with sharp versus round edges at 3 years [20] levels out in the years to follow [21] (Fig. 15.16: by Berger et al. 2008, unpublished data).

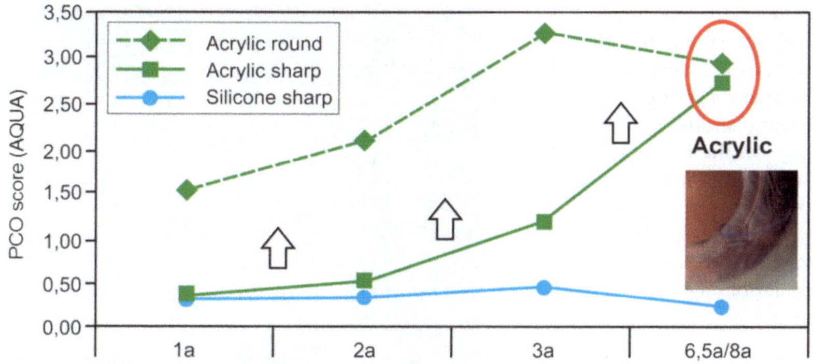

Fig. 15.16 While acrylic IOLs exhibit a propensity for delayed secondary barrier failure after 3 years even if equipped with a sharp posterior edge, silicone IOLs continue to provide an excellent barrier effect even with round-edged optics due to the pronounced collagenous capsular sealing

15.3 ACO and Capsular Shrinkage

While regeneratory PCO takes years to fully develop, fibrotic ACO and PCO reach their final stage during the first 3–6 months [22, 23]. A-LECs may repopulate areas where cells have been scraped off during the surgery. From the edge of the capsulorhexis opening, cells may migrate unto the anterior optic surface ("LEC out- or ongrowth") when overlapping the optic or unto the posterior capsule where such overlap is lacking. With full capsule-optic overlap, fibrotic contraction of the anterior capsule creates a strong encasement ("shrink-wrapping") of the IOL optic, thus maximizing the barrier effect of the posterior optic edge by firmly pulling the posterior capsule around a sharp posterior edge (Fig. 15.17), while the collagen deposited by the A-LECs along the contact line between the capsule leaves adjacent to the optic rim creates a permanent sealing line resisting reopening of the bag by Soemmering's ring formation and delayed E-LEC proliferation when strong enough [2].

Thus, the barrier is *initiated* by the posterior sharp optic edge. It is effectuated by capsular bag fusion and depends on both the speed and completeness of bend formation and the sharpness of the edge. *Later on* capsular bending is *maintained and supplemented* by the collagenous capsular sealing along the capsular fusion line which is a function of A-LEC transdifferentiation catalysed by the optic material. The barrier effect of collagenous capsular sealing then is largely independent of the posterior edge profile of the IOL optic. Three-piece IOLs with sharp silicone optics provide the most sustained barrier effect of all IOLs on the market. Apart from the exquisitely sharp optic edge they can be fitted with, it is the catalysing effect of silicone material on A-LEC transdifferentiation. This results in the formation of a strong collagenous sealing of both capsules along the optic rim which is not readily broken up by Soemmering's ring formation and delayed-onset

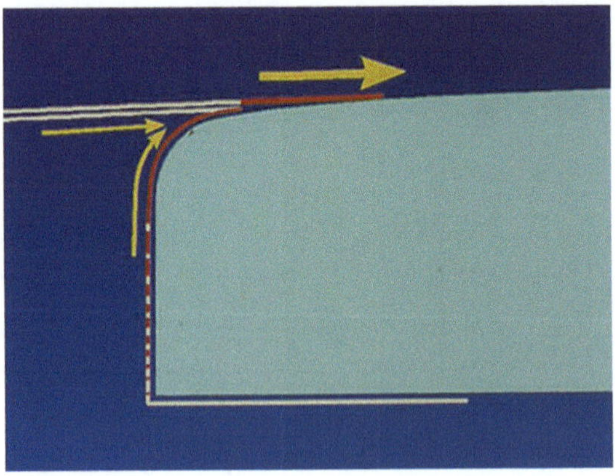

Fig. 15.17 With full capsule-optic overlap, fibrotic contraction of the anterior capsule creates a strong encasement ("shrink-wrapping") of the IOL optic, thus maximizing the barrier effect of the posterior optic edge by firmly pulling the posterior capsule around a sharp posterior edge

E-LEC proliferation. Reducing fibrosis by whatever means will thus inherently interfere with the long-term barrier effect of IOLs [24].

Fibrotic contraction of the anterior capsule leaf, or rhexis opening, depends on the area of contact between the rhexis leaf and the IOL optic, which is not necessarily identical with the amount of capsule-optic overlap when not in full contact. With the anterior capsule fully contacting the optic surface, though, shrinkage will be enhanced as the rhexis diameter decreases and the degree of overlap increases. Whitening and shrinkage also depends on the optic material, and contraction on zonular countertraction, eyes with weak zonules (e.g. pseudoexfoliation—PEX) being prone to excessive capsular contraction ("rhexis phimosis", "capsular contraction syndrome" (Fig. 15.2c), [25]). Shrinkage is more prominent with silicone IOLs: When comparing 3-piece IOLs with identical design, Sacu et al. [9] found contraction with a silicone optic IOL to be significantly more prominent than with its otherwise identical acrylic counterpart. Eyes implanted with monobloc silicone IOLs are especially prone to capsule shrinkage [6] since plate haptics increase the area of IOL-capsule contact. Capsular shrinkage is not different with polypropylene and poly-(methyl methacrylate) loops [26]. However, strong loops with a broad contact angle may counteract the contraction forces of silicone IOLs [27]. Contact of the A-LECs to the IOL initiates a breakdown of the blood-aqueous barrier [28] which may enhance shrinkage and contraction.

15.4 Rationale, Techniques and Instrumentation for Anterior Capsule Polishing

Capsular whitening or shrinkage can be counteracted either by removing the A-LEC layer as the cellular substrate for ACO or by withholding the anterior capsule leaf from establishing contact to the optic surface, thereby preventing the A-LECs from getting into contact with the IOL material as a prerequisite for initiating transdifferentiation. The latter can be achieved by adding an elevated

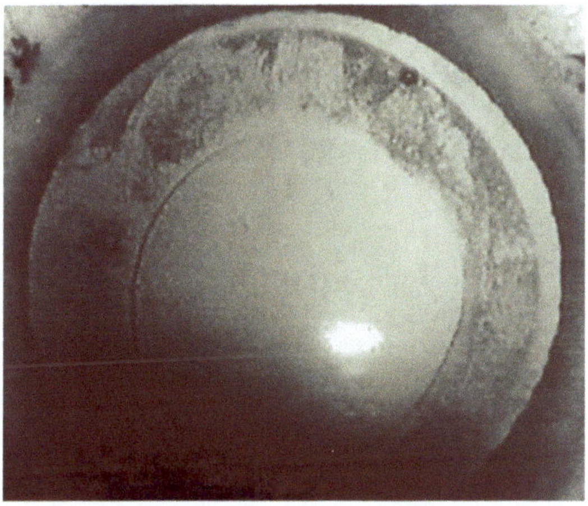

Fig. 15.18 After capsulorhexis and lens removal, because the capsule bag is exposed to infusion fluid and viscoelastic substance, the oedematous A-LEC layer can be seen as a greyish concentric zone and its peripheral border well discriminated

ridge along the anterior optic rim, by additionally implanting a capsular bending ring ("capsular distance ring" [29, 30]) or by using the surgical technique of posterior optic entrapment into a posterior capsulorhexis opening [31, 32]. While the former mechanically holds the anterior capsule at a distance to the anterior optic surface, the latter results in sandwiching of the posterior capsule in between the anterior capsule and anterior optic surface, thus avoiding direct contact to the IOL optic.

When after capsulorhexis and lens removal the capsule bag is exposed to infusion fluid and viscoelastic substance, the oedematous A-LEC layer can be seen as a greyish concentric zone and its peripheral border is well discriminated (Fig. 15.18). This band-shaped area must be fully covered by any polishing instrument used to ascertain complete A-LEC removal. The various techniques and instruments described for anterior capsule polishing, cannulas (Kratz), scrapers (Koch), curettes (Rentsch [33]), sleeved aspiration/irrigation (A/I) tips with or without ultrasound [34] and ultraviolet irradiation probes (Raut [35]), significantly differ in efficiency and safety [36]: Efficiency is defined by the percentage of cell-free areas found on capsule specimen retrieved from the eye after polishing, the amount of remaining cell structures (borders, junctions) on the capsule and the time needed to achieve this. Methods also differ in the capsular lesions they may induce. These vary between lamellar separation of superficial capsule layers and deep almost penetrating lacerations.

Menapace [37] described an aspiration curette with a slit-shaped superior opening with sharp parallel edges and rounded ends and a bypass opening that allows gentle docking unto the undersurface of the anterior capsule (Fig. 15.19), maximizing efficient polishing while avoiding capsular injuries. The slit-shaped opening allows for removing broad bands of A-LECs at a time and reaching behind the iris while still having control of how far the probe tip and slit reach into the periphery.

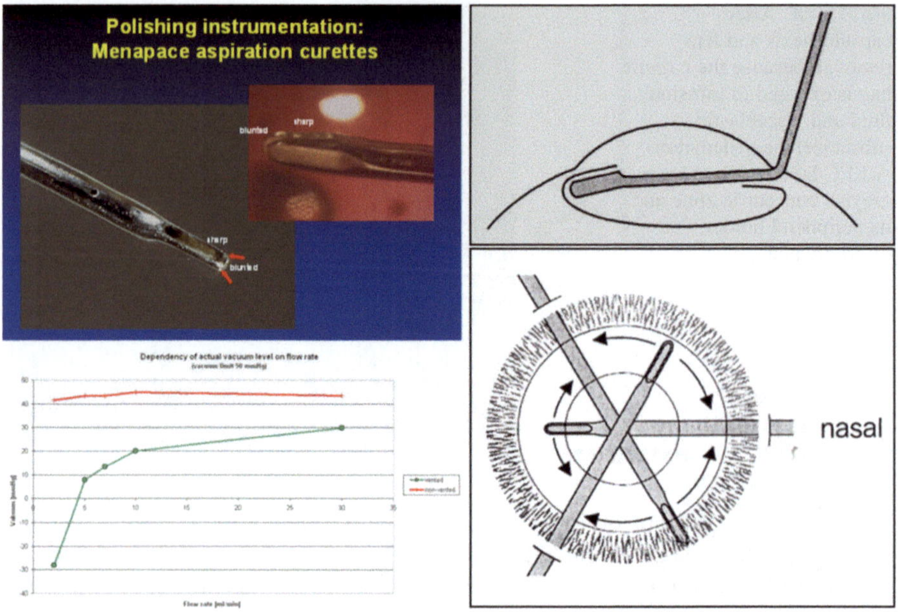

Fig. 15.19 The aspiration curette described by Menapace features a slit-shaped superior opening with sharp parallel edges and rounded ends and a bypass opening that allows gentle docking unto the undersurface of the anterior capsule

15.5 Efficiency and Impact of Anterior Capsule Polishing on Anterior Capsule Opacification

Several studies have been performed to elucidate efficiency and impact of anterior capsule polishing (ACP) by comparing postoperative anterior capsule shrinkage and/or whitening with and without polishing.

Tadros et al. [38] investigated the impact of ACP on capsulorhexis size with a silicone IOL. Six months postoperatively, the size of the capsulorhexis significantly increased in the study group (mean increase 1.07 ± 1.70 mm; $P = 0.01$) and significantly decreased in the control group (mean decrease 3.38 ± 2.37 mm; $P < 0.0001$). The difference in changes in the capsulorhexis areas between the two groups was also highly statistically significant ($P < 0.0001$). Hanson et al. [39] studied the impact of ACP on capsulorhexis aperture after AcrySof (Alcon) acrylic IOL implantation. Three months postoperatively, the mean decrease in aperture was 1.9 % in the polished and 5.6 % in the non-polished group ($P = 0.02$). Grade 2 ACO at 3 months was found in 44 and 61 %, respectively ($P = 0.13$). After careful ACP using biaxial I/A instrumentation for 5 min, Shah et al. [40] found a significant difference in ACO intensity 1 month postoperatively. When disregarding the grade of ACO, only 43 % of the polished eyes developed any grade of ACO compared to 90 % of the controls ($p = 0.001$). At 6 months, however, no statistically significant difference was found ($p = 0.5$). Overall, polished eyes tended to develop lower

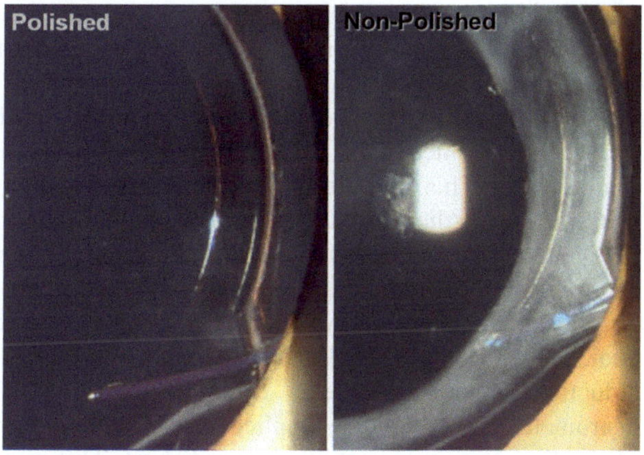

Fig. 15.20 Anterior capsule of an eye polished with a Menapace aspiration curette (*left*) versus the control eye of the same patient

grades of ACO, and grade 4 ACO was only detected in control eyes. In a non-randomized study comparing 180° and 360° ACP with the *Rentsch* curettes, Kruger et al. [41] found no statistical difference between ACO 2 years after surgery. Liu et al. [42] found a statistically significant decrease in fibrotic ACO after aggressive 360° polishing using an I/A probe and including the anterior capsule and equatorial capsule, with a mean ACO score of 1.2 ± 0.5 versus 3.2 ± 0.6 on a scale of 1–5 year. Svancarova et al. [43] also found a significant difference after 2 years. Sacu et al. [44], in a randomized intraindividual comparison study using the *Menapace* aspiration curette for polishing and a standardized slitlamp photographic technique for reproducible documentation [45], also found a statistically significant difference 3 years after surgery. Two silicone IOL models had been implanted: one with a blunted-edge profile (SI-40, Allergan) and the other with a fully rounded rim (Silens6, Bausch & Lomb). The mean ACO score (scale 0–100 %) was 17 % in the polished eyes and 26 % in the control eyes ($P < 0.0001$). In the polished SI-40 eyes mean ACO was 15 % and 26 % in the control SI-40 eyes ($P < 0.01$). In the polished Silens6 eyes mean ACO was 19 % and 26 % in the control Silens6 eyes ($P < 0.003$). As described by Shah et al. [40], the striking initial effect of polishing seen after 1 month (Fig. 15.20) also decreased later on, but was still significant after 3 years. Bolz et al. [46], in a similar trial with silicone IOLs featuring a truncated rim with sharp edges, again found a statistically significant decrease of ACO after 1, 2, 3 and 5 years in eyes that had undergone ACP ($P < 0.01$–0.03).

How can the differences in the findings be explained? ACO results when resident A-LECs on the undersurface of the anterior capsule leaf transdifferentiate upon contact with the IOL optic. When removing these cells, fibrosis should not occur. If it still occurs, this can have two causes: First, LEC abrasion was incomplete, and/or second, the abraded areas have been repopulated by LECs from non-polished areas. The more thoroughly it abrades the LEC layer, and the more peripherally it reaches out, the more effective ACP should be. With the vented slit-cannula designed by Menapace, LEC abrasion is extensive, since the capsule is attracted towards the sharp edge of the slit-shaped aspiration port which removes a broad band of LECs at a time and reaches far out into the capsular periphery behind the iris with still good visual control by viewing the proximal end of the slit. With other aspiration instruments, the small port removes only narrow straits of LECs at a time while the location of the aspiration opening cannot be judged when manipulating behind the iris. With the Rentsch curettes and other non-aspiration devices, power coupling is inadequate since it is only effectuated by elevating the capsule instead of aspirating it. This significantly decreases efficiency of LEC abrasion, leaving LECs behind in the treated areas and/or reducing the treated area itself.

The fact that ACO again increases after month 1 to reach the same grade as the controls after 6 months, as reported by Shah et al. [40] or fails to show any difference, as reported by Kruger et al. [41], while Hansen et al. [39], Liu et al. [42], Sacu et al. [44] and Bolz et al. [46] demonstrated a lasting significant reduction, obviously reflects differences in the efficiency of the polishing instruments and techniques applied. *However, some grade of ACO did develop in all studies even with most efficient polishing*. This may be explained by repopulation of the polished areas either by residual resident A-LECs or by equatorial LECs from the germinative zone. Though Kruger et al. [41] did not specifically address the difference between the polished and non-polished semi-circumference in one and the same eye, the lacking difference in overall ACO in their study indicates that there may be secondary LEC invasion from the equator towards the capsulorhexis edge and even further unto the anterior optic surface. These E-LECs then seem capable of expressing a potential of myofibroblastic transdifferentiation as resident A-LECs. Liu et al. [42] investigated the LEC growth in capsular bags implanted with an IOL in cadaver eyes with the iris removed using inverse phase-contrast light microscopy. One day after surgery, large patches of cells remained under the anterior capsule and the equatorial zone in the unpolished cadaver eyes while the anterior capsule was clear in the polished eyes. After 3 days in culture, many patches of dead cells had formed in the unpolished eyes. After 7 days, cell growth was minimal in the unpolished eyes, while robust cell proliferation was observed in the polished eyes. This corroborates that ACP cannot remove all LECs, particularly those around the equatorial zone, and that ACP itself may stimulate robust proliferation of residual LECs in vitro, while live cells in unpolished eyes tend to die, resulting in less LEC proliferation. In addition, repopulation of the once polished anterior capsule areas by equatorial LECs may be triggered by mechanical trauma to the capsule equator by the polishing instrument. This may be particularly true for bulky non-aspiration curettes like that designed by Rentsch.

15.6 Impact of Anterior Capsule Polishing on Posterior Capsule Opacification and YAG Capsulotomy Rates

The *Vienna IOL Study Group* conducted several clinical studies to elucidate the impact of ACP on PCO using randomized intraindividual comparison.

Menapace et al. published the 3 years results of *regeneratory* PCO formation with and without ACP [47]. They found no statistically significant difference in r-PCO scores as evaluated subjectively and objectively in retroilluminated images.

Svancarova et al., in their study, also did not find differences in PCO rates after 2 or more years [43]. However, when Menapace et al. looked at the YAG rates *after 3 years, 54 % of the eyes that had undergone ACP required YAG-LCT as compared to only 36 % eye without ACP*. During the first year, no YAG-LCT had been necessary in either group. Between years 1 and 3, however, 11 YAG-LCTs were performed in the polished eye group while only 4 in the control group. This suggests early barrier failure caused by lacking collagenous capsule sealing along the optic rim. YAG-LCTs had been performed in the hospital on a non-profit basis based on specified criteria, i.e. a drop of best corrected high-contrast visual acuity to 0.8 or less in eyes with full macular potential *and subjective complaints of blur and reduced contrast vision. Of the latter patients, all patients reported a substantial increase in visual quality even when high-contrast visual acuity had been still good.* The difference between the grade of r-PCO in retroillumination and the quality of vision reported by the patient was particularly pronounced in the polished eye group, suggesting an obvious mismatch between the two parameters. When scrutinizing the interspace between optic and posterior capsule in these eyes before performing the YAG-LCT using oblique, high-intensity slitlamp illumination, the *interspace was seen to be filled with amorphous LEC material lacking well-defined cell borders, which was obviously not adequately picked up by retroillumination photography* [48]. When looking at pairs of eyes with prominent regeneratory PCO, the *capsule bag was typically patent with fusion and bending typically incomplete or absent* in the polished eyes and E-LEC ingrowth typically more profuse without signs of having been halted at the optic edge (Fig. 15.21).

Fibrotic PCO was significantly reduced in the polished eyes [44]: Overall mean f-PCO score was 0.5 in the polished and 1.0 in the unpolished eyes ($P < 0.0007$). In the polished blunted-edge SI-40 eyes mean f-PCO score was 0.4 and 1.1 in the SI-40 control eyes ($P < 0.0006$). In the polished rounded-rim Silens6 eyes, mean f-PCO was 0.6 and 0.9 in the Silens6 control eyes ($P < 0.08$).

15.7 ACP and Impact of the Optic Edge Profile on Regeneratory and Fibrotic PCO

Menapace et al. [47] found a significant increase in YAG-LCT rates after ACP in eyes implanted with silicone IOLs lacking a sharp posterior optic edge. When analysing the YAG-LCT rates of the two subgroups of IOLs, however, they found lower YAC-LCT rates with the IOL featuring blunted edges, profile (SI-40) than with the model equipped with a fully rounded optic rim (Silens6). Bolz et al. [46], in a similar setting, evaluated the impact of ACP on PCO and YAG-LCT rates with a truncated silicone IOL (911A, Pharmacia) with very sharp edges. They again found no difference in r-PCO but also not in YAG-LCT rates, and Sacu et al. [11] found no difference in f-PCO.

How can these differences between the two studies be explained? When addressing f-PCO, the *Vienna IOL Study Group* reported differences dependent upon the optic edge profile: While f-PCO was most accentuated with the IOL with a

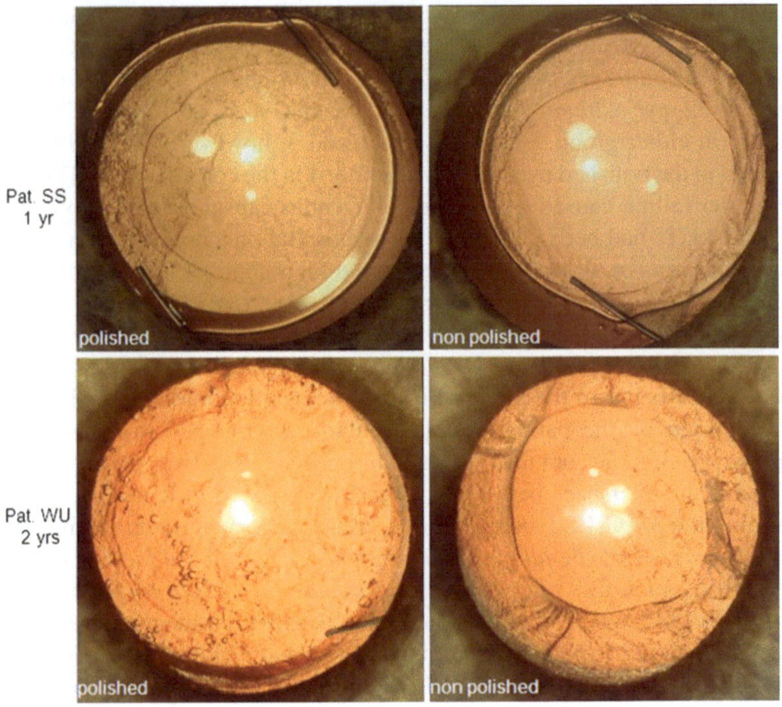

Fig. 15.21 Early barrier failure after ACP (*left*) compared to the control eye (*right*); the rhexis leaf in the ACP eyes is still clear, while that in the control eyes shows fibrotic ACO and f-PCO with radial folds and concentric wrinkling

fully rounded optic rim, it was almost absent with the IOL featuring a truncated rim with sharp edges. With the IOL featuring rounded edges, f-PCO formation was intermediate. In the course of capsular bag closure and capsular fusion, the posterior capsule is typically bent around the posterior optic edge (Fig. 15.6). The sharper the posterior optic edge, the higher the pressure on the capsule along its contact line with the optic edge, and the more acute the bent formed in the posterior capsule. Whatever the prevailing mechanism [13–15], the barrier effect will increase with the sharpness of the posterior optic edge, incrementally preventing migrating A-LEC from the peripheral anterior capsule leaf and/or E-LECs from the capsular bag periphery from accessing the retro-optical space where they quickly transdifferentiate, forming the typical circular zone of f-PCO inside the optic rim. This may explain why the differences observed in f-PCO with and without polishing decreased with increasing sharpness of the posterior optic edge. Since silicone triggers myofibroblastic transdifferentiation more than any other IOL material, these differences have been highlighted by the silicone IOL models used in aforementioned studies by Sacu et al. [11, 44], Menapace et al. [47] (rounded) and by Bolz et al. [46] (sharp edge profile).

15.8 Impact of ACP on Capsular Closure and Sealing

When specifically looking at the interplay between the anterior and posterior capsules along the IOL optic rim, the following can be observed: After polishing, the capsular bag tends to remain open with the two capsule leaves staying at a distance to each other and the capsular fusion line halting more peripherally if at all present. Early LEC ingrowth is seen to invade the posterior capsule behind the IOL optic, and the patent interspace between the capsules and the retro-optical space readily fills up with ill-defined cellular material. With the latter forming, patients start complaining of subjective decrease of visual quality while they still may have good high-contrast visual acuity.

ACP thus not only interferes with collagenous capsular sealing and the formation of a lasting barrier formation by removing the cellular substrate for collagenous sealing and shrink-wrapping of the IOL optic following myofibrillar contraction of the capsules but also interferes with the process of capsule bag closure and capsular bending. This becomes obvious when observing the interplay between the capsular leaves and the formation of the capsular fusion line and its progression towards the IOL edge. ACP also changes the morphological characteristics of the LEC itself. Both may reflect changes in the biochemical environment and physiological characteristics of the residual LECs when abraded or traumatized. As a result, the capsular bag tends to remain patent, reflected by a more peripheral or even lacking fusion line. LECs quickly invade the retro-optical interspace and form poorly discernible layers of ill-defined cells as opposed to the well-defined globular cells seen in non-polished control eyes. While these amorphous cell agglomerates are not picked up by retroillumination, but only by oblique high-intensity slitlamp illumination, they do cause vision-disturbing light scatter. This explains the mismatch between the low PCO readings and the high YAG capsulotomy rates in the polished eyes implanted with IOLs lacking a sharp posterior optic edge. Though capsular fusion and sealing is also compromised with sharp posterior edge optics, the sharp edge seemingly still prevents LECs from invading the retrolental space. This is especially true for IOL models featuring angulated and resilient "capsular-C" haptics that permanently push the optic posteriorly while still preserving the circular shape of the capsule bag like the IOL model used in the study by Bolz et al. [46], thus inducing a lasting circular posterior capsule bend.

15.9 Anterior Capsule Opacification and Polishing with Posterior Optic Buttonholing (POBH)

Menapace [31, 32] looked at the impact of additional ACP in conjunction with his posterior optic buttonholing (POBH) of round-edged open-loop IOLs for PCO eradication. When entrapping the optic of a looped IOL into a posterior capsulorhexis opening, the remaining peripheral posterior capsule slips anteriorly to be sandwiched between the anterior capsule and the optic surface (Fig. 15.22a), thereby precluding direct contact between the anterior capsule leaf and the optic except in the area adjacent to the base of the optic-haptic junction where the posterior capsulorhexis overcrosses the haptic insertion. While the fibrosis of the anterior capsule leaf is

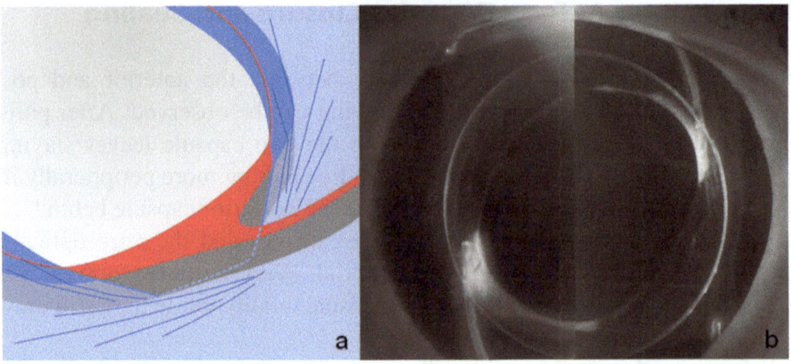

Fig. 15.22 Optic entrapment into a posterior capsulorhexis prevents anterior capsule from contacting the optic except in the area adjacent to the haptic base (**a**); this is where fibrosis primarily forms (**b**)

Fig. 15.23 Since the mechanism of POBH to avoid central opacification is independent of barrier formation at the optic edge, ACP is a valuable adjunct to reduce peripheral LEC reaction without compromising its efficacy: *left* = non-polished, *right* = polished

most prominent in this area of direct contact to the optic adjacent to the haptic insertion and often remains restricted to this area (Fig. 15.22b), it may extend to the sandwiched posterior capsule between the junctions. Fibrosis of the latter originates from the contact line between the rhexis edge of the anterior capsule and the residual peripheral posterior capsule leaf or from the area of direct contact between the anterior capsule leaf and the optic surface adjacent to the haptic-optic junction.

When combining POBH with ACP, a significant further reduction of capsule fibrosis was seen, but again some grade of residual fibrosis often formed. The latter may again emerge from either residual resident A-LECs or from peripheral E-LECs migrating centripetally and taking on the characteristics of A-LECs. While ACP, though decreasing ACO, increases PCO with standard in-the-bag implantation, there is no such negative trade-off with POBH. Since ACP with POBH carries the potential to decrease peripheral LEC reaction (Fig. 15.23) without compromising its efficacy in precluding central capsule opacification, ACP is considered a valuable adjunct to POBH.

References

1. Marcantonio JM, Vrensen GF (1999) Cell biology of posterior capsular opacification. Eye 13:484–488
2. Menapace R (2004) Prevention of PCO. In: Kohnen T, Koch DD (eds) Cataract and refractive surgery, essentials in ophthalmology. Springer, Berlin/Heidelberg/New York, pp 101–122
3. Nishi O, Nishi K, Sakka Y et al (1991) Intercapsular cataract surgery with LEC removal. Part IV: capsular fibrosis by poly(methyl methacrylate). J Cataract Refract Surg 17:471–477
4. Toldos JJM, Roig AA, Benabent EC (1996) Total anterior capsule closure after silicone intraocular lens implantation. J Cataract Refract Surg 22:269–271
5. Dahlhauser KF, Wroblewski KJ, Mader TH (1998) Anterior post cataract (IOL) surgery opacification. J Cataract Refract Surg 24:1216–1219
6. Werner L, Pandey SK, Escobar-Gomez M et al (2000) ACO: a histopathological study comparing different IOL style. Ophthalmology 107:463–471
7. Werner L, Pandey SK, David J, Apple DJ et al (2001) ACO: correlation of pathologic findings with clinical sequelae. Ophthalmology 108:1675–1681
8. Hayashi K, Hayashi H, Nakao F, Hayashi F (1997) Reduction in the area of the anterior capsule opening after polymethylmethacrylate, silicone, and soft acrylic intraocular lens implantation. Am J Ophthalmol 123:441–447
9. Sacu S, Menapace R, Buehl W, Rainer G, Findl O (2004) Effect of intraocular lens optic edge design and material on fibrotic capsule opacification and capsulorhexis contraction. J Cataract Refract Surg 30:1875–1882
10. Kappelhof JP, Vrensen GF (1992) The pathology of after-cataract. A mini review. Acta Ophthalmol Suppl 205:13–24
11. Sacu S, Menapace R, Findl O et al (2004) Influence of optic edge design and anterior capsule polishing on posterior capsule fibrosis. J Cataract Refract Surg 30:658–662
12. Sacu S, Findl O, Menapace R, Buehl W (2005) Influence of optic edge design, optic material, and haptic design on capsular bend configuration. J Cataract Refract Surg 31:1888–1894
13. Nishi O, Nishi K (1999) Preventing posterior capsule opacification by creating a discontinuous sharp bend in the capsule. J Cataract Refract Surg 25:521–526
14. Bhermi GS, Spalton DJ, El-Osta AA, Marshall J (2000) Failure of a discontinuous bend to prevent lens epithelial cell migration in vitro. J Cataract Refract Surg 28:1256–1261
15. Nagamoto T, Fujiwara T (2003) Inhibition of lens epithelial cell migration at the intraocular lens optic edge: role of capsule bending and contact pressure. J Cataract Refract Surg 29:1605–1612
16. Nishi O, Nishi K, Akura J (2002) Speed of capsular bend formation at the optic edge of acrylic, silicone, and poly(methyl methacrylate) lenses. J Cataract Refract Surg 28:431–437
17. Findl O, Menapace R, Sacu S, Buehl W, Rainer R (2005) Effect of optic material on posterior capsule opacification in IOLs with sharp edge optics. Ophthalmology 112:67–72
18. Leydolt C, Davidovic S, Sacu S et al (2007) Long-term effect of 1-piece and 3-piece hydrophobic acrylic intraocular lens on posterior capsule opacification: a randomized trial. Ophthalmology 114:1663–1669
19. Sacu S, Menapace R, Findl O et al (2005) Long-term efficacy of adding a sharp posterior optic edge to a three-piece silicone intraocular lens on capsule opacification: five-year results of a randomized study. Am J Ophthalmol 139:696–703
20. Buehl W, Findl O, Menapace R et al (2005) Long-term effect of optic edge design in an acrylic intraocular lens on posterior capsule opacification. J Cataract Refract Surg 31:954–961
21. Menapace R (2006) Can modern acrylic IOL concur with acrylic IOLs? An up-dated performance comparison. In: Kohnen T, Auffarth GU, Pham DT (eds) Kongress der Deutschsprachigen Gesellschaft für Intraokularlinsenimplantation und Refraktive Chirurgie, 3–4 März 2006. Biermann, Köln, pp 73–82
22. Kimura W, Yamanishi S, Kimura T et al (1998) Measuring the anterior capsule opening after cataract surgery to assess capsule shrinkage. J Cataract Refract Surg 24:1235–1238

23. Park TK, Chung SK, Baek NH (2002) Changes in the area of the anterior capsule opening after intraocular lens implantation. 1. J Cataract Refract Surg 28:1613–1617
24. Menapace R (2007) After-cataract following intraocular lens implantation. Part I: genesis and prevention by optimizing conventional lens implants and surgical techniques. Ophthalmologe 104:253–262
25. Davison JA (1993) Capsule contraction syndrome. J Cataract Refract Surg 19:582–589
26. Gallagher SP, Pavilack MA (1999) Risk factors for anterior capsule contraction syndrome with polypropylene or poly-(methyl methacrylate) haptics. J Cataract Refract Surg 25:1356–1361
27. Sacu S, Menapace R, Findl O (2006) Effect of optic material and haptic design on anterior capsule opacification and capsulorhexis contraction. AJO 141:488–493
28. Tsuboi S, Tsujioka M, Kusube T, Kojima S (1992) Effect of continuous circular capsulorhexis and intraocular lens fixation on the blood-aqueous barrier. Arch Ophthalmol 110:1124–1127
29. Menapace R, Sacu S, Georgopoulos M, Findl O, Rainer G, Nishi O (2008) Efficacy and safety of capsular bending ring implantation to prevent posterior capsule opacification: three-year results of a randomized clinical trial. J Cataract Refract Surg 34:1318–1328
30. Nishi O, Nishi K, Menapace R (1998) Capsule-bending ring for the prevention of capsular opacification: a preliminary report. Ophthalmic Surg Lasers 29:749–753
31. Menapace R (2006) Routine posterior optic buttonholing for eradication of posterior capsule opacification in adults: report of 500 consecutive cases. J Cataract Refract Surg 32:929–943, Erratum in: J Cataract Refract Surg (2006)
32. Menapace R (2008) Posterior capsulorhexis combined with optic buttonholing: an alternative to standard in-the-bag implantation of sharp-edged intraocular lenses? A critical analysis of 1000 consecutive cases. Graefes Arch Clin Exp Ophthalmol 246:787–801
33. Rentsch FJ (1996) Long-term results after lens epithelial cell removal during extracapsular cataract surgery with phacoemulsification. In: Rochels R, Duncker G, Hartmann CH (eds) Kongress der Deutschsprachigen Gesellschaft für Intraokularlinsen Implantation und Refraktive Chirurgie, Springer Berlin/Heidelberg/New York, pp 399–407
34. Nishi O (1987) Removal of lens epithelial cells by ultrasound in endocapsular cataract surgery. Ophthalmic Surg 18:577–580
35. Raut RM (2007) Low-intensity ultraviolet A irradiation of the lens capsule to remove LECs during cataract surgery. J Cataract Refract Surg 33:1025–1032
36. Mathey CF, Kohnen TB, Ensikat HJ, Koch HR (1994) Polishing methods for the lens capsule: histology and scanning microscopy. J Cataract Refract Surg 20:64–69
37. Menapace R, Di Nardo S (2006) Aspiration curette for ACP: laboratory and clinical evaluation. J Cataract Refract Surg 32:1997–2003
38. Tadros A, Bhatt UK, Abdul Karim MN, Zaheer A, Thomas PW (2005) Removal of LECs and the effect on capsulorhexis size. J Cataract Refract Surg 31:1569–1574
39. Hanson RJ, Rubinstein A, Sarangapani S et al (2006) Effect of LEC aspiration on postoperative capsulorhexis contraction with the use of the AcrySof intraocular lens: randomized clinical trial. J Cataract Refract Surg 32:1621–1626
40. Shah SK, Praveen MR, Kaul A et al (2009) Impact of ACP on ACO after cataract surgery: a randomized clinical trial. Eye 23:1702–1706
41. Kruger AJ, Amon M, Schauersberger J et al (2001) ACO and lens epithelial outgrowth on the intraocular lens surface after curettage. J Cataract Refract Surg 27:1987–1991
42. Liu X, Cheng B, Zheng D et al (2010) Role of ACP in residual LEC proliferation. J Cataract Refract Surg 36:208–214
43. Svancarová R, Novák J, Adámková H (2010) The influence of the lens capsule mechanical polishing to the secondary cataract development. Cesk Slov Oftalmol 66:21–55
44. Sacu S, Menapace R, Wirtitsch M et al (2004) Effect of anterior capsule polishing on fibrotic capsule opacification: three-year results. J Cataract Refract Surg 30:2322–2327
45. Sacu S, Findl O, Menapace R et al (2002) Assessment of ACO: photographic technique and quantification. J Cataract Refract Surg 28:271–275

46. Bolz M, Menapace R, Findl O et al (2006) Effect of ACP on the PCO–inhibiting properties of a sharp-edged, 3-piece, silicone intraocular lens: three- and 5-year results of a randomized trial. J Cataract Refract Surg 32:1513–1520
47. Menapace R, Wirtitsch M, Findl O et al (2005) Effect of anterior capsule polishing on posterior capsule opacification and neodymium:YAG capsulotomy rates: three-year randomized trial. J Cataract Refract Surg 31:2067–2075
48. Camparini M, Macaluso C, Reggiani L, Maraini G (2000) Retroillumination versus reflected-light images in the photographic assessment of posterior capsule opacification. Invest Ophthalmol Vis Sci 41:3074–3079

Laser Photolysis System and PCO Prevention

16

Erica Liu, Nick Mamalis, and Liliana Werner

Abstract

Posterior capsule opacification (PCO) is the most common complication following modern cataract surgery. PCO is caused by proliferation of lens epithelial cells (LECs) that remain in the capsular bag following cataract removal. Various methods for preventing PCO by LEC removal have been developed including pharmacological treatments and mechanical methods for LEC removal. A laser photolysis system has been developed for the removal of LECs from the lens capsular bag. A modified Nd:YAG handpiece has been evaluated in human cadaver eyes to document removal of LECs. In addition, extracellular matrix glycoproteins such as laminin and fibronectin, which may play a role in the development of PCO, were shown to be removed from the lens capsular bag by the photolysis system. Preliminary clinical studies have found that the laser was successful in removing the LECs from the anterior portion of the capsular bag with subsequent clear anterior and posterior capsules in the area of treatment for 2 years. In conclusion, the Nd:YAG laser photolysis system shows promise for prevention of PCO by removal of LECs as well as removal of adhesion glycoproteins.

Keywords

Laser photolysis system • Lens epithelial cells • Posterior capsule opacification

16.1 Introduction and Background

Posterior capsule opacification (PCO) and fibrosis is the most common complication following successful cataract surgery due to the proliferation of lens epithelial cells (LECs). The incidence of PCO approaches 30 % in the first several years after

E. Liu, M.D. • N. Mamalis, M.D. • L. Werner, M.D., Ph.D. (✉)
John A. Moran Eye Center, University of Utah, Salt Lake City, UT, USA
e-mail: liliana.werner@hsc.utah.edu

S. Saika et al. (eds.), *Lens Epithelium and Posterior Capsular Opacification*,
DOI 10.1007/978-4-431-54300-8_16, © Springer Japan 2014

cataract surgery [1] and occurs at an even higher percentages of the pediatric population [2].

Significant PCO can be successfully treated by a neodymium:YAG (Nd:YAG) laser capsulotomy. However, there remain significant cost and potential complications associated with the procedure. With one in four cataract patients eventually developing PCO the cost of PCO treatment in the United States is second only to the cost of the initial cataract surgeries [3]. Possible complications include intraocular lens (IOL) dislocation, postoperative intraocular pressure elevation, cystoid macular edema, and an increased incidence of retinal complications, including retinal detachment [1, 4].

PCO is caused by the proliferation of LECs that remain in the capsular bag after phacoemulsification for cataract removal [5]. Residual LECs cloud the capsule by proliferating and migrating along the internal surface of the anterior and posterior lens capsule. Sometimes referred to as a "secondary cataract," PCO causes significant reduction in visual acuity and contrast sensitivity, increased glare, and optical aberrations [6, 7]. LECs can be divided into two main morphological types: anterior epithelial cells and equatorial epithelial cells. When stimulated, anterior epithelial cells tend to proliferate, undergo fibrous metaplasia, and form fibrous tissue. The equatorial epithelial cells are the primary cells responsible for PCO and Soemmering's ring formation. When stimulated, they tend to migrate onto the posterior capsule and proliferate, forming PCO. They can also form regenerative epithelial pearls in the capsular bag fornix termed Soemmering's ring. Occasionally, the LECs can cause fibrosis and contraction with reduced vision and disruption of the position and function of the IOL [8–12]. This phimosis and subsequent need for IOL explantation is of increasing concern with highly flexible thin lenses, particularly those designed for microincision cataract surgery [8].

The development of capsule opacification and fibrosis with contraction potentially becomes an even more disruptive complication with the continued advancement of IOL design. Specialty IOLs, to correct astigmatism and to treat presbyopia with multifocal and accommodating IOLs, are particularly sensitive to position and movement within the capsular bag. Furthermore, patients with these newly developed IOLs may require higher degrees of capsule clarity and require earlier Nd: YAG laser capsulotomies. These new IOL technologies highlight the importance of developing consistent and safe methods to prevent capsule opacification.

16.1.1 Surgical Techniques for PCO Prevention

Techniques used during surgery to prevent PCO rely on two main mechanisms: (1) cleaning and removing the maximum amount of LECs and (2) using barriers to block any proliferation of the remaining LECs. Hydrodissection is a surgical technique that uses a fluid wave to separate the lens cortex from the lens capsule [13, 14]. Fluid is injected directly under the anterior capsule and spreads past the equator to the posterior surface of the lens. This action releases the lens for easier and safer removal and also removes LECs that may contribute to anterior capsule opacification, fibrosis [15], and PCO after surgery. Rotation of the lens is often performed by surgeons to confirm the complete separation of the lens and capsule but can also potentially provide a physical scraping mechanism to remove the remaining LECs from the capsule [14].

Anterior capsule polishing is a controversial technique used to remove LECs from the inner surface of the anterior capsule (more details on its role in PCO prevention can be found in Chap. 15). Multiple studies have shown that removal of LECs by polishing will delay PCO development [16–18]. Other studies, however, have shown either no benefit or even an increase in proliferation of LECs after anterior capsule polishing [19, 20].

In-the-bag fixation involves proper placement of the IOL completely in the capsular bag. This holds the IOL firmly in place and prevents damage to the surrounding uveal tissue. It also keeps the IOL optic flush against the surface of the posterior capsule. Without the potential space between the optic and the posterior capsule, a barrier effect prevents LECs from entering the gap and proliferating, causing PCO [21–23].

Similarly, the capsulorhexis size, shape, and centration matter in acting as a physical barrier to migrating and growing LECs. The ideal size of the capsulorhexis is just smaller than the optic size [24–26] with central alignment in relation to the optic [27]. The shape of the capsulorhexis also plays a role in equal force distribution. Symmetrically round-shaped capsulorhexis shows the least amount of PCO [28]. Large and asymmetric capsulorhexes are associated with increased posterior capsular folds which may play a role in increasing PCO [26, 28]. It is also worthwhile to note, however, that a very small and intact capsulorhexis can lead to increased capsular fibrosis and shrinkage [29, 30]. More details on the role of the capsulorhexis in PCO prevention can be found in Chap. 14.

16.1.2 IOL Design

One of several methods that have been used in an attempt to prevent PCO incorporates changes in IOL material and design (more details on their respective role in PCO prevention can be found in Chap. 17). Based on the theory that apposition of the posterior lens capsule to the IOL may have a barrier effect, successful IOL design modifications have incorporated a sharp posterior optic edge. The sharp edge of the IOL causes the capsular bend, in which the posterior capsule changes configuration to tightly wrap against the posterior, square edge of the IOL [31]. The edge is thought to prevent proliferating LECs from extending across the posterior lens capsule, causing PCO [32, 33]. This configuration, a discontinuous sharp square bend with tight wrapping of the posterior capsule around the sharp posterior lens optic, effectively inhibits LEC migration onto the posterior lens capsule. New IOL designs incorporating this feature must have a sharp posterior edge for 360° without interruption by the haptics. Three-hundred sixty-degree coverage can help delay or prevent PCO proliferation from breaching the barrier at the optic–haptic junction [34].

16.1.3 PCO Prevention by LEC Removal: Pharmacological Treatments

Other methods aim to completely eliminate the LECs from the lens capsular bag using capsular irrigation with drugs, such as antimetabolites, toxic monoclonal antibodies, and other agents. Early attempts to clean the capsule from LECs included irrigation with ethylenediaminetetraacetic acid (EDTA) and trypsin. EDTA is a calcium chelator agent and acts by disrupting the connection between epithelial cells and the basement membrane. Trypsin is a proteolytic enzyme that also disrupts cell bonding. These were both effective but did unacceptable damage to the surrounding structures [35]. Using a closed bag technique with EDTA, acetic acid, 5-fluorouracil, mitomycin C, and distilled water produced similar results in rabbits at the end of the study without eliminating complicating factors of inflammation, fibrosis, and edema [36]. Additional studies used the proteolytic enzyme Dispase to loosen cell junctional complexes in addition to EDTA and minimal irrigation and aspiration; negligible damage to zonules and corneal endothelium was found [37, 38]. Another promising route targets cell signaling at the molecular level. Inhibition of growth and migration modulator, Src kinase, effectively blocked proliferation and migration of LECs, as well as the fibrous metaplasia transition of PCO in chick lens capsular bags [39].

The PerfectCapsule is a device that attaches to and seals the anterior capsule opening and isolates the capsule from the remainder of the anterior chamber. This allows positive inflation of the capsule and a closed system for irrigation of the capsular bag with pharmaceutical solutions to treat residual LECs. Sealed irrigation of the bag would protect the remainder of the anterior segment and minimize damage to additional intraocular structures. It also allows the pharmaceutical solutions to reach the LECs in the fornices of the capsule by expanding the entire bag with positive pressure [40, 41]. However, vacuum seal may not be possible in all patients depending on anterior chamber depth and pupil size [42]. In the first clinical attempts of this device, distilled water was chosen as an irrigant for minimal collateral damage in case of vacuum failure. Distilled water has been shown to lyse LECs in previous studies [43, 44]; however, it was not shown to be effective in clinical use [42].

16.1.4 PCO Prevention by LEC Removal: Mechanical Methods

LECs can also be mechanically removed during cataract surgery. Hydrodissection separates the cortex from the capsule [14]. Standard irrigation/aspiration (I/A) phacoemulsification instruments can be used to polish the lens capsule of LECs. Rough phacoemulsification tips or round cannulas [45] can also manually scrape LECs off the anterior lens capsule and the fornix. Various loop-type devices have also been tried to scrape the LECs from the inside surface of the capsular bag. Ultrasound cleaning is considered to be the most effective technique when mechanically removing the LECs [46]. Another promising development is the

AquaLase, a liquefaction device for the Infiniti Vision System (Alcon Laboratories) that is thought to be safer in extraction of softer cataracts. The AquaLase uses micropulses of heated balanced salt solution to break up the cataract and uses technology that eliminates incision burn [47]. Additionally, the pulse method may physically wash LECs off the capsule [48]. The increased safety of the AquaLase makes breaking the capsule more difficult, allowing for better capsule polishing. Together, these features may help prevent or delay PCO. The challenge remains, however, to reach the regenerative cells in the capsular fornix. During surgery, the iris blocks the visualization of these cells, prohibiting their complete mechanical removal.

16.2 Laser Removal of LECs

A modified Nd:YAG handpiece has been used to remove LECs [49, 50]. The modified Nd:YAG laser photolysis handpiece is coupled with a modification of the laser photolysis system (ARC Laser GmbH). Preliminary porcine studies using the modified laser photolysis system showed confocal laser scanning microscopy and histological evidence of complete ablation of LECs from the anterior lens capsule [51].

16.2.1 Laser and Mechanism

The present design of the laser photolysis unit was developed to treat the inner surface of the lens capsular bag. It is based on a modification of the Dodick Laser PhotoLysis System designed initially for laser cataract phacoemulsification [49] and also used in laboratory experiments to clean the lens capsule. A sheathed fiber-optic bundle leaves the laser console and is connected to the handpiece. Infusion fluid is connected through tubing to the handpiece. The Q-switched Nd:YAG laser has an 8 ns pulse length. The energy per pulse can be set from 4 to 12 mJ, commonly used at 7 mJ. The application frequency of the laser pulses is also variable from 1 to 20 Hz in 1 Hz increments, commonly used at 2 Hz. The laser light is delivered into the handpiece via a fiber-optic cable with a diameter of 283 mm in conjunction with an irrigation channel.

The fiber terminates opposite an oblique titanium plate mounted adjacent to the exit aperture. The laser light travels down the cable, hits the surface of the titanium plate with a spot size of 400–500 nm, is absorbed, and creates plasma with a shock wave that exits the handpiece through a 0.65 mm round opening. The same opening functions as the irrigation port, with the infusion fluid entering the handpiece and flowing around the cable and out the opening. The handpiece tip is 1.05 mm in diameter (Fig. 16.1). The laser photolysis beam itself does not exit the handpiece and does not interact with the tissue directly. The titanium surface, which is hit by the laser radiation, does not come in contact with the tissue. Laboratory experiments with this unit show that with 10 mJ laser power, the shock wave is

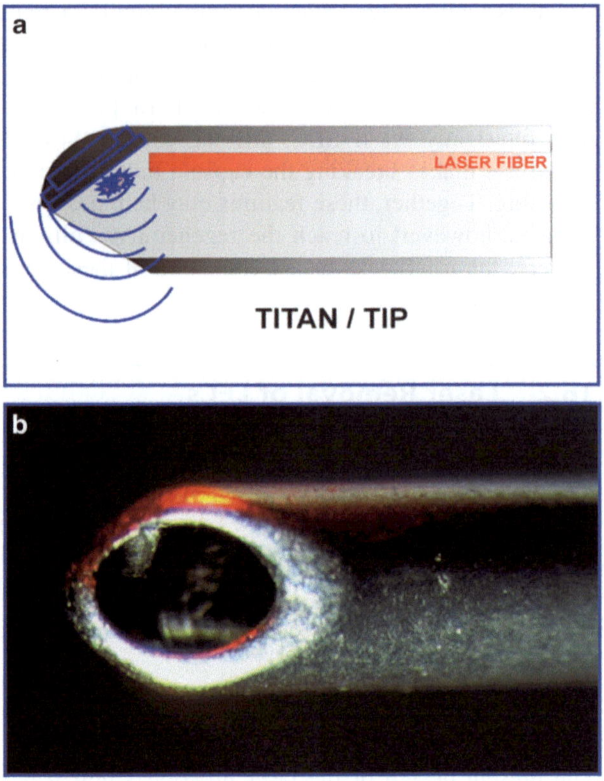

Fig. 16.1 Schematic drawing (**a**) and gross photograph (**b**) showing the handpiece tip of the laser photolysis system (provided by ARC Laser GmbH)

able to impact and remove cells within a 10.0 mm^2 area with the tip held 1.0 mm from the surface. This equals a treatment area of more than 3.0 mm diameter. Thus, when the tip of the instrument is placed through a capsulorhexis beneath the anterior capsule, the shock wave is likely to reach the capsule fornix, where it can destroy the germinal LECs. With less than 4 mJ pulse energy, the shock wave formation is negligible and not efficient. More than 10 mJ laser energy leads to fast destruction of the fiber.

16.2.2 Cadaver Eye and Immunohistochemistry

In a study of human eye-bank eyes [52], the modified Nd:YAG laser photolysis system removed LECs from the inner surface of the anterior lens capsule and the capsular fornix (Fig. 16.2). This was confirmed by light microscopy (LM) and transmission electron microscopy (TEM). In the control-untreated areas, the normal anatomy of LECs and capsular bag were present on both LM and TEM. In areas in which the lens epithelium was treated but not effectively removed by aspiration, small blebs formed within the LECs and cellular debris persisted at the base of the blebs. Similarly, in areas in which the LECs were treated with attempted aspiration

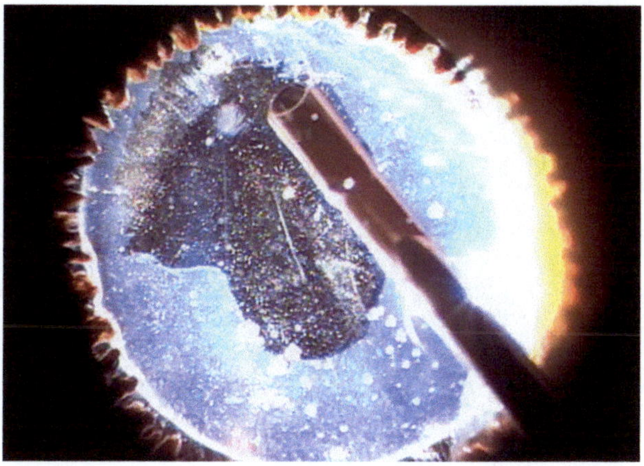

Fig. 16.2 Removal of lens epithelial cells from a human cadaver eye prepared according to the Miyake–Apple technique, by using the laser photolysis system. Published in Mamalis et al. [52]

using standard bimanual I/A handpieces, there was remnant cellular debris from the LECs where the epithelium was not completely removed (Fig. 16.3). These findings show the importance of using laser treatment and I/A in combination to effectively treat and remove LECs.

In contrast, areas of the lens capsule in which the LECs were treated by the laser photolysis system and aspiration were completely clear, with no LECs or cellular debris on LM or on TEM (Figs. 16.4 and 16.5). In areas treated by the Nd:YAG photolysis system, neither laminin nor fibronectin was present on the surface of the lens capsule when evaluated by immunohistochemical analysis (Fig. 16.6). In a clinical setting, this indicates that the cells would have no "carpet" on which to slide and migrate and no adhesion molecules upon which to take up residence [52].

16.2.3 Clinical Studies

Clinical studies of the use of the modified Nd:YAG laser photolysis capsule cleaning device in cataract surgery found the laser was successful in removing LECs from the anterior portion of the capsular bag. The inner surface of the nasal half of the capsule in 17 eyes was treated with the laser after standard phacoemulsification under direct visualization through the dilated pupil; the temporal half of the bag was not treated. When one half of the capsular bag was treated, the anterior and posterior capsule remained clear in most eyes for 2 years; the untreated half of the capsular bag became opaque within a few months after surgery (Fig. 16.7). In the clinical series, treatment with the laser photolysis system prevented not only the capsule from opacifying in the areas that received treatment prevented LEC migration onto the posterior capsule, preventing PCO from spreading from the untreated areas of the capsule to the treated capsule [53].

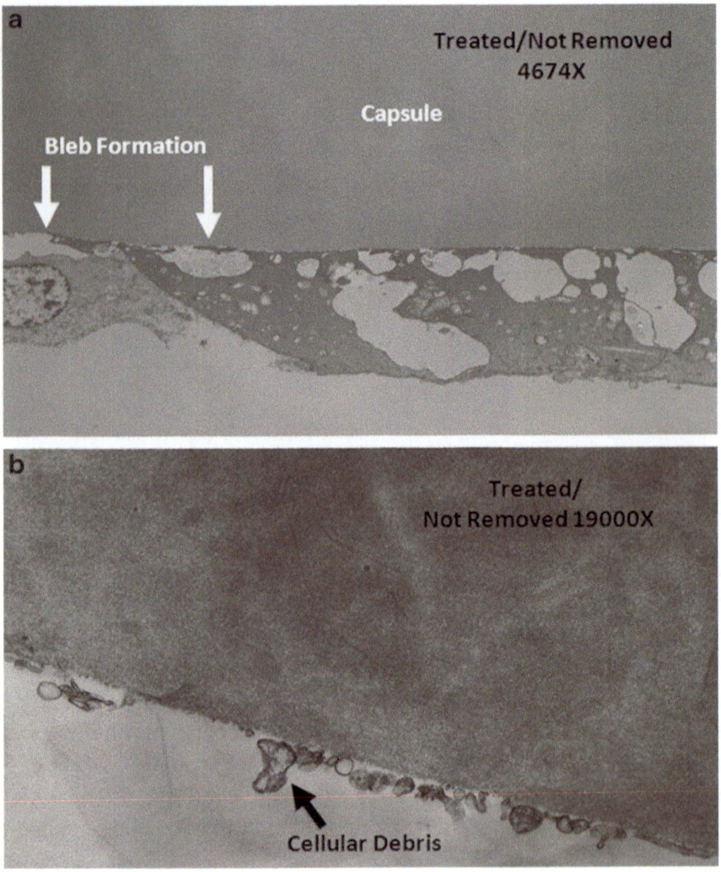

Fig. 16.3 Transmission electron photomicrographs (**a** and **b**) showing the inner surface of capsular bags of a cadaver eyes which were treated with the laser photolysis system. Remnant cellular debris were not effectively removed by aspiration (provided by Hans Grossniklaus, M.D., Emory Eye Center, Emory University, Atlanta, GA)

16.3 Laser Photolysis and PCO Prevention

Posterior migration of LECs after cataract surgery causes opacification of the posterior capsule. Basement membrane and extracellular matrix (ECM) glycoproteins play a role in the development of PCO by providing a substrate for LEC adhesion and may promote LEC migration. Laminin interacts with LECs and is one of the major constituents of the lens basement membrane, the thickest basement membrane in the body. Migration and adhesion of LECs occur in response to lens capsule proteins collagen type IV and laminin. Maximum LEC migration is also promoted by the ECM glycoprotein fibronectin [54], which is not typically found in the adult lens capsule. However, fibronectin can be found in the

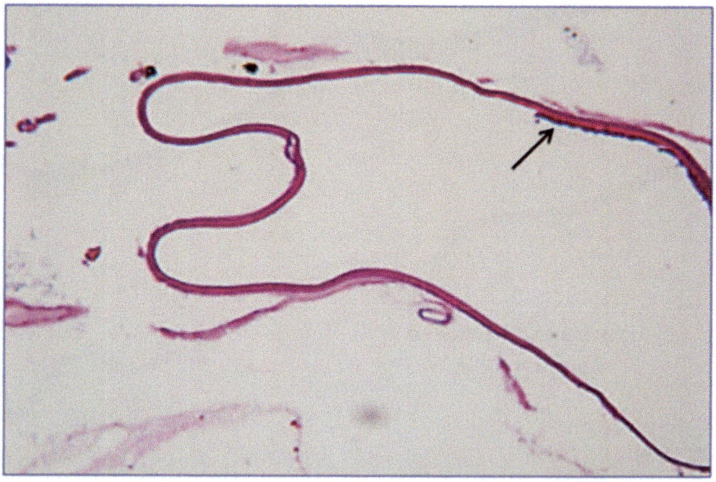

Fig. 16.4 Light photomicrograph of a cadaver eye capsular bag showing successful removal of lens epithelial cells from the fornix by treatment with the laser photolysis system. Persistent cells can be seen in the untreated anterior subcapsular area (*arrow*). Published in Mamalis et al. [52]

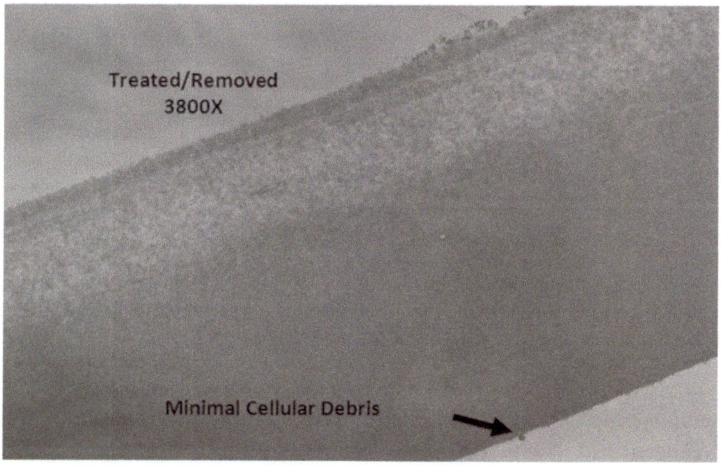

Fig. 16.5 Transmission electron photomicrograph showing the inner surface of a cadaver eye capsular bag which was treated with the laser photolysis system. Remnant cellular debris were effectively removed by aspiration (provided by Hans Grossniklaus, M.D., Emory Eye Center, Emory University, Atlanta, GA)

anterior segment after cataract surgery due to the inflammatory response. Additionally, the inflammatory response in which LECs transform into myofibroblasts can produce additional fibronectin. LEC integrin adhesion receptors can migrate by binding to fibronectin and gaining traction [55].

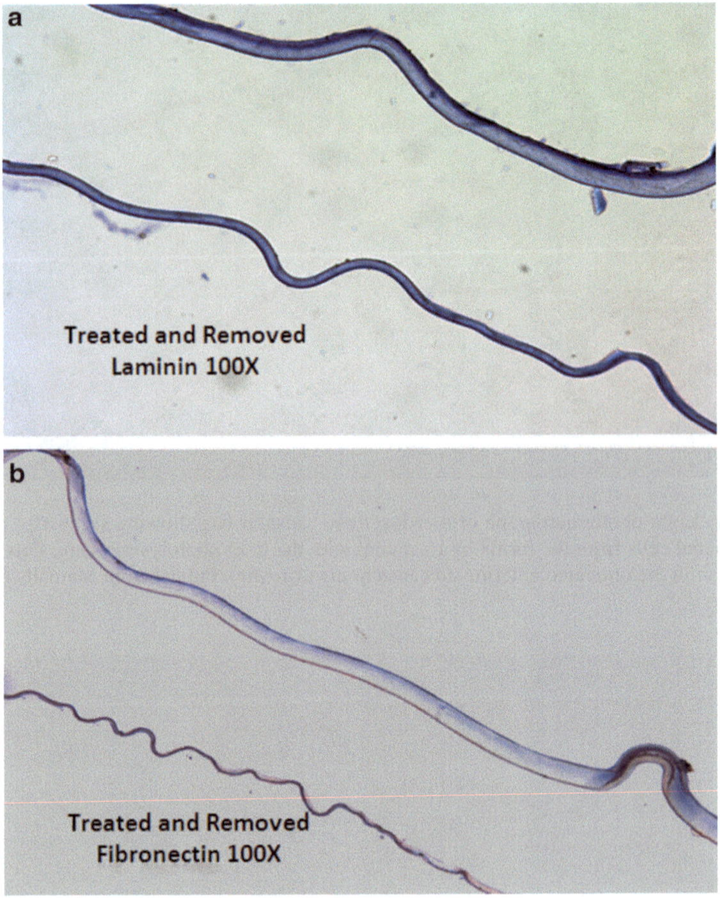

Fig. 16.6 Light photomicrographs of a cadaver eye capsular bag after removal of lens epithelial cells with the laser photolysis system. No immunohistochemical staining is shown for laminin (**a**) or for fibronectin (**b**). (**a** and **b**) published in Mamalis et al. [52]

Collagen type IV is a normal basement membrane component of the lens capsule. Fibronectin has a functional domain that binds to collagen. Because the lens capsule is made of collagen [56], fibronectin could be a mediator for adhesion. Fibronectin, laminin, and collagen type IV also adhere to the surface of IOLs. Furthermore, fibronectin mediates adhesion between IOLs and the lens capsule. Proliferative connective tissue that grows between the capsule and implanted IOLs has shown the presence of these proteins [55]. Interestingly, these adhesion-type molecules are found interposed between the lens capsule and the surface of IOLs, particularly hydrophobic acrylics. The direct adhesion may prevent the potential space that LECs can grow in. This "sandwich theory" may contribute to the decreased PCO noted in the eyes with a hydrophobic acrylic IOL in the capsular bag [57, 58]. The presence of fibronectin and laminin on or in the

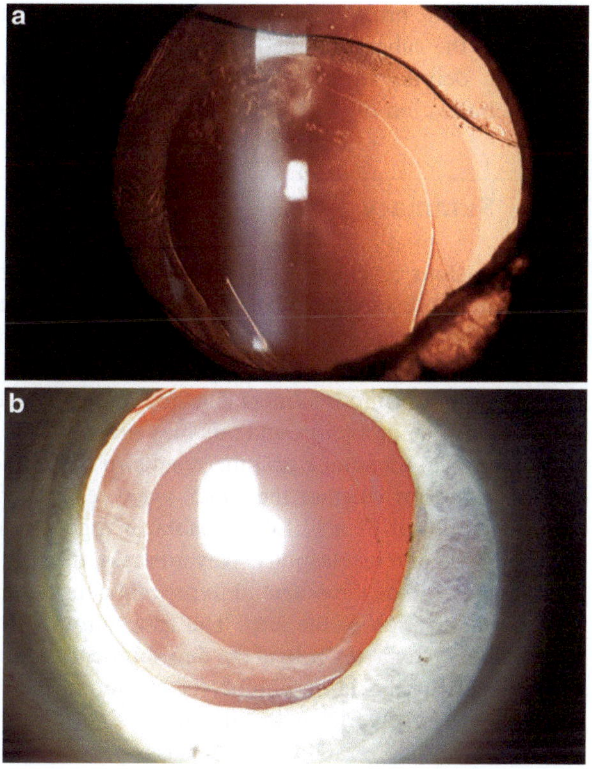

Fig. 16.7 Clinical photographs (**a** and **b**) taken approximately 2 years after surgery show capsule opacities on the untreated side (*left*) and clinically clear anterior and posterior capsules on the treated side (*right*) (provided by Wolfram Wehner, M.D., ARC Laser GmbH)

lens capsular bag may explain how capsule opacification still occurs despite cleaning of the majority of the capsule using I/A. Even a few residual LECs that are not removed from the lens capsule can migrate across an area of previously clear capsule and cause visually significant capsular opacification.

Evaluation of control and untreated areas in the human eye-bank eye study showed positive staining for laminin in the lens capsule basement membrane area with intact LECs. Some epithelial cells themselves also stained positive for laminin. The entire lens epithelium and some zonular fibers stained positive for fibronectin. In areas in which the LECs were treated solely with Nd:YAG photolysis but were not aspirated or removed, the remnant cellular debris stained positive for laminin and fibronectin. In areas in which LEC removal was attempted using solely bimanual I/A, remnant cellular debris also stained positively for fibronectin and laminin. Areas treated by both Nd:YAG photolysis and aspiration were clear of cellular debris and did not stain for fibronectin or laminin.

The results in the cadaver eye study clearly show that the laser photolysis system facilitates complete removal of LECs from the inner surface of the anterior lens capsule as well as the capsular fornix without leaving LEC debris. Immunohistochemical staining did not stain for fibronectin or laminin, confirming the absence of the two adhesive proteins in treated areas. This is thought to be the mechanism of inhibition of the proliferation of residual LECs along the treated lens capsule, thus preventing the subsequent generation of PCO. Use of aspiration without the laser left LEC debris, which contains laminin and fibronectin and may explain re-epithelialization of the capsule and the onset of PCO [52].

16.4 Summary

The conclusions reached in the cadaver eye study are reinforced clinically in the clinical studies discussed above [53]. The halves of eyes treated with Nd:YAG laser photolysis and aspiration remained clear for 2 years post-cataract surgery while the untreated halves opacified within a few months. The Nd:YAG laser photolysis system is a promising treatment for the removal of LECs as well as adhesion and migration of glycoproteins, and this may prevent opacification of the capsular bag.

The surgical technique and the results in the clinical trial were not definitive. Questions remain whether the shock wave reaches the capsule fornix, if there is complete removal of all LECs capable of regeneration, and if the resulting lens capsule will retain its elasticity. Nevertheless, the results indicate that the Nd:YAG photolysis laser has the potential to prevent capsule opacification and fibrosis after phacoemulsification of the lens. The adhesion-type molecules, laminin and fibronectin, likely play a key role in the development and prevention of PCO. The safety and efficacy of the laser photolysis technique can be determined by a formal prospective clinical trial with bilateral design in which the eyes of patients with similar cataracts are randomized to laser photolysis or standard I/A, with photographic comparison of capsule opacities.

References

1. Schaumberg DA, Dana MR, Christen WG et al (1998) A systematic overview of the incidence of posterior capsule opacification. Ophthalmology 105:1213–1221
2. Metge P, Cohen H, Graft F (1989) Intercapsular intraocular lens implantation in children: 35 cases. Eur J Implant Refract Surg 1:169–173
3. Bertelmann E, Kojetinsky C (2001) Posterior capsule opacification and anterior capsule opacification. Curr Opin Ophthalmol 12:35–40
4. Stark WJ, Worthen D, Holladay JT et al (1985) Neodymium:YAG lasers; an FDA report. Ophthalmology 92:209–212
5. Apple DJ, Solomon KD, Tetz MR et al (1992) Posterior capsular opacification. Surv Ophthalmol 37:73–116
6. Sunderraj P, Villada JR, Joyce PW et al (1992) Glare testing in pseudophakes with posterior capsule opacification. Eye 6:411–413
7. Claesson M, KlarCn L, Bechman C et al (1994) Glare and contrast sensitivity before and after Nd:YAG laser capsulotomy. Acta Ophthalmol 72:27–32
8. Zaugg B, Werner L, Neuhann T et al (2010) Clinicopathologic correlation of capsulorhexis phimosis with anterior flexing of single-piece hydrophilic acrylic intraocular lens haptics. J Cataract Refract Surg 36(9):1605–1609
9. Kumar A, Freeman M, Kumar V et al (2008) In the bag IOL dislocation following uncomplicated phacoemulsification. Cont Lens Anterior Eye 31(2):103–106

10. Cavallini GM, Masini C, Campi L et al (2008) Capsulorhexis phimosis after bimanual microphacoemulsification and in-the-bag implantation of the Akreos MI60 intraocular lens. J Cataract Refract Surg 34:1598–1600
11. Qatarneh D, Hau S, Tuft S (2010) Hyperopic shift from posterior migration of hydrophilic acrylic intraocular lens optic. J Cataract Refract Surg 36:161–163
12. Michael K, O'Colmain U, Vallance JH et al (2010) Capsule contraction syndrome with haptic deformation and flexion. J Cataract Refract Surg 36:686–689
13. Fine IH (1992) Cortical cleaving hydrodissection. J Cataract Refract Surg 18:508–512
14. Vasavada AR, Dholakia SA, Raj SM et al (2006) Effect of cortical cleaving hydrodissection on posterior capsule opacification in age-related nuclear cataract. J Cataract Refract Surg 32:1196–1200
15. Peng Q, Visessook N, Apple DJ et al (2000) Surgical prevention of posterior capsule opacification: 3. The intraocular lens barrier effect as a second line of defense. J Cataract Refract Surg 26:198–213
16. Menapace R, Di Nardo S (2006) Aspiration curette for anterior capsule polishing: laboratory and clinical evaluation. J Cataract Refract Surg 32:1997–2003
17. Hanson RJ, Rubinstein A, Sarangapani S et al (2006) Effect of lens epithelial cell aspiration on postoperative capsulorhexis contraction with the use of the AcrySof intraocular lens; randomized clinical trial. J Cataract Refract Surg 32:1621–1626
18. Sacu S, Menapace R, Wirtitsch M et al (2004) Effect of anterior capsule polishing on fibrotic capsule opacification: three-year results. J Cataract Refract Surg 30:2322–2327
19. Menapace R, Wirtitsch M, Findl O et al (2005) Effect of anterior capsule polishing on posterior capsule opacification and neodymium:YAG capsulotomy rates: three-year randomized trial. J Cataract Refract Surg 31:2067–2075
20. Liu X, Cheng B, Zheng D et al (2010) Role of anterior capsule polishing in residual lens epithelial cell proliferation. J Cataract Refract Surg 36(2):208–214
21. Peng Q, Apple DJ, Visessook N et al (2000) Surgical prevention of posterior capsule opacification. Part 2: enhancement of cortical cleanup by focusing on hydrodissection. J Cataract Refract Surg 26:188–197
22. Apple DJ, Peng Q, Visessook N et al (2000) Surgical prevention of posterior capsule opacification: 1. How are we progressing in eliminating this complication of cataract surgery? J Cataract Refract Surg 26:180–187
23. Ram J, Apple DJ, Peng Q et al (1999) Update on fixation of rigid and foldable posterior chamber intraocular lenses: II. Choosing the correct haptic fixation and intraocular lens design to help eradicate posterior capsule opacification. Ophthalmology 106:891–900
24. Aykan U, Bilge AH, Karadayi K et al (2003) The effect of capsulorhexis size on development of posterior capsule opacification: small (4.5 to 5.0 mm) versus large (6.0 to 7.0 mm). Eur J Ophthalmol 13(6):541–545
25. Ravalico G, Tognetto D, Palomba M et al (1996) Capsulorhexis size and posterior capsule opacification. J Cataract Refract Surg 22(1):98–103
26. Hollick EJ, Spalton DJ, Meacock WR (1999) The effect of capsulorhexis size on posterior capsular opacification: one-year results of a randomized prospective trial. Am J Ophthalmol 128(3):271–279
27. Wejde G, Kugelberg M, Zetterström C (2004) Position of anterior capsulorhexis and posterior capsule opacification. Acta Ophthalmol Scand 82:531–534
28. Langwińska-Wośko E, Broniek-Kowalik K, Szulborski K (2011) The impact of capsulorhexis diameter, localization and shape on posterior capsule opacification. Med Sci Monit 17:CR577–CR582
29. Hayashi K, Hayashi H, Nakao F, Hayashi F (1997) Reduction in the area of the anterior capsule opening after polymethylmethacrylate, silicone, and soft acrylic intraocular lens implantation. Am J Ophthalmol 123:441–447
30. Kimura W, Yamanishi S, Kimura T et al (1998) Measuring the anterior capsule opening after cataract surgery to assess capsule shrinkage. J Cataract Refract Surg 24:1235–1238

31. Nishi O, Nishi K, Akura J (2002) Speed of capsular bent formation at the optic edge of acrylic, silicone and poly(methyl methacrylate)lenses. J Cataract Refract Surg 28:431–437
32. Auffarth GU, Golescu A, Becker KA et al (2003) Quantification of posterior capsule opacification with round and sharp edge intraocular lenses. Ophthalmology 110:772–780
33. Nishi O (1999) Posterior capsule opacification. Part I: experimental investigations. J Cataract Refract Surg 25:106–117
34. Nixon DR, Apple DJ (2006) Evaluation of lens epithelial cell migration in vivo at the haptic-optic junction of a one-piece hydrophobic acrylic intraocular lens. Am J Ophthalmol 142:557–562
35. Humphry RC, Davies EG, Jacob TJC et al (1988) The human anterior lens capsule – an attempted chemical debridement of epithelial cells by ethylenediaminetetracetic acid (EDTA) and trypsin. Br J Ophthalmol 72:406–408
36. Fernandez V, Fragoso MA, Billotte C et al (2004) Efficacy of various drugs in the prevention of posterior capsule opacification: experimental study of rabbit eyes. J Cataract Refract Surg 30:2598–2605
37. Nishi O, Nishi K, Hikida M (1991) Removal of lens epithelial cells by dispersion with enzymatic treatment followed by aspiration. Ophthalmic Surg 22:444–450
38. Nishi O, Nishi K, Hikida M (1993) Removal of lens epithelial cells following loosening of the junctional complex. J Cataract Refract Surg 19:56–61
39. Walker JL, Wolff IM, Zhang L et al (2007) Activation of SRC kinases signals induction of posterior capsule opacification. Invest Ophthalmol Vis Sci 48(5):2214–2223
40. Maloof AJ, Neilson G, Milverton EJ et al (2003) Selective and specific targeting of lens epithelial cells during cataract surgery using sealed-capsule irrigation. J Cataract Refract Surg 29:1566–1568
41. Maloof AJ, Pandey SK, Neilson G et al (2005) Selective death of lens epithelial cells using demineralized water and Triton-X-100 with PerfectCapsule sealed-capsule irrigation; a histologic study in rabbit eyes. Arch Ophthalmol 123:1378–1384
42. Rabsiber TM, Limberger IJ, Reuland AJ et al (2007) Long-term results of sealed capsule irrigation using distilled water to prevent posterior capsule opacification: a prospective clinical randomized trial. Br J Ophthalmol 91:912–915
43. Crowston JG, Maloof AJ, Healey PR et al (2003) Water modulation of lens epithelial cells during cataract surgery. J Cataract Refract Surg 29:2464–2465
44. Crowston JG, Healey PR, Hopley C et al (2004) Water-mediated lysis of lens epithelial cells attached to lens capsule. J Cataract Refract Surg 30:1102–1106
45. Isakov I, Madjarov B, Bartov E (1995) Safe method for cleaning the posterior lens capsule. J Cataract Refract Surg 21(4):371–372
46. Mathey CF, Kohnen TB, Ensikat HJ et al (1994) Polishing methods for the lens capsule: histology and scanning electron microscopy. J Cataract Refract Surg 20(1):64–69
47. Nekolová J, Jirásková N, Pozlerová J et al (2009) Three-year follow-up of posterior capsule opacification after AquaLase and NeoSoniX phacoemulsification. Am J Ophthalmol 148 (3):390–395
48. Mackool RJ, Brint SF (2004) AquaLase: a new technology for cataract extraction. Curr Opin Ophthalmol 15:40–43
49. Dodick JM, Christiansen J (1991) Experimental studies on the development and propagation of shock waves created by interaction of short Nd:YAG laser pulses with a titanium target; possible implications for Nd:YAG laser phacolysis of the cataractous human lens. J Cataract Refract Surg 17:794–797
50. Dodick JM, Sperber LTD, Lally JM et al (1993) Neodymium-YAG laser phacolysis of the human cataractous lens. Arch Ophthalmol 111:903–904
51. Pollhammer M, Meiller R, Rummelt C et al (2007) In situ ablation of lens epithelial cells in porcine eyes with the laser photolysis system. J Cataract Refract Surg 33:697–701
52. Mamalis N, Grossniklaus HE, Waring GO 3rd et al (2010) Ablation of lens epithelial cells with a laser photolysis system: histopathology, ultrastructure, and immunochemistry. J Cataract Refract Surg 36(6):1003–1010

53. Wehner W, Waring GO 3rd, Mamalis N et al (2010) Prevention of lens capsule opacification with ARC neodymium:YAG laser photolysis after phacoemulsification. J Cataract Refract Surg 36(6):881–884
54. Olivero DK, Furcht LT (1993) Type IV collagen, laminin, and fibronectin promote the adhesion and migration of rabbit lens epithelial cells in vitro. Invest Ophthalmol Vis Sci 34:2825–2834
55. Linnola RJ, Sund M, Ylönen R, Pihlajaniemi T (1999) Adhesion of soluble fibronectin, laminin, and collagen type IV to intraocular lens materials. J Cataract Refract Surg 25:1486–1491
56. Kappelhof JP, Pameyer JH, DeJong PTVM (1986) The proteinaceous coating and cytology of implant lenses in rabbits. Am J Ophthalmol 102:750–758
57. Linnola RJ, Werner L, Pandey SK et al (2000) Adhesion of fibronectin, vitronectin, laminin, and collagen type IV to intraocular lens materials in pseudophakic human autopsy eyes. Part I: histologic sections. J Cataract Refract Surg 26:1792–1806
58. Linnola RJ, Werner L, Pandey SK et al (2000) Adhesion of fibronectin, vitronectin, laminin, and collagen type IV to intraocular lens materials in pseudophakic human autopsy eyes. Part 2: explanted intraocular lenses. J Cataract Refract Surg 26:1807–1818

Part V

Intraocular Lense/Devices and PCO

PCO Prevention: IOL Material Versus IOL Design

17

Caleb Morris, Liliana Werner, and Manfred Tetz

Abstract

Intraocular lens characteristics may influence the outcome of posterior capsule opacification. Different studies have described the possible roles of material and design on the prevention of this complication, which are discussed in this chapter. These include studies evaluating material properties, particularly adhesive properties leading to different patterns of protein adsorption on the lens surfaces. The most important intraocular lens feature in posterior capsule opacification prevention with in-the-bag fixated intraocular lenses was found to be the presence of a square edge on the posterior optic surface. Nevertheless, studies evaluating the microstructure of the edges of currently available foldable intraocular lenses found that all square edges in the market are not equally square. As a group, hydrophilic acrylic lenses were found to have less square edges than hydrophobic acrylic and silicone lenses. Furthermore, animal and clinical studies demonstrated that the square edge should be present for 360 around the lens optic, for maximal efficiency in terms of prevention of posterior capsule opacification.

Keywords

Fibronectin • Hydrophobic acrylic • Intraocular lens • Posterior capsule opacification • Square edge

C. Morris, M.D. • L. Werner, M.D., Ph.D. (✉)
John A. Moran Eye Center, University of Utah, Salt Lake City, UT, USA
e-mail: liliana.werner@hsc.utah.edu

M. Tetz, M.D.
Eye Center at the Spreebogen and Berlin Eye Research Institute (BERI), Berlin, Germany

S. Saika et al. (eds.), *Lens Epithelium and Posterior Capsular Opacification*,
DOI 10.1007/978-4-431-54300-8_17, © Springer Japan 2014

17.1 Introduction and Background

Apple et al. identified 6 factors considered instrumental in the prevention of posterior capsule opacification (PCO) after evaluating more than 5,000 pseudophakic human eyes obtained postmortem [1]. Three of the factors were related to the surgical technique, while the other 3 were related to the intraocular lens (IOL) choice. The IOL-related factors were as follows: biocompatible IOL to reduce stimulation of cellular proliferation, maximal IOL optic-posterior capsule contact (angulated haptic; "bioadhesive" biomaterial to create a "shrink wrap"), and IOL optic geometry with a square, truncated optic edge. It appears therefore that both, IOL material and IOL design, may influence the outcome of PCO. However, it is important to understand their relative roles and participation in the prevention of this complication.

17.2 IOL Material in the Prevention of PCO

After the introduction of the hydrophobic acrylic AcrySof lens (Alcon Laboratories) in the market, with reported low rates of PCO and neodymium: YAG (Nd:YAG) laser posterior capsulotomy, great attention was placed on the adhesive characteristic of this hydrophobic acrylic material. Oshika et al. found that soft hydrophobic acrylic IOLs adhere better to the lens capsule than polymethyl-methacrylate (PMMA) IOLs; silicone IOLs did not adhere at all in their study [2]. Linnola et al. found the same attachment pattern in a rabbit corneal tissue culture study in which hydrophobic acrylic AcrySof IOLs attached better than PMMA, heparin-surface-modified (HSM) PMMA, silicone, or hydrophilic acrylic IOLs [3]. In an in vitro study by Nagata et al., IOLs were pressed against a collagen film in a fortified balanced salt solution; the pulling out force was measured at the moment the lens detached from the film as it was pulled up [4]. Hydrophobic acrylic IOLs adhered more strongly to a collagen film than PMMA IOLs.

17.2.1 "Sandwich" Theory

The "sandwich" theory postulated by Dr. Linnola stated that an IOL made of a bioadhesive material allows a single layer of lens epithelial cells (LECs) to bond to the IOL and the posterior capsule at the same time, producing a sandwich pattern that includes the IOL, a cell monolayer, and the posterior capsule [3, 5–7]. The sealed sandwich structure would prevent further LEC proliferation. Therefore, the degree of bioactivity/bioadhesivity could explain the differences in PCO and Nd: YAG rates among IOL materials.

Different patterns of protein adhesion to the surfaces of different IOLs were found to be related to adhesive properties of certain materials. In an in vitro study, Linnola et al. used 75 PMMA, HSM PMMA, silicone, hydrophobic acrylic, and hydrophilic acrylic IOLs, which were incubated with radioactive iodine-labeled soluble fibronectin, laminin, or collagen type IV [5]. The amount of absorbed protein was measured with a gamma counter. Hydrophobic acrylic IOLs (AcrySof) showed the highest binding to fibronectin; the difference between the hydrophobic acrylic and the other materials, except PMMA, was significant. The adsorption of fibronectin to the IOL surfaces was confirmed in another in vitro study by Johnston et al. [8]. They found that surface adsorption of fibronectin was significantly greater on hydrophobic acrylic IOLs than on PMMA IOLs after 1 day and 1 week.

In two follow-up studies using pseudophakic human eyes obtained postmortem, Linnola et al. also found that the protein fibronectin is the major extracellular protein involved in the adhesion between the hydrophobic acrylic AcrySof IOL and the capsular bag [6, 7]. In the first study, 38 autopsy eyes containing PMMA, silicone, hydrophobic acrylic, or hydrophilic acrylic IOLs were assessed [6]. Histological sections were prepared from each eye, and immunohistochemical analyses were performed for fibronectin, vitronectin, laminin, and collagen type IV. One hundred fifty-two specimens (sections) were analyzed. A sandwich-like structure (anterior or posterior capsule/fibronectin/1 cell layer/fibronectin/IOL surface) was seen in 12 of 14 autopsy eyes with hydrophobic acrylic lenses, 3 of 10 with a PMMA IOL ($P = 0.0094$), 1 of 10 with a silicone IOL ($P = 0.0022$), and 0 of 4 with a hydrophilic acrylic IOL ($P = 0.0041$). The thicker fibrocellular tissue on the inner surface of the anterior or posterior capsule that was in contact with silicone IOLs was lined with collagen type IV. Vitronectin and laminin were not found at the fibrocellular tissue-IOL interface in any specimen.

In the second study, 32 autopsy eyes containing PMMA, silicone, hydrophobic acrylic, or hydrophilic acrylic IOLs were assessed [7]. The IOLs were explanted from the capsular bag, and both sides of the IOLs were immunohistochemically stained for fibronectin, vitronectin, laminin, or collagen type IV. The number of cells on the IOL surfaces was also counted. Hydrophobic acrylic IOLs had significantly more fibronectin adhering to their surfaces than PMMA ($P < 0.01$) or silicone ($P < 0.01$) IOLs. Silicone IOLs had more collagen type IV adhesion than the other IOLs ($P < 0.05$–0.06). Collective protein adhesion differed significantly between hydrophobic acrylic IOLs and PMMA and silicone IOLs, but not between PMMA and silicone IOLs. Hydrophobic acrylic IOLs also had significantly more LECs on their surfaces. This is consistent with the finding that these IOLs adhere to the capsule better than PMMA or silicone IOLs, confirming that adhesion is likely mediated and promoted through LEC production of fibronectin or comparable proteins that bind the lens to the capsular bag (Fig. 17.1).

Collagen type IV lined the cells and tissue on the surface of silicone IOLs. This might be a factor in the higher rate of anterior capsule contraction (phimosis) with silicone IOLs than with hydrophobic acrylic IOLs [9–11]. The fibrocellular tissue under the anterior capsule was encapsulated from the IOL surface by collagen type IV in eyes with silicone IOLs. This probably does not produce the same kind of capsule-IOL surface attachment as fibronectin does.

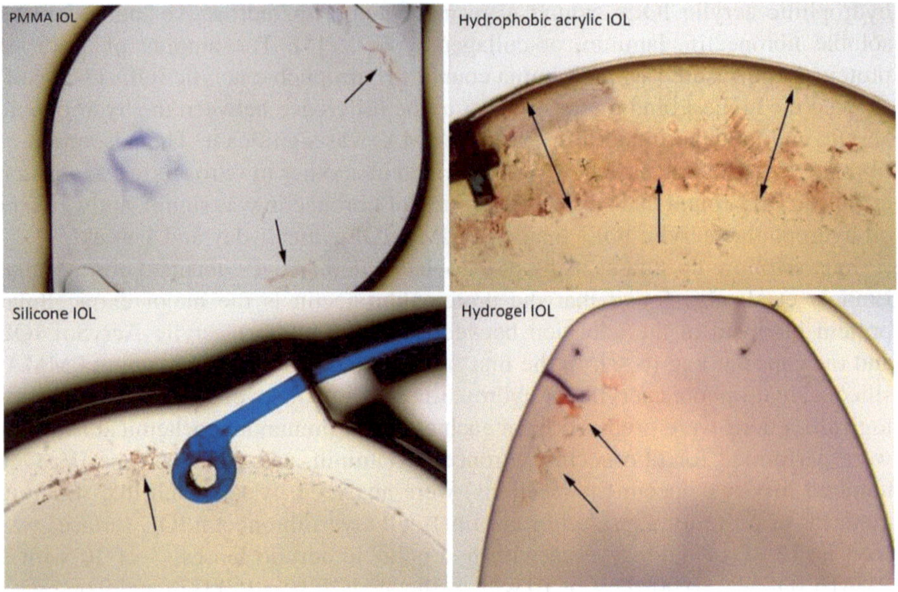

Fig. 17.1 Gross photographs of intraocular lenses explanted from cadaver eyes, manufactured from different materials. Immunohistochemical stain for fibronectin (*arrows*) on the surface of the lenses shows the presence of a more significant amount of the protein attached on the surface of the hydrophobic acrylic lens. Published in Linnola et al. [7]

17.3 IOL Design in the Prevention of PCO

After evaluation of adhesive properties of IOL materials and their effect on PCO, the focus shifted towards the possible role of the lens design in the prevention of this complication. Different studies suggested a more significant role of the square posterior optic edge as a barrier to LEC proliferation behind the lens optic, a design feature that was present in the AcrySof lens [12]. Later, other studies have shown that modern posterior chamber IOLs with a square posterior optic edge, regardless of the material used in their manufacture, have been associated with better results in terms of PCO prevention [13–17]. According to experimental studies, this may be due to the mechanical barrier effect exerted by the square edge [18, 19], contact inhibition of migrating LECs at the capsular bend created by the sharp optic edge [20, 21], higher pressures exerted by IOLs with a square-edged optic profile on the posterior capsule [22, 23], or perhaps to various mechanism combinations. This IOL design feature has generally been incorporated into modern foldable IOL designs, and it can be appropriately assessed in morphological studies using scanning electron microscopy (SEM). A series of SEM studies done in Germany at the BERI aimed to evaluate and describe different IOLs at the microedge level.

17.3.1 Preliminary Study on the IOL Microedge Structure

Tetz and Wildeck made the first attempt to evaluate and quantify, at the microscopic level, how sharp the optic edge must be to effectively prevent LECs from growing onto the posterior capsule [24]. Plano +0.0 D PMMA IOLs with 11 defined edge designs were especially manufactured for use in this in vitro preliminary study. Different edge designs were obtained by removal from the tumble-polishing machine at different times. The edge's ability to stop cell growth was evaluated by placing each IOL into cell culture and observing bovine LEC growth over 18 days on average. Experimental PMMA IOLs with different edge profiles were imaged under SEM with a standard magnification of ×500, and the area above the edge, representing the deviation from an ideal square, was calculated with a digital system based on the Evaluation of Posterior Capsule Opacification System (EPCO 2000 program). Only three groups of PMMA IOLs, those with the sharpest edge design, prevented the growth of LECs onto the visual axis of the IOL. The edge design that effectively stopped cell growth was characterized by an area above the edge, measured with the EPCO system, of 13.5 μm^2 at most.

17.3.2 Microedge Structure of Commercially Available Hydrophobic IOLs

In follow-up studies, Werner et al. used an improved methodology to evaluate the optic microedge structure of currently available IOLs manufactured from different materials and marketed as square-edged IOLs [25, 26]. In the first study, commercially available lenses manufactured from hydrophobic acrylic and silicone materials were obtained through letters sent to IOL manufacturers [25]. Generally, 2 IOLs of each design were evaluated: a +20.0 D and a +0.0 D whenever available for a particular design. In case a +0.0 D was not available, the lowest dioptric power was used for that particular design. The methodology used in that study was as follows: Each IOL was carefully removed from its original packaging with a toothless forceps and mounted on a support for SEM analysis. During SEM examination, the analysis of each optic edge was done from a perpendicular view. Photographs of the optic edge of each IOL were obtained at three magnifications: ×25 or ×100, ×300, and ×1,000. The first two magnifications were used to document the overall orientation of the specimen, and the ×1,000 magnification photographs were used for the microedge analysis. The SEM photographs of each IOL were saved as electronic, high-resolution JPEG files. They were then imported into the AutoCAD LT 2000 system (Autodesk). This program, which is commonly used in engineering and architecture, allows accurate area calculations. The first step was to adjust the scale of the photograph into the program using the reference bar incorporated on the right bottom corner of each SEM photograph. After the scale on each photograph was confirmed by measuring

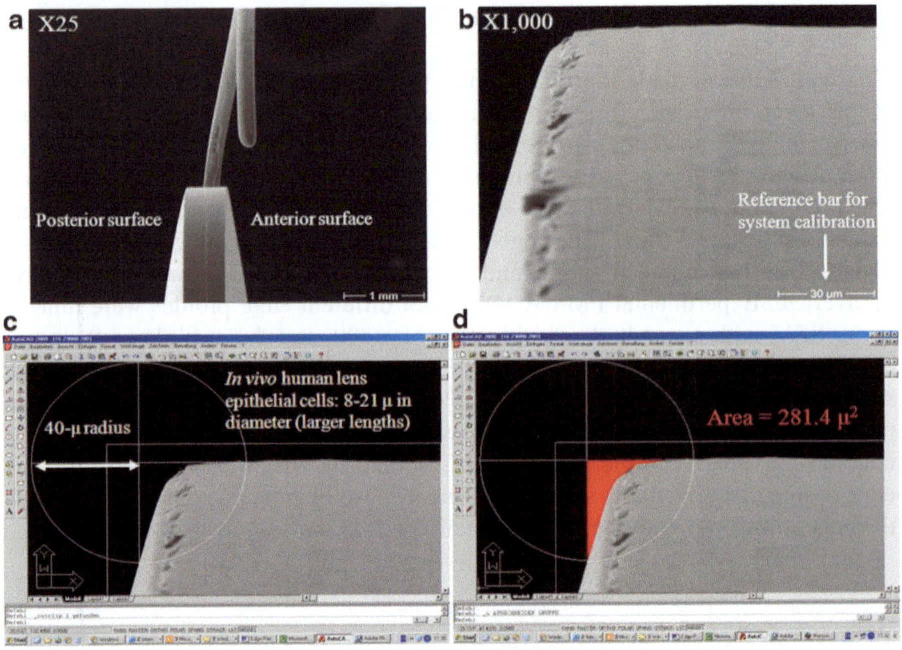

Fig. 17.2 Scanning electron microscopy and AutoCAD analyses of one intraocular lens used in the evaluation of the microedge study of commercially available lenses. (**a**) Perpendicular view of the optic edge obtained with a magnification of ×25. All IOLs were oriented with the lateral edge up and the anterior and posterior surfaces on the *right* and *left* sides, respectively. The scanning electron microscopy photographs of ×25 and ×300 helped to control the orientation of the specimens. In this case, the silicone lens is equiconvex; that is, the distance between the *right edge* and the anterior surface and between the *left edge* and the posterior surface (*bottom* of photograph) is the same. (**b**) Perpendicular view of the lateral-posterior optic edge obtained with ×1,000 magnification. The 30-μm bar was used to adjust the scale of the photograph into the AutoCAD program. (**c** and **d**) AutoCAD screens of the analyses of the photograph in B using a 40-μm radius circle. The magnification of the photographs on the screens was adjusted to incorporate the entire *bottom-right* quadrant of each *circle*. The area in *red* corresponds to the deviation from the ideal square. Published in Werner et al. [25]

the reference bar and obtaining the corresponding value, a reference circle of known radius, divided into 4 quadrants by 2 perpendicular lines passing through its center, was projected onto the photograph. The position of the circle was adjusted so that the end of both perpendicular lines touched the lateral and posterior IOL optic edges. The area of the lateral-posterior IOL edge deviating from a perfect square defined by the 2 perpendicular lines inside the reference circle was easily delineated using the computer mouse. The measurement of the area was then calculated by the program and provided in square microns. The minimum circle radius size of 40 μm was chosen as a function of the size of the human LEC (Fig. 17.2).

The commercially available IOLs were compared with an experimental square-edged PMMA IOL (reference IOL) manufactured for use in the preliminary in vitro

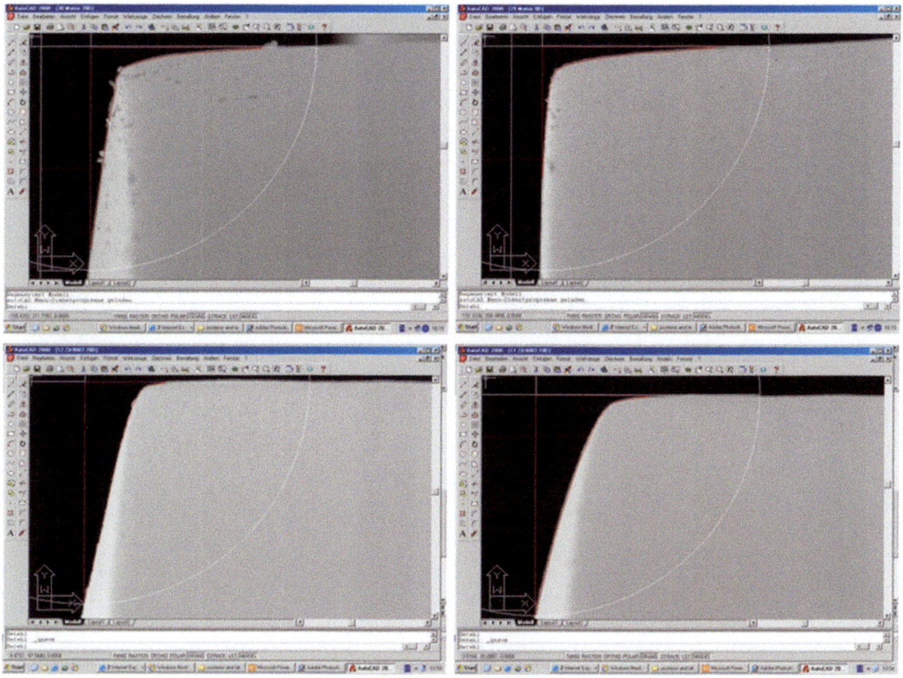

Fig. 17.3 Photographs from AutoCAD screens of the analyses of scanning electron microscopy photographs of four different hydrophobic acrylic intraocular lenses. The area delineated in *red* shows the deviation from the ideal square, which varies among the lenses. Published in Werner et al. [25]

study [24]. The edge design of the experimental IOL effectively stopped cell growth in culture. Two silicone IOLs (+20.0 D and +0.0 D) manufactured with round optic edges were used as controls. For the square-edged PMMA IOL, the value of the area measured with the AutoCAD system with the 40-μm radius circle was 34.0 μm^2. The respective value for the +20.0 D control silicone IOL was 729.3 and for the +0.0 D control silicone IOL, 727.3 μm^2.

There was a large variation in the deviation area from a perfect square, not only between different IOL designs but also between different powers of the same design (Figs. 17.3 and 17.4). Considering the measurements done with the 40-radius circle, the values for hydrophobic acrylic ($N = 19$) and silicone ($N = 11$) lenses were 183.38 ± 82.18 and 74.39 ± 88.54 μm^2, respectively (all dioptric powers evaluated included). The hydrophobic IOLs used, labeled as square-edged IOLs, had an area of deviation from a perfect square ranging from 4.8 to 338.4 μm^2. Of the 30 commercially available square-edged, hydrophobic IOLs evaluated, only seven silicone lenses of five designs had area values that were smaller than, or close to, those of the reference square-edged PMMA IOL [25].

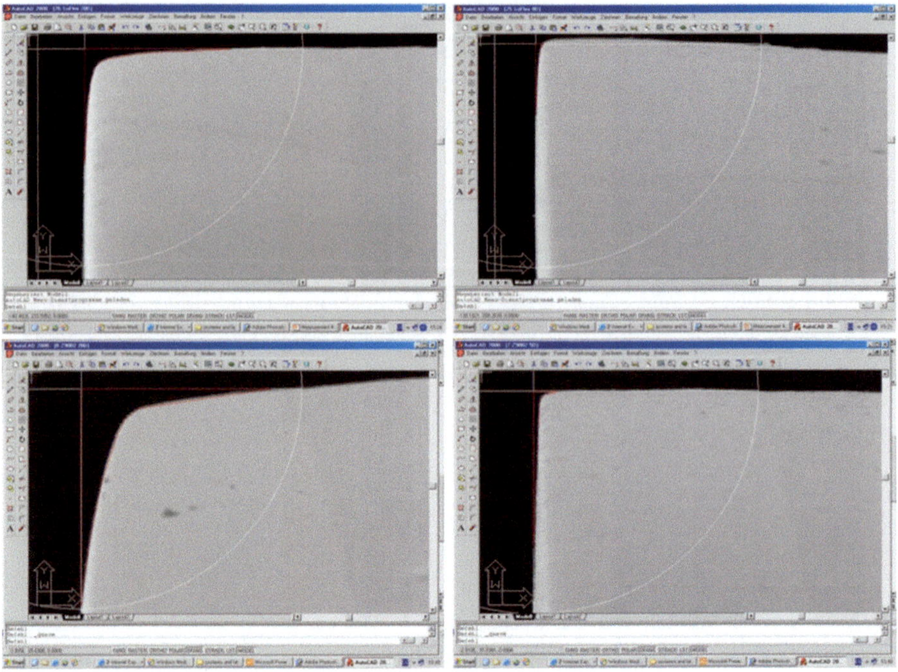

Fig. 17.4 Photographs from AutoCAD screens of the analyses of scanning electron microscopy photographs of four different silicone intraocular lenses. The area delineated in *red* shows the deviation from the ideal square, which varies among the lenses. Published in Werner et al. [25]

17.3.3 Microedge Structure of Commercially Available Hydrophilic IOLs

In the second study on commercially available IOLs marketed as square-edged lenses, hydrophilic acrylic lenses were evaluated using the same methodology as described above [26]. However, it is important to highlight that an environmental SEM technique was used for the hydrophilic acrylic lenses in order to evaluate them under low vacuum conditions, preventing dehydration and thus artifactual alteration of the edge. The microedge structure of modern hydrophilic IOLs, most of which have water content in the vicinity of 26 %, may be significantly modified during the vacuum required in standard SEM procedures.

The study of hydrophilic acrylic lenses had an area of deviation from a perfect square ranging from 60.84 to 871.51 μm^2 for the +20 D lenses (379.01 ± 188.26; $N = 24$), and from 35.52 to 826.55 μm^2 for the low diopter lenses (281.71 ± 241; $N = 23$), as measured with the 40-μm circle ($P = 0.12$; not significant). The area of deviation from a perfect square ranged from 35.52 to 826.55 μm^2 for the single-piece lenses (280.44 ± 189.85; $N = 33$), and from 130.2 to 871.51 μm^2 for the three-piece lenses (451.51 ± 242.29; $N = 14$), as measured with the 40-μm circle ($P = 0.01$; significant). Considering all lenses included in the study ($N = 47$), the

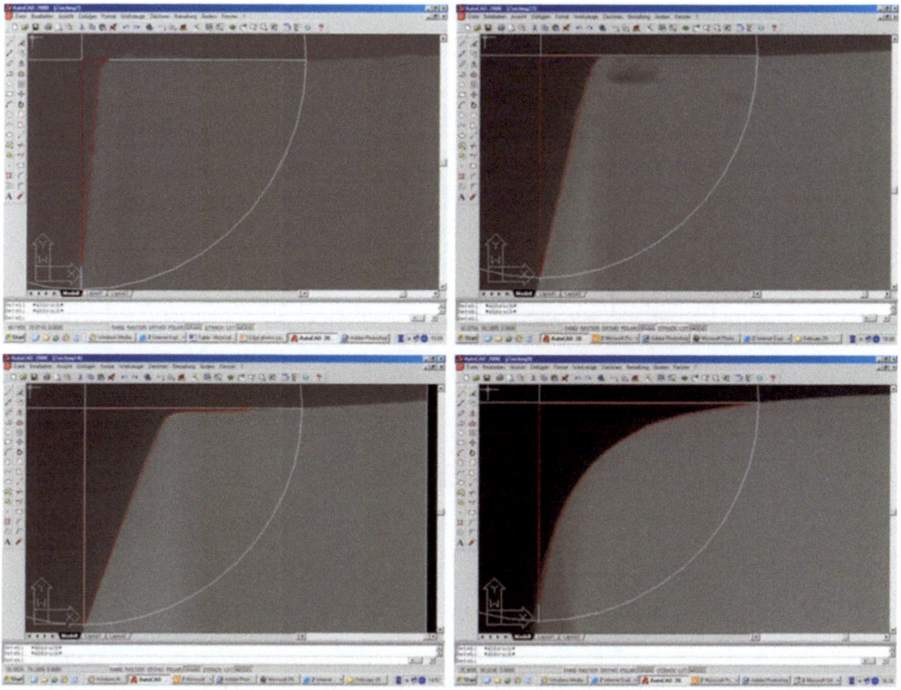

Fig. 17.5 Photographs from AutoCAD screens of the analyses of environmental scanning electron microscopy photographs of four different hydrophilic acrylic intraocular lenses. The area delineated in *red* shows the deviation from the ideal square, which varies among the lenses. Published in Werner et al. [26]

area of deviation from a perfect square ranged from 35.52 to 871.51 μm^2 (331.39 ± 218.90) (Fig. 17.5). The area measurement values of hydrophilic acrylic lenses, as a group, were found to be higher than the values reported for hydrophobic acrylic or silicone lenses in the first study on commercially available lenses. As a group, however, hydrophilic acrylic lenses also showed the largest variations (largest standard deviations). The differences among the three groups of materials were found to be statistically significant (Fig. 17.6).

Nanavaty et al. also performed a SEM study comparing the edge profile of commercially available square-edged IOLs [27]. Their study included a total of 17 square-edged designs of +20.0 D, with five hydrophobic acrylic, seven hydrophilic acrylic, and five silicone lenses. Perpendicular images with a magnification of ×500 were obtained and analyzed by using a purpose-designed software to produce a line tracing of the edge profile of the lenses. The sharpness of the edge profile was then quantified by measuring the local radius of curvature at the point on the posterior edge with the smallest radius. Their conclusions are similar in that as a group, hydrophilic acrylic lenses appear to have relatively rounder edges in comparison to hydrophobic acrylic and silicone lenses. This is probably due to the

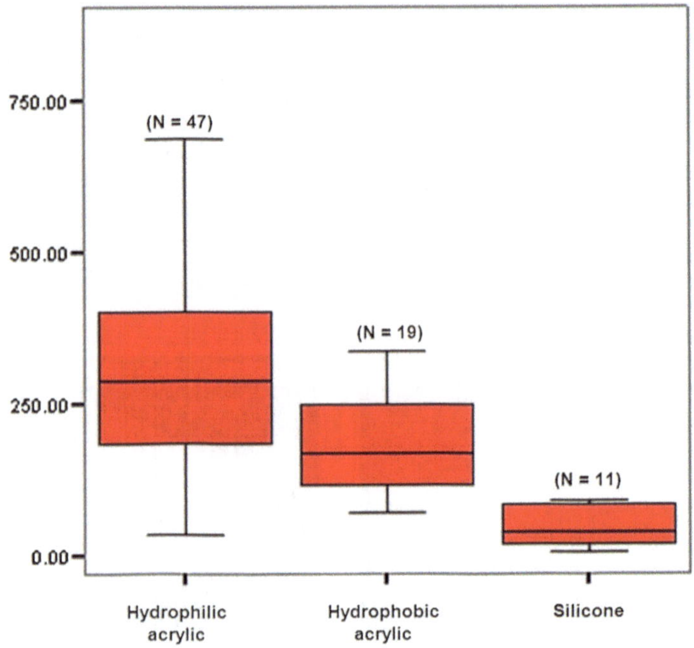

Fig. 17.6 *Graph* showing the variations in the area of deviation from a perfect square (measured in μm^2) among three groups of intraocular lenses: hydrophilic acrylic, hydrophobic acrylic, and silicone. All lenses included were commercially available and marketed as having a square posterior optic edge. As a group, silicone lenses showed less deviation from a perfect square (therefore, *squarer edges*), with less variation among the lenses evaluated

manufacturing process of hydrophilic acrylic lenses, which involves lathe cut from dehydrated blocks, which are then rehydrated. Water absorption by the IOL material may render the final aspect of the edge rounder as the IOL swells.

17.3.4 Clinical Significance of Microedge Structure Studies

The factor that may play the most important clinical role in evening out the differences in the microedge profiles observed in the abovementioned studies is shrink wrapping of the IOL by the capsular bag, which enhances contact between the posterior IOL surface and the posterior capsule. However, this factor may not even out large differences in edge profile.

The results of microedge structure studies are interesting in the light of some clinical studies comparing square-edged IOLs manufactured from different materials, reporting higher rates of PCO with hydrophilic acrylic lenses. As an example, Kugelberg et al. evaluated 120 eyes of 120 patients who had phacoemulsification and were prospectively randomized to receive a square-edged hydrophobic acrylic (Alcon SA60AT) or a square-edged hydrophilic acrylic

IOL (Bausch & Lomb BL27) [28]. Both IOLs are 1-piece designs with an optic diameter of 6.0 mm and no optic-haptic angulation. At the 1-year follow-up, there was a statistically significant difference in PCO between the two groups, with the hydrophilic acrylic group having a greater area and severity of PCO as analyzed with POCOman software. At the 2-year follow-up of the same study, patients with the SA60AT hydrophobic acrylic IOL still had less PCO as well as better high- and low-contrast visual acuity than patients with the BL27 hydrophilic acrylic IOL [29].

Richter-Mueksch et al. evaluated the uveal and capsular biocompatibility of 86 eyes in 78 patients with pseudoexfoliation who had cataract surgery [30]. In a nonrandomized protocol, the eyes received 1 of the following squared-edged IOLs: OphthalMed Injectacryl F3000 (hydrophilic acrylic), Alcon AcrySof MA60MB (hydrophobic acrylic), or Pharmacia CeeOn 911 (silicone). PCO was statistically greater in the hydrophilic acrylic group, according to semiquantitative analyses.

In a meta-analysis including 23 prospective randomized controlled clinical trials, the authors concluded that hydrophilic acrylic IOLs were significantly associated with higher rates of PCO and Nd:YAG laser posterior capsulotomies than IOLs of other materials [31]. However, the majority of the studies included in the analysis hydrophilic acrylic IOLs not marketed as square-edged IOLs. Finally, in a prospective randomized study, Heatley et al. evaluated 106 eyes of 53 patients with bilateral cataract. One eye was implanted with an SA60AT IOL and the other with a Rayner Centerflex 570H IOL (1-piece design with square edges and no optic-haptic angulation) [32]. The percentage PCO area, measured with the POCOman system, was higher in the hydrophilic IOL group than in the hydrophobic IOL group at 1 month ($P < 0.05$), 6 months ($P < 0.001$), and 12 months ($P < 0.001$). At 1 year, the median PCO values were 50.3 % in the hydrophilic IOL group and 4.9 % in the hydrophobic IOL group. Incorporation of an "enhanced" square edge into this IOL design clearly improved PCO results, as shown in more recent rabbit and clinical studies [33–35].

As shown above, in many instances the authors concluded that higher PCO rates with hydrophilic acrylic lenses were related to a "material" effect; however, the edges of the lenses included in the clinical studies were perhaps just not comparable. Evaluation of the microedge structure of commercially available lenses showed that all square edges in the market are not the same, and perhaps large variations in edge profile may account for differences in clinical outcomes of postoperative PCO.

17.3.5 Optic-Haptic Junctions of Single-Piece Designs

Most available prospective, randomized clinical studies have shown no statistically significant difference between 1-piece and 3-piece hydrophobic acrylic IOL in terms of PCO rates [36–38]. Rabbit studies performed in our laboratory demonstrated that whenever PCO started with single-piece lenses, it had a tendency to start at the level of the optic-haptic junctions [19, 34]. This was observed with single-piece hydrophobic or hydrophilic acrylic lenses. It has been hypothesized that the sharp, square posterior optic edge of those single-piece lenses was

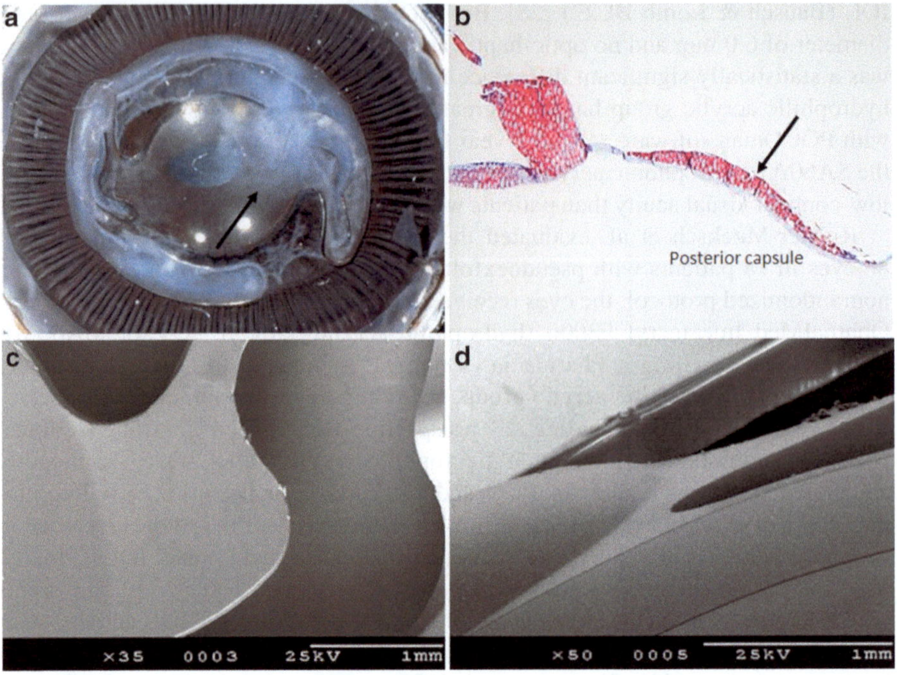

Fig. 17.7 Gross photograph (Miyake-Apple view; **a**) and corresponding histopathological correlation (**b**) of a rabbit eye implanted with a looped, single-piece hydrophilic acrylic intraocular lens with a smooth transition at the optic-haptic junctions (shown in the scanning electron microscopy photographs; **c** and **d**). Cell ingrowth causing peripheral and central posterior capsule opacification formation appears to have started at the optic-haptic junction (*arrows*). Published in Werner et al. [34]

interrupted at the optic-haptic junction, allowing proliferating, migrating LECs to enter the area between the posterior capsule and the optic at that site. These findings led to a design modification of a single-piece hydrophilic acrylic lens, incorporating an enhanced edge, which improved the barrier effect at the level of the optic-haptic junctions (Figs. 17.7 and 17.8). This has been confirmed in rabbit [33, 34] and clinical studies [35] with the same hydrophilic acrylic lens. In a study using pseudophakic human eyes obtained postmortem also performed in our laboratory, 75 % of eyes implanted with 1-piece hydrophobic acrylic lenses with initial peripheral PCO had the initiation site at the optic-haptic junctions [39]. A new hydrophobic acrylic IOL design, which has recently entered the US market, addressed the lack of barrier effect at the junctions by creating a full 360° square posterior optic edge while maintaining a 1-piece configuration. We have not yet received in our laboratory cadaver eyes implanted with this design. However, a recently published paired-eye clinical study showed significantly less PCO with this new design in comparison with a 1-piece interrupted-edge hydrophobic acrylic IOL 2 years after implantation [40].

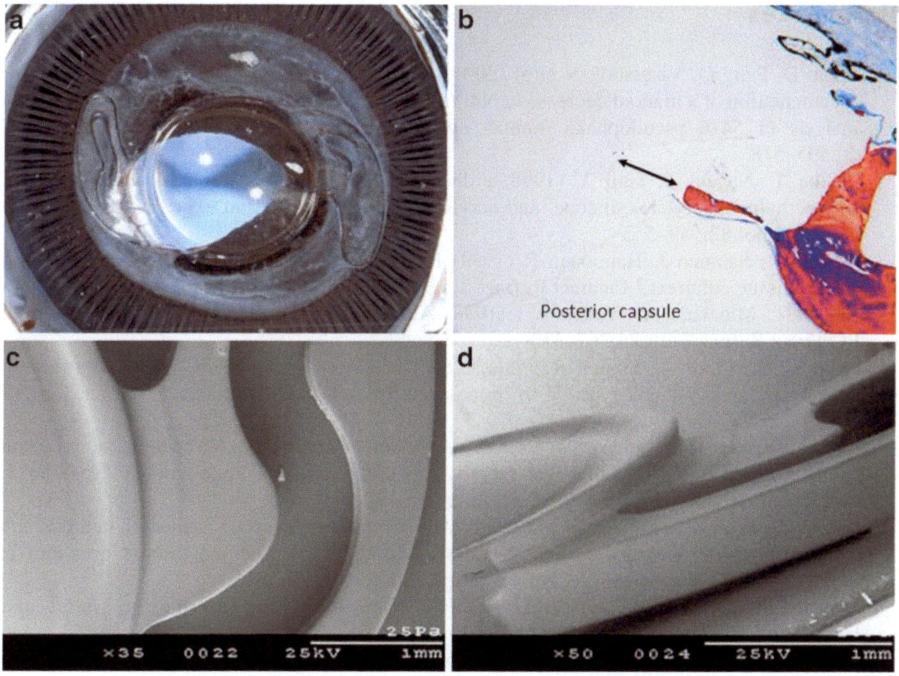

Fig. 17.8 Gross photograph (Miyake-Apple view; **a**) and corresponding histopathological correlation (**b**) of a rabbit eye implanted with a looped, single-piece hydrophilic acrylic intraocular lens with an enhanced *square edge* at the optic-haptic junctions (shown in the scanning electron microscopy photographs; **c** and **d**). The edge profile of the optic-haptic junction of the lens can be seen in the histological section. The material that originated from Soemmering's ring was blocked at the extra ridge of the optic-haptic junction of this lens design (*arrow*). Published in Werner et al. [34]

17.4 Summary

As described above, there is convincing evidence that both material and design may play a role in preventing PCO formation. Some hydrophobic acrylic IOLs appear to form the best bioactive bond with the posterior capsule, and these lenses also exhibit a square optic edge. Both of these factors are likely to contribute to the success of this IOL type in preventing PCO. However the studies listed above have shown that silicone, PMMA, and hydrophilic acrylic IOLs, which have less favorable bioactive bond formation, can also achieve low levels of PCO formation if they incorporate a square optic edge. These findings provide evidence that while material type and bioactive bond formation can play an important role in PCO prevention, a sharp (true to square)-edged optic may play a more preponderant role.

References

1. Apple D, Peng Q, Visessook N et al (2001) Eradication of posterior capsule opacification: documentation of a marked decrease in Nd:YAG laser posterior capsulotomy rates noted in an analysis of 5416 pseudophakic human eyes obtained postmortem. Ophthalmology 108 (3):505–518
2. Oshika T, Nagata T, Ishii Y (1998) Adhesion of lens capsule to intraocular lenses of polymethylmethacrylate, silicone, and acrylic foldable materials: an experimental study. Br J Ophthalmol 82:549–553
3. Linnola R, Salonen J, Happonen R (1999) Intraocular lens bioactivity tested using rabbit corneal tissue cultures. J Cataract Refract Surg 25:1480–1485
4. Nagata T, Minakata A, Watanabe I (1998) Adhesiveness of AcrySof to a collagen film. J Cataract Refract Surg 24:367–370
5. Linnola RJ, Sund M, Ylönen R, Pihlajaniemi T (2003) Adhesion of soluble fibronectin, vitronectin, and collagen type IV to intraocular lens materials. J Cataract Refract Surg 29 (1):146–152
6. Linnola R, Werner L, Pandey S et al (2000) Adhesion of fibronectin, vitronectin, laminin, and collagen type IV to intraocular lens materials in pseudophakic human autopsy eyes Part 1: histological sections. J Cataract Refract Surg 26:1792–1806
7. Linnola R, Werner L, Pandey S et al (2000) Adhesion of fibronectin, vitronectin, laminin, and collagen type IV to intraocular lens materials in pseudophakic human autopsy eyes Part 2: Explanted intraocular lenses. J Cataract Refract Surg 26:1807–1818
8. Johnston R, Spalton D, Hussain A, Marshall J (1999) In vitro protein adsorption to 2 intraocular lens materials. J Cataract Refract Surg 25:1109–1115
9. Ursell P, Spalton D, Pande M (1997) Anterior capsule stability in eyes with intraocular lenses made of poly(methyl methacrylate), silicone, and AcrySof. J Cataract Refract Surg 23:1532–1538
10. Werner L, Pandey S, Escobar-Gomez M et al (2000) Anterior capsule opacification: a histopathological study comparing different IOL styles. Ophthalmology 107:463–471
11. Werner L, Pandey SK, Apple DJ et al (2001) Anterior capsule opacification: correlation of pathological findings with clinical sequelae. Ophthalmology 108:1675–1681
12. Nishi O, Nishi K, Sakanishi K (1998) Inhibition of migrating lens epithelial cells at the capsular bend created by the rectangular optic edge of a posterior chamber intraocular lens. Ophthalmic Surg Lasers 29:587–594
13. Schauersberger J, Amon M, Kruger A et al (2001) Comparison of the biocompatibility of 2 foldable intraocular lenses with sharp optic edges. J Cataract Refract Surg 27:1579–1585
14. Buehl W, Findl O, Menapace R et al (2002) Effect of an acrylic intraocular lens with a sharp posterior optic edge on posterior capsule opacification. J Cataract Refract Surg 28:1105–1111
15. Prosdocimo G, Tassinari G, Sala M et al (2003) Posterior capsule opacification after phacoemulsification: silicone CeeOn Edge versus acrylate AcrySof intraocular lens. J Cataract Refract Surg 29:1551–1555
16. Auffarth G, Golescu A, Becker K, Volcker H (2003) Quantification of posterior capsule opacification with round and sharp edge intraocular lenses. Ophthalmology 110:772–780
17. Nixon DR (2004) In vivo digital imaging of the square-edged barrier effect of a silicone intraocular lens. J Cataract Refract Surg 30:2574–2584
18. Peng Q, Visessook N, Apple DJ et al (2000) Surgical prevention of posterior capsule opacification. Part 3. Intraocular lens optic barrier effect as a second line of defense. J Cataract Refract Surg 26:198–213
19. Werner L, Mamalis N, Izak AM et al (2005) Posterior capsule opacification in rabbit eyes implanted with single-piece and three-piece hydrophobic acrylic intraocular lenses. J Cataract Refract Surg 31:805–811
20. Nishi O, Nishi K (1999) Preventing posterior capsule opacification by creating a discontinuous sharp bend in the capsule. J Cataract Refract Surg 25:521–526

21. Nishi O, Yamamoto N, Nishi K, Nishi Y (2007) Contact inhibition of migrating lens epithelial cells at the capsular bend created by a sharp-edged intraocular lens after cataract surgery. J Cataract Refract Surg 33:1065–1070
22. Bhermi G, Spalton D, El-Osta MJ (2002) Failure of a discontinuous bend to prevent lens epithelial cell migration in vitro. J Cataract Refract Surg 28:1256–1261
23. Boyce J, Bhermi G, Spalton D, El-Osta A (2002) Mathematic modeling of the forces between an intraocular lens and the capsule. J Cataract Refract Surg 28:1853–1859
24. Tetz M, Wildeck A (2005) Evaluating and defining the sharpness of intraocular lenses: part 1: influence of optic design on the growth of the lens epithelial cells in vitro. J Cataract Refract Surg 31:2172–2179
25. Werner L, Muller M, Tetz M (2008) Evaluating and defining the sharpness of intraocular lenses. Microedge structure of commercially available square-edged hydrophobic lenses. J Cataract Refract Surg 34:310–317
26. Werner L, Tetz M, Feldmann I et al (2009) Evaluating and defining the sharpness of intraocular lenses. Microedge structure of commercially available square-edged hydrophilic lenses. J Cataract Refract Surg 35:556–566
27. Nanavaty M, Spalton D, Boyce J et al (2008) Edge profile of commercially available square-edged intraocular lenses. J Cataract Refract Surg 34:677–686
28. Kugelberg M, Wejde G, Jayaram H, Zetterström C (2006) Posterior capsule opacification after implantation of a hydrophilic or a hydrophobic acrylic intraocular lens: one-year follow-up. J Cataract Refract Surg 32:1627–1631
29. Kugelberg M, Wejde G, Jayaram H, Zetterström C (2008) Two-year follow-up of posterior capsule opacification after implantation of a hydrophilic or hydrophobic acrylic intraocular lens. Acta Ophthalmol 86:533–536
30. Richter-Mueksch S, Kahraman G, Amon M et al (2007) Uveal and capsular biocompatibility after implantation of sharp-edged hydrophilic acrylic, hydrophobic acrylic, and silicone intraocular lenses in eyes with pseudoexfoliation syndrome. J Cataract Refract Surg 33:1414–1418
31. Cheng J, Wei R, Cai J et al (2007) Efficacy of different intraocular lens materials and optic edge designs in preventing posterior capsular opacification: a meta-analysis. Am J Ophthalmol 143:428–436
32. Heatley CJ, Spalton D, Kumar A et al (2005) Comparison of posterior capsule opacification rates between hydrophilic and hydrophobic single-piece acrylic intraocular lenses. J Cataract Refract Surg 31:718–724
33. Vargas LG, Izak AM, Apple DJ et al (2003) Implantation of a single-piece, hydrophilic, acrylic, minus-power foldable posterior chamber intraocular lens in a rabbit model: clinicopathologic study of posterior capsule opacification. J Cataract Refract Surg 29:1613–1620
34. Werner L, Mamalis N, Pandey S et al (2004) Posterior capsule opacification in rabbit eyes implanted with hydrophilic acrylic intraocular lenses with enhanced square edge. J Cataract Refract Surg 30:2403–2409
35. Nishi Y, Rabsilber TM, Limberger IJ et al (2007) Influence of 360-degree enhanced optic edge design of a hydrophilic acrylic intraocular lens on posterior capsule opacification. J Cataract Refract Surg 33(2):227–231
36. Sacu S, Finle O, Menapace R et al (2004) Comparison of posterior capsule opacification between the 1-piece and 3-piece AcrySof intraocular lenses; two-year results of a randomized trial. Ophthalmology 111:1840–1846
37. Leydolt C, Davidovic S, Sacu S et al (2007) Long-term effect of 1-piece and 3-piece hydrophobic acrylic intraocular lens on posterior capsule opacification; a randomized trial. Ophthalmology 114:1663–1669
38. Bender L, Nimsgern C, Jose R et al (2004) Effect of 1-piece and 3-piece AcrySof intraocular lenses on the development of posterior capsule opacification after cataract surgery. J Cataract Refract Surg 30:786–789

39. Ness PJ, Werner L, Maddula S et al (2011) Pathology of 219 human cadaver eyes with 1-piece or 3-piece hydrophobic acrylic intraocular lenses: capsular bag opacification and sites of square-edged barrier breach. J Cataract Refract Surg 37:923–930
40. Nixon DR, Woodcock MG (2010) Pattern of posterior capsule opacification models 2 years postoperatively with 2 single-piece acrylic intraocular lenses. J Cataract Refract Surg 36 (6):929–934

Capsular and Uveal Biocompatibility of Different IOLs in Eyes With and Without Associated Conditions

18

Michael Amon and Guenal Kahraman

Abstract

High biocompatibility is one of the main features expected of an intraocular lens (IOL) implant. In terms of anatomical position an implant is close to, or in contact with, uveal tissue and the capsule. Therefore, uveal reactions of the eye must be distinguished from capsular reactions. As both reactions are influenced by the implant, one should rightly address the subjects of uveal biocompatibility and capsular biocompatibility separately. The iris, the ciliary body, and the choroid are composed of vascularized tissue and are close to the implant. In cases of iridolenticular synechiae or sulcus position of the IOL, portions of the lens come into direct contact with uveal tissue. Changes in blood-aqueous barrier due to surgical trauma and the implanted foreign body cause leakage of proteins and macrophages from blood.

The main parameters of capsular biocompatibility are lens epithelial cell (LEC) outgrowth, anterior capsule opacification (ACO), posterior capsule opacification (PCO), and contraction of the capsule; IOLs that perform well in these respects may be said to possess high capsular biocompatibility.

The foreign-body cell reaction is the most important parameter of uveal biocompatibility; IOLs causing a very mild foreign-body reaction could be rightly referred to as IOLs with uveal biocompatibility. Implants with high uveal biocompatibility are especially suitable for eyes with a compromised blood-aqueous barrier (BAB).

Keywords

Anterior capsule opacification • Capsular biocompatibility • Lens epithelial cells • Posterior capsule opacification

M. Amon, M.D. (✉) • G. Kahraman, M.D.
Department of Ophthalmology, Academic Teaching Hospital of St. John, Vienna, Krankenhaus der Barmherzigen Brüder, Johannes von Gott Platz 1, 1020 Vienna, Austria
e-mail: amon@augenchirurg.com; kahraman@gmx.at

S. Saika et al. (eds.), *Lens Epithelium and Posterior Capsular Opacification*,
DOI 10.1007/978-4-431-54300-8_18, © Springer Japan 2014

18.1 Capsular and Uveal Biocompatibility

Biocompatibility is one of the most important prerequisites of an intraocular implant [1]. This term is frequently used to describe specific characteristics of intraocular lenses (IOLs).

Biocompatibility is the capability of a prosthesis implanted in the body to exist in harmony with tissue without causing deleterious changes [2]. As an implant reacts with different tissues in the eye, a specific IOL may prove to be favorable with reference to a specific tissue reaction of the body, but may perform less effectively in respect of a different reactive pattern. Therefore, it has become necessary to systematize the terminology of biocompatibility in a tissue-dependent manner.

In terms of their anatomic position, intraocular implants are mainly in the immediate vicinity of, or in contact with, uveal tissue and lens tissue. The implant itself causes a specific pathophysiological reaction in the uvea and in lens epithelial cells (LECs) [3–5]. Therefore, a distinction should be made between uveal and capsular reactions of the eye [6]. As both of these reactive processes are influenced by the implant, it would be appropriate to introduce the terms uveal and capsular biocompatibility. The reactive pattern of both types of tissues influences blood-aqueous barrier (BAB) changes with all of its consequences [7].

A further tissue-specific subdivision into corneal and vitreal biocompatibility would be plausible, but appears to be of secondary importance for IOLs compared to the significance of the two above mentioned types of tissue. The term "retinal biocompatibility" would be meaningful in connection with IOLs.

18.1.1 Capsular Biocompatibility

In most cases the capsular bag and the LECs are the only ocular tissues in direct contact with the IOL. Contact between the LECs and lens material causes proliferation, myofibroblastic metaplasia, and the production of cytokines (IL-1, IL-6, IL-8, FGF, TGF-β, TNF-α, PGE2, etc.) [5]. The secretion of cytokines may compromise the BAB, which, in turn, influences LEC metaplasia. Due to proliferation, LECs may spread onto the "capsule-free" anterior lens surface [8, 9]. The severity of this reaction appears to be mainly dependent on IOL material [10–12]. However, the principal complications of capsular reactions to the implant are the development of opacities of the anterior (anterior capsule opacification, ACO) and posterior capsule (posterior capsule opacification, PCO) and contraction of the capsular bag [13]. ACO and PCO appear to be influenced by lens material as well as lens design [5].

Thus, as the main parameters of capsular biocompatibility are LEC outgrowth, ACO, PCO, and capsular contraction, IOLs that perform well in these respects could be said to possess high capsular biocompatibility.

The terms *uveal* and *capsular biocompatibility* of intraocular implants would be useful to make more specific statements about IOLs and select the most suitable implant for a specific indication.

18.1.2 Uveal Biocompatibility

The iris, the ciliary body, and the choroid are composed of vascularized tissue and are close to the implant. In cases of iridolenticular synechiae or sulcus position of the IOL, portions of the lens come into direct contact with uveal tissue. Changes in BAB due to surgical trauma and the implanted foreign body cause leakage of proteins and macrophages from blood [3, 4].

Protein adsorption occurs on the lens surface shortly after implantation of an IOL [14]. This membrane is composed of various proteins (albumin, fibrinogen, fibronectin, etc.). Its composition is also influenced by lens material [15]. The membrane, in turn, influences cell adhesion and activation on the implant. Initially, small round cells and fibroblast-like cells adhere to this membrane. Later on, as a result of fusion of macrophages, they develop into epithelioid cells and foreign-body giant cells [16–18].

Macrophages themselves produce various cytokines (IL-1, TGF-β, TGF-α, PDGF, etc.), which again influence the subsequent course of inflammation and result in BAB changes of varying intensity. The foreign-body reaction described above was found to be influenced by lens material and lens design [3, 16].

Since the foreign-body cell reaction is the most important parameter of uveal biocompatibility, IOLs causing a very mild foreign-body reaction could be rightly referred to as those with uveal biocompatibility.

Implants with high uveal biocompatibility are especially suitable for eyes with a compromised BAB (uveitis, pseudoexfoliation—PEX, diabetes, etc.).

Selection of the best IOL, suited to a patient's initial situation, is of fundamental importance for the outcome of surgery. This is especially true in critical cases. Even today—several decades after the development of intraocular implants—this decision is a major challenge for the ophthalmological surgeon. In addition to a large number of objective criteria (such as various lens materials and designs), the surgeon is confronted with a number of subjective criteria associated with the marketing of lenses.

The influence of different IOL designs and IOL materials on capsular biocompatibility, and the importance of preexisting conditions such as PEX or uveitis, will be discussed in this section.

We will also try to establish a system that will serve as an aid in decision-making with regard to selecting the optimal lens for various initial situations.

18.2 Intraocular Lens Selection

Basically, every type of lens reacts differently in the eye because of its material and/or design. Nevertheless, it was found that mutually associated lens materials and similar lens types react in a similar way. Hence it would be meaningful to divide IOLs into groups of material and lens designs because lens material as well as lens design influence biocompatibility [3].

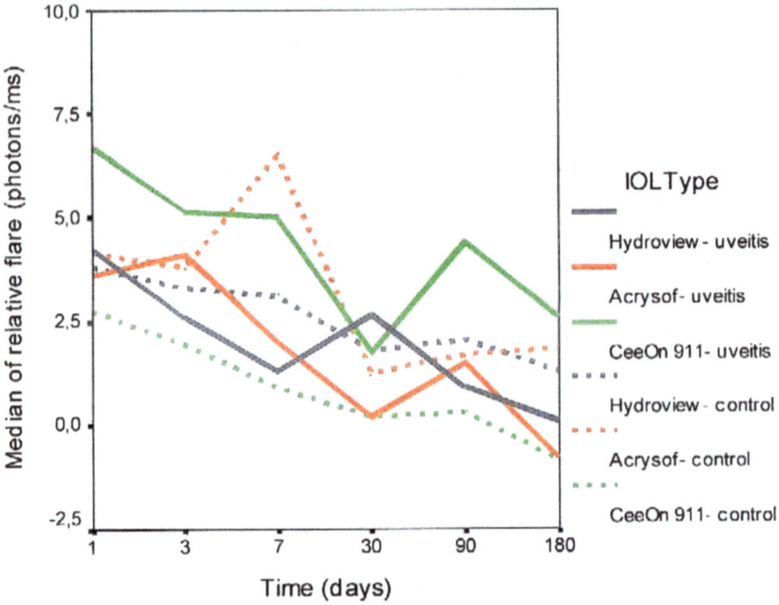

Fig. 18.1 Course of median relative flare values in the three IOL groups in uveitic and control eyes (hydrophilic acrylic (Hydroview, Bausch & Lomb), hydrophobic acrylic (AcrySof, Alcon), or silicone (CeeOn911, Pharmacia))

In this chapter, silicones (silicone of the new generation), hydrophobic acryls, and the very large group of hydrophilic acryls (including heparin-coated foldable lenses; only thoroughly tested and evaluated IOLs without lens opacification) will be differentiated and their intraocular behavior discussed.

The results of several studies have shown that silicone lenses lead to ACO and the development of rhexis phimosis [17]. Hydrophobic acryl demonstrates a stronger uveal reaction while hydrophilic acryl causes a greater capsular reaction [18] (Figs. 18.1 and 18.2).

With reference to lens design, all three groups of materials perform better in respect of PCO rates when the optic edge is sharp [8] and the optic-haptic junction narrow. Uveal reaction is mainly influenced by lens material [19].

Prospective studies focusing on lens biocompatibility in senile cataract, cataract with PEX, and cataract with uveitis showed that the uveal as well as capsular biocompatibility of implants reduce as their inflammatory disposition rises (Fig. 18.3). Furthermore, it was found that lens-specific differences become evident only in eyes at risk [20].

In summary, hydrophilic acryl and silicone are associated with greater uveal biocompatibility for "compromised eyes," whereas hydrophobic lens material (silicone, acryl) with a sharp-edged optic and a narrow optic-haptic junction shows greater capsular biocompatibility [17].

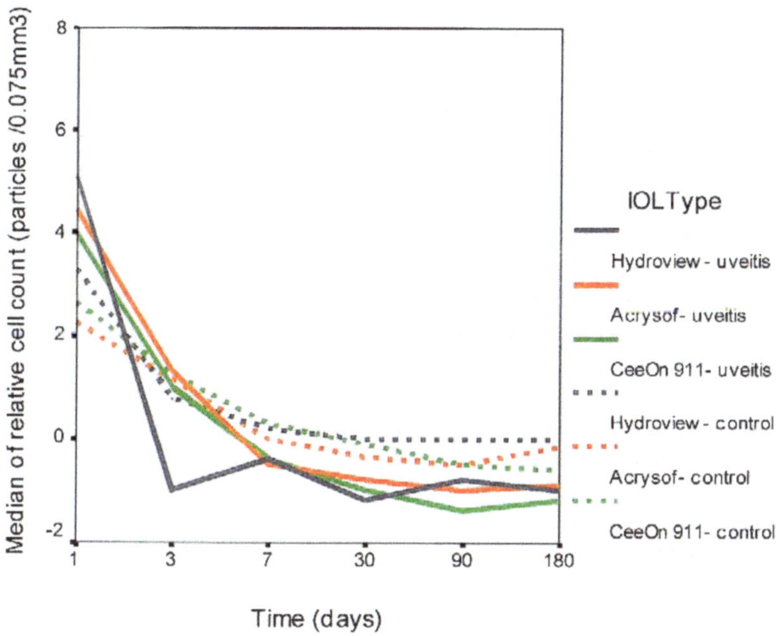

Fig. 18.2 Course of median relative cell counts in the three IOL groups in uveitic and control eyes

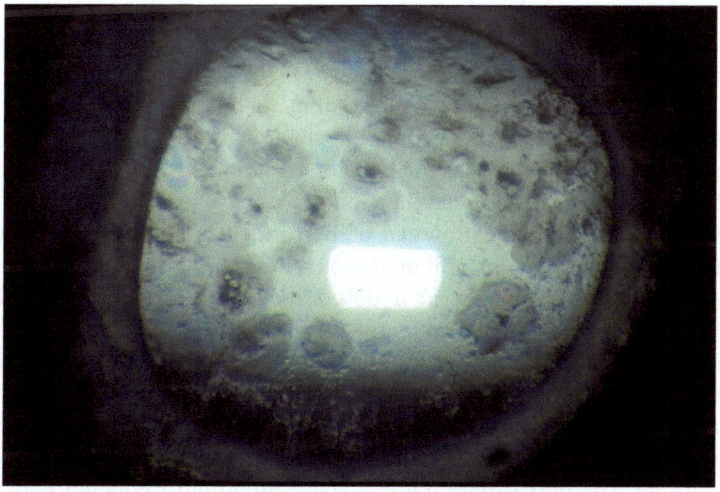

Fig. 18.3 Foreign-body giant cell reaction on a hydrophobic acrylic IOL after 4 months of implantation (PEX eye)

The laser flare cell measurement (LFCM) shows significant differences between patients with senile cataract, PEX, or uveitis, but lens-specific differences have not been observed with the LFCM [12]. This fact underlines the unspecific nature of quantifying "flare" for the evaluation of biocompatibility.

Based on the results presented here in summarized form, a few interesting aspects and selection criteria may be derived with regards to the selection of the optimal implant for a specific situation. It should be emphasized that, in addition to the intraocular implant, the postoperative outcome is influenced by the patient's underlying disease, perioperative therapy, and surgical trauma. The use of a minimally traumatic and precise surgical technique utilizing the entire armamentarium of ophthalmic surgery (meticulous small incision surgery; MSIS) is a prerequisite for achieving satisfactory results. The implant itself cannot compensate for surgical or therapeutic errors. The success of treatment depends on the optimal interplay of all of these factors. These aspects will be discussed in the following with reference to specific baseline situations.

18.3 Uveitis

The optimal treatment of complex cataract is a major challenge for the ophthalmologist. Accurate preoperative diagnostic investigations are more important here than in any other setting. Further crucial aspects include the establishment of the indication for surgery at the correct point in time, an appropriate perioperative therapy concept, the use of a nontraumatic surgical technique, finding the optimal lens implant, and monitoring the patient very carefully postoperatively in order to treat potential complications on time.

In patients with chronic recurrent uveitis, surgery should be performed during the disease-free interval. However, even in these cases and especially in chronic forms of the disease, local and even systemic anti-inflammatory therapy should be started on a timely basis before surgery.

As uveitis is associated with a preexisting disorder of the BAB, it is most important to minimize surgical trauma and thus avoid additional complications. The selected lens type determines the incision technique. Basically, one may select a scleral incision or even a corneal incision.

As long as phacoemulsification can be performed in a controlled manner, one should not perform iatrogenic pupil dilatation. However, if the pupil is very narrow, techniques to dilate the pupils in a dosed manner (high-viscosity ophthalmic viscosurgical devices—OVDs, iris retractors, pupil dilatation) should be given preference over iridotomy. Pupillary membranes and posterior synechiae should be exposed and removed completely as far as possible.

The surgeon should perform an exactly dimensioned capsulorhexis (circumferential overlap of the IOL optic periphery). Trypan blue dye may be used to stain the anterior capsule. The rhexis margin is less stable after high-frequency capsulotomy.

The technique of phacoemulsification does not differ from the standard technique, but one may anticipate higher rates of miosis and zonulopathy in uveitic patients.

Selection of the implant is especially important in uveitic patients, as they frequently develop recurrent foreign-body giant cell membranes and posterior synechiae. Since these cell membranes produce cytokines, which then lead to cystoid macular edema (CME) and massive functional impairment, and they also cause massive optical impairment, uveal biocompatibility should be given prime importance when selecting the lens.

18.3.1 Lens Material

Thoroughly tested and evaluated hydrophilic lens surfaces (no superficial or "intrastromal" calcifications; heparin-coated PMMA, hydrogel) and modern silicones are more biocompatible and therefore indicated in uveitic patients. Foreign-body giant cells develop much less frequently on hydrophilic lenses and silicone than on hydrophobic acryl [18, 21]. The implantation of heparin-coated PMMA lenses was regarded as the gold standard until recently. As these implants do demonstrate similar uveal biocompatibility as foldable hydrophilic acryl lenses (with or without a heparin coating), but are not foldable, the foldable lens should now be given preference because of the smaller incision.

18.3.2 Lens Design

In addition to lens material, lens design is obviously of enormous significance in uveitic patients. The creation of a sufficient distance between the iris and the lens is especially important to avoid synechiae. Therefore, haptic angulation and a sharp-edged optic with a narrow optic-haptic junction are essential to minimize PCO rates, which are basically higher in uveitic patients.

Finally, due to advances in the treatment of the underlying disease, cataract surgery, and IOL technology, the prognosis of cataract surgery has been improved in uveitic patients. As a result, the indication for cataract surgery can be established earlier in many cases.

18.4 Pseudoexfoliation Syndrome (PEX)

Eyes with PEX are susceptible to more severe postoperative inflammation (less than in the presence of uveitis) and more pronounced alterations of the capsule than eyes with senile cataract alone. For this reason, uveal and capsular biocompatibility of the implant is of great significance in this setting as well. As these patients are inclined to develop zonulopathies and massive capsular bag contraction, it is very important to perform optimal surgery without straining the zonular apparatus and by creating an ideal rhexis (360° overlap of the optic periphery, not too small).

18.4.1 Lens Material

In this situation the surgeon may select any of the three lens materials. As rhexis phimosis has been frequently observed in connection with silicone, it is important to size the rhexis correctly [22]. One may additionally use a capsular tension ring.

18.4.2 Lens Design

Lens design should be similar to that used in uveitic patients. The creation of a sufficient distance between the iris and the lens appears to be important here as well (it is also important in combined surgery for cataract and glaucoma). Angulation of the haptic and a sharp-edged optic with a narrow optic-haptic junction is needed to minimize the high PCO rates associated with PEX. Besides, the lens design should permit the use of a nontraumatic implantation technique.

18.5 Vitreoretinopathy, Silicone Oil Filling

The visibility of the fundus after lens implantation is especially important in these patients. Capsular biocompatibility is clearly of prime importance in this setting.

18.5.1 Lens Material

Silicone and hydrophobic acryl are given preference because they have yielded better results in respect of PCO. However, as silicone oil is spread over a large area on a silicone lens and adheres to its surface, silicone lenses are contraindicated in this setting. Yet, the contraindication is not absolute because a silicone lens can be easily replaced if necessary.

18.5.2 Lens Design

A large optic diameter (6.5–7 mm) improves the visibility of the fundus and would theoretically be associated with a plane or concave posterior optic surface in the silicone-filled eye, thus influencing refraction to a lesser degree. A sharp-edged optic and a narrow optic-haptic junction ensure greater capsular biocompatibility.

18.6 Sulcus Fixation

As the entire capsular bag or at least central portions of the capsule are no longer present in eyes requiring IOL sulcus fixation, and the IOL is in direct contact with the uvea (fixed with sutures or by portions of the capsule in the ciliary sulcus), only uveal biocompatibility is important in these cases.

18.6.1 Lens Material

Hydrophilic acryl is the lens material of choice in this specific situation. Heparin-surface-modified (HSM)-PMMA would also be an alternative, but due to the larger incision required for this lens it would be more reasonable to give preference to a foldable lens. Silicone is associated with greater uveal biocompatibility; however, if one needs to apply silicone oil, the latter could be in direct contact with the silicone lens.

18.6.2 Lens Design

The lens haptic should be of sufficient overall diameter (more than 13 mm) to prevent decentration and rotation. To avoid iridolenticular synechiae it should be angulated. The optic edge should be rounded or matted to minimize glare phenomena.

18.7 Trauma

In the heterogeneous group of trauma patients, cataract is frequently associated with other complex pathologies, such as coloboma, aniridia, zonulopathy, changes in the vitreous body, glaucoma, etc. Therefore, selection of the best IOL in these cases depends on the additional functions of the implant (aniridia lens, eyelet for the sulcus suture, etc.). Lens material and lens design should be aligned to the indication and the respective situation. The range of special products available in the market is now quite extensive.

18.8 Congenital Cataract

A review of recent studies data showed that, in these patients, the surgical technique is the prime factor that influences the success of the operation. Particularly in children less than 5 years of age, a surgical technique combined with posterior capsulorhexis, vitrectomy, and if necessary "posterior optic capture" should be used.

The time of lens implantation and selection of the appropriate refraction should be considered with great care. As refraction is subject to enormous change especially in the first 12 months of a child's life, and the eye's uveal and capsular reaction to the implant are more intensive during this time, in most cases it would be advisable to implant an IOL after the age of 1 year. Due to the preexisting strong postoperative inflammation and massive capsule alterations in children, the implant should possess high uveal as well as capsular biocompatibility.

The use of a blue-light filter is controversially discussed at the present time. The existing body of data is not sufficient to make a conclusive statement in this regard.

Obviously, one should expect the implant to show a high degree of retinal biocompatibility (protects the retina from phototoxicity), because children have to keep the implant for several decades. Assuming that the additional blue-light protection has no disadvantages (color vision, contrast vision, etc.), the use of implants with better light protection in these young patients is worthy of discussion ("in dubio pro reo strategy").

Due to the changes in refraction that occur in the course of a child's life, it may be necessary to replace lenses or use additional IOLs. The use of a capsule spacer ring would be meaningful in this setting. It would facilitate explantation as well as renewed capsular bag implantation. However, the currently available models are not suitable for use in children.

18.8.1 Lens Material

Hydrophilic acryl (a thoroughly tested and evaluated product with no demonstrable opacification) may be used because one has to perform posterior capsulorhexis and vitrectomy in any case. To balance the lower capsular biocompatibility of this product, a posterior optic capture would be useful when this material is used. Hydrophobic acryl is employed today in the majority of cases and yields satisfactory results. Basically one may also use silicone, but one can never entirely rule out the possibility of vitreoretinopathy in children at a later point in time. It may be necessary to use silicone oil in these cases.

18.8.2 Lens Design

The haptic should not be too rigid because the diameter of the capsular bag is usually much smaller than it is in adults. The optic should have a sharp edge and be of sufficient diameter (6 mm), because young patients usually have a wide pupil. A narrow optic-haptic junction facilitates "optic capture."

18.9 Senile Cataract and Maculopathy

Cataract surgery as such is liable to worsen preexisting maculopathy. Therefore, it is very important to use a nontraumatic surgical technique (MSIS) and minimize intraoperative exposure to light. Since uveal as well as capsular reactions are associated with the production of cytokines, which in turn are important for the propagation of maculopathy, high uveal and capsular biocompatibility are essential for the optimal outcome of surgery. Whether IOLs with a blue filter have a protective effect in patients with preexisting damage of the macula has not been proven yet. However, an "in dubio pro reo strategy" could be adopted in these patients as well.

18.9.1 Lens Material

Basically IOLs may be selected from all three groups of materials, provided they have been thoroughly tested and evaluated.

18.9.2 Lens Design

To ensure optimal capsular biocompatibility and avoid secondary interventions in the lens capsule in this risk group, the lens optic should be sharp-edged and the optic-haptic junction narrow. Addressing concepts of prismatic optics and "magnifying lenses" would exceed the limitations of this chapter.

18.10 Clear Lens Extraction

Obviously, one should use implants with high uveal and capsular biocompatibility in patients without senile cataract as well, because one never knows whether additional pathologies may develop at a later point in time (endophthalmitis, diabetic retinopathy, uveitis, etc.). Particularly in these patients one may select special additional optical functions (multifocal lens, toric IOL, aberration-corrected IOLs, etc.) and/or other properties (microincision lens, for instance). Generally it should be noted that there is a clear trend towards injector systems, which have many advantages (less contamination of the implant, length of the incision, etc.)

18.10.1 Lens Material

Basically, IOLs may be selected from any of the three materials, provided they have been thoroughly tested and evaluated.

18.10.2 Lens Design

To ensure optimal capsular biocompatibility the lens optic should be sharp-edged. The optic design should be selected to ensure minimal dysphotopsia (configuration of the edge, refraction index, surface radiuses, deposits, etc.). When no alternative strategy is used ("enhanced square edge," capsular spacer ring) to compensate the weak point of the optic-haptic junction, the latter should be as narrow as possible. The lens haptic should not be angulated because this will ensure more rapid postoperative refractive stability.

18.11 Conclusion

In summary, given the development of modern lens technology, the surgeon is able to achieve satisfactory postoperative results in complicated cases provided the surgeon uses thoroughly tested implants. However, lens biocompatibility is not sufficient in all cases. Further effort in basic research, implantology, and ophthalmology is required to achieve optimal results in patients with any associated conditions in the future.

References

1. Scales J (1953) Discussion on metals and synthetic materials in relation to soft tissues: tissue reaction to synthetic materials. Proc R Soc Med 46:647
2. Becker EL, Landau SI (1986) International dictionary of medicine and biology, vol 1. Churchill Livingstone, New York/Edinburgh/London/Melbourne
3. Amon M, Menapace R (1994) In vivo documentation of cellular reactions on lens surfaces for assessing the biocompatibility of different intraocular implants. Eye 8 (Pt 6):649–656
4. Wolter JR (1983) Morphology of the capsule-like portion of the reactive membranes on intraocular lens implants. Graefes Arch Clin Exp Ophthalmol 220:58–65
5. Wolter JR (1982) Cell life on the surface of lens implants. Graefes Arch Clin Exp Ophthalmol 218:244–249
6. Amon M (2001) Uveal and capsular Biocompatibility of Intraocular Implants. J Cataract Refract Surg 27(2):178–179
7. Miyake K, Asakura M, Kobayashi H (1984) Effect of intraocular lens fixation on the blood-aqueous barrier. Am J Ophthalmol 98:451–453
8. Spalton DJ, Shah SM, Kerr Muir MG (1993) Specular microscopy of the anterior intraocular lens surface. Eye 7:707–710
9. Nishi O (1999) Posterior capsule opacification. Experimental investigations. J Cataract Refract Surg 25:106–117
10. Apple D, Solomon K, Tetz M (1991) Posterior capsule opacification. Surv Ophthalmol 37:73–116
11. Schauersberger J, Kruger A, Schild G, Amon M (2001) Lens epithelial cell outgrowth on three different intraocular lenses. J Cataract Refract Surg 27(6):850–854
12. Schild G, Schauersberger J, Amon M, Abela-Formanek C, Kruger A (2005) Lens epithelial cell ongrowth: comparison of 6 types of hydrophilic intraocular lens models. J Cataract Refract Surg 31(12):2375–2378
13. Wolter JR (1993) Continuous sheet of lens epithelium on an intraocular lens: pathological confirmation of specular microscope. J Cataract Refract Surg 19:789–792
14. Linnola R, Werner L, Pandey S, Escobar-Gomez M, Znoiko S, Apple D (2000) Adhesion of fibronectin, vitronectin, laminin, and collagen type IV to intraocular lens materials in pseudophakic human autopsy eyes. Part 1: histological sections. J Cataract Refract Surg 26:1792–1806
15. Wenzel M, Reim M, Heinze M, Böcking A (1988) Cellular invasion on the surface of intraocular lenses. In vivo cytological observations following lens implantation. Graefes Arch Clin Exp Ophthalmol 226:449–454
16. Versura P, Caramazza R (1992) Ultrastructure of cells cultured onto various intraocular lens materials. J Cataract Refract Surg 18:58–64
17. Kruger A, Schauersberger J, Abela C, Schild G, Amon M (2000) Two year results: sharp versus rounded optic edges on silicone lenses. J Cataract Refract Surg 26:566–570

18. Abela-Formanek C, Amon M, Schauersberger J, Kruger A, Nepp J, Schild G (2002) Results of hydrophilic acrylic, hydrophobic acrylic and silicone intraocular lenses in uveitic eyes with cataract: comparison to a control group. J Cataract Refract Surg 28(7):1141–1152
19. Ravalico G, Baccara F, Lovisato A, Tognetto D (1997) Postoperative cellular reaction on various intraocular lens materials. Ophthalmology 104:1084–1091
20. Amon M (1992) The importance of in-vivo documentation of cellular reactions on lens-surfaces for assessing the biocompatibility of different intraocular implants. Spektrum der Augenheilkunde 6(7): Suppl 7
21. Abela-Formanek C, Amon M, Kahraman G, Schauersberger J, Dunavoelgyi R (2011) Biocompatibility of hydrophilic acrylic, hydrophobic acrylic, and silicone intraocular lenses in eyes with uveitis having cataract surgery: long-term follow-up. J Cataract Refract Surg 37(1):104–112. doi:10.1016/j.jcrs.2010.07.038
22. Richter-Mueksch S, Kahraman G, Amon M, Schild-Burggasser G, Schauersberger J, Abela-Formanek C (2007) Uveal and capsular biocompatibility after implantation of sharp-edged hydrophilic acrylic, hydrophobic acrylic, and silicone intraocular lenses in eyes with pseudoexfoliation syndrome. J Cataract Refract Surg 33(8):1414–1418

Capsule-Bending Ring for the Prevention of Posterior Capsule Opacification

19

Okihiro Nishi, Kayo Nishi, and Rupert Menapace

Abstract

In 64 patients, a capsule-bending ring (CBR), an open poly(methyl methacrylate) (PMMA) ring with a truncated edge profile that should create a sharp bend in the equatorial capsule, was implanted in one eye of patients with a hydroxy(ethyl methacrylate) (HEMA) intraocular lens (IOL). The contralateral eye received only the IOL as control. Anterior capsule opacification (ACO) and shrinkage were significantly reduced in the eyes with the ring. Posterior capsule opacification (PCO) score (EPCO) was 0.235 ± 0.215, 0.287 ± 0.200, and 0.398 ± 0.248 with the ring and 0.530 ± 0.190, 0.670 ± 0.225, and 1.111 ± 0.298 without the ring, at 6 months, 1 year, and 2 years, respectively ($P < 0.01$ at each period). Nd:YAG laser posterior capsulotomy was performed in 4 eyes with ring and 17 eyes without the ring, respectively, after 2 years ($P < 0.01$). The CBR significantly reduced anterior capsular fibrosis and shrinkage as well as PCO. The ring may be useful for those patients who are at high risk of developing eye complications from capsular opacification requiring Nd:YAG laser capsulotomy, for those who are expected to undergo vitreoretinal surgery and photocoagulation by facilitating better fundus visualization, and for those who have pediatric cataracts.

Keywords

Anterior capsule opacification • Capsule-bending ring • Capsule tension ring • Discontinuous capsular bend • Posterior capsule opacification

Proprietary Interest: The authors have no proprietary interest in the method and products described in this paper.

O. Nishi, M.D. (✉) • K. Nishi, M.D.
Nishi Eye Hospital, 4-14-26, Nakamichi, Higashinari-ku, Osaka 537-0025, Japan
e-mail: okihiro@nishi-ganka.or.jp

R. Menapace, M.D.
Department of Ophthalmology, Universitäts-Augenklinik, Vienna, Austria

S. Saika et al. (eds.), *Lens Epithelium and Posterior Capsular Opacification*,
DOI 10.1007/978-4-431-54300-8_19,

19.1 Introduction

To prevent posterior capsule opacification (PCO) by removing or killing lens epithelial cells (LECs), various mechanical, physical, or immunological methods have been developed, but none of the methods proved to be satisfactorily practical, effective, and safe as a routine clinical procedure [1, 2]. The introduction of intraocular lenses (IOLs) with posterior sharp edges has, however, significantly reduced the formation of PCO and the subsequent need for neodymium:YAG (Nd: YAG) laser capsulotomy. To explain the effect, we proposed a new concept based on our experimental and clinical observations that the migration of LEC is inhibited [3, 4] at the sharp, "discontinuous"[1] capsular bend which is created by the sharp edge of the optic, due to contact inhibition [5].

However, the sharp edge effect of IOLs, in other words, the formation of a sharp, discontinuous capsular bend, can fail in several circumstances, which may explain the reason that the sharp edge is not able to prevent PCO in 100 % of the cases. The first is that the discontinuous capsular bend at the posterior optic edge is formed only when there is circumferential overlap of the optic by the anterior capsule leaf, which is not always achieved. Second, LECs migrate posteriorly in 1–2 weeks [6] and the capsular bend is formed in 2–4 weeks after surgery [7], i.e., some LECs can migrate posteriorly over the capsular bend, before it is formed and will proliferate causing later PCO. Once the bend is formed, the subsequent migration of LECs will be stopped. Third, LECs at the capsular bend were found to be in the G0 phase of the cell cycle, indicating that they were contact inhibited [5], while other LECs apart from the capsular bend proliferate, forming Soemmering's ring. This padded, increasingly stuffed after cataract can break up the capsular bend years after surgery by spreading the capsular bag, and may awake contact-inhibited LECs at the bend from the G0 phase, which will then commence to re-proliferate, migrating posteriorly. This may explain the high 10-year cumulative rate of Nd:YAG laser capsulotomy for the most widely used hydrophobic acrylic IOL [8]. In a long-term follow-up, PCO appears not to be prevented but rather delayed by the IOL with sharp edges.

These observations suggest the need for a reevaluation of the sharp edge effect of the IOL optic. Such a bend can be obtained by an optic with sharp edges like that of an acrylic IOL or a capsule tension ring with a certain width and sharp edges. Such a ring can create the capsular bend immediately at the time of cataract surgery, as opposed to the sharp optic edge that requires 2–4 weeks for the bend formation. The formation of Soemmering's ring could, therefore, be reduced and delayed, because

[1] "Discontinuous" is a mathematical concept describing a curve that is not continuous; that is, a curve that, at some point, is abruptly broken or interrupted and can no longer continue to be continuous. In other words, at that point, tangential curves cannot be continuously drawn. With regard to the posterior capsule, this point of curve interruption or "discontinuity" is where the posterior capsule and its contact with the sharp edge of the optic are abruptly bent, thus creating a capsular bend.

the bend formation takes place much more peripherally toward the equatorial zone of the lens capsule, so that it does hardly break up the bend in the late postoperative period. Pursuing the latter possibility, we confirmed in an experimental study using rabbits [9] that the capsule-bending ring (CBR), a modified capsule tension ring with a sharp edge, significantly inhibited the LEC migrating on the posterior capsule and prevented PCO.

19.2 Animal Experiments

19.2.1 Capsule-Bending Ring for Animal Experiments

The CBR [9] is an open, band-shaped poly(methyl methacrylate) (PMMA) ring measuring 11 mm in diameter with pre-tension (13-mm diameter when the ring is open), 0.2 mm in thickness, and 1.0 mm in width (Fig. 19.1a, b). The ring is polished minimally in order to keep the edges sharp and rectangular, thus facilitating the creation of a sharp, discontinuous bend in the equatorial capsule. A crooked eyelet is provided at both ring ends to avoid possible spearing of the capsular fornix and to facilitate manipulation during insertion.

An animal study investigated the inhibitory effect of a discontinuous capsular bend created by a CBR on migrating LECs. This CBR was implanted into the capsular bag in the rabbit eyes. In the histopathological examinations 8 weeks after surgery [9], LECs accumulated at the equatorial corner outside the ring, showing the inhibition of LEC migration. No LECs were found on the posterior capsule. In some eyes, a remarkable finding was observed. Besides LECs at the equatorial corner outside the ring, other LECs accumulated inside the ring forming a mass of LECs, whereby no LECs migrated onto the posterior capsule, although these LECs were in a wide contact with the posterior capsule (Fig. 19.2).

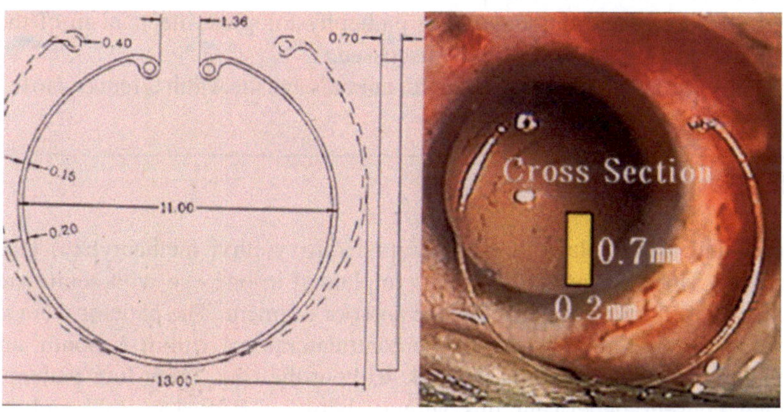

Fig. 19.1 *Left*: Schematic of capsule-bending ring type 14E. *Right*: The figure shows a type 14E ring for clinical use. Note the square-edged cross section (Reprinted with permission [10])

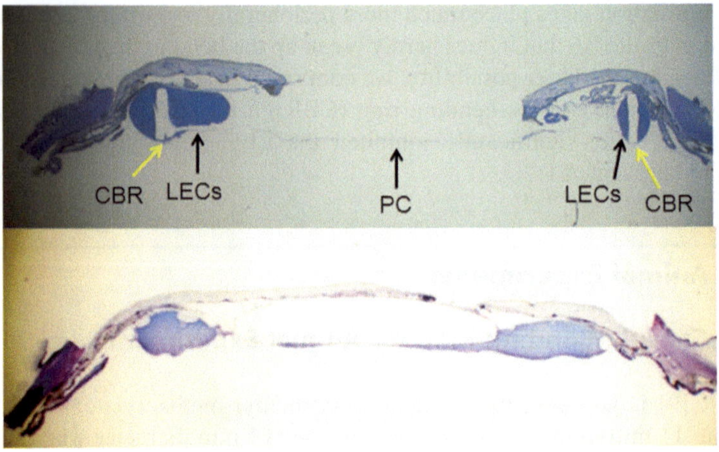

Fig. 19.2 Histopathological sections. *Top*: Rabbit lens capsule section containing a capsule-bending ring (CBR) 4 weeks after surgery. The posterior capsule is clean. The LECs (*black arrows*) proliferated on the CBR (*yellow arrows*), yet they never migrated onto the posterior capsule. LECs are found in a rectangular angle between the CBR and anterior and posterior capsules, indicating that the sharp discontinuous capsule bend induced contact inhibition as in vitro on a well bottom. *Bottom*: In the control eye, the LECs formed a Soemmering's ring cataract and migrated onto the posterior capsule (Reprinted with permission [9])

19.3 Capsule-Bending Ring for Humans

Based on the animal study, Nishi and Menapace modified the ring for humans [10] (CBR type 14E, Morcher, Germany) (Fig. 19.1a, b) and performed a clinical trial with the ring at two institutions: Nishi Eye Hospital, Osaka, and Department of Ophthalmology, University of Vienna. The two-year clinical results in Nishi Eye Hospital and 3-year results in Department of Ophthalmology, University of Vienna, are compiled here, and the underlying pathophysiological mechanism of the preventive effect of the ring on PCO is discussed.

Basically, the ring is the same as that for animals, but the width is reduced to 0.7 mm.

19.4 Patient Selection

The ring and a foldable IOL made from hydroxy(ethyl methacrylate) (HEMA) (Hydroview H60M, Storz, USA) were implanted in one eye with senile cataract without any ocular abnormalities in the anterior segment. The patients were scheduled to undergo similar surgery in the contralateral eye within 1 month, and the contralateral eye received only the IOL without the ring. Sixty-four patients with informed consent underwent this surgery between February 1997 and August 1997 at Nishi Eye Hospital. In the Vienna Study, 60 patients were involved. Inclusion criteria were non-high myopic, bilateral senile cataract, no ocular abnormalities, immature nuclear or cortical cataract, and good pupil dilation.

19.5 Surgical Technique

After phacoemulsification following a self-sealing corneoscleral incision 3 mm in length and continuous curvilinear capsulorhexis (CCC) approximately 4.5 mm in diameter, Healon GV® was injected into the capsular bag and anterior chamber. To ensure that the optic edge was entirely and securely covered by the anterior capsular margin, we created a CCC with a rather small diameter. One end of the ring was inserted into the capsular bag, and the ring was inserted slowly using a forceps, so that the ring end glided along the capsular fornix until the outer end reached the corneoscleral incision. Then the edge of the ring was grasped by a Simcoe IOL forceps and placed into the capsular bag by compression. As an alternative technique, a special inserter that we designed can be used (Geuder, Heidelberg, Germany). Then a foldable Hydroview was implanted into the capsular bag.

19.6 Results

The results shown here are from the Osaka Study [10, 11], unless they are indicated to be from the Vienna Study [8].

19.6.1 Posterior Capsule Opacification

PCO was classified as Elschnig pearl type in all cases. PCO in the eyes with the ring was much less marked in general, compared to that in the eyes without the ring (Fig. 19.3).

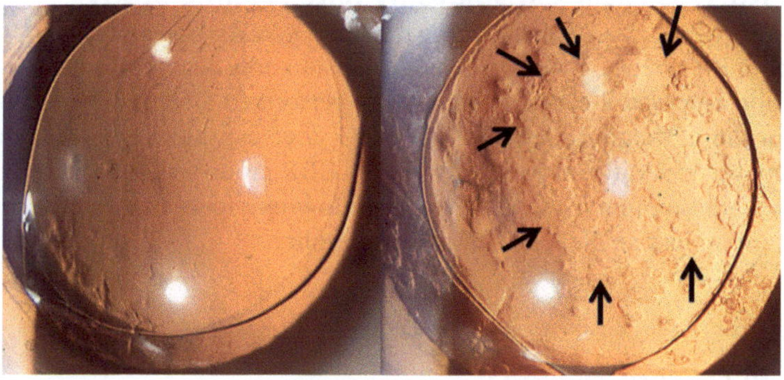

Fig. 19.3 Retroillumination photographs of anterior capsule fibrosis and PCO in an eye with the ring (*left*) and the contralateral eye without the ring (*right*) after 2 years. The capsule with the ring is clear. The capsule without the ring is opacified by Elschnig pearls. The anterior capsule fibrosis in the eye without the ring (indicated by the *arrows*) is difficult to be appreciated under retroillumination (Reprinted with permission [11])

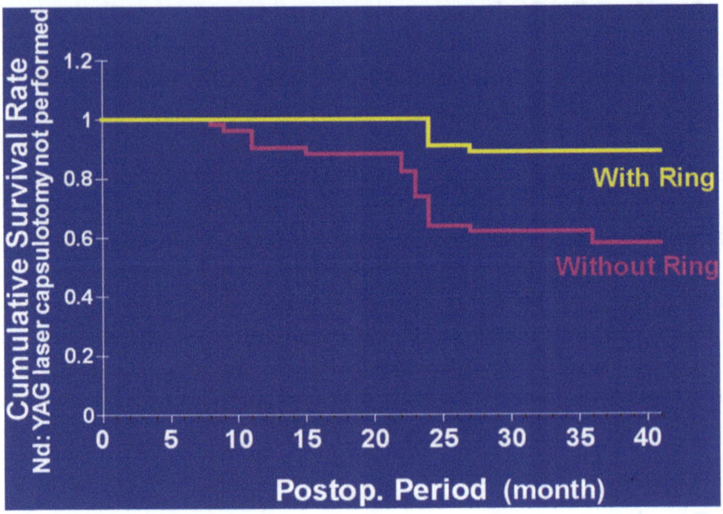

Fig. 19.4 Survival curve of Nd:YAG laser capsulotomy. The difference was significant at 1 and 2 years ($P < 0.01$) (Reprinted with permission [11])

19.6.2 Nd:YAG Laser Capsulotomy Rate

The cumulative number of patients who underwent capsulotomy was 0 with the ring and 5 without the ring at 1 year and 4 and 17 at 2 years, respectively. The difference was significant at both 1 year and 2 years ($P < 0.01$) (Fig. 19.4).

19.6.3 Nd:YAG Laser Capsulotomy Rate (Vienna Study)

In the no-CBR group, 1 Nd:YAG laser capsulotomy was performed before the 1-year follow-up and 2 capsulotomies were performed after the 1-year follow-up; no eye in the CBR group required Nd:YAG laser capsulotomy. After the 2-year follow-up, seven additional patients in the no-CBR group and one in the CBR group required a capsulotomy. At the 3-year follow-up, four additional eyes in the no-CBR group had an Nd:YAG laser capsulotomy.

19.6.4 PCO Score by EPCO

The score did not significantly differ among eyes with the ring, but did significantly differ among eyes without the ring. The score also significantly differed between eyes with and without the ring at each time point (Table 19.1).

Table 19.1 PCO score evaluated by EPCO ($n = 22$)

Post-op period	Eyes with ring	Eyes without ring
6 mons.	0.235 ± 0.215	0.530 ± 0.190
	**	
1 yr.	0.287 ± 0.200	0.670 ± 0.225
	**	
2 yr.	0.398 ± 0.248	1.111 ± 0.298
	**	

(Eyes without ring: 6 mons. vs 2 yr. **; 1 yr. vs 2 yr. **)

There was a significant difference in the score between the two groups at each period. The score was significantly different among eyes without the ring. * indicates $P < 0.01$ (Reprinted with permission [11])

19.6.5 Postoperative Inflammation

On the evaluation of postoperative inflammation by laser flare cellmetry, there was no significant difference in aqueous flare intensity between the subject and the control eyes at day 3, 2 weeks, and 1 month after surgery (Mann–Whitney *U* test).

19.6.6 Pupil Diameter (Vienna Study)

At all examinations, except preoperatively, there was a statistically significant difference between groups in pupil diameter under full medical dilation (tropicamide 1 %, phenylephrine 10 %). The mean diameter was 6.1 mm in the CBR group and 6.6 mm in the no-CBR group.

19.6.7 Anterior Capsular Fibrosis

Fibrosis was seen in 100 % of eyes without the ring after 1 year, whereas nearly 70 % of the eyes with the ring did not show any fibrosis up to 2 years. There was a significant difference between the eyes with and without a ring at each time point ($P < 0.01$, Mann–Whitney *U* test).

19.6.8 Anterior Capsular Shrinkage

Table 19.2 shows the results.

The preventive effect of the ring on anterior capsular shrinkage was evaluated by measuring the area within the CCC using a Scheimpflug camera (EAS-1000, Nidek, Gamagori, Japan). The area of anterior capsular opening was significantly

Table 19.2 Anterior capsular shrinkage (area within CCC) ($n = 22$)

Post-op period	Eyes with ring	Eyes without ring
6 mons.	0.996 ± 0.058 (27.3 mm^2)	0.607 ± 0.122 (17.2 mm^2)
	**	
1 yr.	0.935 ± 0.073 (26.4 mm^2)	0.575 ± 0.113 (16.2 mm^2)
	**	
2 yrs.	0.906 ± 0.071 (23.2 mm^2)	0.529 ± 0.102 (14.5 mm^2)
	**	

The score indicates the index of the CCC area to 1 that corresponds to the area within the 6 mm optic, i.e., 28.3 mm^2. Accordingly, the area of each mean index is given with each parenthesis. Between the two groups, there was a significant difference at each period ($P < 0.01$). Within the same group there was no difference. *indicates $P < 0.01$ (Reprinted with permission [11])

greater in the eyes with the ring at 1 and 3 months after surgery ($P < 0.01$, Mann–Whitney U test).

As capsular fibrosis was noted in around 30 % of the eyes with the ring and 100 % without the ring, anterior capsular shrinkage was generally much less marked in the eyes with the ring. The difference in area was not significant at 6 months, 1 year, and 2 years within the eyes with the ring as well as within those without the ring, but was significant between the eyes with the ring and those without the ring at each time point.

19.6.9 Capsulorhexis-Optic Clearance

There was no space noted in any eye without the ring except for one eye after 6 months and that eye eventually lost the space after 1 year. However, nearly 80 % of the eyes with the ring showed still a distance between the rhexis and the IOL optic after 2 years.

19.6.10 Posterior Capsular Stress Folds

19.6.10.1 Osaka Study

The folds that were seen at 6 months disappeared with time in many eyes without the ring. In the only eye with the ring in which a capsular fold was seen, the fold disappeared at the 1 year examination (Table 19.3).

Table 19.3 Posterior capsular folds

Post-op period	Eyes with ring	Eyes without ring
6 months	1/52	25/52
1 year	0/48	15/48
2 years	0/42	12/42

Reprinted with permission [11]

Table 19.4 Lens epithelial cell outgrowth onto the IOL optic

Post-op period	Eyes with ring	Eyes without ring
6 months	0/52	20/52
1 year	0/48	16/48
2 years	0/48	11/42

Reprinted with permission [11]

19.6.10.2 Vienna Study

One week postoperatively, 2 eyes (3.3 %) in the CBR group and 47 eyes (78.3 %) in the no-CBR group had traction folds in the posterior capsule; the difference was statistically significant. Of the 40 eyes evaluated after 1 year, none in the CBR group and 21 (42 %) in the no-CBR group had haptic-induced stress folds in the posterior capsule ($P < 0.001$).

19.6.11 Lens Epithelial Cell Outgrowth onto the IOL Optic

No outgrowth was noted in the eyes with the ring during the whole postoperative period, whereas a significant number of eyes without the ring showed LEC outgrowth (Table 19.4).

19.6.12 Distance of Eyelets (Vienna Study)

Eyelet distance measurements were obtained in 23 eyes (46 %) in the CBR group evaluated at 1 week and 1 year by gonioscopy. At 1 week, close eyelet apposition (distance 0.0–0.5 mm) was found in 6 eyes (26 %). A distance of 0.5–1.0 mm was found in 7 eyes (30 %) and a distance of 1.0–2.0 mm in 9 eyes (39 %). One eye had eyelet apposition. By 1 year, the number of eyes with close eyelet attachment had risen to 16 (70 %) and the number of eyes with a distance greater than 1.0 mm had decreased to 3 (13 %). Three eyes had eyelet overlap of less than 0.5 mm. In the only eye with an overlap exceeding 0.5 mm, extensive capsulorhexis-optic contact had caused substantial fibrotic shrinkage of the capsulorhexis.

19.6.13 Intraocular Pressure (Vienna Study)

Postoperatively, intraocular pressure decreased by approximately 4 mmHg in both groups but did not change thereafter. There was no statistically significant difference at any time between the two groups.

19.6.14 Best Corrected Visual Acuity (Vienna Study)

There was no difference between the CBR group and the no-CBR group in mean best corrected visual acuity (BCVA) 1 week after surgery. From 1 to 3 years, BCVA was better in the CBR group; the difference was statistically significant at 1 and 2 years.

19.7 Comments and Discussions

19.7.1 Posterior Capsule Opacification

Because the criteria for YAG laser capsulotomy were mostly subjective, and there was no assurance that the criteria were applied consistently, we evaluated PCO by EPCO which should be more objective. PCO in terms of Nd:YAG laser capsulotomy rate as well as EPCO score was also significantly reduced by the ring.

PCO, however, could not be completely prevented by the ring, though the PCO seen in these eyes was not very marked. This may be due to several factors. The LECs most distal from the anterior capsule center could remain within the ring and later proliferate onto the posterior capsule. The possibility of this condition may depend on the anatomical structure of the individual anterior segment and the width of the ring. A wider ring might affect all the peripheral LEC outside the posterior capsular bend that is created by the posterior ring edge, but a wider ring may cause iris or ciliary body chafing.

Another possibility is that the two ends of the open ring were not overlapped or closely apposed within the capsular bag due to the discontinuous nature of the ring, and the LEC migrated between the two ring ends onto the posterior capsule. The condition of the ring ends could be observed by gonioscopy, and we found such a condition in some eyes. Therefore, Nishi and Menapace have modified the ring; the eyelet at one end was abolished and the end slightly elongated, so that this end can overlap the crooked eyelet of the other ring end in the capsular bag.

19.7.2 Anterior Capsular Fibrosis and Shrinkage, Capsular Folds, and LEC Outgrowth

Our results clearly showed that anterior capsular fibrosis and shrinkage, capsular folds, and LEC outgrowth were significantly prevented by the CBR compared to those in the control eyes at 2 years in the Osaka Study and 3 years following the Vienna Study, respectively. Anterior capsular fibrosis and shrinkage can cause a decrease in the visual acuity, IOL decentration, or capsular block syndrome. Although posterior capsular folds are mostly visually insignificant, they can be regarded as PCO causing visual disturbance under certain circumstances.

We evaluated LEC outgrowth, because hydrogel IOLs as used in this study should show an enhanced response to LECs that grow over the anterior optic surface to a much greater extent [12, 13]. Although the functional consequences of anterior LEC outgrowth are as yet unclear and the visual acuity or contrast sensitivity are not affected when the LECs do not cover the visual axis, IOLs that are associated with increased LEC outgrowth could be anticipated to have worse PCO [12, 13]. In fact, hydrogel IOLs were found to develop significantly more PCO compared with PMMA and silicone IOLs [13].

The 0.7 mm width and stretching effect of the ring on the capsule prevented the anterior capsule from coming into contact with the IOL. This open capsule effect was first described by Hara [14] and recently by other authors [15, 16]. As a result, LEC outgrowth was prevented and LECs underneath the anterior capsule were not induced to undergo fibrous pseudometaplasia. Anterior capsular fibrosis and subsequent shrinkage were, thus, prevented. As for capsular shrinkage in terms of the area of capsular opening, it decreased significantly in the eyes without ring, whereas the decrease in capsular opening area in the ring eyes was far less marked, as the early results (up to 3 months) showed. The results showed also that capsular shrinkage progressed up to 3–6 months after surgery and then remained rather stable in both groups. The ring effect appeared to be almost persistent as the rate of capsular fibrosis, shrinkage, and positive distance formed almost a plateau 1–2 years after surgery in the eyes with the ring, compared to control eyes.

19.7.3 Complications

We did not encounter any serious complications that were related to the ring implantation in the early postoperative period in both Osaka and Vienna studies. Postoperative inflammation was not significantly different, and intraocular pressure was also within the normal range compared to those in the control group. In the subsequent 2-year follow-up and 3-year follow-up in the Vienna Study, we never observed any adverse complications. The ring was well tolerated by the eyes.

19.7.4 Indication

Though both anterior and posterior capsule opacifications were significantly inhibited, routine use of the ring may be questionable. There may be potential complications; implantation requires additional surgical time and there is concern regarding cost. However, those patients who are at high risk of developing complications from capsular opacification requiring Nd:YAG laser capsulotomy might benefit from ring implantation, such as younger patients with high myopia, patients with retinitis pigmentosa which tends to promote anterior capsular shrinkage, and children. For pseudoexfoliation syndrome, we do not recommend ring implantation. The ring may effectively prevent capsular fibrosis and subsequent capsular shrinkage, but ring implantation may be risky for those eyes with weak and defective zonules that might be damaged by the surgical maneuvers during implantation of a very rigid ring. The ring could spear the capsule or cause further damage to zonules leading to possible luxation. Dick et al. [17] modified the ring for children by decreasing the overall diameter and implanting it in pediatric eyes. The results may be encouraging, but longer follow-up will be necessary. Those who are going to or expected to undergo vitreoretinal surgery also might benefit from the ring implantation. Vitreous surgery or photocoagulation may be facilitated by providing better visualization of the ocular fundus.

19.7.5 Pathophysiology of LECs on the Preventive Effect of the CBR

There appears to be two pathophysiological mechanisms for the inhibition of migrating LECs: sharp edge effect and open capsule effect.

19.7.5.1 Sharp Edge Effect

In a large number of our experimental studies on the preventive effect of an IOL with sharp edges [3, 4, 18, 19] on PCO, the IOL, whether acrylic, silicone, or PMMA, created a sharp, discontinuous capsular bend at its sharp edge, and the LEC migration was stopped at the capsular bend (Fig. 19.5). Likewise, the present capsule-bending ring created a sharp discontinuous capsular bend which is located more peripherally toward the equatorial zone of a lens capsule, as shown in the animal study (Fig. 19.2). This phenomenon is comparable to cell cultures. In cell culture, cells cease to proliferate when they reach the rectangularly standing wall on a well bottom, the condition being called "confluent" (Fig. 19.6). These cells never climb on the wall, but occasionally form cell layers in long-term culture. Conceivably, such a cell within the confluent cell layer is pressed out from the cell row due to the increasing pressure within the cell layer (by the growth of each cell, not by cell division that is suppressed by contact inhibition). The cell is pushed out on the confluent original cell layer and begins to proliferate forming another layer. This process is repeated so that the cultured cells can eventually form several layers on the bottom of a well (Fig. 19.6). But even in such a long-term culture forming layers, LECs never climb onto and ascend the wall. This phenomenon can explain the reason that the cells found on the central side of the ring do not migrate onto the posterior capsule (Fig. 19.2). The cells in the corner formed by the CBR and both anterior and posterior capsules are analogous to those on the cell culture forming cell layers, whereby the cells never climb on the cell wall. We have

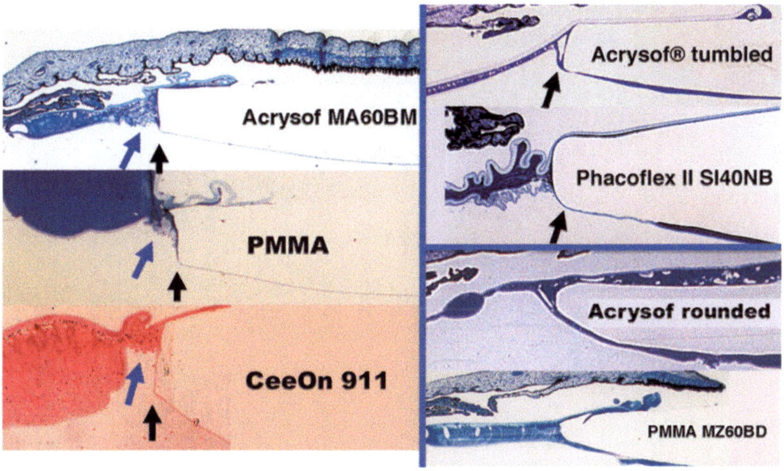

Fig. 19.5 Capsular bend formation of various degrees. The sharp bend is formed at the sharp optic edge (*left column*) regardless of the material composition. The truncated optic rim with round edges (*top right*) creates a bend, but one that is less sharp; the rounded optic rim (two figures on the *bottom right*) is not able to form a bend (Reprinted with permission [16])

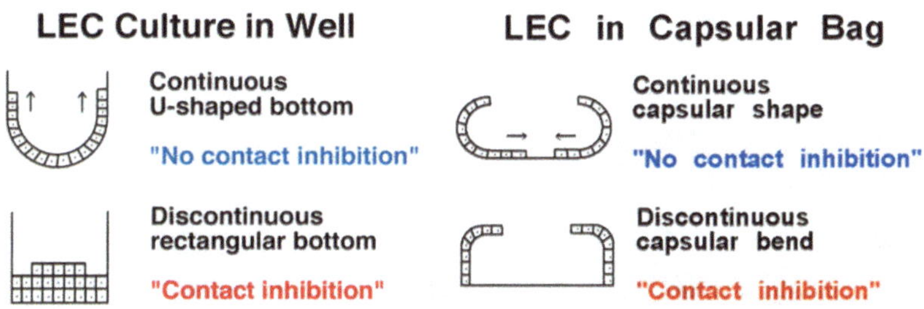

Fig. 19.6 *Left*: Migration and contact inhibition of LECs cultured in a well. *Right*: Migration and contact inhibition in the capsular bag, which is analogous to that observed in culture (drawing on the *left*) (Reprinted with permission [4])

immunohistochemically shown that the sharp discontinuous bend of the capsule on which LECs migrate induced contact inhibition of the migrating LECs: They were in the G0 phase of the cell cycle [5] (Fig. 19.7).

19.7.5.2 Open Capsule Effect (Avoidance of the Contact Between Anterior Capsule and IOL)

This effect is analogous to that reported by Hara [14]. In the more recent reports, Nagamoto [15] and Werner [16] described the same effect. In this effect, two mechanisms can be considered.

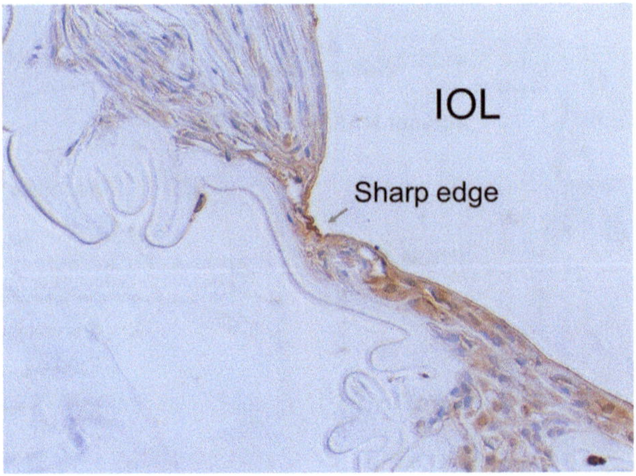

Fig. 19.7 Lens epithelial cells at the capsular bend 7 weeks after surgery in a rabbit eye that received a CeeOn Edge IOL. All cells before the capsular bend were Ki-67 negative, indicating these LECs were contact inhibited. At the bend, there were no LECs. In contrast on the posterior capsule after the bend, there were ample LECs that stained *brownish*, thus positive for Ki-67, indicating they were proliferating. These LECs must have migrated before the formation of the capsular bend (Reprinted with permission [5])

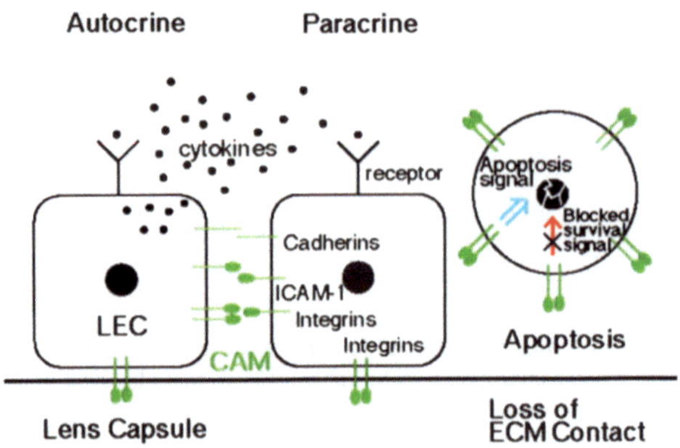

Fig. 19.8 Two modalities in the intercellular signaling and apoptosis of the cell. While secreted cytokines (*dots*) (IL-1, 6, 8, TGF-β, b-FGF, etc.) act in an autocrine or paracrine manner, cell adhesion molecules (CAMs) transduce signals as plasma membrane-bound molecules. Any detachment of the LEC from the underlying ECM, i.e., lens capsule, leads to apoptosis of the cells by activating apoptosis signals or blocking survival signals (Reprinted with permission [1])

19.7.5.3 Decreased Production and Dilution of Cytokines Which Stimulate LEC Proliferation

LECs beneath the anterior capsule can first proliferate when they come into contact with the posterior capsule or IOL optic. Upon the contact, these LECs produce various cytokines [20, 21] which stimulate and promote cell proliferation [22] working in an autocrine or paracrine manner [1] (Fig. 19.8). This stimulus will be further spread toward the equatorial zone (paracrine manner, i.e., to the neighboring cell), so that the neighboring cells are further stimulated to proliferate. When the capsular bag is open so that the anterior capsule does not come into contact with the IOL, LECs cannot proliferate, as cytokine production may be insufficient due to the lack of the contact between the anterior capsule and the IOL. LEC proliferation may be also suppressed, because the cytokines they produce will be diluted and cleared by continuous aqueous humor circulation due to the open capsular bag. Thus, the cytokine production that is required to LEC proliferation is reduced and diluted.

19.7.5.4 Inhibitory Effect of Aqueous Humor

It is speculated that an as yet unknown factor is present that inhibits LEC proliferation. Its detection awaits future studies.

19.8 Conclusion

In conclusion, the capsule-bending ring effectively inhibited ACO and PCO. The concept that a discontinuous capsular bend may induce contact inhibition of migrating LECs and reduce PCO appeared to be proven by the ring. The ring cannot be recommended for routine use due to its potential risks during its insertion, but may be clinically useful in the eyes of children or in the eyes anticipated to require retinal laser or vitrectomy where fundus visualization is of paramount importance.

References

1. Nishi O (1999) Posterior capsule opacification. Part I: experimental investigations. J Cataract Refract Surg 25:106–117
2. Tetz M, Nimsgern C (1999) Posterior capsule opacification. Part II: clinical findings. J Cataract Refract Surg 25:1662–1674
3. Nishi O, Nishi K, Sakanishi K (1998) Inhibition of migrating lens epithelial cells at the capsular bend created by the rectangular optic edge of a posterior chamber intraocular lens. Ophthalmic Surg Lasers 29:587–594
4. Nishi O, Nishi K (1999) Preventing posterior capsule opacification by creating a discontinuous sharp bend in the capsule. J Cataract Refract Surg 25:521–526
5. Nishi O, Yamamoto N, Nishi K, Nishi Y (2007) Contact inhibition of migrating lens epithelial cells at the capsular bend created by a sharp-edged intraocular lens after cataract surgery. J Cataract Refract Surg 33:1065–1070
6. Nishi O (1986) Incidence of posterior capsule opacification in eyes with and without posterior chamber intraocular lenses. J Cataract Refract Surg 12:519–522
7. Nishi O, Nishi K, Akura J (2002) Speed of capsular bend formation at the optic edge of acrylic, silicone, and poly(methyl methacrylate) lenses. J Cataract Refract Surg 28:431–437
8. Menapace R, Sacu S, Georgopoulos M et al (2008) Efficacy and safety of capsular bending ring implantation to prevent posterior capsule opacification. Three-year results of a randomized clinical trial. J Cataract Refract Surg 34:1318–1328

9. Nishi O, Nishi K, Mano C et al (1998) The inhibition of lens epithelial cell migration by a discontinuous capsular bend created by a band-shaped loop or a capsule bending ring. Ophthalmic Surg Lasers 29:119–225
10. Nishi O, Nishi K, Menapace R (1998) Capsule-bending ring for the prevention of capsular opacification: a preliminary report. Ophthalmic Surg Lasers 29:749–753
11. Nishi O, Nishi K, Menapace R, Akura J (2001) Capsular bending ring to prevent posterior capsule opacification: 2 year follow-up. J Cataract Refract Surg 27:1359–1365
12. Amon M, Menapace R (1991) Cellular invasion on hydrogel and poly(methyl methacrylate) implants: an in vivo study. J Cataract Refract Surg 17:774–779
13. Hollick EJ, Spalton DJ, Ursell PG (1999) Surface cytologic features on intraocular lenses. Arch Ophthalmol 117:872–878
14. Hara T, Hara T, Sakanishi K, Yamada Y (1995) Efficacy of equator rings in an experimental rabbit study. Arch Ophthalmol 113:1060–1065
15. Nagamoto T, Tanaka N, Fujiwara T (2009) Inhibition of posterior capsule opacification by a capsular adhesion-preventing ring. Arch Ophthalmol 127:471–474
16. Kavoussi SC, Werner L, Fuller SR et al (2011) Prevention of capsular bag opacification with a new hydrophilic acrylic disk-shaped intraocular lens. J Cataract Refract Surg 37:2194–2200
17. Dick HB, Schwenn O, Pfeiffer N (1999) Implantation of the modified endocapsular bending ring in pediatric cataract surgery using a viscoadaptive viscoelastic agent. J Cataract Refract Surg 25:1432–1436
18. Nishi O, Nishi K (2000) Wickströmk. Preventing lens epithelial cell migration using intraocular lenses with sharp rectangular edges. J Cataract Refract Surg 26:1543–1549
19. Nishi O, Nishi K (2003) Effect of the optic size of a single-piece acrylic intraocular lens on posterior capsule opacification. J Cataract Refract Surg 29:348–353
20. Nishi O, Nishi K, Ohmoto Y (1996) Synthesis of interleukin-1, interleukin 6, and basic fibroblast growth factor by human cataract lens epithelial cells. J Cataract Refract Surg 22:852–858
21. Nishi O, Nishi K, Fujiwara T et al (1996) Effects of the cytokines on the proliferation of and collagen synthesis by human lens epithelial cells. Br J Ophthalmol 80:63–68
22. Nishi O, Nishi K, Imanishi M (1992) Synthesis of interleukin-1 and prostaglandin E2 by lens epithelial cells of human cataracts. Br J Ophthalmol 76:338–341

PCO Prevention with Endocapsular Equator Rings

20

Tsutomu Hara

Abstract

An endocapsular equator ring (E-ring), a closed ring with a square edge, is made of flexible silicone. The loops of the intraocular lens (IOL) are fixed in the inner groove of the E-ring. The ring prevents development of posterior capsular opacification (PCO) by stopping the posterior movement of postoperative metamorphosed lens epithelial cells at the equator. There is no contact between the IOL and the posterior capsule. The posterior capsule retains transparency without touching the IOL. The ring, which has a 9.5-mm outer edge diameter, fits most eyes except those of patients with high myopia. Besides the high success of the device in preventing PCO, the E-ring has another important potential, that is, it facilitates late IOL exchanges. In addition, the developmental process of PCO is shown through histochemical observation.

Keywords

Endocapsular equator ring (E-ring) • IOL exchange • Lens epithelial cells (LECs) • Posterior capsular opacification

Lens epithelial cells (LECs) cause posterior capsular opacification (PCO). After routine phacoemulsification, the LECs begin to change in the pericapsulorhexis area earlier than in the periphery [1] (Fig. 20.1). One layer of static LECs gradually develops into multiple layers of mobile fibroblastic-like cells [2] (Figs. 20.2 and 20.3).

Using an in situ hybridization technique, Azuma et al. found that the postoperative histochemical changes at the anterior lens capsule (ALC) and equator were similar to the changes associated with conventional wound healing [3]

T. Hara, M.D. (✉)
Hara Eye Hospital, Nishi 1-1-1, Utsunomiya 320-0861, Japan
e-mail: office@haraganka.com

S. Saika et al. (eds.), *Lens Epithelium and Posterior Capsular Opacification*,
DOI 10.1007/978-4-431-54300-8_20,

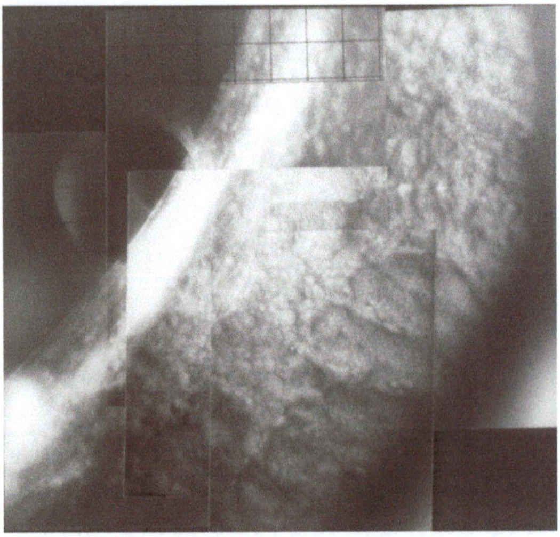

Fig. 20.1 Panoramic specular microscopy of LECs in a human eye 14 days postoperatively. Original magnification × 29. The vertical line in each square is 100 μm. The actual size of the hole in the IOL is 400 μm in diameter, and the distance between the outer rim of the hole and the rim of the IOL optic is 300 μm. The cell density is 1,480 cells/mm^2 near the margin of the continuous curvilinear capsulorhexis and 4,030 cells/mm^2 near the equator

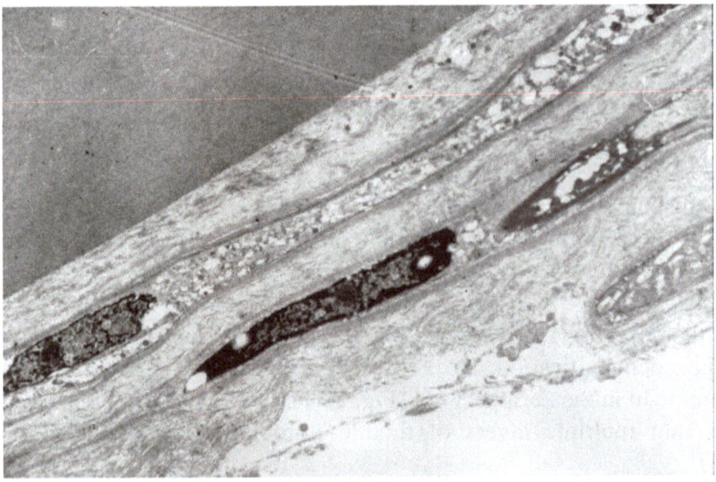

Fig. 20.2 Transmission electron microscopic view of an opacified anterior capsule excised from a human eye. Multilayered cells surrounded by fibers are seen on the inner surface of the anterior capsule (original magnification × 1,200). (Reprinted from Hara et al. [2]. Reproduced with permission of SLACK Incorporated)

(Figs. 20.4 and 20.5). This means that the postoperative changes under the ALC are similar to the normal wound-healing process, indicating that this is a physiologic process and almost impossible to prevent completely. Three possible procedures to prevent PCO can be considered, i.e., removal of all LECs intraoperatively,

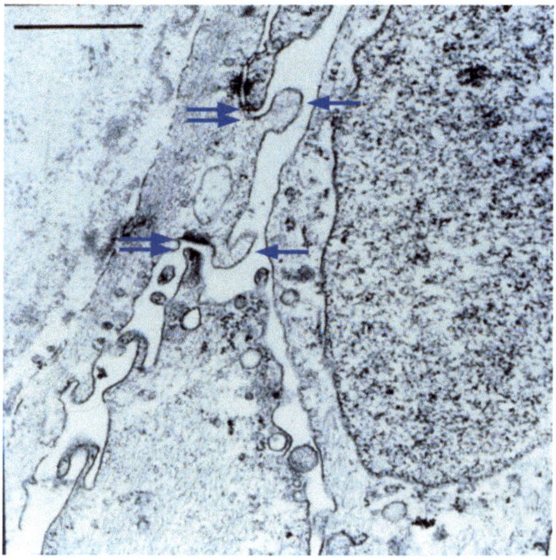

Fig. 20.3 Transmission electron microscopic view from static LECs to mobile fibroblast-like cells. Fibroblast-like cells transformed from the LECs and fibril formation in a human eye 25 months postoperatively. The cells contain many cytoplasmic microtubules and underwent amoeboid processes at their surfaces (*arrows*). In some places, tight junctions are seen (*double arrows*) (original magnification × 250). Bar = 1,000 μm. (Reprinted from Hara et al. [2]. Reproduced with permission of SLACK Incorporated)

mediation of the proliferative process as much as possible if it is impossible to prevent, and allowing the process to continue and stopping it before it reaches the posterior lens capsule (PLC). The endocapsular equator ring (E-ring) was designed based on the last possibility. We previously addressed these three possibilities. In this chapter, we briefly describe the results.

20.1 LEC Removal

Many trials have been performed to remove the LECs from under the ALC. Besides the studies of the E-ring, from 1985, we conducted experiments with Drs. Nishi, Sakka, Hayashi, and Iwata and Menicon Co to restore accommodation using lens refilling after cataract removal; ultimately about six diopters of accommodation were obtained [4–6]. In those studies, preventing opacification in the ALC and PLC was indispensable. We tried many procedures to remove LECs by simple aspiration, aspiration with ultrasound [7], and double cryopexy [8, 9]. However, it was impossible to remove all the LECs, and dense ALC opacification developed. The reason for the failure is shown in Fig. 20.6. After cataract surgery, the LECs become activated. We have to understand that numerous LECs were posterior to the geographic equator. Clinically, it is almost impossible to remove all LECs from that location, especially superiorly. If some cells remain, these activated cells can proliferate more freely in the newly provided cell-free area (Fig. 20.7).

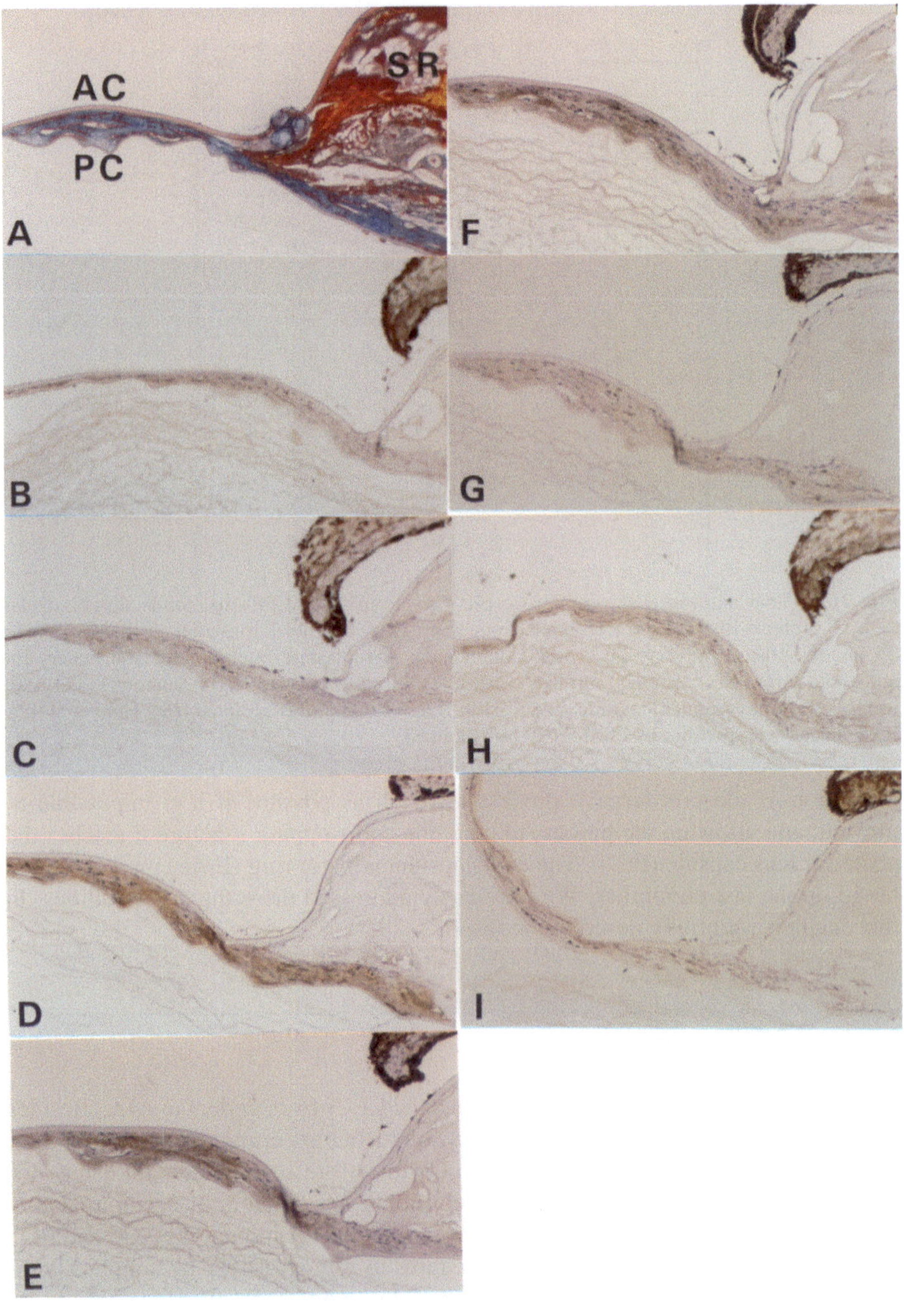

Fig. 20.4 Postoperative proliferative tissue in the peripheral area. (**a**) Fibrous tissue between the anterior capsule (AC) and the posterior capsule (PC) after extracapsular cataract extraction in a control autopsy eye stained with Mallory-Azan. Soemmering's ring (SR) resulted from proliferation of the LECs. The fibrous tissue is immunohistochemically positive for (**b**) chondroitin-6-sulfate, (**c**) dermatan sulfate, (**d**) keratin sulfate, (**e**) type I and (**f**) type III collagen, (**g**) cytokeratin, (**h**) vimentin, and (**i**) α-smooth muscle actin. (Original magnification ×20). (Reprinted from Azuma et al. [3]. Reproduced with permission of Springer-Verlag)

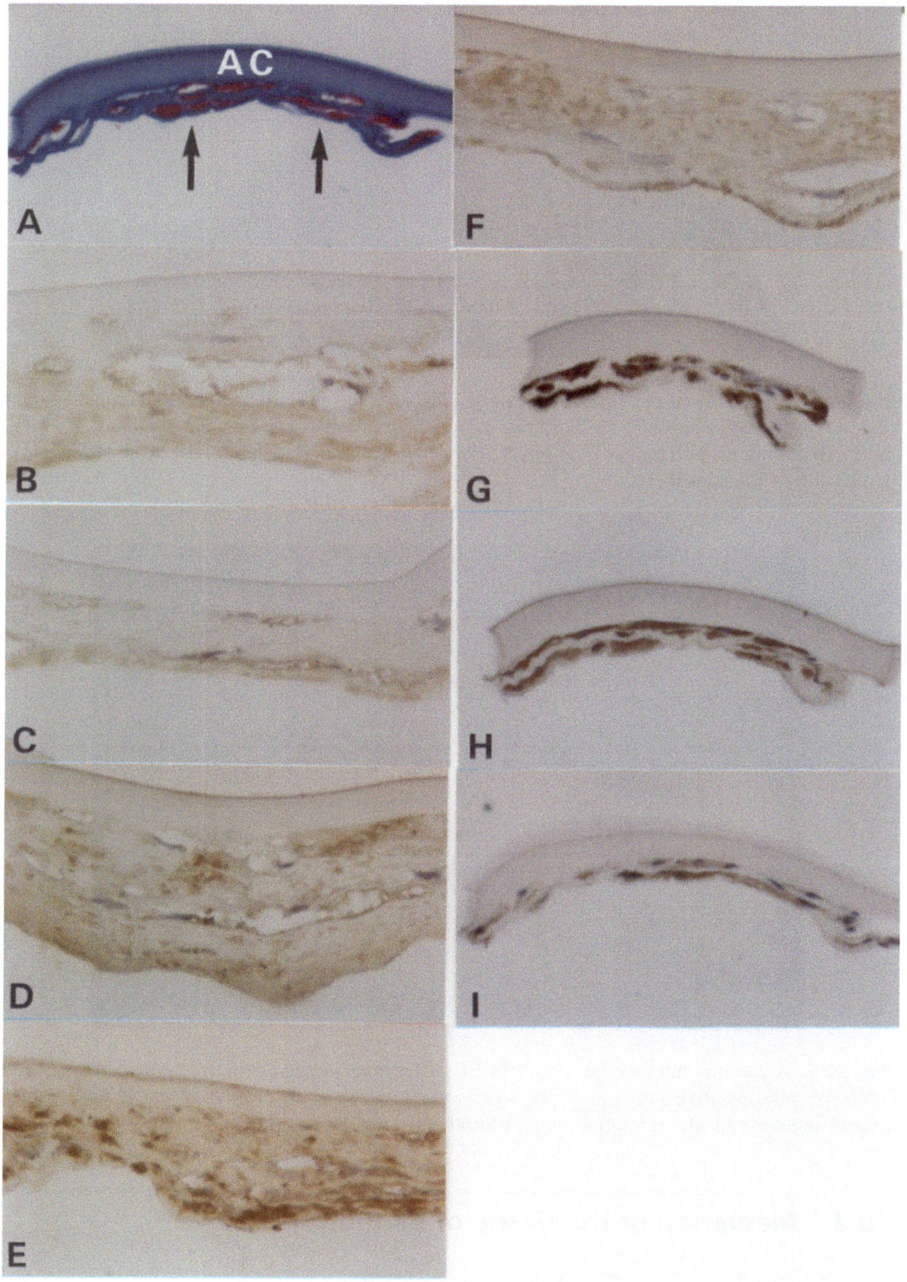

Fig. 20.5 Postoperative proliferative tissue in the central area. An opacified central anterior capsule after endocapsular phacoemulsification and aspiration. (**a**) Mallory-Azan staining. The fibrous tissue (arrows) is underneath the anterior capsule (AC). The opacity is immunohistochemically positive for (**b**) chondroitin-6-sulfate, (**c**) dermatan sulfate, (**d**) keratin sulfate, (**e**) type I and (**f**) type III collagen, (**g**) cytokeratin, (**h**) vimentin, and (**i**) α-smooth muscle actin. (Original magnification ×20). (Reprinted from Azuma et al. [3]. Reproduced with permission of Springer-Verlag)

Fig. 20.6 LECs around the equator. Residual LECs remain posterior to the geographic equator (*arrow*). *AC* anterior capsule, *PC* posterior capsule. (Reprinted from Hogan, Alvarado, Weddell. W. B. Histology of the Human Eye. p 649, Happer & Row, Philadelphia [10]. Reproduced with permission of Elsevier Inc.)

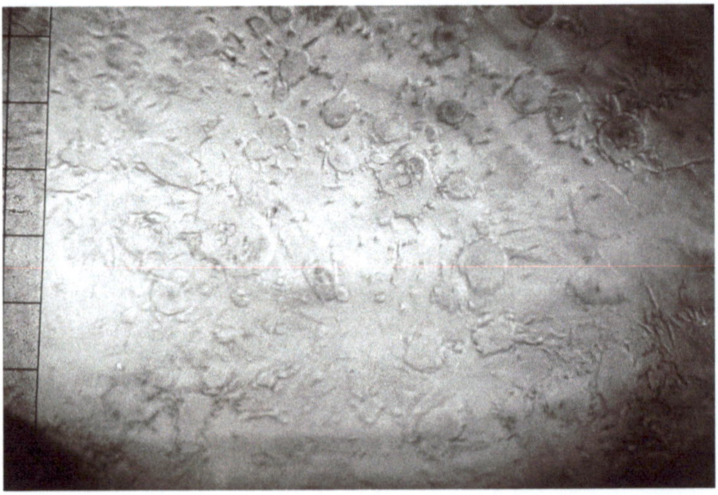

Fig. 20.7 A specular microscopic view of LECs after intensive removal from a human eye. After intensive intraoperative removal of the LECs, the cells begin to proliferate under the retained central anterior capsule. (Original magnification ×18)

20.2 Mediation of the Grade of Metamorphosis

20.2.1 Contact with Fresh Aqueous Humor

Nagamoto et al. later invented a special ring that allowed the LECs to be in contact with fresh aqueous humor [11]. Although we did not notice this initially, retrospectively our E-ring already had adequate space for this purpose. Miyata et al. also reported a study on the use of a fravonoid compound to inhibit LEC proliferation [12]; however, we were not involved in those experiments.

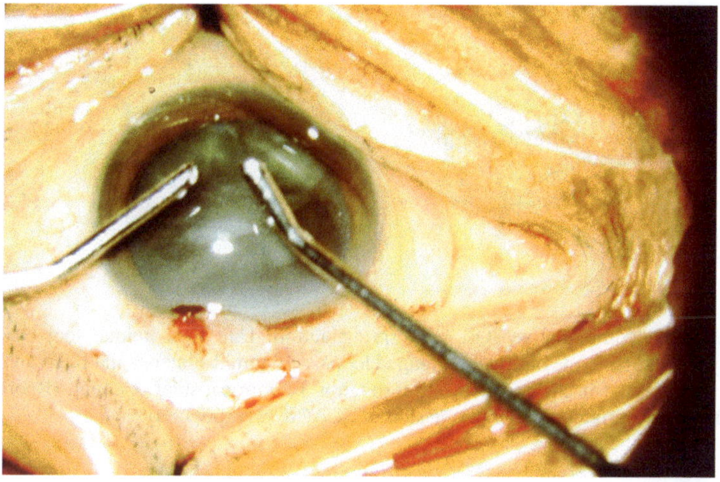

Fig. 20.8 Pioneering technique of endocapsular phacoemulsification with a sleeveless bent ultrasound tip and sleeveless bent irrigation/aspiration tip in 1989. The two tips are used. (*Right*) A 0.9-mm bent fragmatome. (*Left*) A bent 20-gauge May infusion needle. A human cataract is removed through a 1.5-mm scleral incision and 1.5-mm anterior capsular opening. (Reprinted from Hara and Hara [14]. Reproduced with permission of SLACK Incorporated)

20.2.2 Compression of LECs (Compression Inhibition)

At one time, we had been interested in pursuing mechanical restriction to inhibit development of LECs in order to maintain the capsular transparency. We hypothesized that space limitation was one reason for the slow and steady proliferation of the normal LECs inside the capsule throughout life. It was commonly observed that after most LECs were removed, the residual LECs proliferated more and irregularly [13]. During an accommodation study, to prepare for future lens refilling to inject a flexible material into a void capsular bag through a 1.5-mm anterior capsular opening and later seal it, we completed a new technique to remove a hard human cataract through both a 1.5-mm scleral incision and a 1.5-mm anterior capsular opening using a bent sleeveless ultrasound tip and separate bent infusion needle (Fig. 20.8). This was the beginning of the current small-incision procedure. At that time, we often observed the fate of ALCs that were almost completely retained. At that time, it was too premature to inject an experimental material inside an empty human capsular bag. Then, we used conventional hard intraocular lens (IOL). When we implanted a conventional IOL into a void capsular bag, most ALCs later opacified. The adhesion between the capsules and IOL optics was loose. However, when an IOL was not implanted, the ALC and PLC adhered more tightly than their adhesion with the IOL optic. Then, in many cases the ALC remained relatively transparent [2, 13]. In the early 1990s, in other experiments of postoperative accommodation, we invented a flexible double-optic accommodative IOL with spring action (referred to as the spring IOL) [15, 16] (Fig. 20.9) and compared

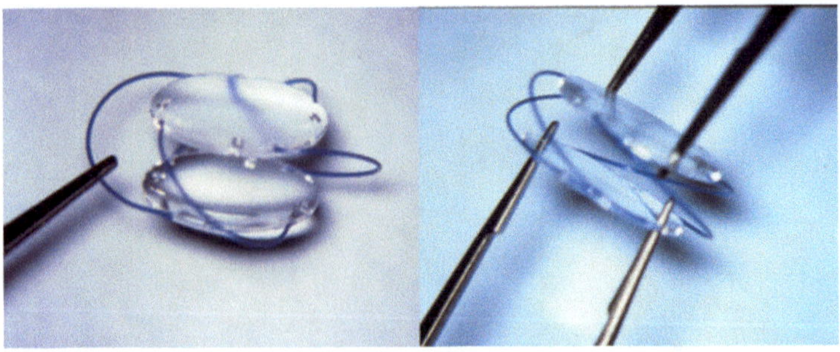

Fig. 20.9 An accommodative IOL with spring action. Due to a unique loop structure, by torsionally compressing the obliquely arranged loops, a 10 mm 3-dimensional IOL can be reduced to 6 mm and inserted in the capsular bag through a 3.5 mm capsular opening. (Reprinted from Hara et al. [15]. Reproduced with permission of SLACK Incorporated)

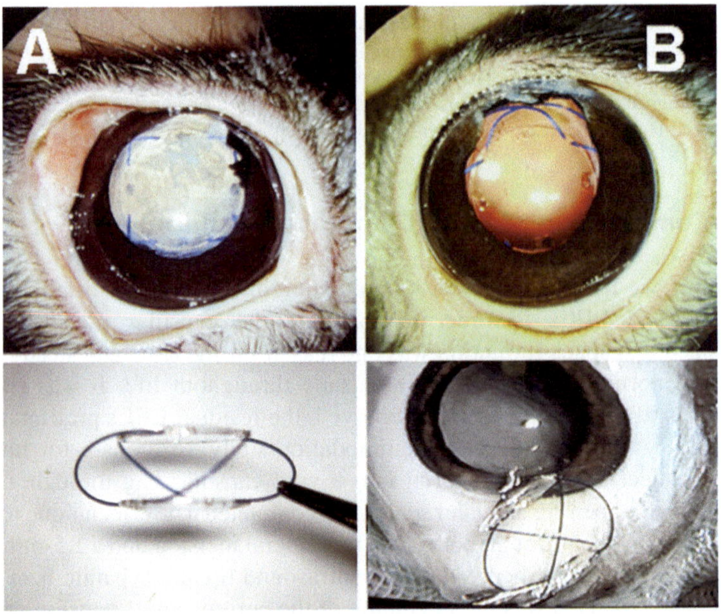

Fig. 20.10 PCO based on the grade of internal compression. (**a**) Severe PCO is seen in the eye in which a small spring IOL was implanted. (**b**) Transparency is maintained in the eye that received a large spring IOL. (This figure was published in Hara et al. [13], Copyright European Society of Cataract and Refractive Surgeons)

two types of spring IOLs. When the capsule was pressed tightly from inside by a large expanding IOL, the ALC and PLC remained transparent. However, later cellular proliferation occurred from the free equatorial area. When the internal compression was weak because of a loose small spring IOL, the entire capsule opacified (Fig. 20.10). We did not implant the IOL in the human eye at the time.

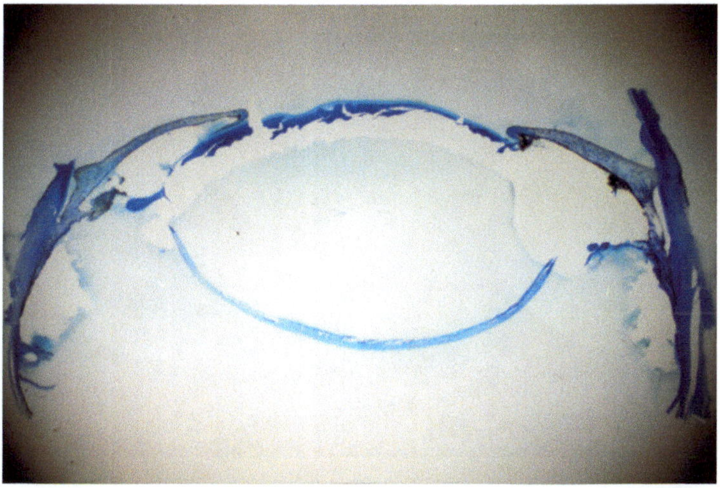

Fig. 20.11 PCO in eyes with an endocapsular balloon. Proliferative tissue in the space between the capsule and the balloon. The balloon filled with silicone was lost during tissue preparation (original magnification ×7). (Reprinted from Hara et al. [5]. Reproduced with permission of Elsevier Inc.)

However, the basic concept was succeeded to a dual-optic accommodative IOL, which was implanted in the human eye and is currently undergoing clinical trials [17]. An endocapsular balloon, which also was used for lens refilling to restore accommodation, seemed ideal for achieving even compression of the entire capsular bag from inside. However, the results were unsatisfactory [5] (Fig. 20.11). It was difficult to conduct further trials in humans at the time.

20.3 Preventing the Process at the Equator: E-Ring

In 1991, we introduced our new idea for the E-ring.

20.3.1 E-ring Structure

The device is a flexible silicone closed ring with a square edge (Fig. 20.12). The height and the width are both 1 mm (1,000 μm). Through subsequent repeated trials, it became clear that the outer diameter of the 9.5-mm ring fit all human eyes except those with high myopia. The ring has a groove in the inner surface where both IOL loops are engaged. The total weight of the 9.5-mm E-ring (23.6 mg) and IOL (10.6 mg) is 34.2 mg, which is only about one-sixth of the 230 mg of the human crystalline lens in an 80-year-old patient [19].

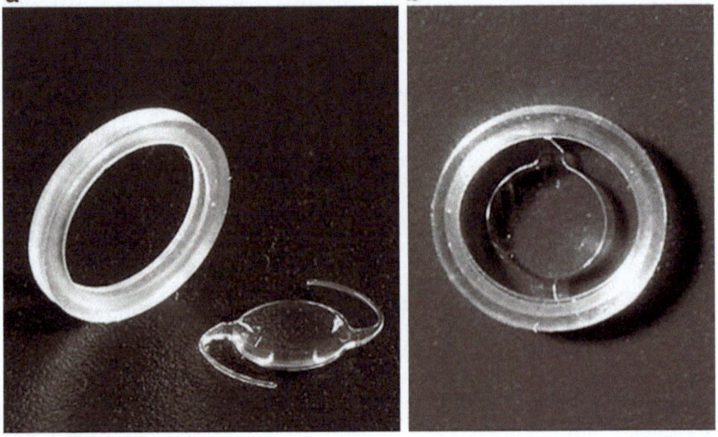

Fig. 20.12 The E-ring. (**a**) The endocapsular equator ring is a closed silicone ring with a square edge, an outer diameter of 9.5 mm, and width and thickness of 1.0 mm respectively. (**b**) A groove on the inner surface facilitates fixation of the IOL loops. (Reprinted from Hara et al. [18]. Reproduced with permission of American Medical Association)

20.3.2 Surgical Technique

Using a conventional IOL injector, the E-ring can be inserted into an empty capsular bag through a 3.2-mm sutureless limbal incision.

20.3.3 History of the E-Ring

After the first report of the E-ring [20] was published, rabbit eyes were studied [21]. The E-ring maintained the circular shape of the equator and successfully prohibited development of PCO (Figs. 20.13 and 20.14). The results were confirmed in monkey eyes [22]. The anterior chamber angle remained wide, in contrast to the narrow anterior chamber angle in the rabbit eyes (Fig. 20.15). Ultimately, the first case report of a young man with atopic eye was published in 2007 (Fig. 20.16) [23]. In 2011, we reported the results achieved with 14 patients with 2–7 years of follow-up [18]. These patients received an E-ring and IOL in one eye and only an IOL in the fellow eye. The mean ± standard deviation PCO values in the eyes with an E-ring centrally, nasally, and temporally were 4.4 ± 2.47, 2.0 ± 0.95, and 3.3 ± 5.75, respectively. The values in the control eyes were 11.4 ± 5.42, 25.5 ± 13.31, and 20.0 ± 10.11, respectively. The eyes with an E-ring had significantly ($P = 0.005$, $P = 0.001$, and $P = 0.001$, respectively) lower values. No eyes with an E-ring required Nd:YAG laser posterior capsulotomy postoperatively compared with six (43 %) of 14 control eyes. The study proved the ability of the E-ring to prevent PCO.

a b

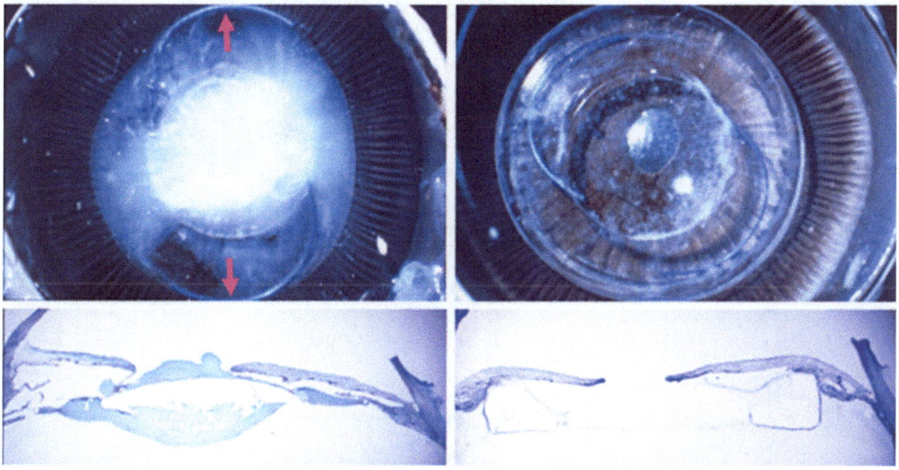

Fig. 20.13 An E-ring in rabbit eyes. (**a**) A Miyake view of a rabbit eye implanted with a 13.0-mm IOL alone. The superior and inferior equator is pushed slightly outward by the loop (*arrow*). Both the ALC and PLC are severely opacified. (**b**) An eye with an E-ring. The *circular shape* of the equator is well retained. Proliferation is prohibited and capsular transparency is retained. The iris root is elevated. The follow-up period was 2.2 months (original magnification ×7). (Reprinted from Hara et al. [21]. Reproduced with permission of American Medical Association)

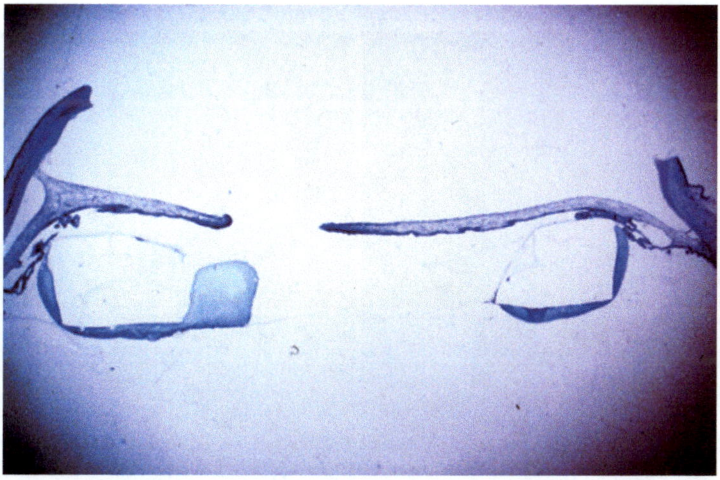

Fig. 20.14 A failed case. On the *left*, the equatorial cells proliferated under the ring onto the PLC (*toluidine blue*, original magnification ×7). (Reprinted from Hara et al. [21]. Reproduced with permission of American Medical Association)

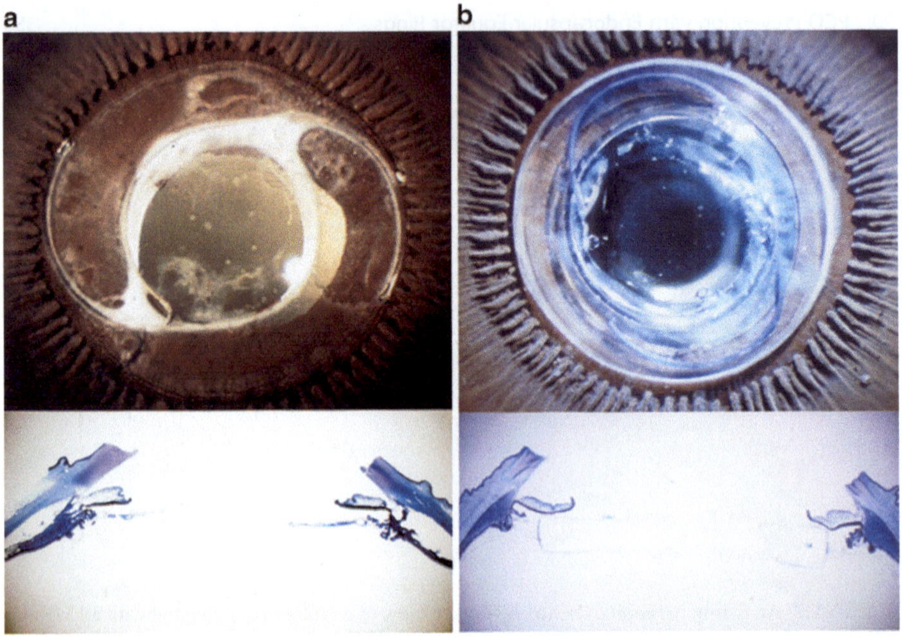

Fig. 20.15 E-ring in a monkey eye. (**a**) A monkey eye with only an IOL. The PLC is moderately opacified as a result of LEC proliferation, and the capsular bag is deformed by compression of the IOL loops. (**b**) A monkey eye with an E-ring and an IOL. The PLC is clear, and the capsular equator is round. The IOL loops are in the groove, and there is an open space between the IOL and the PLC. The anterior chamber angle remains wide. (Reprinted from Hashizoe et al. [22]. Reproduced with permission of Springer-Verlag)

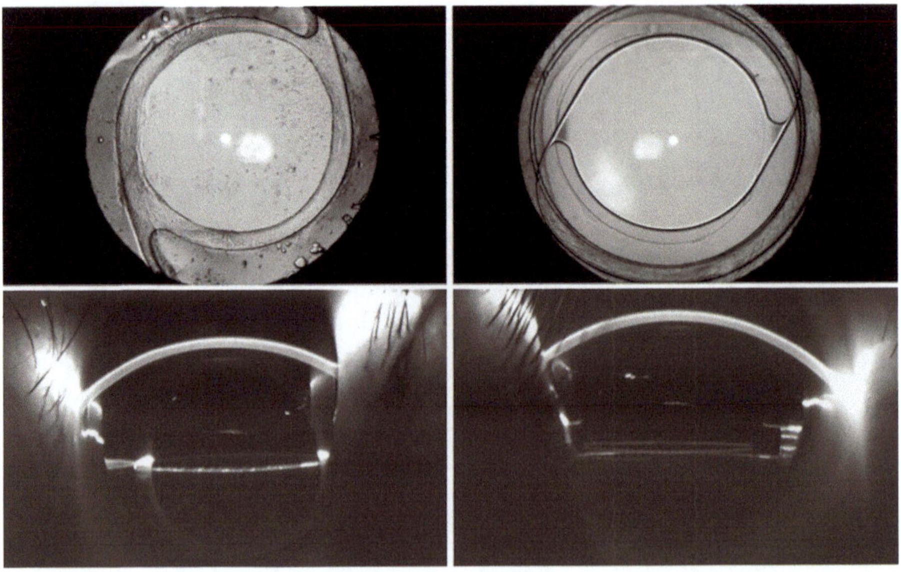

Fig. 20.16 E-ring in a human eye. (*Right*) This is a view of an E-ring in a 22-year-old man with an atopic eye, obtained 2 years after implantation. (*Left*) A control eye. In the eye with the E-ring, the device and the IOL remain in the correct position and the PLC is clear over an area wider than the IOL optic. There is no contact between the ring and the iris and ciliary body. The IOL optic and loops are not in contact with the PLC. No PCO has developed. (Reprinted from Hara et al. [23]. Reproduced with permission of American Medical Association)

20.3.4 Mechanisms of the E-Ring for Preventing PCO

The benefits of the E-ring include the highly preventive effect of the square edge; at a narrow equator the effect doubled by the presence of an anterior and posterior square edge; the sufficient width (1,000 μm) of the device to suppress 71 LECs (14 μm each) around the equator (Fig. 20.6); and the maintenance of adequate space in the bag for easy contact between the postoperative LECs and fresh aqueous fluid.

20.3.5 E-Ring Features

The device retains capsular transparency without touching the IOL optic. Late IOL exchange becomes a possibility. The device has a possible application for pediatric cataract.

Because the circular shape of the bag equator is almost completely retained, postoperative capsular bag contraction is minimized, which consequently prevents IOL dislocation and retinal detachment. Thus, it is highly indicated for young patients with or without atopy.

Despite being a closed ring, the 9.5-mm outer diameter ring fits most eyes, except highly myopic eyes. The device can be implanted through a 3.2-mm sutureless incision and can be used during most routine cataract procedures without zonular and PLC disorders.

The E-ring makes possible the exact adjustment and easy modification of the axis orientation of a toric IOL.

Besides preventing PCO, the E-ring also facilitates late IOL exchanges. In light of this, we believe that the IOL and the E-ring should remain separate and not be combined into one product.

The E-rings for animal experiment were produced by Menicon Co., Nagoya, Japan, and those for human eyes were produced by Morcher GmbH, Stuttgart, Germany. The E-ring is available commercially from Morcher GmbH.

References

1. Hara T, Hara T (1992) Panoramic specular microscopy of lens epithelial cells in the postoperative pseudophakic eye. J Cataract Refract Surg 18(6):589–593
2. Hara T, Azuma N, Chiba K et al (1992) Anterior capsular opacification after endocapsular cataract surgery. Ophthalmic Surg 23(2):94–98
3. Azuma N, Hara T, Hara T (1998) Extracellular matrix of opacified anterior capsule after endocapsular cataract surgery. Graefes Arch Clin Exp Ophthalmol 236(6):531–536
4. Nishi O, Hara T, Hara T et al (1989) Further development of experimental techniques for refilling the lens of animal eyes with a balloon. J Cataract Refract Surg 15(5):584–588
5. Hara T, Sakka Y, Sakanishi K et al (1994) Complication associated with endocapsular balloon implantation in rabbit eyes. J Cataract Refract Surg 20(5):507–512
6. Sakka Y, Hara T, Yamada Y et al (1996) Accommodation in primate eyes after implantation of refilled endocapsular balloon. Am J Ophthalmol 21(2):210–212

7. Nishi O (1987) Removal of lens epithelial cells by ultrasound in endocapsular cataract surgery. Ophthalmic Surg 18(8):577–580
8. Caldwell D (1984) Freezing of the posterior capsule in extracapsular extraction. Highl Ophthalmol 12(7):1–14
9. Hara T, Hara T (1988) Observations on lens epithelial cells and their removal in anterior capsule specimens. Arch Ophthalmol 106(12):1683–1687
10. Hogan MJ, Alvarado JA, Weddell JE (1971) Histology of the human eye. W. B. Saunders Company, Philadelphia, p 649
11. Nagamoto T, Tanaka N, Fujiwara T (2009) Inhibition of posterior capsule opacification by a capsular adhesion–preventing ring. Arch Ophthalmol 127(4):471–474
12. Miyata Y, Oshitari T, Okuyama Y (2013) Polymethoxyflavones as agents that prevent formation of cataract: nobiletin congeners show potent growth inhibitory effects in human lens epithelial cells. Bioorg Med Chem Lett 23(1):183–187
13. Hara T, Hara T, Yasuda A et al (1993) Retaining postoperative anterior capsular transparency without lens epithelial cell removal. Eur J Implant Ref Surg 5(1):42–45
14. Hara T, Hara T (1989) Endocapsular phacoemulsification and aspiration (ECPEA)-Recent surgical technique and clinical results. Ophthalmic Surg 20:469–475
15. Hara T, Hara T, Yasuda A et al (1990) Accommodative intraocular lens with spring action. Part 1. Design and placement in an excised animal eye. Ophthalmic Surg 21(2):128–133
16. Hara T, Hara T, Yasuda A et al (1992) Accommodative intraocular lens with spring action. Part 2. Fixation in the living rabbit. Ophthalmic Surg 23(9):632–635
17. Ossma IL, Galvis A, Vargas LG et al (2007) Synchrony dual-optic accommodating intraocular lens. Part 2: pilot clinical evaluation. J Cataract Refract Surg 33(1):47–52
18. Hara T, Hara T, Narita M et al (2011) Long-term study of posterior capsular opacification prevention with endocapsular equator rings in humans. Arch Ophthalmol 129(7):855–863. doi:10.1001/archophthalmol.2011.38
19. Worgul BV (1982) Lens. In: Jakobiec FA (ed) Ocular anatomy, embryology, and teratology. Harper & Row, Philadelphia, p 360
20. Hara T, Hara T, Yamada Y (1991) "Equator ring" for maintenance of the completely circular contour of the capsular bag equator after cataract removal. Ophthalmic Surg 22(6):358–359
21. Hara T, Hara T, Sakanishi K et al (1995) Efficacy of equator rings in an experimental rabbit study. Arch Ophthalmol 113(8):1060–1065
22. Hashizoe M, Hara T, Ogura Y (1998) Equator ring efficacy in maintaining capsular bag integrity and transparency after cataract removal in monkey eyes. Graefes Arch Clin Exp Ophthalmol 236(4):375–379
23. Hara T, Hara T, Hara T (2007) Preventing postoperative posterior capsule opacification with an endocapsular equator ring in a young human eye: 2-year follow-up. Arch Ophthalmol 125 (4):483–486

PCO Prevention with IOLs Maintaining an Open or Expanded Capsular Bag

21

Anne Floyd, Liliana Werner, and Nick Mamalis

Abstract

Postoperative capsular opacification is a multifactorial physiological consequence of cataract surgery. Opacification may involve the anterior or posterior capsules and may have a significant impact on visual function. Capsular opacification, composed of cortical/pearl and/or fibrotic components, can disrupt the proper functioning of an intraocular lens, particularly specialized ones such as accommodating lenses. It has been hypothesized that intraocular lens designs that maintain an open or expanded capsular bag are associated with better bag clarity. This may be due to mechanisms that include mechanical compression of residual lens epithelial cells within the capsular bag by a relatively bulky device/intraocular lens with overall inhibition of residual lens epithelial cells metaplasia and migration/proliferation of these cells. Another factor may be the mechanical stretch of the bag at the level of the equatorial region, maintaining the overall bag contour. Lastly, constant irrigation of the inner capsular bag compartment by the aqueous humor may also have an influence on the prevention of proliferation of residual cells. This chapter discusses devices that have been designed to minimize or prevent the development of anterior and/or posterior capsule opacification. Devices described include capsular rings and different designs of intraocular lenses which are intended for implantation within the capsular bag after phacoemulsification. The commonality among these devices is to maintain an open or expanded capsular bag, which effectively prevents capsular opacification. While some of the devices are already in clinical use, others are currently under investigation in animal studies.

Keywords

Capsular ring • Hydrophilic acrylic • Hydrophobic acrylic • Intraocular lens • Posterior capsule opacification

A. Floyd, M.D. • L. Werner, M.D., Ph.D. (✉) • N. Mamalis, M.D.
John A. Moran Eye Center, University of Utah, Salt Lake City, UT, USA
e-mail: liliana.werner@hsc.utah.edu

S. Saika et al. (eds.), *Lens Epithelium and Posterior Capsular Opacification*,
DOI 10.1007/978-4-431-54300-8_21, © Springer Japan 2014

21.1 Introduction and Background

Posterior capsule opacification (PCO) is the most common long-term complication of cataract surgery resulting in visual impairment and necessitating additional procedures, such as Nd:YAG laser posterior capsulotomy [1, 2]. The material opacifying the posterior capsule may have cortical/pearl and/or fibrotic components [1, 2], whereas anterior capsule opacification (ACO) is essentially a fibrotic entity [3, 4]. Prevention of overall capsular bag opacification has become one of the primary goals of intraocular lens (IOL) design and development, particularly in specialized IOLs such as multifocal or accommodating lenses [5, 6]. Prevention of any form of fibrosis within the capsular bag is particularly important for accommodating IOLs, which are generally designed to move within the bag or have their optical shapes altered in response to accommodating stimuli [6]. The capsular rings and IOLs discussed in this chapter are designed to maintain an open or expanded capsular bag and have demonstrated a relative lack of PCO and ACO.

21.2 Previously Described Intraocular Devices Maintaining an Open or Expanded Capsular Bag

21.2.1 Capsular Bending Ring

A specially designed capsular tension ring (CTR), developed by Nishi and Menapace, has proven efficacious at decreasing ACO and PCO formation [7, 8]. This open capsular ring implant, named capsular bending ring (CBR), is made from polymethylmethacrylate (PMMA) and has a rectangular profile with a thickness of 0.15–0.20 mm, a height of 0.7 mm, and a diameter of 11.0 mm. The CBR is placed in the bag, such that the capsular bend is created at the equator. This placement prevents the capsular bag from collapsing and the anterior capsule from contacting the IOL optic or posterior capsule. One way this device prevents capsular bag opacification is by minimizing or entirely preventing growth or migration of the equatorial lens epithelial cells (LECs) onto the posterior capsule or around the IOL optic. The other mechanism is by minimizing or preventing any fibrosis or shrinkage of the anterior capsule. The ring also prevents formation of striae or folds in the posterior capsule. These features work together to maintain a transparent capsular bag after cataract surgery with appropriate in-the-bag IOL placement. More details on the CBR can be found in Chap. 19.

21.2.2 Endocapsular Equator Ring

Hara and coauthors designed the endocapsular equator ring to be compatible with modern small-incision surgical techniques and to maintain the shape of the capsular bag after cataract extraction [9–13]. The equator ring, named the E-ring, is a closed, silicone ring that is 1.0 mm wide and 1.0 mm thick with an outer diameter of 9.0 mm.

The edge of the ring is square and there is a groove on its inner aspect that allows for IOL loop fixation. This device was able to effectively maintain the circular contour of the capsular bag equator and prevent the invading LECs from migrating toward the center along the posterior capsule. This design validated that the square edge preserves capsular transparency by preventing posterior movement of the stimulated LECs. When the E-ring is properly placed at the equator, the anterior and posterior capsules are separated and ample space is retained in the capsule. In rabbit and monkey studies, eyes receiving the IOL and the ring had significantly less capsular bag opacification than eyes receiving the IOL only, and the circular contour of the bag was maintained by the ring [10, 11]. Dimensions of the ring were later optimized for human implantation, and clinical studies confirmed its PCO prevention effect. Postoperative Scheimpflug photography performed in the clinical studies clearly showed no contact between the IOL and the posterior capsule in the presence of the ring [12, 13], demonstrating that contact between the IOL optic and the posterior capsule may not be necessary for PCO prevention in this circumstance.

Because this CTR maintains an open capsular bag, it allows for sufficient endocapsular flow of the aqueous humor. The aqueous humor contains some inhibitory growth factors to prevent the transformation and subsequent proliferation of the LECs. Studies have shown that maintaining contact between fresh aqueous humor and the LECs and maintaining an ample endocapsular aqueous humor exchange have been effective at decreasing the development of ACO and PCO [14–20]. The E-ring effectively maintains the circular contour of the capsular bag and inhibits the cells at the capsular bag equator from proliferating toward the center of the posterior capsule. More details on endocapsular equator rings can be found in Chap. 20.

21.2.3 Capsular Adhesion-Preventing Ring

Nagamoto and coauthors introduced a thin, open capsular adhesion-preventing ring (CAPR) made of PMMA that measures 2.0 mm in height with an inner diameter of 6.5 mm and outer diameter of 8.5 mm [21]. This CTR contains four grooves for IOL loop fixation and four distinct holes to allow enhanced endocapsular circulation of the aqueous humor. It holds the posterior capsule away from the anterior capsule, preventing adhesion of the capsules and further allowing the aqueous humor to circulate into the capsular bag through the grooves and holes. When the CAPR was evaluated in animal studies there was a reduction in PCO formation when the appropriate placement was achieved within the bag.

It is still unknown whether enhanced endocapsular flow alone with increased exposure to inhibitory growth factors is sufficient to prevent PCO development. Achieving complete anteroposterior capsular separation may be a critical component to inhibit the formation of PCO as well. A combination of anteroposterior capsular separation and increased endocapsular aqueous humor circulation with associated growth-inhibitory factors may be required for the reduction of PCO. Although aqueous humor at the early postoperative period was shown to stimulate proliferation of LECs in rabbit models [14, 15] the growth-promoting effect of postoperative aqueous humor gradually decreased and disappeared approximately 1 month after the surgery [14]. Results in previous studies [16, 17] indicate that transforming growth factor beta2 (TGF-β2) in the normal aqueous humor inhibits proliferation of LECs and corneal endothelial cells. In a recent letter to the editor, Nishi [18] stated that constant irrigation by the aqueous humor may prevent certain cytokines involved in stimulating LEC proliferation from reaching a threshold concentration level in the bag compartment. According to a study by Nishi and coauthors [18–20] interleukin-1 would be one such cytokine. Therefore, increasing the endocapsular flow and exposure to fresh aqueous humor may have a role in the prevention of LEC proliferation and subsequent capsular bag opacification formation.

The use of the CAPR effectively reduced the development of PCO in the rabbits when the complete anteroposterior capsular separation and increased aqueous humor circulation were achieved [21].

21.2.4 Spring-Loaded IOL

Hara and coauthors also developed a spring-loaded lens to provide sufficient accommodation through dual optics and flexible loops [22]. Hara's lens had two 6.0 mm PMMA optics connected with obliquely arranged poly(vinylidene fluoride) (PVDF) loops, with a horizontal length of 10.0 mm. Lenses with anteroposterior lengths of 4.0 and 8.0 mm were implanted in rabbit eyes. The surgical technique involved endocapsular phacoemulsification through a small upper central anterior capsulotomy. After complete evacuation of the capsular bag, cryopexy was applied to the entire anterior capsule in an attempt to remove the LECs. The anterior capsule opening was extended to 6.0 mm on both horizontal sides for the IOL implantation; thus, almost complete capsular bags were retained. Rabbit eyes receiving the thin (4.0 mm) spring-loaded lenses had opacified anterior capsules over its entire area, while rabbit eyes receiving thick (8.0 mm) spring-loaded lenses retained transparent anterior capsules. The authors postulated that the anterior capsules retained their transparence in the eyes implanted with thick spring-loaded lenses because of mechanical compression against the LECs [22, 23].

21.3 Newly Described IOLs Maintaining an Open or Expanded Capsular Bag

21.3.1 Concept 360

The Concept 360 (Corneal Laboratoire) is a single-piece IOL manufactured from a foldable hydrophilic acrylic material with water content of 26 %. The lens has an optic diameter of 6.0 mm, an overall diameter of 11.5 mm, and square optic and haptic edges. Its overall design is that of a disc-shaped lens, with six haptic

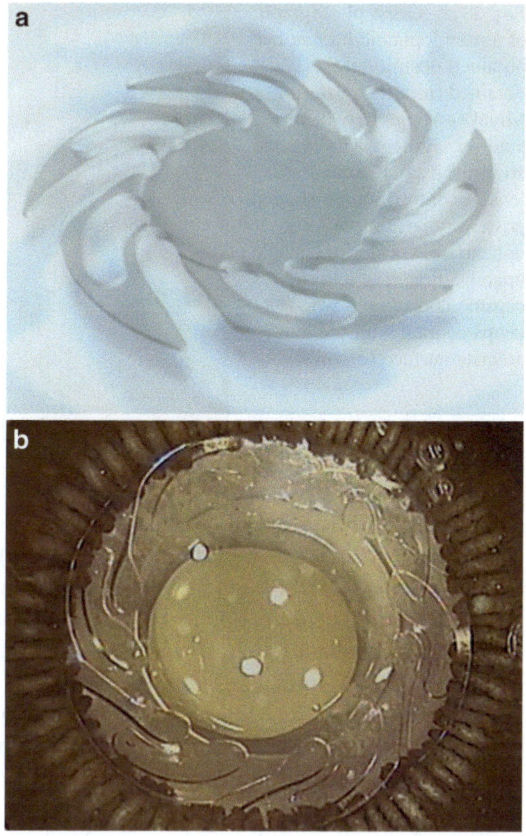

Fig. 21.1 Concept 360 intraocular lens. (**a**) Schematic drawing showing the design of the lens. The overall design is that of a disc-shaped lens with the appearance of a propeller; the space between the haptic components will decrease as a function of the diameter of the capsular bag. (**b**) Gross photograph of the lens experimentally implanted in a cadaver eye (Miyake–Apple view). Published in: Werner et al. [26]

components having a 10° posterior optic–haptic angulation that gives the lens the appearance of a propeller. The design of this IOL keeps the anterior capsule away from the optic surface and when the haptic components contact each other without any gaps between them, there is also a CTR peripheral effect (Fig. 21.1). A preliminary clinical study of this IOL found that the lack of anterior capsule–optic contact prevented anterior capsule-related complications [24, 25].

It has been demonstrated that the anterior capsule opacifies postoperatively where it remains in contact with the anterior IOL surface [3, 4]. This has been particularly observed with silicone plate lenses, because of the large area of contact between the anterior surface of the lens and the inner surface of the anterior capsule (Fig. 21.2). The configuration of the haptic components of the Concept 360 simulates the presence of a broad, band-shaped capsular ring, which helps to keep the anterior capsule away from the anterior optic surface. This configuration, associated with the 10° posterior optic–haptic angulation, prevented contact between the anterior lens surface and the anterior capsule in four cadaver eyes in a preliminary study, in which the capsulorhexis margin was at distance from the

Fig. 21.2 Gross photograph of a pseudophakic human eye obtained postmortem, obtained from the posterior or Miyake–Apple view of the anterior segment. The eye had an in-the-bag silicone plate lens. A central, square-shaped posterior capsulotomy was done to treat posterior capsule opacification. The anterior capsule is opacified where it keeps contact with the lens' anterior surface (*arrow*)

anterior lens surface for 360° [26]. An IOL with a CTR effect may have significant advantages in the prevention of capsular bag opacification. As described in Sect. 21.2.1, a CTR designed to prevent opacification in the capsular bag was evaluated in two centers, one in Japan and the other in Austria [7, 8]. Both centers reported a significant reduction in ACO and PCO with the rings compared with the contralateral eyes implanted with the same lens design. Equatorial CTRs have the ability to maintain the contour of the capsular bag and to stretch the posterior capsule [27]. In high-resolution laser interferometric studies, a space between the IOL and the posterior capsule was found with different lens designs [28]. With a CTR in place, this space was found to be smaller or nonexistent. CTRs also produce a circumferential stretch on the capsular bag, with the radial distention forces equally distributed. Formation of traction folds in the posterior capsule, which may be used as an avenue for cell ingrowth, is thus avoided. The absence of capsular striae with the Concept 360 was confirmed in the preliminary study using cadaver eyes [26].

In smaller eyes, the distance between the haptics of the Concept 360 is minimized as they come close together to conform to the capsular bag diameter. In this situation, the contact between the periphery of the lens and the equatorial region of the capsular bag may produce the effect of a complete CTR. However, in capsular bags with larger diameters, the haptics of the Concept 360 are not close together. The gaps between them may represent avenues for migration or proliferation of any residual cells or cortical material in the capsular bag equator, eventually forming PCO. The influence of excess overlapping between the haptic components of the lens (possibly observed in very small eyes, also creating gaps) on the CTR effect needs to be assessed.

Association of various factors within the design of the Concept 360 may help in PCO prevention. Spontaneous rotational movements of this disc-shaped lens during irrigation/aspiration may help in the dislocation of residual cortical material and cells out of the equatorial region of the capsular bag. This would make the material more accessible to the surgeon's view and therefore promote a more complete cortical cleanup. The square haptic edge profile, in association with contact of the haptic periphery around the equatorial region of the capsular bag (CTR effect), may promote a significant barrier effect against cell proliferation. The contact between the square posterior optic edge of the lens and the posterior capsule, further enhanced by the posterior optic–haptic angulation, is likely to represent another significant barrier against cell proliferation and thus PCO formation [26].

Evaluation of new lenses in cadaver-eye studies with the Miyake–Apple view is useful to assess their fit within the human capsular bag and their possible interaction with capsular bag structures. With respect to aspects such as prevention of ACO and PCO and the absence of posterior capsule striae, the preliminary clinical results with this lens are encouraging [25], but only further long-term clinical studies can confirm these findings.

21.3.2 Synchrony

The Synchrony (Visiogen, Inc./Abbott Medical Optics) is a foldable, single-piece, dual-optic IOL manufactured from silicone. The IOL has two optic components (anterior and posterior); each has the general design of a plate-haptic silicone IOL and the two are connected by a bridge through the haptics with a spring function. The posterior aspect is designed with a significantly larger surface area than the anterior aspect and it has two tabs to maintain stability in the capsular bag during the accommodation/unaccommodation process. The anterior optic has two expansions oriented parallel to the haptic component that lift the capsulorhexis edge up, preventing complete contact between the anterior capsule and the anterior surface of the IOL (and therefore preventing opacification of the anterior capsule). In this dual-optic IOL system, the anterior IOL has a high plus power beyond that required to produce emmetropia, while the posterior IOL has a minus power to return the eye to emmetropia [29–35]. The degree of capsular bag expansion provided by the Synchrony IOL may have played a role in the maintenance of capsular bag transparency and the prevention of capsular bag contraction and capsulorhexis phimosis, as observed in rabbit and clinical studies (Fig. 21.3).

The presence of two IOL optics in the capsular bag raises concerns about the possibility of ingrowth of regenerative/proliferative crystalline lens material between them, with formation of interlenticular opacification (ILO) [36–41]. It is expected that the thickness of the interlenticular space will experience dynamic changes, as a function of efforts of accommodation. Whether this will influence the outcome of postoperative ILO formation still needs to be assessed in long-term clinical studies. In a series of 25 patients implanted with the Synchrony lens and followed for 12 months, no ILO formation was observed [33].

During the clinical follow-up of rabbits implanted with the dual-optic accommodating IOL, pairs of silicone plate lenses, or pairs of hydrophobic acrylic lenses, we observed ILO formation in the hydrophobic acrylic control lenses, which started at the level of the optic–haptic junctions [35]. It has to be considered that the rabbit is an accelerated model for PCO. Progressive regrowth of the crystalline lens

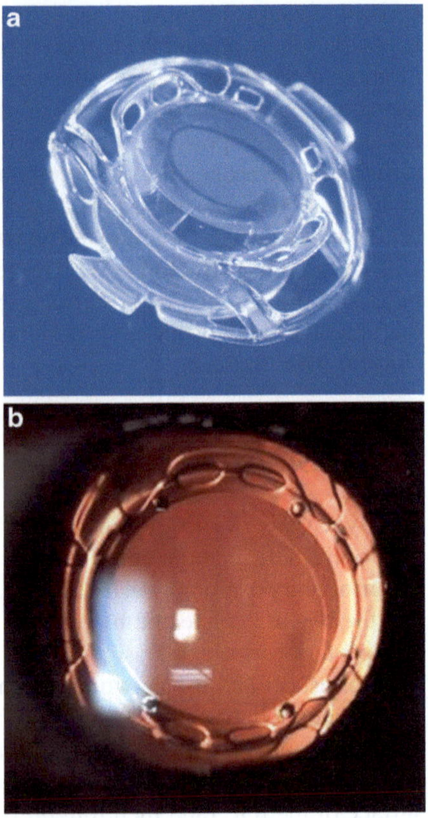

Fig. 21.3 Synchrony intraocular lens. (**a**) Gross photograph of the lens showing its different components (provided by Visiogen/AMO). (**b**) Clinical photograph of a Synchrony lens taken approximately 2 years after implantation. The capsular bag remained remarkably clear (provided by Ivan L. Ossma, MD, MPH, Bucaramanga, Columbia)

material is observed in the postoperative period; thus, Soemmering's formation is generally abundant with this animal. The ILO development confirms previous observations in rabbit studies, in which the optic–haptic junctions of a single-piece hydrophobic acrylic design and those of a single-piece hydrophilic acrylic design were the sites of the beginning of ingrowth of material causing PCO [42–44] (Fig. 21.4). The same study also showed that ILO was not an issue with the dual-optic accommodating lens.

21.3.3 Zephyr

The Zephyr IOL (Anew Optics, Inc.) is a single-piece, disc-shaped hydrophilic acrylic IOL with an overall diameter of 10.02 mm [45, 46]. The lens is suspended between two complete haptic rings connected by a pillar of the haptic material, such that the anterior ring rests against the anterior capsule at some distance from the capsular equator and the posterior ring rests against the posterior capsule also at some distance from the capsular equator. This particular disc-shaped IOL has been

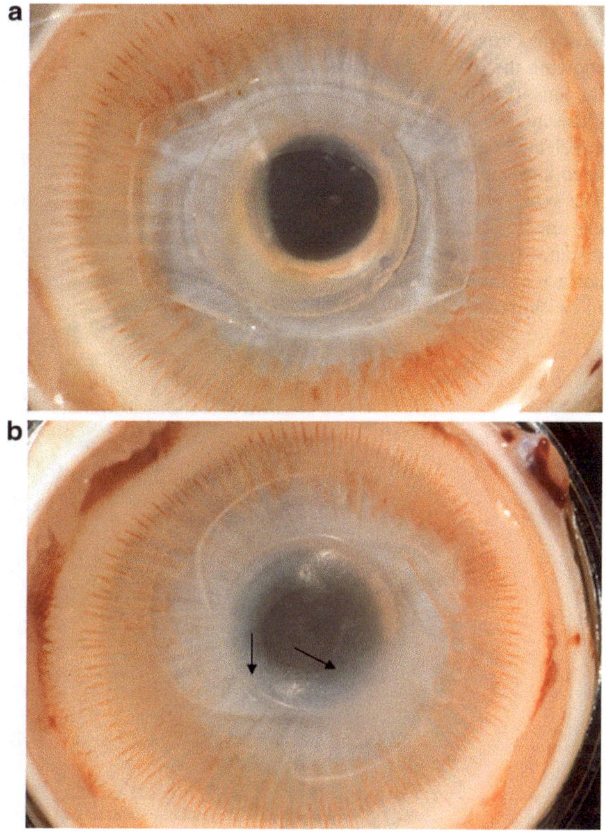

Fig. 21.4 Gross photographs of enucleated rabbit eyes taken 6 weeks postoperatively (Miyake–Apple view). (**a**) The eye was implanted with a Synchrony lens and exhibits an overall clear capsular bag. (**b**) The eye was implanted with a pair of single-piece hydrophobic acrylic lenses. Besides posterior capsule opacification, the *arrows* show interlenticular opacification, generally starting at the level of optic–haptic junctions. Published in Werner et al. [35]

modified to incorporate features to enhance the endocapsular flow of aqueous humor. The haptic perforations allow circulation of the aqueous throughout the capsule, both anteriorly and posteriorly to the optic as well as in the fornix of the capsular bag. In rabbit studies, the peripheral rings of the disc-shaped IOL prevented ACO and PCO by expanding the capsular bag and preventing IOL surface contact with the anterior capsule. Soemmering's ring formation was observed in localized areas around the equatorial region of the capsular bag between the peripheral rings. In these studies, the anterior capsule showed a distinct separation from the anterior IOL surface, which was associated with complete clarity of the anterior capsule throughout the studies [45, 46] (Figs. 21.5 and 21.6).

In the same abovementioned rabbit studies, some of the Zephyr lenses exhibited the optics bulging anteriorly instead of toward the posterior capsule, with a lack of contact between the optic and the posterior capsule. This could have been caused by the fact that the lens diameter was slightly too large for the capsular bags of the rabbits. In any event, capsular bag opacification scoring in these eyes was no different than in the eyes exhibiting posterior bulging of the optic, and the preventative PCO effect was maintained. This fact provides further evidence that contact

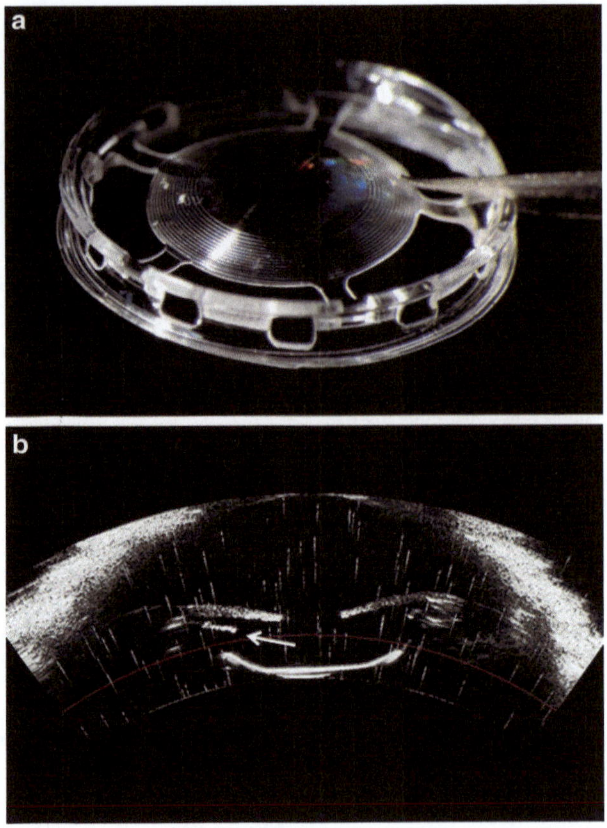

Fig. 21.5 Zephyr intraocular lens. (**a**) Gross photograph showing the overall design of the lens. (**b**) High-frequency ultrasound scan of an enucleated rabbit eye that had been implanted with the lens. The *arrow* shows the anterior capsule, which is kept at distance from the anterior surface of the lens. Published in Leishman et al. [46]

between the optic and the posterior capsule is not required for PCO prevention, providing the overall IOL design configuration is disc-shaped with a large CTR effect around the equator, keeping the bag open and/or expanded.

21.3.4 FluidVision

The FluidVision (PowerVision) is a new accommodating lens composed of a hollow, fluid-filled hydrophobic acrylic optic and oversized, hollow, fluid-filled haptics. The relatively large haptic elements of this lens keep the anterior and posterior capsules apart [47]. The fluid in the optic and haptics is an index-matched silicone oil which flows back and forth between the haptics and optic to change curvature and hence the power of the optic (Fig. 21.7). An initial clinical evaluation of an earlier prototype of this lens conducted in a limited number of sighted eye subjects demonstrated the potential to achieve more than five diopters of power change [48].

Cellular proliferation within the capsular bag after implantation of an accommodating IOL could potentially impair its function. Postoperative fibrosis with

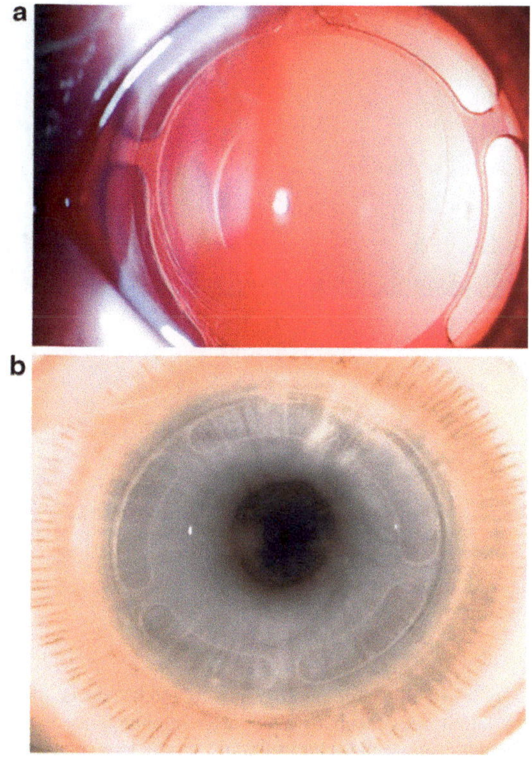

Fig. 21.6 Slit-lamp photograph (**a**) and gross Miyake–Apple view photograph (**b**) of a rabbit eye taken 5 weeks after implantation of the Zephyr lens. The eye has clear anterior and posterior capsules. Published in Kavoussi et al. [45]

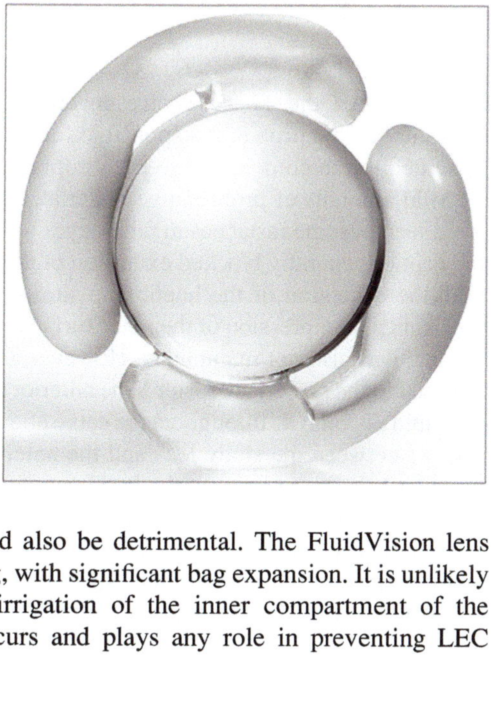

Fig. 21.7 Photograph showing the overall design of the FluidVision intraocular lens (provided by PowerVision)

contraction of the capsular bag could also be detrimental. The FluidVision lens practically fills the entire capsular bag, with significant bag expansion. It is unlikely in this configuration that constant irrigation of the inner compartment of the capsular bag by aqueous humor occurs and plays any role in preventing LEC

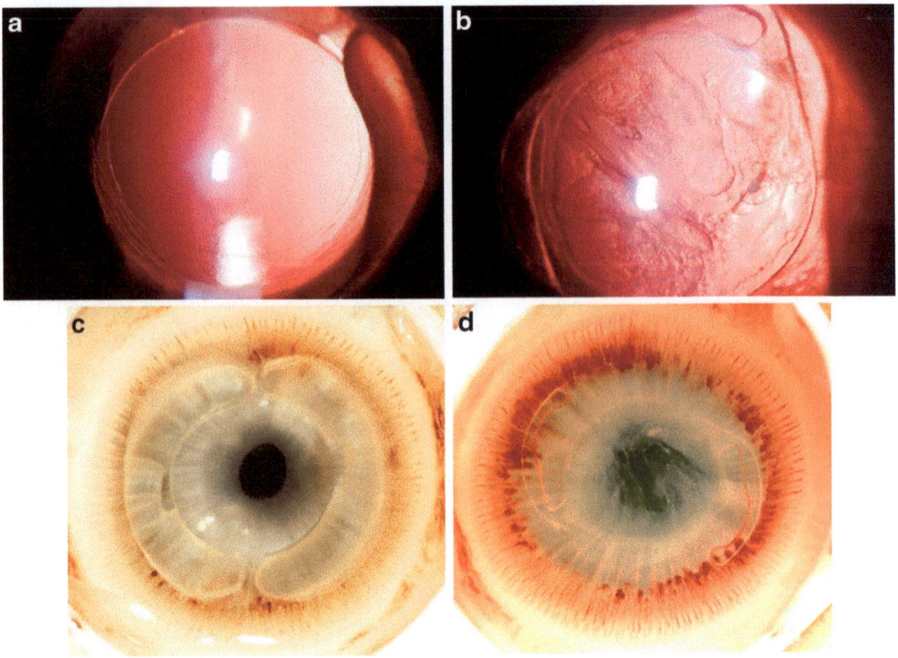

Fig. 21.8 Slit-lamp photographs (**a** and **b**) and gross Miyake–Apple view photographs (**c** and **d**) of rabbit eyes taken 6 weeks after implantation of the FluidVision lens (**a** and **c**) or a control standard single-piece hydrophobic acrylic lens (**b** and **d**). The eye with the FluidVision shows clear anterior and posterior capsules, while the eye with the control IOL developed diffuse posterior capsule opacification, starting at the optic–haptic junctions. Published in Floyd et al. [47]

proliferation. Other factors such as mechanical compression and/or stretching of the capsular bag are probably responsible for the significant prevention of capsular bag opacification in comparison to eyes implanted with control, standard lenses.

Mild amounts of proliferative material, limited to the fornix of the capsular bag, were seen in some instances in rabbit eyes with the FluidVision [47]. The presence of the haptics generally blocked extension of the proliferative material toward the optic, with the exception of the haptic gap sites. In those two areas, there was a lack of mechanical compression of the inner surface of the capsular bag; the shape/contour of the bag was also not maintained. However, the optic edge blocked the access of the material to the posterior capsule. The anterior capsule remained remarkably clear with the FluidVision IOL throughout the same abovementioned rabbit study [47]. The only contact between the study lens and the anterior capsule was at the periphery of the capsular bag, at the level of the haptic components. A fine wrinkling of the anterior capsule was observed in some instances at that area. The anterior capsule at and around the capsulorhexis edge was generally devoid of any fibrosis, as it was kept at a distance from the anterior IOL surface (Figs. 21.8 and 21.9). The clarity of the capsular bag with the silicone oil-filled IOL 6 weeks postoperatively was remarkable

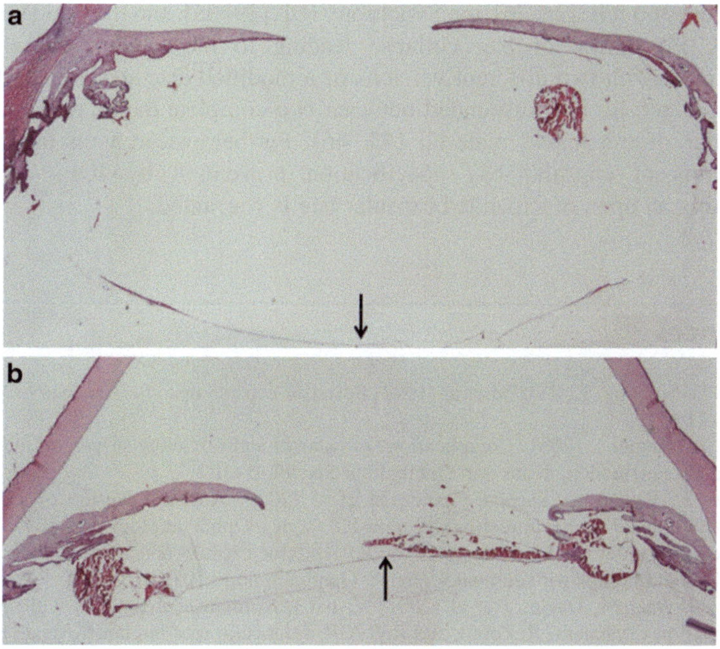

Fig. 21.9 Representative light photomicrographs from histopathologic sections cut from both eyes of the same rabbit. (**a**) Eye implanted with the FluidVision lens. (**b**) Eye implanted with a control IOL. Note the expanded capsular bag and limited Soemmering's ring formation in the study eye compared to the control eye. Significant PCO is observed with the control IOL, whereas PCO is absent with the study IOL. The *arrows* in the photographs show the posterior capsule. Published in Floyd et al. [47]

in the rabbit model, both clinically and pathologically [47]. This was in marked contrast to the control standard IOL (single-piece looped design), which showed a large amount of proliferative material with significant PCO, generally starting at the optic–haptic junctions.

21.4 Summary

Prevention of opacification of the capsular bag after cataract extraction and IOL implantation, especially the fibrotic types, appears to be even more important now with the increased interest in the development of accommodating IOLs. There are concerns that late postoperative capsular bag fibrosis might prevent long-term functioning of these lenses. This may be associated with not only PCO but also ACO. Prevention of capsular bag collapse with maintenance of bag clarity has been described with various devices and IOLs, such as the CBR of Nishi and Menapace [7, 8], Hara's equator ring [9–13], Nagamoto's acrylic CAPR [21], a PMMA-PVDF dual-optic accommodating IOL with spring action also designed by Hara [22, 23], the Concept 360 IOL [25, 26], the Synchrony IOL [29–35], and the FluidVision lens [47, 48]. Prevention of bag collapse leading to bag clarity has also been demonstrated with two different versions of a modified disc-shaped, single-piece, hydrophilic acrylic IOL suspended between two complete haptic rings connected by a pillar of the haptic material [45, 46]. Further research on the proposed mechanisms of capsular bag opacification prevention by IOLs or devices maintaining an open of expanded capsular bag is warranted.

References

1. Apple D, Solomon K, Tetz M et al (1992) Posterior capsule opacification. Surv Ophthalmol 37:73–116
2. Apple D, Werner L (2001) Complications of cataract and refractive surgery: a clinicopathological documentation. Trans Am Ophthalmol Soc 99:95–109
3. Werner L, Pandey S, Escobar-Gomez M et al (2000) Anterior capsule opacification: a histopathological study comparing different IOL styles. Ophthalmology 107:463–471
4. Werner L, Pandey S, Apple D et al (2001) Anterior capsule opacification: correlation of pathological findings with clinical sequelae. Ophthalmology 108:1675–1681
5. Pepose J, Qazi M, Davies J et al (2007) Visual performance of patients with bilateral vs combination Crystalens, ReZoom, and ReSTOR intraocular lens implants. Am J Ophthalmol 144:347–357
6. Alió J, Piñero D, Plaza-Puche A (2010) Visual outcomes and optical performance with a monofocal intraocular lens and a new-generation single-optic accommodating intraocular lens. J Cataract Refract Surg 36:1656–1664
7. Nishi O, Nishi K, Menapace R (1998) Capsule-bending ring for the prevention of capsular opacification: a preliminary report. Ophthalmic Surg Lasers 29:749–753
8. Menapace R, Sacu S, Georgopoulos M et al (2008) Efficacy and safety of capsular bending ring implantation to prevent posterior capsule opacification; three-year results of a randomized clinical trial. J Cataract Refract Surg 34:1318–1328
9. Hara T, Hara T, Yamada Y (1991) "Equator ring" for maintenance of the completely circular contour of the capsular bag equator after cataract removal. Ophthalmic Surg 22:358–359
10. Hara T, Hara T, Sakanishi K et al (1995) Efficacy of equator rings in an experimental rabbit study. Arch Ophthalmol 113:1060–1106
11. Hashizoe M, Hara T, Ogura Y et al (1998) Equator ring efficacy in maintaining capsular bag integrity and transparency after cataract removal in monkey eyes. Graefes Arch Clin Exp Ophthalmol 236:375–379
12. Hara T, Hara T, Hara T (2007) Preventing posterior capsular opacification with an endocapsular equator ring in a young human eye; 2-year follow-up. Arch Ophthalmol 125:483–486
13. Hara T, Hara T, Narita M et al (2011) Long-term study of posterior capsular opacification prevention with endocapsular equator rings in humans. Arch Ophthalmol 129:855–863
14. Wallentin N, Wickström K, Lundberg C (1998) Effect of cataract surgery on aqueous TGF beta and lens epithelial cell proliferation. Invest Ophthalmol Vis Sci 39:1410–1418
15. Reddan J, Weinsieder A, Wilson D (1979) Aqueous humor from traumatized eyes triggers cell division in the epithelia of cultured lenses. Exp Eye Res 28:267–276
16. Kurosaka D, Nagamoto T (1994) Inhibitory effect of TGF-beta2 in human aqueous humor on bovine lens epithelial cell proliferation. Invest Ophthalmol Vis Sci 35:3408–3412
17. Chen K-H, Harris D, Joyce N (1999) TGF-beta2 in aqueous humor suppresses S-phase entry in cultured corneal endothelial cells. Invest Ophthalmol Vis Sci 40:2513–2519

18. Nishi O (2012) Other factors in PCO prevention [letter]. J Cataract Refract Surg 38:924–925, reply by Werner L, Mamalis N, Kavoussi S, 925
19. Nishi O, Nishi K, Imanishi M (1992) Synthesis of interleukin-1 and prostaglandin E2 by lens epithelial cells of human cataracts. Br J Ophthalmol 76:338–341
20. Nishi O, Nishi K, Ohmoto Y (1994) Effect of interleukin 1 receptor antagonist on the blood-aqueous barrier after intraocular lens implantation. Br J Ophthalmol 78:917–920
21. Nagamoto T, Tanaka N, Fujiwara T (2009) Inhibition of posterior capsule opacification by a capsular adhesion-preventing ring. Arch Ophthalmol 127:471–474
22. Hara T, Hara T, Yasuda A et al (1990) Accommodative intraocular lens with spring action. Part 1. Design and placement in an excised animal eye. Ophthalmic Surg 21:128–133
23. Hara T, Hara T, Yasuda A et al (1992) Accommodative intraocular lens with spring action–Part 2. Fixation in the living rabbit. Ophthalmic Surg 23:632–635
24. Werner L (2010) IOL designs maintaining an open or expanded capsular bag. Presented at 2010 ESCRS clinical research symposium: new IOL materials and micro design, Paris, France, 4 Sept 2010
25. McGrath D (2003) New IOL tackles anterior-capsule-related complications. Euro Times 8:12, http://www.escrs.org/eurotimes/September2003/5.asp
26. Werner L, Hickman M, LeBoyer R et al (2005) Experimental evaluation of the Corneal Concept 360 intraocular lens with the Miyake-Apple view. J Cataract Refract Surg 31:1231–1237
27. Menapace R, Findl O, Georgopoulos M et al (2000) The capsular tension ring: designs, applications, and techniques. J Cataract Refract Surg 26:898–912
28. Findl O, Drexler W, Menapace R et al (1998) Accurate determination of effective lens position and lens-capsule distance with 4 intraocular lenses. J Cataract Refract Surg 24:1094–1098
29. McLeod S, Portney V, Ting A (2003) A dual optic accommodating foldable intraocular lens. Br J Ophthalmol 87:1083–1085
30. McLeod S (2006) Optical principles, biomechanics, and initial clinical performance of a dual-optic accommodating intraocular lens (an American Ophthalmological Society thesis). Trans Am Ophthalmol Soc 104:437–452
31. McLeod S, Vargas L, Portney V et al (2007) Synchrony dual-optic accommodating intraocular lens. Part 1: optical and biomechanical principles and design considerations. J Cataract Refract Surg 33:37–46
32. Ossma I, Galvis A, Vargas L et al (2007) Synchrony dual-optic accommodating intraocular lens. Part 2: pilot clinical evaluation. J Cataract Refract Surg 33:47–52
33. Ossma-Gomez I et al (2005) Synchrony dual-optic accommodating IOL: one year follow up. Presented at ASCRS symposium on cataract, IOL and refractive surgery, Washington, DC, April 2005
34. Werner L, Pandey S, Izak A et al (2004) Capsular bag opacification after experimental implantation of a new accommodating intraocular lens in rabbit eyes. J Cataract Refract Surg 30:1114–1123
35. Werner L, Mamalis N, Stevens S et al (2006) Interlenticular opacification: dual-optic versus piggyback intraocular lenses. J Cataract Refract Surg 32:655–661
36. Shugar J, Schwartz T (1999) I Interpseudophakos Elschnig pearls associated with late hyperopic shift: a complication of piggyback posterior chamber intraocular lens implantation. J Cataract Refract Surg 25:863–867
37. Gayton J, Apple D, Peng Q et al (2000) Interlenticular opacification: a clinicopathological correlation of a complication of posterior chamber piggyback intraocular lenses. J Cataract Refract Surg 26:330–336
38. Shugar J, Keeler S (2000) Interpseudophakos intraocular lens surface opacification as a late complication of piggyback acrylic posterior chamber intraocular lens implantation. J Cataract Refract Surg 26:448–455
39. Werner L, Shugar J, Apple D et al (2000) Opacification of piggyback IOLs associated with an amorphous material attached to interlenticular surfaces. J Cataract Refract Surg 26:1612–1619

40. Werner L, Apple D, Pandey S et al (2002) Analysis of elements of interlenticular opacification. Am J Ophthalmol 133:320–326
41. Spencer T, Mamalis N, Lane S (2002) Interlenticular opacification of piggyback acrylic intraocular lenses. J Cataract Refract Surg 28:1287–1290
42. Werner L, Mamalis N, Izak A et al (2005) Posterior capsule opacification in rabbit eyes implanted with 1-piece and 3-piece hydrophobic acrylic intraocular lenses. J Cataract Refract Surg 31:805–811
43. Vargas L, Izak A, Apple D et al (2003) Implantation of a single-piece, hydrophilic, acrylic, minus-power foldable posterior chamber intraocular lens in a rabbit model; clinicopathologic study of posterior capsule opacification. J Cataract Refract Surg 29:1613–1620
44. Werner L, Mamalis N, Pandey S et al (2004) Posterior capsule opacification in rabbit eyes implanted with hydrophilic acrylic intraocular lenses with enhanced square edge. J Cataract Refract Surg 30:2403–2409
45. Kavoussi S, Werner L, Fuller S et al (2011) Prevention of capsular bag opacification with a new hydrophilic acrylic disk-shaped intraocular lens. J Cataract Refract Surg 37:2194–2200
46. Leishman L, Werner L, Bodnar Z et al (2012) Prevention of capsular bag opacification with a modified hydrophilic acrylic disk-shaped intraocular lens. J Cataract Refract Surg 38:1664–1670
47. Floyd A, Werner L, Liu E et al (2013) Capsular bag opacification with a new accommodating intraocular lens. J Cataract Refract Surg 39:1415–1420
48. Roux P, Nichamin L (2010) Early implantation results of shape-changing accommodating IOL in sighted eyes. Presented at 2010 ASCRS meeting, Boston, 12 April 2010

Lens Epithelium and Posterior Capsular Opacification: Prevention of PCO with the Bag-in-the-Lens (BIL)

22

Marie-José Tassignon and Sorcha Ní Dhubhghaill

Abstract

The bag-in-the-lens (BIL) cataract surgery technique is a unique approach to prevent posterior capsular opacification. Insertion of the lens requires a primary posterior continuous curvilinear capsulorhexis. The lens is positioned entirely suspended by the anterior and posterior capsules. This unique placement confers stability and predictability, but most significantly it results in a 0 % rate of PCO. Lens epithelial cells may continue to proliferate between the two capsules, but the design of the BIL implant seals any point of entry. We conclude that BIL technique of cataract extraction is an optimal approach to prevent PCO in all patients.

Keywords

Anterior interface • Bag-in-the-lens (BIL) • Posterior capsule opacification (PCO) • Primary posterior circular continuous capsulorhexis (PPCCC) • Ring caliper

22.1 Introduction

From the earliest days of lens implantation in cataract surgery, posterior capsular opacification (PCO) has been the most frequently encountered postoperative complication [1]. Progress in intraocular lens (IOL) design has focused on lens shape

M.-J. Tassignon, M.D., Ph.D. (✉)
Department of Ophthalmology, Universitair Ziekenhuis Antwerpen, Antwerp University Hospital, Wilrijkstraat 10, 2650, Edegem (Antwerp), Belgium

Department of Medicine, Antwerp University, Wilrijk, Belgium
e-mail: marie-jose.tassignon@uza.be

S. Ní Dhubhghaill, M.B., Ph.D.
Department of Medicine, Antwerp University, Wilrijk, Belgium

S. Saika et al. (eds.), *Lens Epithelium and Posterior Capsular Opacification*,
DOI 10.1007/978-4-431-54300-8_22,

and material composition predominantly, but despite high expectations, these lenses deliver modest benefit over their predecessors, and PCO still occurs frequently [2]. The advances and limitations of IOL material and design are covered elsewhere in this volume (Chap. 18). The most optimized lens-in-the-bag approach seems to delay PCO development rather than entirely prevent it. It appears therefore that while lens epithelial cells (LEC) can freely proliferate behind the IOL, there is still a risk of PCO.

22.1.1 Nd:YAG Capsulotomy

The mainstay of PCO treatment is currently neodymium-doped yttrium aluminum garnet (Nd:YAG) laser capsulotomy where a series of high-intensity laser pulses are directed at the posterior capsule. The laser energy perforates the capsule, which results in an opening behind the IOL. Although this technique is simple to learn and easy to tolerate, it is by no means a complication free intervention. In fact, there are significant variations on how the technique is performed and a lack of evidencebased clinical guidelines [3]. The rate of Nd:YAG laser capsulotomy is approximately 24 % only 5 years after phacoemulsification [4]. This rate, however, only represents eyes where PCO formation has resulted in reductions in visual acuity. It is likely that an additional proportion of patients experience deterioration in quality of vision due to posterior capsular changes, though this was not detected by standard visual acuity assessment.

Complications observed with Nd:YAG capsulotomy range from the common to the extremely rare but severe [5–8]. The more frequently seen complications include lens pitting, cystoid macular edema (CME), and retinal breaks and detachment which are increased, particularly in younger and myopic patients [9–12]. Moreover, proliferation of residual LECs and subsequent closure of the capsulotomy have also been reported [13, 14].

Even uncomplicated capsulotomies can undermine a previously optimal surgical procedure. The original reports on the negligible effect of Nd:YAG capsulotomy on vision used standard visual acuity as a parameter. As technology has improved, the capacity to detect visual aberrations not seen by previous means has shed a new light on the eye post Nd:YAG. Aberrometric analysis has been performed comparing PCO patients post a 2.5 mm Nd:YAG capsulotomy and a control group of pseudophakic patients who had no evidence of PCO [15]. While the best-corrected visual acuity was the same in both groups, mean total higher-order aberrations were higher in the post-Nd:YAG groups. Persistent wave-front aberrations have also been detected particularly in hydrophilic IOLs even after a larger 4.5 mm to 5.0 mm capsulotomy [16]. Both studies indicate that, despite treatment, visual outcomes post Nd:YAG laser capsulotomy are not ideal and not comparable to outcomes if PCO had never developed.

High-quality refractive outcomes may also be compromised after capsulotomy. Vrijman et al. reported a cohort of patients with multifocal apodized diffractive pseudophakia in whom 7 % had a significant change in subjective refraction after capsulotomy [17]. There is also an additional cost and patient burden associated with Nd:YAG and access to the equipment is limited in the third world, so the ideal cataract surgery would not routinely require a secondary intervention [3]. Finally, in patients who require subsequent IOL exchange, the presence of an Nd:YAG capsulotomy increases the difficulty of the procedure and the risk of vitreous loss [18].

22.1.2 The Posterior Capsule

Gimbel and Neuhann originally described a technique to remove a portion of the posterior capsule as a means of preventing PCO [19]. Removing the posterior capsule via this primary posterior continuous curvilinear capsulorhexis (PPCCC) was thought to remove the scaffold structure whereby LECs could proliferate and cause PCO. This does not appear to be the case [20]. Although PPCCC openings may be relatively stable in uncomplicated elderly patients [21], a two-year follow-up study showed a PCO rate of 2 % over 2 years [22]. In high-risk PCO cases such as children, young adults, and uveitic patients, 23–40 % will experience partial or complete closure of the PPCCC [23, 24]. In vitro evidence supports this observation by showing that LECs cells retain the capacity to proliferate even in the absence of a capsular scaffold [25]. It is unclear whether the LECs grow using anterior hyaloid as support or whether it is on a matrix secreted by the cells themselves. Regardless, it appears that simply performing a PPCCC is not sufficient to entirely prevent PCO.

Gimbel and DeBroff subsequently described a technique that combined PPCCC with optic buttonholing ("posterior optic buttonholing" or POBH) to reduce the rate of visual axis reopacification in the pediatric population [26, 27]. In this technique, the lens haptics remain in the bag, but the optic itself is positioned behind the posterior capsule with the margin of the posterior capsule overlying the anterior surface of the optic [28]. This should prevent ingress of migrating LECs into the retrolenticular space. The seal is not complete, however, and LECs may enter the retrolenticular space through the haptic crossover points. The anterior capsule also lies free over the optic and may be prone to LEC accumulation over the anterior surface of the lens in the long term. The concern in regard to rotation behind the posterior rhexis made this approach unsuitable for standard toric IOLs [28].

22.2 The Bag-in-the-Lens Technique

The BIL technique represents a significant change in the approach to PCO prevention [29]. Standard PCO prevention strategies have focused on surgical technique, IOL shape and material, pharmacological reduction of LEC, and use of additional implants such as capsular tension rings [2]. In the BIL technique, the IOL itself forms the barrier to PCO development. The lens is composed of a monofocal

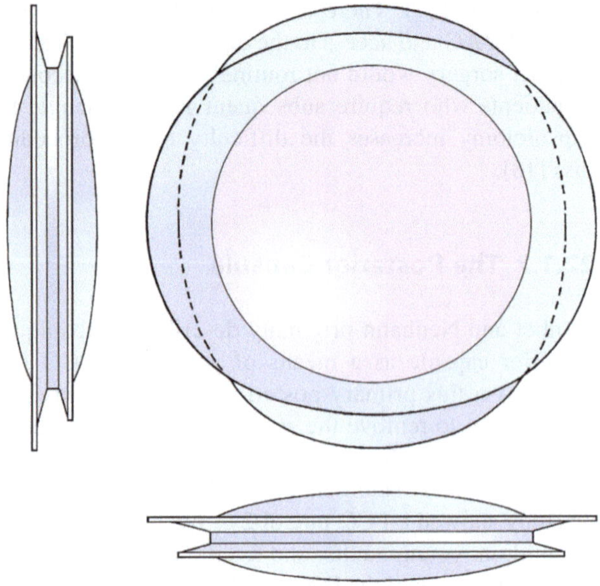

Fig. 22.1 Structure and profile of the BIL intraocular lens [30]

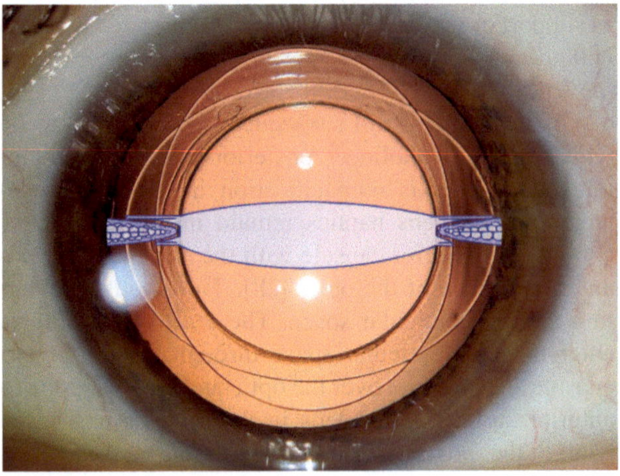

Fig. 22.2 Centrally sited IOL with side-on schematic view

spherical hydrophilic material and designed as a biconvex optic with two elliptical flanged haptics (Fig. 22.1). The plane haptics are aligned perpendicularly, the posterior haptics lying on the horizontal meridian, and the anterior oriented on the vertical (Fig. 22.2). Between the two haptics lies a channel, which when sited/ positioned contains the anterior and posterior capsules. The haptics are 0.15 mm thick and the groove between them is 0.25 mm. The total diameter ranges from 6.5 to 8.5 mm depending on the case requirements.

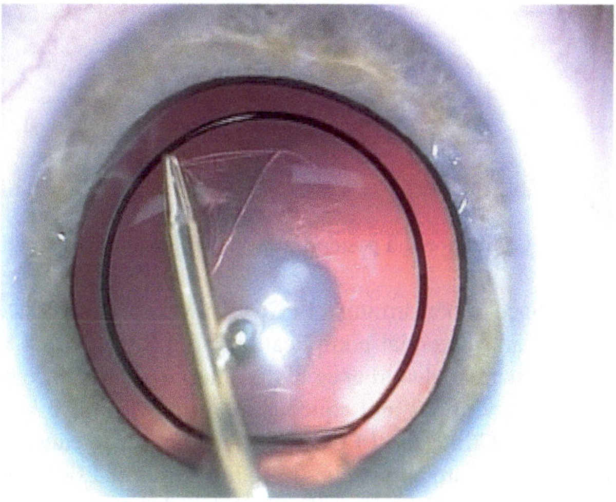

Fig. 22.3 Ring caliper for accurate sizing of anterior rhexis

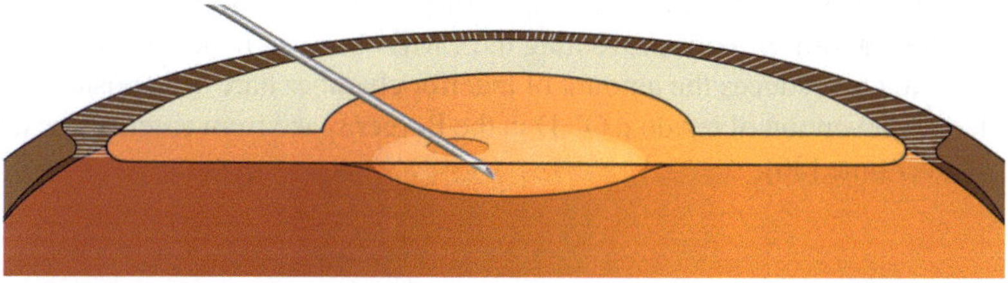

Fig. 22.4 Injection of OVD into Berger's space through small perforation in PC

22.2.1 The Anterior Rhexis

The creation of the anterior and posterior calibrated continuous curvilinear capsulorhexes is a crucial step in the BIL technique. The success of the lens is dependent on an accurately centered, well-fitting anterior rhexis which can be assisted by the use of a ring-shaped caliper (Fig. 22.3) [31]. The phacoemulsification is performed as standard. The size of the PPCCC may be then gauged using the anterior rhexis as a guide. A slightly undersized rhexis allows a firm seal on the BIL.

22.2.2 The Posterior Rhexis

The PPCCC is commenced with a small incision into the capsule. Ophthalmic viscosurgical device (OVD) is injected into the small defect and under the capsule into Berger's space (Fig. 22.4) [32]. This is a gentle technique and at the one-year follow-up of 60 patients, not a single case experienced vitreous loss [32].

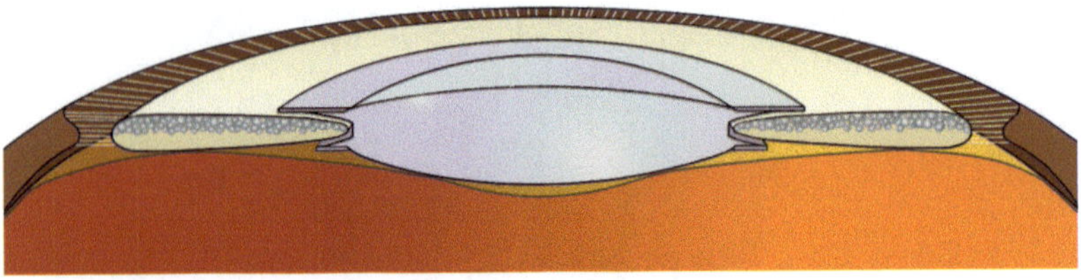

Fig. 22.5 Anterior and posterior capsular edges positioned in the lens optic groove

Reluctance in performing a PPCCC may stem from concern for disturbing the anterior hyaloid and potential complications. A planned PPCCC, unlike accidental capsular rupture, with no disturbance of the anterior hyaloid face retains the diffusion properties across the vitreous and aqueous interface of an eye with an intact posterior capsule (PC) [33]. Clinically, PPCCC does not appear to cause any higher postoperative risks than the standard approaches [34]. In a series of 1,000 cases of POBH, no case of CME was reported and the rate of retinal detachment was not higher than standard cataract surgery [35]. Menapace suggests that the use of OVD in Berger's space combined with a watertight seal between the rhexis and optic may impede the access of cytokines and stabilizes the anterior vitreous preventing the CME observed in accidental PC rupture [28]. The seal also reduces the amount of anterior chamber flare seen postoperatively, likely due to prevention of residual OVD in the Berger space from entering the anterior chamber over time [36].

22.2.3 Insertion of the Lens

The foldable BIL is inserted through a 2.8 mm corneal incision into the anterior chamber. It is maneuvered into a position on top of the anterior capsule with the lens apposed to the capsule at the six o'clock position. The lens is gently manipulated temporally and nasally to slide the posterior haptic under the posterior rhexis. The lens haptic design prevents lens tilting and subluxation and when it is moved into position, the anterior and posterior capsules are aligned and apposed. The OVD is removed from the anterior chamber and acetylcholine is injected into the anterior chamber to prevent iris capture.

The BIL functions to prevent PCO in two ways. Firstly, the posterior capsule in the visual axis is removed and any residual LECs in that region are eradicated acting much in the same way as a PPCCC. Secondly, once the lens is correctly sited, the anterior and posterior capsules are sandwiched together (Fig. 22.5). The only opening through which LECs could enter the anterior hyaloid space is sealed within the interhaptic groove of the lens. This highlights some crucial points of the surgical technique. Although it is important to maintain the anterior chamber stability with OVD, during the surgery, it is important not to fill the bag itself with OVD at any stage. Filling the bag creates a concave shape in the posterior rhexis which increases the risk of PPCCC ripping and puncturing the anterior hyaloid. Separating out the anterior and posterior capsule edges also significantly hampers the correct insertion of the lens.

In addition, both the anterior and posterior rhexes should be sized to fit snugly around the lens optic. The inherent elasticity of the capsule allows for undersizing of both rhexes. Positioning the 5 mm optic into a 4.5–5 mm rhexis allows for a very firm and secure apposition of the capsule margin and the lens groove. A loose capsule-lens apposition would result in IOL instability and openings in the bag through which LECs may migrate. It is important not to size the capsulorhexes too small however as excessive pressure and manipulation in positioning the lens can cause stress to the zonular fibers. The lens is suspended entirely by the anterior and posterior capsular edges. The capsule provides considerable support with less than 0.3 mm decentration over time consistent with that seen in standard lens-in-the-bag surgery without excessive capsular fibrosis and contraction [37]. In cases of weak zonular fibers and large axial lengths ($\geq$26 mm), the positioning of a capsular tension ring can add stability to the anterior vitreo-capsular interface.

The potential for lens rotation made the POBH a poor approach for insertion of toric lenses. The unique position of the BIL confers a high resistance to rotation (1°), making it one of the most rotationally stable lenses available [38]. Therefore, unlike POHB, BIL is a very viable option for toric lens correction [39–41]. Results for the first 52 eyes treated with a toric BIL implant have been reported and shown that 82 % of patients achieved correction of the astigmatism [42].

22.3 Bag-in-the-Lens PCO Results

Initial histological examination of human donor eyes that had been implanted in vitro with the BIL technique showed that even after 6 weeks culture LECs did not proliferate over the anterior or posterior surface of the IOL [43]. Similarly, in the rabbit animal model, no evidence of LEC proliferation was found outside of the residual capsule [43].

The follow-up results of the subsequent patients are equally promising. A direct comparison of the BIL with the standard lens-in-the-bag (LIB) technique for 100 cases illustrates the efficacy of the lens to prevent PCO in 100 % of cases [44]. In this study, uncomplicated age-related cataracts were treated and, in both the BIL and LIB groups, lens material was matched, both being comprised of hydrophilic acrylic biomaterial (Morcher). Visual acuity outcomes were similar at 6 months in both groups. Nd:YAG capsulotomy was only performed if Snellen visual acuity dropped by two lines or more and at this threshold, 28 % of standard LIB cases required treatment within 48 months with a median interval to treatment of 20 months. Over a comparable 3-year period, no patient who had undergone BIL surgery required capsulotomy.

The problem of PCO is significantly higher in the pediatric population, but the results of the BIL approach in these patients are similarly encouraging. The results of 34 eyes in 22 children have been reported [45]. The mean age of the children treated was 6 years and 2 months (range 2 months to 14 years). A wide range of cataract type was treated including nuclear fetal cataract and spherophakia. In patients with persistent fetal vasculature (PFV), an anterior vitrectomy was performed at the time of cataract surgery until the vascular cord was retracted. Patients were followed for up to 68 months postoperatively and the visual axis maintained its clarity in all patients who had a successful BIL procedure [45].

22.4 Complications

The largest cohort of BIL outcomes has recently been reported [46]. In a report of 807 cases of cataract treated with the BIL technique, retinal detachment was reported in 10 eyes (1.24 %). Iris was captured in 19 (2.35 %), hypopyon in 3 (0.37 %), and toxic anterior segment syndrome in one patient (0.12 %). CME was found in one patient, which is a rate significantly lower than reported rates in standard cataract surgery. Most crucially, there were no reports of PCO in cases of uncomplicated primary surgery. The main reason for the development of secondary cataract was due to incorrect placement of the lens. In cases where both capsular edges are not positioned within the interhaptic groove, a communication is formed whereby the LECs may gain access to the posterior chamber resulting in proliferation and opacification (Fig. 22.6). These cases were treated by a secondary surgery with anterior subluxation of the lens, aspiration of proliferative material, and repositioning of the lens. There were three pediatric patients in whom the lenses had not been optimally sited. In all three cases visual axis reproliferation did occur within a time span of 6 months postoperatively. These were treated in an approach similar to the adult surgery. Pupillary block was observed in one child (Fig. 22.7). The iris may become caught around the optic edges and the resulting pupil block requires urgent intervention. Based on experience, we recommend a primary iridectomy as a matter of course in children (Fig. 22.8). This is not, however, routinely required in adults.

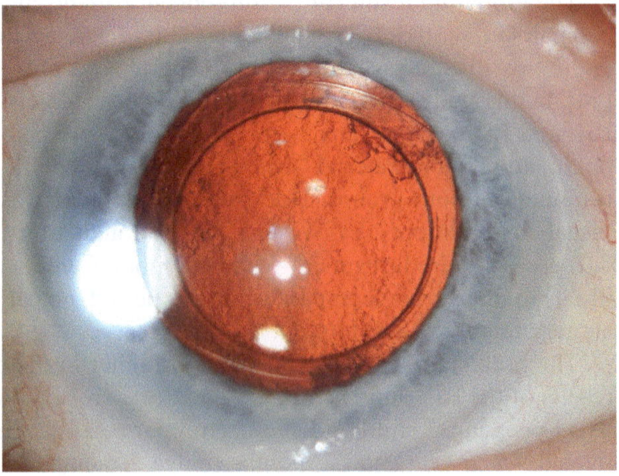

Fig. 22.6 Secondary cataract in a pediatric case in which the PPCCC was not performed and the BIL positioned within the anterior capsulorhexis. PCO occurred within the first 6 months postoperatively

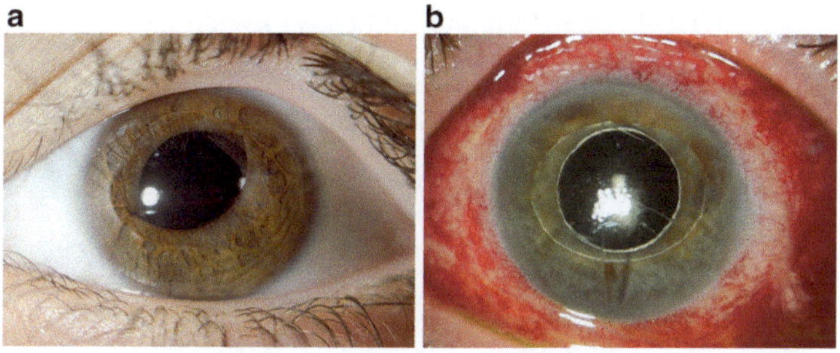

Fig. 22.7 Iris capture. (**a**) Partial. (**b**) Complete

Fig. 22.8 Iridotomy in a pediatric case

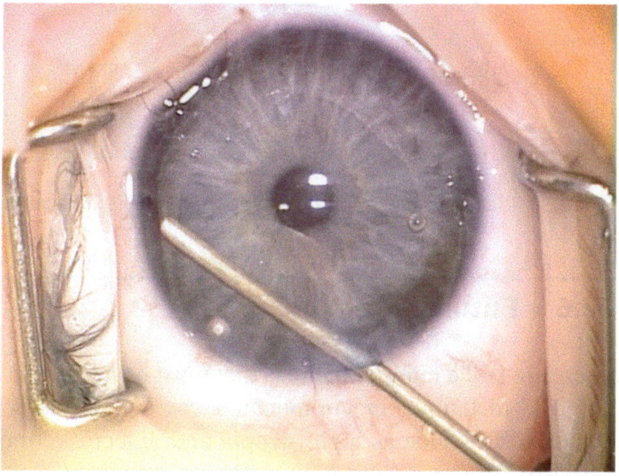

22.5 Histology of the BIL Approach

Immunohistochemical staining has illustrated some fundamental differences between capsular bags of the BIL and the LIB techniques [47]. Capsules incubated with lenses consistent with an in vitro model of LIB show significant staining for α-SMA, a marker of LEC transformation. Conversely, the comparative BIL model showed no staining indicating while LEC proliferation was occurring, no evidence of transformation was found. Subsequent analysis of the first postmortem eye donated from a patient who had undergone the BIL procedure showed that the LECs were confined between the anterior and posterior capsules as expected though this lens had been in situ for only 5 months [48]. Analysis of further examples confirmed that the rhexis openings stayed perfectly clear [49]. LEC proliferation and transformation was confirmed by the presence of Soemmering's ring formation and opacification of the capsule around the optics (Fig. 22.9). The most significant finding, however, was that all of this regenerative material was completely confined by the edges of the capsules sealed by the groove of the BIL implant (Fig. 22.10).

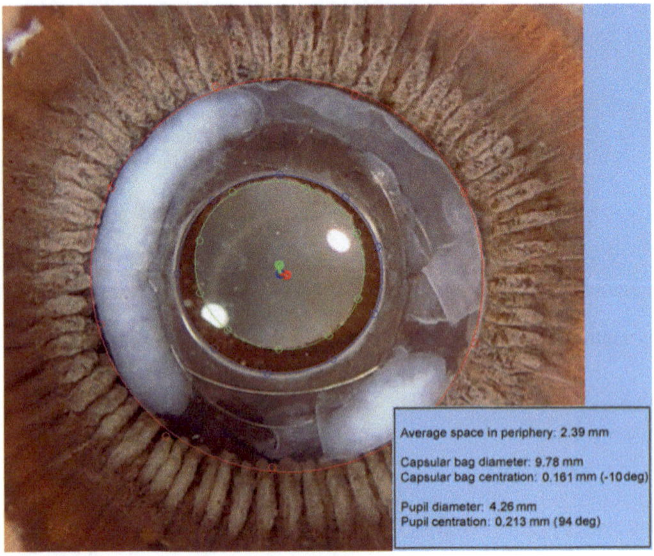

Fig. 22.9 Soemmering's ring forming peripherally but a centrally clear optic [49]

22.6 Future Directions

There are still a number of areas that must be addressed in this technique. The use of the BIL technique in complicated cases has not yet been fully validated and comorbidities such as high myopia and diabetic retinopathy that complicate standard LIB surgery are currently under assessment. Cataract surgery is also associated with a decline in corneal endothelial cells as a result of phacoemulsification, surgical time, and IOL type [50]. The effect of the BIL implant on long-term corneal endothelial cell density is currently under investigation. CME is also a complication seen relatively frequently after phacoemulsification surgery, though the exact rate is not yet known. Rates of postoperative CME will soon be reported as part of a large-scale multicenter investigation, the ESCRS PREvention of Macular EDema (PREMED). This study will allow direct comparison of the BIL postoperative rates with standard surgery.

22.7 Conclusion

The BIL surgical technique was originally designed to combat the challenges posed by PCO in the high-risk pediatric population; however, over time it has been found to have a number of advantages over standard LIB cataract surgeries in adults. The lens design is the first, and currently only, that allows the surgeon to control IOL centration with long-term certainty. LEC proliferation over time occurs in a concentric 360° pattern ensuring that secondary LEC growth will not influence BIL

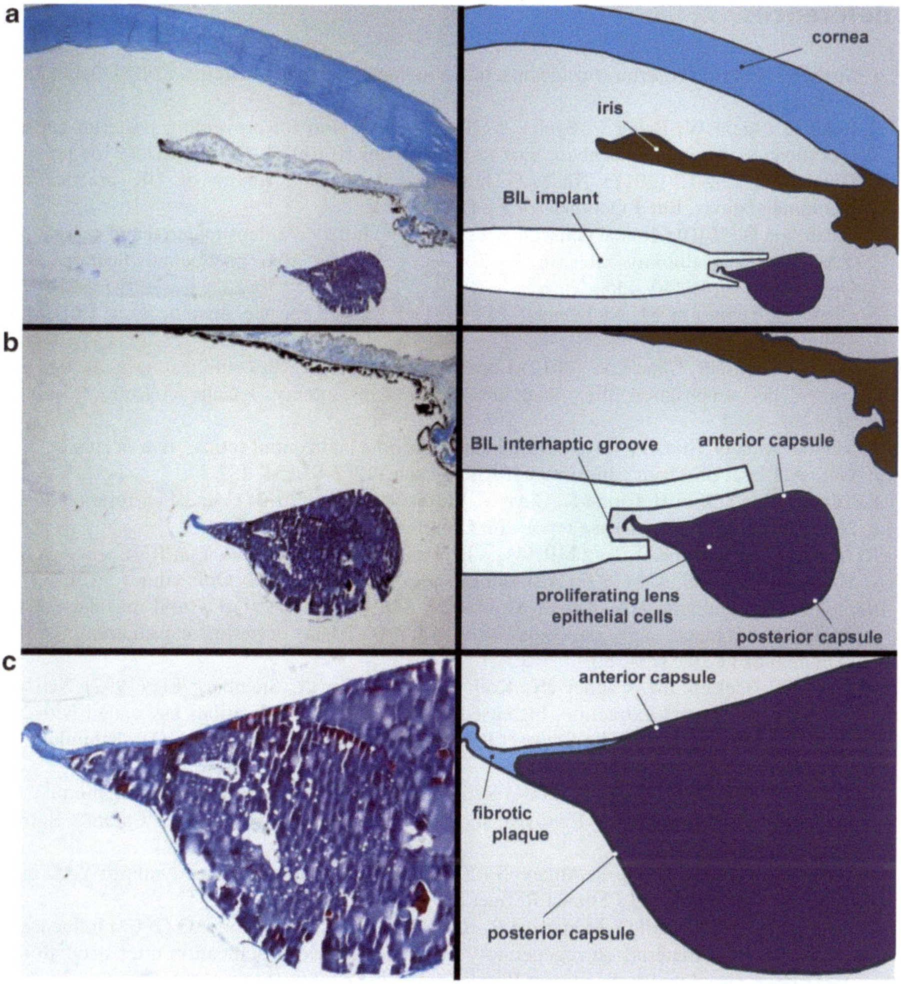

Fig. 22.10 Histology of BIL 3 years after implantation with corresponding schematic (**a**) magnification (×20) showing the BIL in situ. (**b**) Magnification (×40) demonstrating restriction of proliferation within the interhaptic groove. (**c**) Magnification (×100) showing fibrotic plaque at the interhaptic groove sealing the proliferation within anterior and posterior capsules [49]

centration and stability in the posterior chamber. In complicated cataract cases where capsular ripping or zonular dehiscence occurs, this stability may naturally be compromised otherwise stability is assured. Because of these centration capabilities, the lens alignment can be considered objective and independent, and the BIL approach can be considered ideal for complex optics (the so-called premium IOLs).

Lastly, although the BIL technique requires a high degree of surgical accuracy, the skill levels required should not be considered prohibitive. However, the application of newer femtosecond laser techniques may provide an even further level of accuracy. All surgical steps in the BIL technique that can be performed with higher precision will likely improve long-term stability and outcomes.

References

1. Spalton D (2013) Posterior capsule opacification: have we made a difference? Br J Ophthalmol 97(1):1–2
2. Findl O, Buehl W, Bauer P, Sycha T (2010) Intervention for preventing posterior capsule opacification. Cochrane Database Syst Rev 17(2). doi:10.1002/14651858.CD003738.pub3
3. Gomaa A, Liu C (2011) Nd:YAG laser capsulotomy: a survey of UK practice and recommendations. Eur J Ophthalmol 21(4):385–390
4. Johansson B (2010) Clinical consequences of acrylic intraocular lens material and design: Nd: YAG-laser capsulotomy rates in 3 × 300 eyes 5 years after phacoemulsification. Br J Ophthalmol 94(4):450–455
5. Ohadi C, Hamilton M, McDonnell PJ (1991) Posterior capsule opacification. Curr Opin Ophthalmol 2:46–52
6. Ho JC, Yuen NS, Chan CW (2012) Capsule membrane growth complicating glaucoma and corneal decompensation after neodymium:YAG capsulotomy. J Cataract Refract Surg 38 (8):1492–1494
7. Malukiewicz G, Stafiej J (2011) Malignant glaucoma and central retinal vein occlusion after Nd:YAG laser posterior capsulotomy. Klin Oczna 113(7–9):254–257
8. Türkcü FM, Yüksel H, Cingü K, Cınar Y, Murat M, Caça I (2013) Corneal perforation during Nd:YAG capsulotomy: a case report. Int Ophthalmol 33(1):99–101
9. Apple DJ, Solomon KD, Tetz MR, Assia EI, Holland EY, Legler UF, Tsai JC, Castaneda VE, Hoggatt JP, Kostick AM (1992) Posterior capsule opacification. Surv Ophthalmol 37(2):73–116
10. Steinert RF, Puliafito CA, Kumar SR, Dudak SD, Patel S (1991) Cystoid macular edema, retinal detachment, and glaucoma after Nd:YAG laser posterior capsulotomy. Am J Ophthalmol 112(4):373–380
11. Javitt JC, Tielsch JM, Canner JK, Kolb MM, Sommer A, Steinberg EP (1992) National outcomes of cataract extraction. Increased risk of retinal complications associated with Nd: YAG laser capsulotomy. The Cataract Patient Outcomes Research Team. Ophthalmology 99 (10):1487–1497
12. Ranta P, Tommila P, Kivela T (2004) Retinal breaks and detachment after neodymium:YAG laser posterior capsulotomy. Five-year incidence in a prospective cohort. J Cataract Refract Surg 30:58–66
13. Oshika T, Santou S, Kato S, Amano S (2001) Secondary closure of neodymium: YAG laser posterior capsulotomy. J Cataract Refract Surg 27(10):1695–1697
14. Georgopoulos M, Findl O, Menapace R, Buehl W, Wirtitsch M, Rainer G (2003) Influence of intraocular lens material on regeneratory posterior capsule opacification after neodymium: YAG laser capsulotomy. J Cataract Refract Surg 29(8):1560–1565
15. Casprini F, Balestrazzi A, Tosi GM, Lazzarotto M, Malandrini A, Lepri F, Martone G, Caporossi T, Caporossi A (2008) Optical aberrations in pseudophakic eyes after 2.5-mm Nd: YAG laser capsulotomy for posterior capsule opacification. J Refract Surg 24(7):702–706
16. Rozema JJ, Koppen C, De Groot V, Tassignon MJ (2009) Influence of neodymium: YAG laser capsulotomy on ocular wavefront aberrations in pseudophakic eyes with hydrophilic and hydrophobic intraocular lenses. J Cataract Refract Surg 35(11):1906–1910
17. Vrijman V, van der Linden JW, Nieuwendaal CP, van der Meulen IJ, Mourtis MP, Lapid-Gortzak R (2012) Effect of Nd:YAG laser capsulotomy on refraction in multifocal apodized diffractive pseudophakia. J Refract Surg 28(8):545–550
18. Leysen I, Bartholomeeusen E, Coeckelbergh T, Tassignon MJ (2009) Surgical outcomes of intraocular lens exchange: Five year study. J Cataract Refract Surg 35(6):1013–1018
19. Gimbel HV, Neuhann T (1990) Developments, advantages, and methods of the continuous circular capsulorhexis technique. J Cataract Refract Surg 16:31–37
20. Georgopoulos M, Menapace R, Findl O, Pettenel V, Kiss B, Rainer G (2003) After-cataract in adults with primary posterior capsulorhexis: comparison of hydrogel and silicone intraocular lenses with round edges after 2 years. J Cataract Refract Surg 29:955–960

21. Van Tenten Y, De Groot V, Wuyts FL, Tassignon MJ (2000) Quantitative measurement of the PCCC area in the postoperative period. Br J Ophthalmol 84:1117–1120
22. Yazici AT, Bozkurt E, Kara N, Yildirim Y, Demirok A, Yilmaz OF (2012) Long-term results of phacoemulsification combined with primary posterior curvilinear capsulorhexis in adults. Mid East Af J Ophthalmol 19(1):115–119
23. Tassignon MJ, De Groot V, Smets RM, Tawab B, Vervecken F (1996) Secondary closure of posterior continuous curvilinear capsulorhexis. J Cataract Refract Surg 22(9):1200–1209
24. Tassignon MJ, De Groot V, Vervecken F, Van Tenten Y (1998) Secondary closure of posterior continuous curvilinear capsulorhexis in normal eyes and eyes at risk for postoperative inflammation. J Cataract Refract Surg 24(10):1333–1338
25. DeGroot V, Vrensen GFJM, Willekens B, Van Tenten Y, Tassignon MJ (2003) In vitro study on the closure of posterior capsulorrhexis in the human eye. IOVS 44(5):2076–2083
26. Gimbel HV, DeBroff BM (1994) Posterior capsulorhexis with optic capture: maintaining a clear visual axis after pediatric cataract surgery. J Cataract Refract Surg 20:658–664
27. Gimbel HV (1997) Posterior continuous curvilinear capsulorhexis and optic capture of the intraocular lens to prevent secondary opacification in pediatric cataract surgery. J Cataract Refract Surg 23(Suppl 1):652–656
28. Menapace R (2008) Posterior capsulorhexis combined with optic buttonholing: an alternative to standard in-the-bag implantation of sharp-edged intraocular lenses? A critical analysis of 1000 consecutive cases. Graefes Arch Clin Exp Ophthalmol 246(6):787–801
29. Tassignon MJ, DeGroot V, Vrensen GF (2002) Bag-in-the-lens implantation of intraocular lenses. J Cataract Refract Surg 28(7):1182–1188
30. Tassignon MJBR, inventor, Morcher GmbH, assignee (2000) Intraocular lens and method for preventing secondary opacification. US patent 6,027,531. http://patft.uspto.gov/netacgi/nph-Parser?Sect1=PTO1&Sect2=HITOFF&d=PALL&p=1&u=%2Fnetahtml%2FPTO%2Fsrchnum.htm&r=1&f=G&l=50&s1=6027531.PN.&OS=PN/6027531&RS=PN/6027531
31. Tassignon MJ, Rozema JJ, Gobin L (2006) Ring-shaped caliper for better anterior capsulorhexis sizing and centration. J Cataract Refract Surg 32(8):1253–1255
32. DeGroot V, Leysen I, Neuhann T, Gobin L, Tassignon MJ (2006) One-year follow-up of bag-in-the-lens intraocular lens implantation in 60 eyes. J Cataract Refract Surg 32 (10):1632–1637
33. DeGroot V, Hubert M, VanBest JA, Engelen S, Van Aelst S, Tassignon MJ (2003) Lack of fluorophotometric evidence of aqueous-vitreous barrier disruption after posterior capsulorhexis. J Cataract Refract Surg 29(12):2330–2338
34. Galand A, van Cauwenberge F, Moosavi J (1996) Posterior capsulorhexis in adult eyes with intact and clear capsules. J Cataract Refract Surg 22(4):458–461
35. Stifter E, Menapace R, Neumayer T, Luksch A (2008) Macular morphology after cataract surgery with primary posterior capsulorhexis and posterior optic buttonholing. Am J Ophthalmol 146(1):15–22
36. Stifter E, Menapace R, Kriechbaum K, Vock L, Luksch A (2009) Effect of primary posterior continuous curvilinear capsulorhexis with and without posterior optic buttonholing on postoperative anterior chamber flare. J Cataract Refract Surg 35:480–484
37. Verbruggen KHM, Rozema JJ, Gobin L, Coeckelbergh T, De Groot V, Tassignon MJ (2007) Intraocular lens centration and visual outcomes after bag-in-the-lens implantation. J Cataract Refract Surg 33(7):1267–1272
38. Rozema JJ, Gobin L, Verbruggen K, Tassignon MJ (2009) Changes in rotation after implantation of a bag-in-the-lens intraocular lens. J Cataract Refract Surg 35(8):1385–1388
39. Gobin L, Tassignon MJ, Mathysen D (2011) Spherotoric bag-in-the-lens intraocular lens: Power calculation and predictive misalignment nomogram. J Cataract Refract Surg 37 (6):1020–1030
40. Tassignon MJ, Gobin L (2010) Toric bag-in-the-lens implant: why and how to implant. Exp Rev Ophthalmol 5(2):135–141

41. Gobin L, Tassignon MJ, Wouters K, De Groot V (2011) Surgically induced astigmatism after intraocular lens implantation using the bag-in-the-lens technique. J Cataract Refract Surg 37 (6):1015–1019
42. Tassignon MJ, Goblin L, Mathysen D, Van Looveren J (2011) Clinical results after spherotoric intraocular lens implantation using the bag-in-the-lens technique. J Cataract Refract Surg 37 (5):830–834
43. DeGroot V, Tassignon MJ, Vrensen GFJM (2005) Effect of bag-in-the-lens implantation on posterior capsule opacification in human donor eyes and rabbit eyes. J Cataract Refract Surg 31 (2):398–405
44. Leysen I, Coeckelbergh T, Gobin L, Smet H, Daniel Y, DeGroot V, Tassignon MJ (2006) Cumulative neodymium: YAG laser rates after bag-in-the-lens and lens-in-the-bag intraocular lens implantation: comparative study. J Cataract Refract Surg 32(12):2085–2090
45. Tassignon MJ, De Veuster I, Godts D, Kosec D, Van den Dooren K, Gobin L (2007) Bag-in-the-lens intraocular lens implantation in the pediatric eye. J Cataract Refract Surg 33 (4):611–617
46. Tassignon MJ, Goblin L, Mathysen D, Van Looveren J, De Groot V (2011) Clinical outcomes of cataract surgery after the bag-in-the-lens intraocular lens implantation following ISO standard 11979-7:2006. J Cataract Refract Surg 37(12):2120–2129
47. De Keyzer K, Leysen I, Timmermans JP, Tassignon MJ (2008) Lens epithelial cells in an in vitro capsular bag model: lens-in-the-bag versus bag-in-the-lens technique. J Cataract Refract Surg 34(4):687–695
48. Werner L, Tassignon MJ, Gobin L, Rozema JJ, Davis D, Brubaker J (2008) Bag-in-the-lens: first pathological analysis of a human eye obtained postmortem. J Cataract Refract Surg 34 (12):2163–2165
49. Werner L, Tassignon MJ, Zaugg BE, De Groot V, Rozema J (2010) Clinical and histopathological evaluation of six human eyes implanted with the bag-in-the-lens. Ophthalmology 117(1):55–62
50. Ventura AC, Walti R, Bohnke M (2001) Corneal thickness and endothelial density before and after cataract surgery. Br J Ophthalmol 85(1):18–20

Posterior Capsule Opacification with Microincision (MICS) IOLs

23

David Spalton

Abstract

The history of cataract has been of continual evolution towards smaller and smaller incision sizes, but there is, however, little point in reducing incision size unless intraocular lenses (IOLs) can be designed to be inserted without the need to enlarge the incision. The demands are great because the microincision cataract surgery (MICS) IOLs must perform as well as a conventional IOL. They must be able to withstand the pressures and forces required to compress it enough to withstand the trauma of injection. They must also be rigid enough to withstand the forces of postoperative bag contraction. IOL design and materials, as well as inserter technology, using coatings to facilitate injection, have advanced so that IOLs can now be inserted through 1.8 mm incisions or even 1.4 mm using wound-assisted techniques. Early MICS IOLs used hydrophilic polymers as these are more compressible and have a plate-haptic configuration, but clinical experience has shown these suffered from high rates of posterior capsule opacification (PCO). More recently hydrophobic materials and open-loop haptic designs have become available but still have increased PCO compared to conventional IOLs, possibly due to their reduced thickness.

MICS IOL design continues to advance and the importance of objective comparable fellow-eye-controlled PCO studies cannot be overemphasized if we are to achieve a good MICS IOL design.

Keywords

Comparative PCO studies • IOL edge design • Microincision IOL • MICS

D. Spalton, F.R.C.S., F.R.C.P., F.R.C.Ophth. (✉)
Professor of Ophthalmology and Emeritus Consultant Ophthalmic Surgeon, St Thomas' Hospital,
The London Clinic Eye Centre, 119 Harley St, London W1G 6AU, UK
e-mail: profspalton@gmail.com

S. Saika et al. (eds.), *Lens Epithelium and Posterior Capsular Opacification*,
DOI 10.1007/978-4-431-54300-8_23, © Springer Japan 2014

23.1 Introduction

One of the major themes in the development of cataract surgery over the last 30 years has been the ever-decreasing incision size required to perform surgery. Intracapsular surgery required a 10 mm incision; with extracapsular surgery and PMMA intraocular lenses (IOLs), this decreased to 6 mm and with foldable IOL materials to less than 3 mm. The reduction in incision size has been driven by the benefits of quicker visual rehabilitation, reduced blood aqueous barrier damage, less wound-related complications and reduced astigmatism so that now virtually all phaco surgery is performed through less than a 3 mm incision. Microincision cataract surgery (MICS) is defined as an incision size of less than 2 mm and this can be achieved by separating the infusion from the phaco and aspiration hand piece, wound burn from the phaco hand piece being prevented by using short pulsed phaco bursts so that the temperature never increases enough to damage and shrink collagen. MICS proponents cite advantages of quicker visual rehabilitation, virtually no surgically induced astigmatism and technical advantages from separating the aspiration and infusion, for example, in floppy iris or small pupil cases. Conversely, MICS demands a change to a biaxial technique and coaxial forceps for capsulorhexis and generates wound problems from 'oar locking' whereby the pivoting movement of the hand pieces in the cornea to reach lenticular fragments traumatizes the wound making it more difficult to self-seal at the end of surgery. The future role of MICS therefore still remains a matter of debate.

Biaxial MICS makes no sense unless the IOL can be delivered through the surgical wound without enlargement and a number of European companies now make such IOLs. Aspheric, toric, multifocal and toric multifocal designs are all available. In general these IOLs are made from hydrophilic polymers and have variations on a plate-type platform design (Fig. 23.1), the reason being that hydrophilic polymers are more compressible than hydrophobic materials and the plate-haptic design allows easier injection without tearing or damaging the IOL. Good visual outcomes have been reported with all designs; however, the demands on these IOLs are high because they must perform as well or better than a traditional IOL:

- The IOLs are thin and fragile but must be able to resist tearing during insertion.
- These IOLs are thinner and more flexible than conventional IOLs but must be rigid enough to withstand the decentering and distorting effects of capsular bag fibrosis and shrinkage postoperatively. This proved to be a problem with one of the initial MICS IOL designs. The ThinOptX UltraChoice 1.0 IOL (ThinOptX Inc.) (Fig. 23.2) had reduced thickness by incorporating a Fresnel prism-type design to reduce optic thickness, but unfortunately postoperative bag fibrosis caused the lens to buckle in the bag and it was withdrawn from the market.
- PCO performance should be equivalent to standard IOL designs.

Posterior capsule opacification (PCO) remains the most common complication of cataract surgery. It is a multifactorial problem influenced by patient factors (e.g., age, concurrent ocular disease), surgical technique [1] and IOL design. The contribution of lens design to the development of PCO is well established; factors include

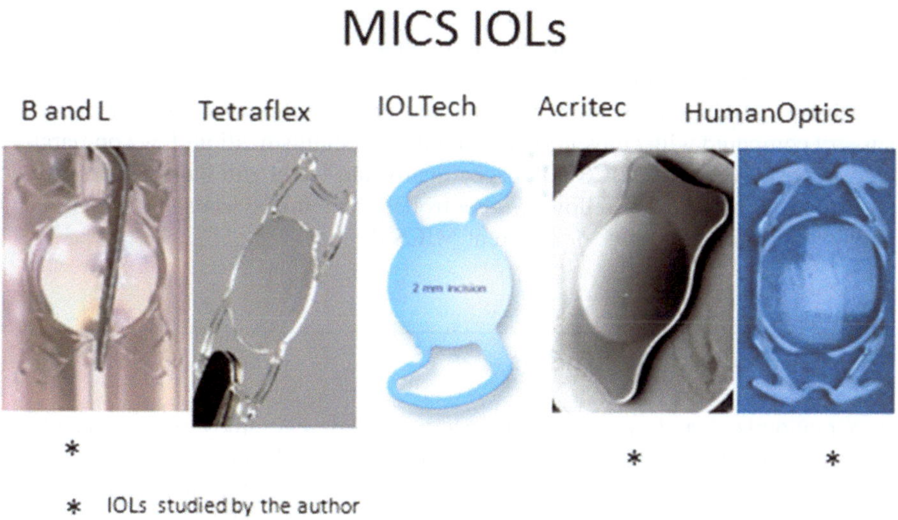

Fig. 23.1 Early designs of MICS IOLs were based on a plate-haptic platform

Fig. 23.2 The ThinOptX IOL was the first IOL specifically developed for MICS. It used Fresnel rings to reduce thickness for insertion but unfortunately tended to buckle in the capsular bag with postoperative fibrosis. The photo shows a scanning electron photomicrograph of the lens

optic diameter, haptic design and biomaterial, but the most critical factor is the square posterior edge design [2–5]. Edge design and biomaterial are particularly relevant to MICS IOLs. Many studies suggest that PCO is increased with hydrophilic IOLs [6]; however, although hydrophilic IOLs have a 'square edge' when in the dry state, environmental scanning electron microscopy shows this is inferior to hydrophobic edges [7]. The reason for this is that hydrophilic IOLs are machined from the dehydrated polymer which is then rehydrated and tumble polished to remove lathing marks and burr but also blunt the edge in the process. Newer techniques can prevent this.

Most of MICS studies have concentrated on the surgical aspects of bimanual surgery compared with conventional coaxial phacoemulsification; few comparative studies have evaluated MICS IOLs specifically and fewer still have used objective techniques in prospective fellow-eye-controlled randomized trials to compare PCO results with a standard IOL or different MICS designs.

23.2 Comparative Studies

In a randomized and controlled prospective fellow eye study we evaluated the HumanOptics MC611MI IOL in comparison to the Alcon AcrySof MA60AC which have very different designs [8]. Three-piece AcrySof IOLs have a long history of excellent PCO performance. The HumanOptics IOL was a hydrophilic plate-designed lens with a complex haptic designed to absorb capsular shrinkage postoperatively (Fig. 23.3a–c). An identical coaxial phacoemulsification and a similar wound were utilized for both groups; thus, the only variable was the two different IOLs. Interestingly, the haptic of the HumanOptics MC611MI could be seen to fold towards or under the optic, although this did not appear to have either an effect on vision or subsequent PCO (Fig. 23.3b). Although both IOLs have good PCO performance overall, our results demonstrated that the PCO performance of the HumanOptics MC611MI IOL was not as good as that of the Alcon MA60AC. Percentage area PCO increased steadily in the MC611MI group and was greater than in the MA60AC group from 3 months of follow-up onwards, achieving statistical significance at 12 and 24 months. The difference in PCO performance between these IOLs is likely to reflect their differences in biomaterial and design. The HumanOptics IOL had a square edge design, but the haptics of the IOL were each joined to the optic by two broad-based attachments. The absence of the square edge barrier at the optic–haptic junction of an IOL has been described as the

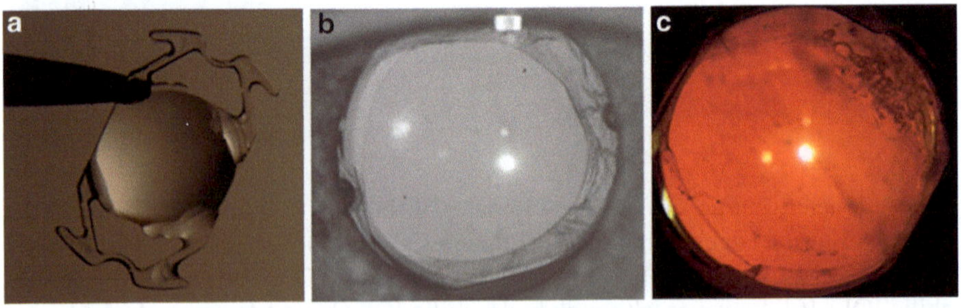

Fig. 23.3 (**a–c**) The haptics of the HumanOptics MC611M1 MICS IOL (**a**) were designed to collapse and absorb the force of capsular contraction, but these tended to buckle under the IOL (**b**). Lens epithelial cells can be seen migrating into the posterior capsule along the plate-type haptic axis (**c**)

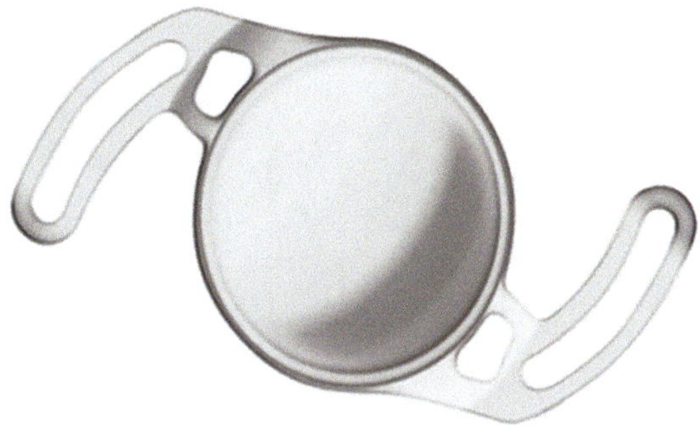

Fig. 23.4 The Croma Idea 613XC MICS IOL has a more conventional open-loop haptic design

'Achilles heel' for PCO prevention as it provides unhindered migration for lens epithelial cells (LECs) from the equatorial region of the capsule onto the central posterior capsule [9]. A 'through haptic' pattern of PCO was observed in 55 % of eyes implanted with the HumanOptics MC611MI IOL and was the most likely cause of greater PCO in this group (Fig. 23.3c). As a corollary to greater PCO in the HumanOptics MC611MI group, we also observed better visual acuity in the MA60AC group which was significant at 3, 6 and 12 months postoperatively. One patient in the MC611MI group required an Nd:YAG laser capsulotomy, whilst no MA60AC eyes required a capsulotomy. Visual acuity was observed to progressively improve in the MA60AC group over the period of follow-up and this may relate to regression of PCO membranes frequently observed with AcrySof IOLs when sequential retroillumination images are assessed subjectively.

Two prospective studies have demonstrated poor PCO performance with another MICS IOL, the ThinOptX UltraChoice 1.0 (Fig. 23.2). In a non-comparative study, 64 % of 50 eyes required an Nd:YAG capsulotomy for visually significant PCO after 15 months of follow-up [10]. A comparative study found that the ThinOptX IOL demonstrated worse PCO and consequently visual performance than the Alcon AcrySof MA30AC 1 year postoperatively [11]. Poor PCO performance thus outweighed the benefits of a smaller incision with these MICS IOLs.

In another prospective randomized trial, which is of particular interest because the MICS IOL used had arguably a better design of a hydrophilic IOL with a hydrophobic surface and 360° square edge barrier, Gangwani et al. compared PCO between the MICS IOL and a standard single-piece AcrySof IOL [12]. Each patient received the MICS IOL (Idea 613 XC, Croma Pharma, Germany) in 1 eye and an AcrySof SA60AT (Alcon Laboratories, USA) as a control in the contralateral eye to allow intraindividual comparison. The MICS IOL was a single-piece hydrophilic/hydrophobic copolymer IOL with a water content of 25 %, an optic diameter of 6.0 mm and overall diameter of 13.0 mm. It had broad open-loop haptics with a broad optic–haptic junction, creating a monoblock-like appearance (Fig. 23.4). The optic had a 360° double-square edge and 9° stepped angulated haptics for increased pressure of the optic against the posterior capsule. PCO 2 years after surgery was greater in the MICS IOL group than in the AcrySof IOL group (mean AQUA score 2.6 ± 2.0 versus 1.9 ± 1.9, $p = 0.02$). Four eyes in the MICS IOL group and no eye

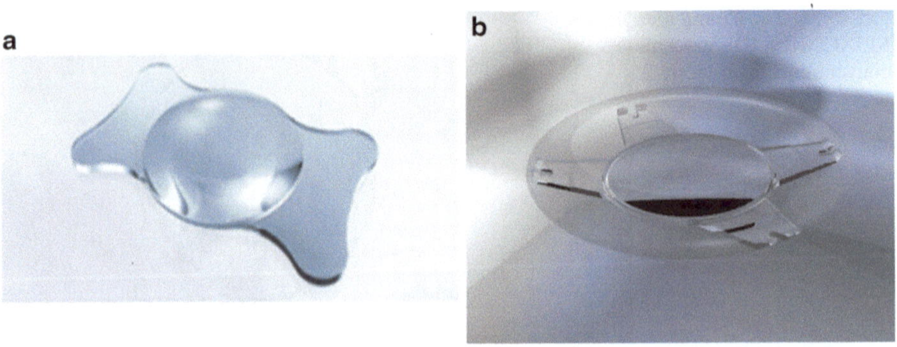

Fig. 23.5 The AcriSmart 36A IOL has a plate-haptic design (**a**). The Akreos MI60 has a novel 4 haptic configuration (**b**)

in the AcrySof IOL group required Nd:YAG capsulotomy ($p = 0.03$). One patient in the MICS IOL group had severe bag contraction, with both haptics folding inward. There was no difference in anterior capsular opacification between the two IOLs and interestingly no difference in average tilt or decentration either, suggesting the MICS IOL was stable in the bag and resisted capsular contraction.

In another prospective, randomized, fellow-eye comparison [13] we evaluated the difference in PCO performance between two aspheric MICS IOLs: one of plate-haptic design and the other of a rather different design although still based on a plate principle. The lenses used were the AcriSmart 36A (Carl Zeiss Meditec, Germany, now renamed as the CT ASPHINA® 509M) and the Akreos MI60 (Bausch & Lomb, Rochester, Minnesota, USA) (Fig. 23.5). The AcriSmart 36A IOL is a hydrophilic IOL with a hydrophobic surface and is used as a platform for widely used toric, multifocal and multifocal toric designs (so it is likely that these variations will have a similar PCO performance). The Akreos MI60 IOL is also a hydrophilic MICS IOL with four haptics designed to be resistant to decentration from capsular fibrosis. Both IOLs can be implanted through a 1.8 mm incision. We also compared the PCO performance of these two MICS IOLs to that of a conventional single-piece hydrophobic acrylic IOL (AcrySof SN60AT Alcon Fort Worth, Texas, USA) using information from our database. Mean percentage PCO score was significantly less with AcriSmart 36A at 1, 3 and 12 months compared to Akreos MI60 ($p = 0.03$, 0.02 and 0.05, respectively). At 1 month, AcrySof SN60AT showed more PCO, whereas at 12 months Akreos MI60 IOL showed more PCO (Fig. 23.6). At 24 months, mean PCO score remained under 11 % with AcrySof SN60AT, whereas mean PCO score increased linearly with time in both the AcriSmart 36A and Akreos MI60 groups, with a maximum of up to 16 % and 23 %, respectively.

Spyridaki and Höh [14] also found high Nd:YAG capsulotomy rates for patients implanted with hydrophilic MICS IOLs, with rates as follows: for AcriSmart (Zeiss) 20 %, Thinlens (Technomed) 33 %, Acriflex (Oculentis) 11 % and Careflex (Medizintechnik) 32 % at 850 days postoperatively.

MICS IOL platforms continue to progress and open-loop hydrophobic designs are becoming available. A recent study found no difference in PCO between a Hoya

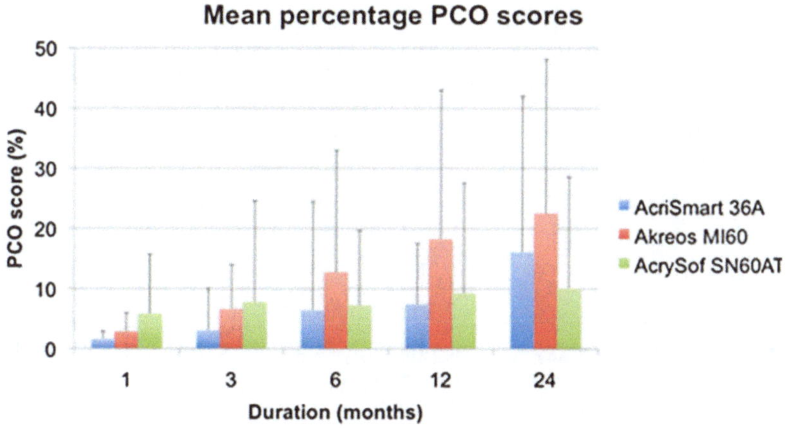

Fig. 23.6 Graph comparing PCO scores of the two MICS IOLs with that of a conventional AcrySof single-piece IOL—this has significantly lower PCO

3-piece or single-piece hydrophobic MICS IOL at 1 year after surgery; furthermore, PCO rates appear to be lower than those found for hydrophilic IOLs although these results can only be preliminary as follow-up was only for 1 year [15]. Interestingly, both IOLs had a high incidence of buttonholing through the rhexis, something which remains unexplained. Another study compared an open-loop MICS IOL to a standard single-piece Alcon AcrySof IOL [16]. One hundred patients were randomized in a fellow-eye-controlled study to implantation of an iMics1 NY-60 IOL (Hoya Corp.) in one eye and an AcrySof SN60WF IOL (Alcon Laboratories) in the other eye. PCO was assessed objectively at 3 years after surgery. The objective PCO score (mean $\pm$ standard deviation) was 3.0 ± 2.0 for the iMics1 NY-60 IOL and 1.9 ± 1.4 for the AcrySof SN60WF IOL ($P < 0.001$). Three years after surgery 35.6 % of patients underwent a capsulotomy in the iMics1 NY-60 eye and 13.7 % underwent a capsulotomy in the AcrySof SN60WF eye ($P = 0.001$). There was no statistically significant difference in best-corrected visual acuity, rhexis–IOL overlap, capsular folds or anterior capsule opacification between the IOLs. PCO scores were similar to those found using the same technology to measure PCO as Gangwani's study [12]. This is a well-controlled long-term study with objective PCO measurement and suggests that a well-designed hydrophobic MICS IOL still has inferior PCO performance than a gold standard IOL. The reasons for this are unclear but could imply that IOL thickness is a significant factor.

23.3 Conclusion

Quite clearly, whilst MICS IOLs can have a good visual performance, quantitative PCO studies show that hydrophilic MICS IOLs of various designs have more PCO in comparison to a 'gold standard' hydrophobic IOL. Possible explanations are the plate-haptic design which allows LEC migration through the broad optic–haptic

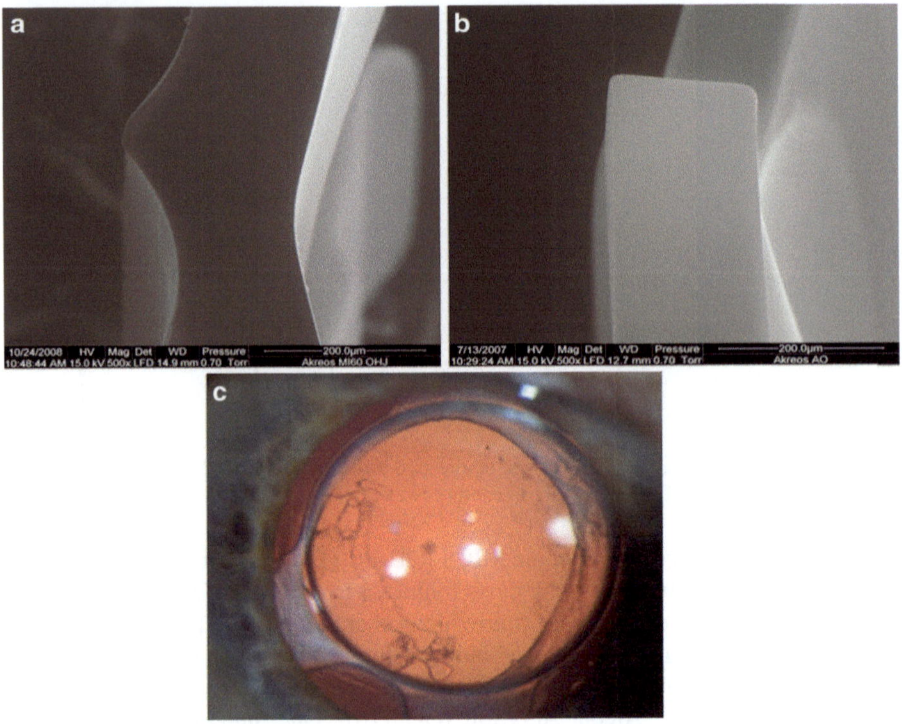

Fig. 23.7 (**a–c**) The Bausch & Lomb Akreos MI60 has poor-quality 'square' edge profiles at the optic–haptic junction (**a**) and at the optic edge (**b**). Lens epithelial cells can be seen migrating through the optic–haptic junction (**c**)

junction, the hydrophilic material or the quality of the square edge. Using scanning environmental electron microscopy, which does not distort hydrophilic materials, we were able to image the IOL edge and quantify the sharpness of this by measuring the local radius of curvature [4]. Furthermore, we were able to correlate the edge sharpness with clinical studies and determine the parameters for PCO protection [17].

The development of MICS IOLs demonstrates again many of the principles and problems of IOL design. The Bausch & Lomb MI60 IOL used in one of our studies (Fig. 23.7), described in this chapter [13], was found to have excessively high PCO, both by ourselves and other authors, and retroillumination images show LECs migrating through the optic–haptic junctions (Fig. 23.7c). This IOL has now been completely reengineered and remarketed as the Incise IOL using a lower water content polymer and now has a superb 360° square edge of equal quality to those seen with the best hydrophobic IOLs (Fig. 23.8). Clinical experience with this lens should answer the question of whether it is the edge or the polymer that is responsible for high PCO rates with hydrophilic IOLs.

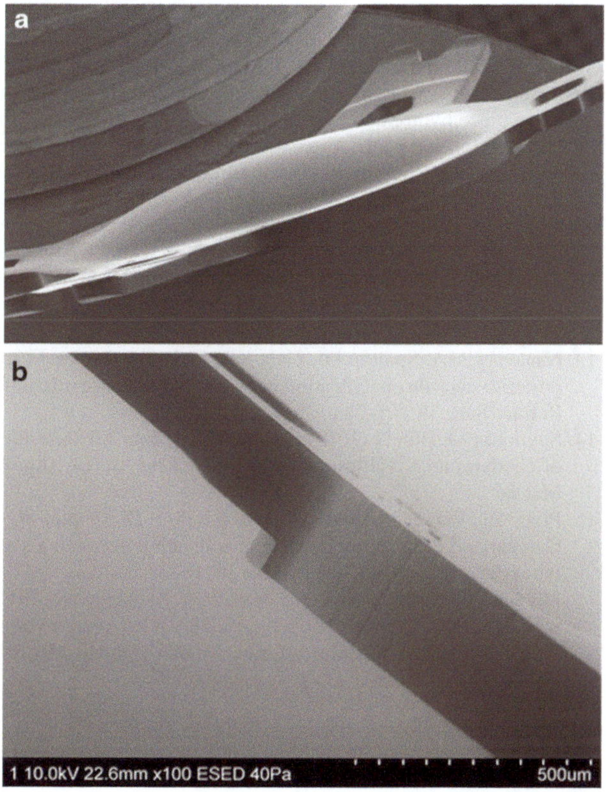

Fig. 23.8 (**a**, **b**) The Akreos MI60 has been reengineered using a different polymer (**a**) and now has a high-quality sharp posterior edge profile extending through the optic–haptic junction (**b**)

References

1. Hollick EJ, Spalton DJ (1999) The effect of capsulorhexis size on posterior capsule opacification. Amer J Ophthal 128:271–279
2. Meacock WR, Spalton DJ, Boyce JF, Jose RJ (2001) Effect of optic size on posterior capsular opacification with 5.5 mm and 6.0 mm optic diameter Acrysof intraocular lenses. J Cataract Refract Surg 27:1194–1198
3. Meacock WR, Spalton DJ (2001) Effect of intraocular lens haptic compressibility on the posterior lens capsule after cataract surgery. J Cataract Refract Surg 27:1366–1371
4. Nanavaty M, Spalton DJ, Boyce J, Brain A, Marshall J (2008) Edge profile of commercially available square-edged intraocular lenses. J Cataract Refract Surg 34:677–686
5. Hancox J, Spalton D, Cleary G, Boyce J, Nanavaty MA, Thyagarajan S, Marshall J (2008) Fellow eye comparison of posterior capsule opacification with AcrySof SN60AT and AF-1 YA-60BB blue blocking intraocular lenses. J Cataract Refract Surg 34:1489–1494
6. Findl O, Buehl W, Bauer P, Sycha T (2007) Interventions for preventing posterior capsule opacification. Cochrane Database Syst Rev 18(3):CD003738
7. Werner L, Tetz M, Feldmann I, Bücker M (2009) Evaluating and defining the sharpness of intraocular lenses: microedge structure of commercially available square-edged hydrophilic intraocular lenses. J Cataract Refract Surg 35:556–566

8. Cleary G, Spalton DJ, Hancox J, Boyce J, Marshall J (2009) Randomized intraindividual comparison of posterior capsule opacification between a microincision intraocular lens and a conventional intraocular lens. J Cataract Refract Surg 35:265–272
9. Nixon DR (2006) Evaluation of lens epithelial migration in vivo at the optic haptic junction of a one piece hydrophobic intraocular lens. Amer J Ophthalmol 142:557–562
10. Prakash P, Kasaby HE, Aggarwal RK, Humfrey S (2007) Microincision bimanual phacoemulsification and Thinoptx implantation through a 1.70 mm incision. Eye 21:177–182
11. Kaya V, O¨ ztu¨rker ZK, O¨ ztu¨rker C, Yas ARO¨, Sivrikaya H, Agca A, Yilmaz O¨F (2007) ThinOptX vs AcrySof: comparison of visual and refractive results, contrast sensitivity, and the incidence of posterior capsule opacification. Eur J Ophthalmol 17:307–314
12. Gangwani V, Hirnschall N, Koshy J, Crnej A, Nishi Y, Maurino V, Findl O (2011) Posterior capsule opacification and capsular bag performance of a microincision intraocular lens. J Cataract Refract Surg 37:1988–1992
13. Nanavaty MA, Spalton DJ, Gala KB, Dhital A, Boyce J (2013) A fellow eye comparison of posterior capsule opacification between two aspheric microincision intraocular lens. J Cataract Refract Surg 39:705–711
14. Spyridaki M, Höh H (2010) Comparison of four MICS intraocular lenses regarding their rates of neodymium:YAG laser capsulotomy. Klin Monbl Augenheilkd 227:208–214, Epub 2010 Mar 16
15. Prinz A, Vecsei-Marlovits PV, Sonderhof D, Irsigler P, Findl O, Weingessel B (2013) Comparison of posterior capsule opacification between a 1-piece and a 3-piece microincision intraocular lens. Brit J Ophthalmol 97:18–22
16. Leydolt C, Schriefl S, Stifter E, Haszcz A, Menapace R (2013) Posterior capsule opacification with the iMics1 NY-60 and AcrySof SN60WF 1-piece hydrophobic acrylic intraocular lenses: 3-year results of a randomized trial. Amer J Ophthal 156:1317–1318

Part VI

Special Cases

PCO and the Pediatric Eye

24

Abhay R. Vasavada, Sajani K. Shah, Vaishali Vasavada, and M.R. Praveen

Abstract

Posterior capsule opacification (PCO) still remains one of the biggest deterrents to visual rehabilitation in pediatric eyes, particularly in infants and young children. Different patient-related factors, ocular and systemic conditions, and surgical techniques influence the development of PCO. In children, the management of the posterior capsule and anterior vitreous during cataract surgery plays a vital role in ensuring a clear visual axis and thereby successful technical and functional outcomes. Meticulous use of surgical techniques and appropriate IOLs remains the mainstay for retarding the development of PCO in children. Alternative approaches to retard or delay PCO include bag-in-the-lens implantation, optic capture of IOL without anterior vitrectomy, and devices like sealed capsule irrigation. Further, timely and age-appropriate management of this visual axis opacification is essential to reduce amblyopia and allow good visual development.

Keywords

Anterior vitrectomy • Intraocular lens (IOL) implantation • Pediatric cataract surgery • Posterior capsule management • Posterior capsule opacification

24.1 Introduction

The primary goal of pediatric cataract surgery is to ensure an anatomically safe eye and achieve binocular single vision that lasts for the child's lifetime. Visual axis obscuration (VAO) is a significant obstacle to visual rehabilitation after pediatric

None of the authors have any financial interest in this chapter.

A.R. Vasavada, M.S., F.R.C.S. (✉) • S.K. Shah, M.S. • V. Vasavada, M.S. • M.R. Praveen, D.O.
Iladevi Cataract and IOL Research Centre, Gurukul Road, Memnagar,
Ahmedabad 380 052, Gujarat, India
e-mail: icirc@abhayvasavada.com

S. Saika et al. (eds.), *Lens Epithelium and Posterior Capsular Opacification*,
DOI 10.1007/978-4-431-54300-8_24, © Springer Japan 2014

cataract surgery [1–4]. In the last decade, there have been a number of noteworthy changes in the field of pediatric cataract extraction with intraocular lens (IOL) implantation. Despite these improvements in techniques and technologies, management of pediatric cataract poses a challenge to cataract surgeons. Posterior capsule opacification (PCO) is the most frequent complication after pediatric cataract surgery and its onset deters visual rehabilitation. If PCO is left untreated, it obscures the central visual axis, especially during the critical period of postnatal growth. This can lead to stimulus deprivation amblyopia and can hamper the development of binocular single vision. It is reported that correlated binocular vision is essential for successful recovery from experimentally induced amblyopia [5–8] and that the absence of correlated binocular vision may play a critical role in the development of amblyopia [6].

Therefore, while surgeons are planning the management of the posterior capsule in pediatric eyes, special attention needs to be paid to the maintenance of a clear visual axis. In recent times, primary posterior capsulectomy (with or without anterior vitrectomy) is considered a "routine surgical step" to prevent PCO, especially in young children.

Apart from PCO, another factor that has a great bearing on the outcome of pediatric cataract surgery is the management of the anterior vitreous [9–11]. It is an accepted fact that the anterior vitreous face (AVF) is more "reactive" in infants and young children. The AVF provides a scaffold across which the lens epithelial cells (LECs) migrate, resulting in opacification of the intact AVF. This can lead to VAO even when the posterior capsule is not present.

Ideally, the aim of pediatric cataract surgery is to delay the onset of VAO including PCO and to treat it promptly. This chapter describes the risk factors for PCO in pediatric eyes, its impact on the growing eye, strategies to minimize or delay the onset of PCO and VAO, clinical presentation, and treatment modalities.

24.2 Risk Factors for PCO in Pediatric Eyes

Unlike cataract surgery in adults, an intact posterior capsule predisposes children to a higher rate of PCO following cataract surgery even without coexisting ocular comorbidities [12]. In children, the high activity of LECs and exaggerated postoperative inflammation induce a greater propensity for PCO [13].

An in vitro experiment of human capsules with a central hole of 5 mm (posterior capsulotomy) placed in a culture showed that even in the absence of the posterior capsule which is their natural substrate, the LECs that remain after cataract extraction have the potential to proliferate and form a monolayer of LECs on a basal lamina of vitreous origin. These LECs are able to close the posterior capsulorhexis partially or totally in approximately one-third of the cases [14].

A traumatic surgery, an incomplete cortical aspiration, inappropriate management of the posterior capsule in accordance with the age of child at the time of cataract surgery, and inadequate anterior vitrectomy increase the risk of VAO. Eyes with associated ocular anomalies (e.g., anterior segment dysgenesis, iris hypoplasia, or persistent fetal vasculature) are at a ninefold higher risk of developing VAO compared with eyes without associated ocular anomalies [2].

There are several other contributing factors for PCO formation after cataract surgery [15–22].

These include:

- Age of the patient at the time of cataract surgery (the younger the age at the time of surgery, the faster is the onset, and the more severe is the PCO)
- Associated ocular anomalies
- Surgical technique
- IOL (design, material, and site of fixation)

24.3 Magnitude of the Problem

PCO occurs more frequently when the posterior capsule is left intact [23–31] (Fig. 24.1). With an intact posterior capsule, various articles have reported that PCO ranges from 14.7 to 100 % (average 25.1 %; excluding eyes with 100 % PCO) in children younger than 4 years of age. Further, PCO is much more common when an IOL is implanted, and when surgery is performed in the first year of life, despite posterior capsulotomy and vitrectomy being performed [2, 32]. In eyes that have

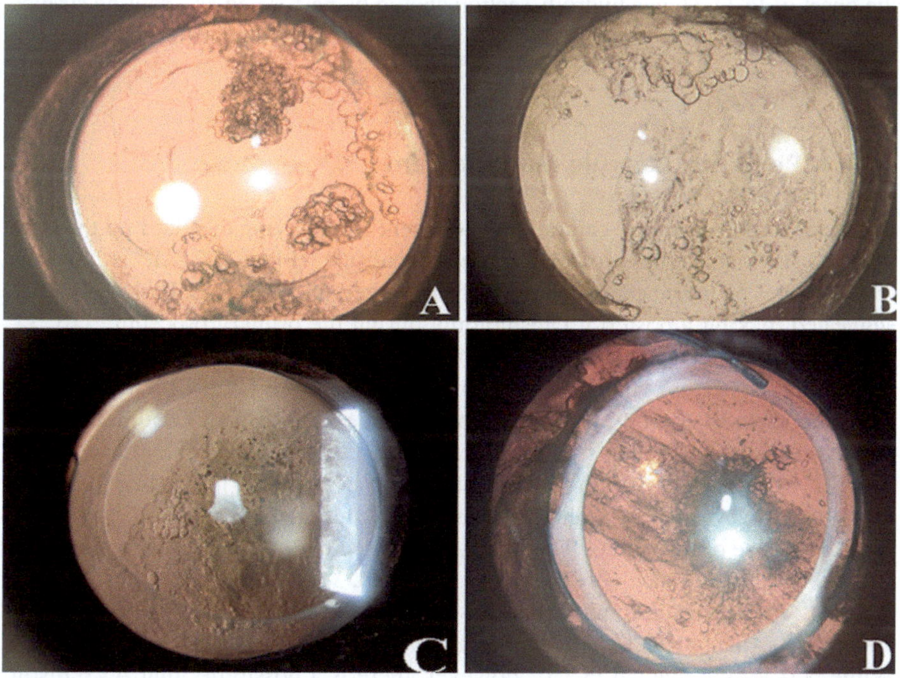

Fig. 24.1 Postoperative images of PCO with intact posterior capsules. (**a**) At 1 year. (**b**) At 2 years. (**c**) At 3.5 years. (**d**) At 5 years

been operated upon in infancy, PCO tends to occur within the first 6 months after cataract surgery [2]. Using hydrophobic acrylic IOLs, the rate of PCO was reported to range from 8.1 % when children under 2 years of age were reviewed to 80 % when children below 6 months of age were included [2, 30, 32, 33], the average PCO rate being 44 %. In children older than 2 years of age at the time of cataract surgery, the PCO rate after primary posterior capsulectomy and vitrectomy varied from 0 to 20.6 % (average 5.1 %). The average rate of secondary intervention in these eyes was 13.8 % (range 0–68 %). Thus PCO is one of the biggest deterrents of a successful visual outcome following an excellent technical outcome.

24.4 Strategies to Reduce or Delay PCO in Children

Creating an anterior continuous curvilinear capsulorhexis (ACCC) allows a stable and predictable in-the-bag IOL fixation. An ACCC smaller than the size of the optic allows the anterior capsule leaflet overlying the optic edge to fuse well with the optic's anterior surface. This results in the formation of a closed environment, which can restrict the migration of LECs toward the visual axis. Cortical cleaving hydrodissection should be done in multiple quadrants if a breach in the posterior capsule is not suspected, as it ensures a cleavage between the lens capsule and the cortex [19, 21, 22]. Multiple quadrant cortical cleaving hydrodissection can delay the onset of PCO. Hence hydrodissection should be performed as a routine surgical step during pediatric cataract surgery if no breach in the posterior capsule is suspected.

24.4.1 Posterior Capsule Management

A crucial step in pediatric cataract management for delaying or retarding the onset of PCO is management of the posterior capsule (Fig. 24.2). It is this that differentiates pediatric cataract surgery from surgery in adult eyes. Parks and others suggested making a posterior capsulotomy during the primary procedure [16, 34–36]. The posterior capsule opening can be created using various approaches including manual posterior continuous curvilinear capsulorhexis (PCCC), vitrectorhexis, radiofrequency diathermy, and Fugo plasma blade. Manual PCCC, although technically challenging, remains the gold standard for posterior capsule management. A PCCC of around 3.5–4 mm is aimed at so as to allow stable in-the-bag fixation of the IOL. However, in these highly elastic capsules, it is not always possible to control the size of the PCCC. The use of a femtosecond laser has great potential for successful performance of precisely centered, circular, continuous anterior and posterior capsulotomies in these eyes [37]. Currently, our preferred approach is manual PCCC, even when we are planning to perform vitrectomy. While performing manual PCCC, however, there is a possibility that AVF disturbance may occur due to the attachment of the posterior capsule to the AVF through the hyaloidopatellar ligament [38, 39]. However, AVF disruption often goes

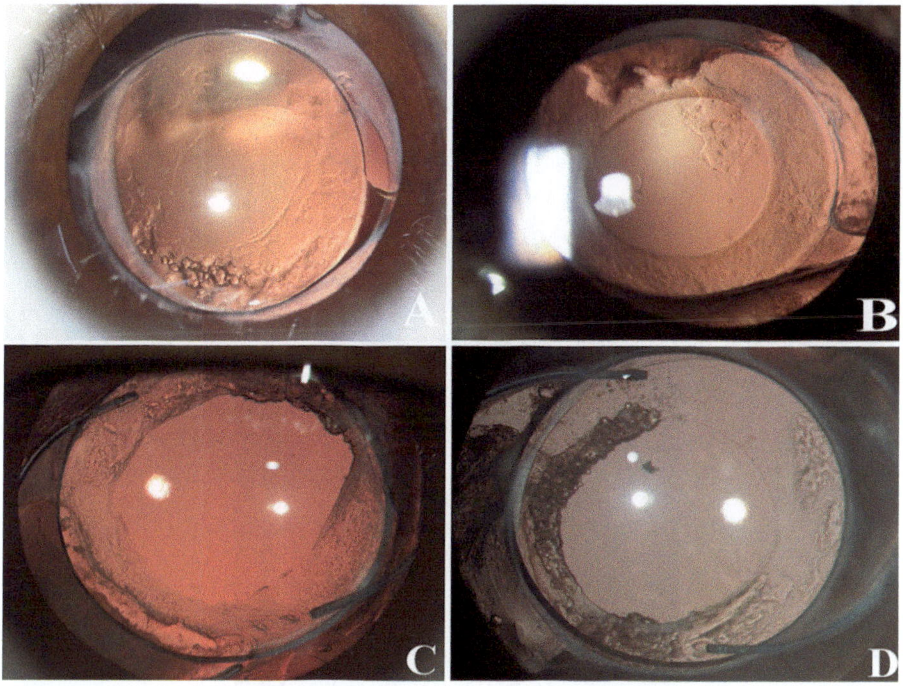

Fig. 24.2 Postoperative images of clear visual axis with PCCC. (**a**) At 1.5 years. (**b**) At 2 years. (**c**) At 3 years. (**d**) At 4 years

unnoticed because of the transparent nature of the vitreous and because the surgeon is not critically looking for AVF disruption. AVF disruption should be dealt with, even in older children where a PCCC is planned without anterior vitrectomy. We have described a few subtle signs to identify disruption of the AVF during PCCC [39]. Moreover, the use of preservative-free triamcinolone acetonide injected intracamerally helps in identifying AVF disruption and the presence of residual vitreous strands in the anterior chamber [40]. An AVF disturbance is generally accompanied by a postoperative scaffold response [10], in addition to the risks for glaucoma and posterior segment complications.

24.4.2 Anterior Vitrectomy

While PCCC can delay the onset of VAO, it cannot eliminate it [17, 41]. Even when the central posterior capsule is removed, the AVF provides a scaffold across which LECs and their products migrate, resulting in opacification. This response also provides a scaffold to inflammatory debris and exudates, leading to VAO. The rapid and severe obscuration of the visual axis would require an early surgical intervention, which is undesirable. Performing an anterior vitrectomy eliminates the possibility of a scaffold response of the AVF, subsequently delaying central VAO.

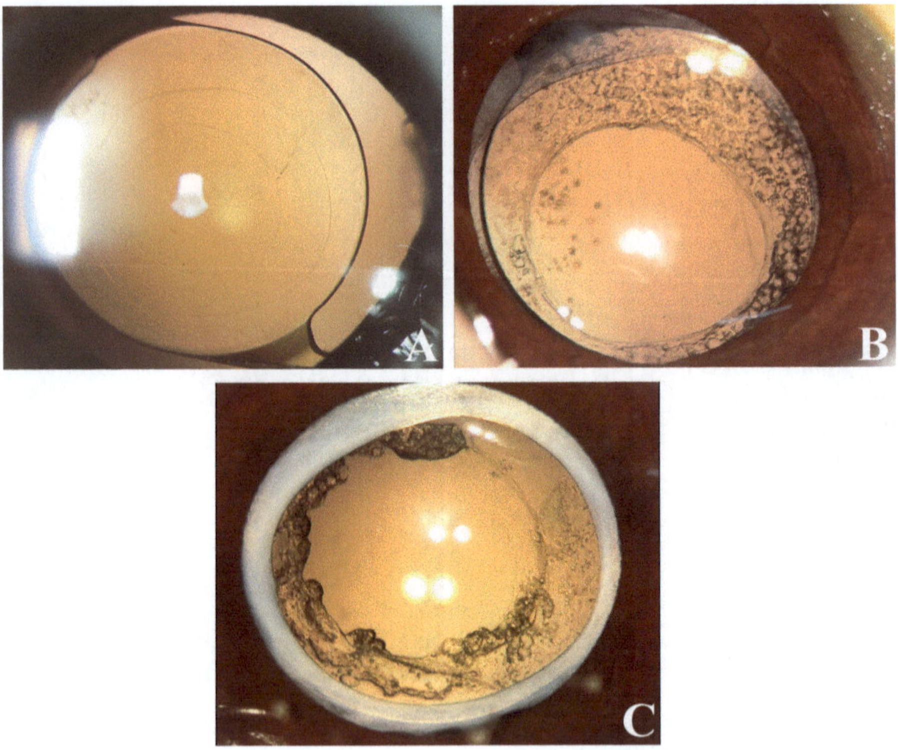

Fig. 24.3 Images of a clear visual axis after PCCC and anterior vitrectomy during congenital cataract surgery: (a) 1 year postoperatively, (b) 2 years postoperatively, and (c) 4.5 years postoperatively

This also effectively negotiates the duration when the chances of amblyopia setting in are the maximum. Surgical removal of the posterior capsule and anterior vitreous has long been considered the gold standard in pediatric cataract surgery. It is essential to perform an anterior vitrectomy during the primary procedure in children, especially those under 3 years of age, since the proliferation and migration of LECs are extremely fast and aggressive, and fibrous membranes may form on the intact AVF. A study of existing literature has revealed that reported rates of VAO after vitrectomy are less than 6 % (Fig. 24.3). Anterior vitrectomy can be performed via the limbus after posterior capsulotomy or through the pars plana following IOL implantation. Lensectomy with anterior vitrectomy is also a popular approach when IOL implantation is not planned.

The occurrence of VAO following PCCC and anterior vitrectomy is often attributed to an inadequate vitrectomy. This throws up the question: "how can the endpoint of vitrectomy be determined?" Residual vitreous strands can lead to all kinds of complications in the eye. They can cause irregularities in the pupil, glaucoma, and IOL displacement and may even lead to VAO. Therefore identification and removal of residual vitreous strands play a crucial role in ensuring successful outcomes after pediatric cataract surgery.

Performing an anterior vitrectomy is not as important in older children as in young ones and also when considering AcrySof® IOL implantation [30]. However, the concerns while performing anterior vitrectomy include possibility of glaucoma [42], increased inflammation, and occurrence of posterior segment complications like late-onset retinal detachment and vitreous traction. Occurrence of macular edema, however, appears to be less in children. In our surgical approach, we have stratified posterior capsule management according to the age of the child [17, 30, 31]. Children under 3 years are subjected to PCCC and anterior vitrectomy; children between 3 and 6 years are subjected to PCCC but not to vitrectomy; in children over 6 years, PCCC is not performed.

24.4.2.1 Choosing Between the Limbal and the Pars Plana/Plicata Approaches for Anterior Vitrectomy

While carrying out vitrectomy, the goal is to remove only the central anterior vitreous without attempting to remove the peripheral or posterior vitreous. This kind of limited vitrectomy is performed either through the limbal approach or the pars plana approach. Although it is commonly referred to as pars plana approach, the majority of surgeons make an entry closer to the limbus through the "pars plicata" region. However, in small eyes or in eyes of young children, it is recommended that pars plana capsulectomy or capsulorhexis along with anterior vitrectomy be performed.

24.4.2.2 Approaches for Posterior Capsulotomy

Different surgeons prefer different techniques for performing posterior capsulotomy. A survey done by the American Society of Cataract and Refractive Surgeons revealed that manual PCCC and posterior vitrectorhexis were preferred by an almost equal number of respondents (41.3 and 43.2 %, respectively) [43]. However, the anatomy of the pars plana may not be well defined in these young eyes.

24.5 IOL Implantation Techniques

24.5.1 In-the-Bag IOL Fixation

By maintaining the anatomical compartments of the eye, bag fixation reduces the risk of chronic inflammation and pupil capture. Additionally, it enhances the barrier effect offered by the sharp, square-edged IOLs in preventing the development of PCO [44]. In young children, in whom anterior vitrectomy is planned, an in-the-bag IOL may be implanted following manual PCCC and limbal anterior vitrectomy. Alternatively pars plana vitrectorhexis may be performed after implantation of the IOL in the capsular bag.

24.5.2 Optic Capture

Although anterior vitrectomy is part of the surgical strategy in very young children, vitrectomy has its own complications, including retinal breaks, retinal detachment, and higher incidence of glaucoma [42]. Therefore, an alternative technique proposed to solve the problem of PCO without performing anterior vitrectomy is posterior capsulorhexis with optic capture of the IOL. Gimbel et al. [45–48] first described the technique of posterior capsulorhexis with optic capture in pediatric eyes. They were of the opinion that placing the optic behind the primary PCCC and fixing the haptics in the bag (optic capture through posterior capsulorhexis) minimize the risk of VAO. This, they believed, may allow surgeons to avoid planned anterior vitrectomy.

It has been hypothesized that obliteration of the capsular bag and the posterior location of the IOL optic prevent migration of the LECs along the vitreous face. Optic capture fuses the anterior and posterior leaflets of the capsular bag almost completely in a 360 ° position, except at the haptic-optic junctions. Theoretically, capsular fusion anterior to the IOL optic might reduce central LEC migration or at least direct the cell movement anteriorly over the lens optic, which is presumably an unsuitable substrate for LEC survival. During pediatric cataract surgery, the main advantage of performing optic capture through PCCC is its ability to achieve a well-centered IOL and prevent or retard VAO (Fig. 24.4).

Although optic capture through PCCC is believed to prevent the need for vitrectomy, the role of anterior vitrectomy is controversial [49–51]. In a few studies, the authors have proven that posterior capsulorhexis and optic capture without vitrectomy cannot eliminate secondary opacification completely, especially in patients more than 5 years old [50, 51]. However, further studies with acrylic IOLs of suitable designs may improve the results with this technique.

24.5.3 Posterior Capsulorhexis Combined with Optic Buttonholing

Recently Menapace and coauthors introduced posterior optic buttonholing (POBH). This is a remarkably safe and effective technique offering several advantages. It not only excludes retro-optical opacification, but also withholds capsular fibrosis by obviating direct contact between the anterior capsular leaf and the optic surface [52].

24.5.4 Posterior Vertical Capsulectomy with Optic Entrapment

Grieshaber and coauthors [53] introduced primary posterior vertical capsulotomy with optic entrapment. After lens aspiration, a small hole is made in the posterior capsule at the 12 o'clock position followed by gentle injection of sodium

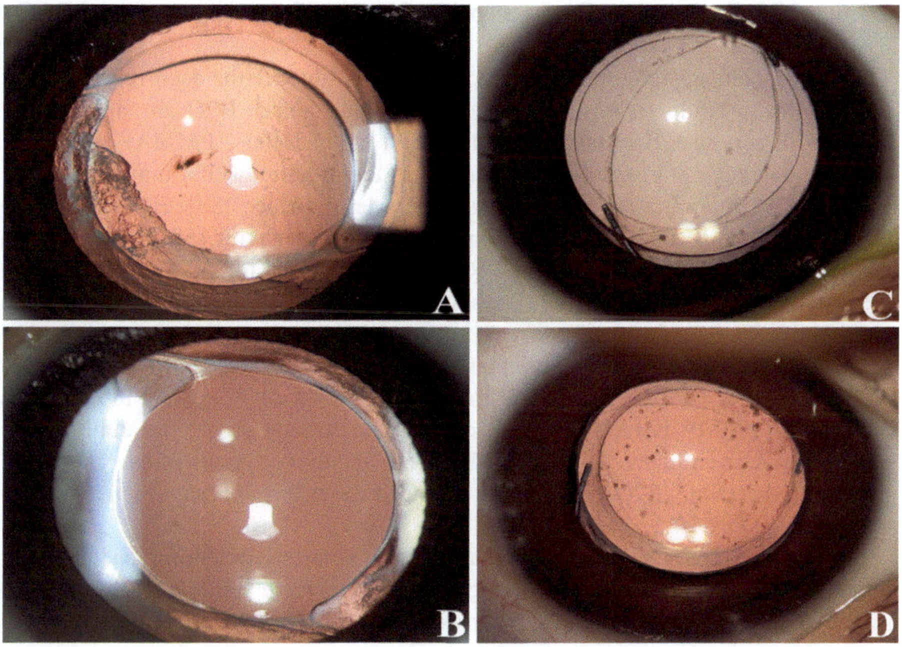

Fig. 24.4 (**a**) Image of a mild PCO after performing optic capture without vitrectomy at the 5.5 years postoperative follow-up with single-piece PMMA IOL. (**b**) A central clear visual axis after performing optic capture with vitrectomy at the 6 years postoperative follow-up with single-piece PMMA IOL. (**c** and **d**) A central clear visual axis after performing optic capture at 1-month and 1-year postoperative follow-up with three-piece hydrophobic acrylic IOLs

hyaluronate through the initial hole to separate the posterior capsule from the anterior hyaloid face. A straight Sutherland microscissor is used to perform posterior vertical capsulotomy along the 12 to 6 o'clock meridian. The IOL is implanted with the inferior haptics inserted in the bag with a Sinskey hook, followed by insertion of the optic and superior haptic. The optic is captured through a vertical posterior capsulotomy. The haptic-optic junction is positioned to be in line with the vertical capsulotomy in a 12 to 6 o'clock position.

24.5.5 Bag-in-the-Lens Implantation

Tassignon and colleagues have crafted a surgical procedure for pediatric cataractous eyes, which they have dubbed "bag in the lens." Here the anterior and posterior capsules are placed in the groove of a specially designed IOL after a capsulorhexis of the same size is created in both capsules. The principle behind this IOL design is to ensure a clear visual axis by mechanically tucking the two capsules into the IOL, thereby preventing any migration of proliferating LECs [54–57].

24.6 IOL Factors to Reduce or Delay the Development of PCO

The material and design of the IOL implanted can influence the occurrence of PCO in children. In adults, it is widely recognized that rates of PCO are significantly lower with acrylic IOLs than with polymethyl methacrylate (PMMA) or silicone IOLs. It has been suggested that heparin surface-modified IOLs may reduce the incidence of inflammatory precipitates on the lens surface [58]. However, these IOLs do not show a favorable response in capsular behavior [58, 59]. Although PMMA and surface-modified PMMA IOLs have been used for several years, during the last decade, many reports have supported the use of hydrophobic acrylic IOLs in children [2, 30, 31, 33, 60–63]. When primary PCCC and anterior vitrectomy were performed, the onset of PCO was later in eyes with AcrySof® IOLs as compared with those with PMMA IOLs (6–8 months in eyes with AcrySof® IOLs versus 3 months in PMMA IOLs). This duration is sufficient to negotiate the amblyogenic period in young children. In older children who are above 6 years of age, implantation of an AcrySof® IOL may obviate the need to perform an anterior vitrectomy [31, 64, 65] (Fig. 24.5).

Reports from published literature suggest that when the posterior capsule is left intact, pediatric eyes with hydrophobic acrylic IOLs have a similar or lower PCO rate than eyes with PMMA IOLs. However, PCO after acrylic IOL implantation is

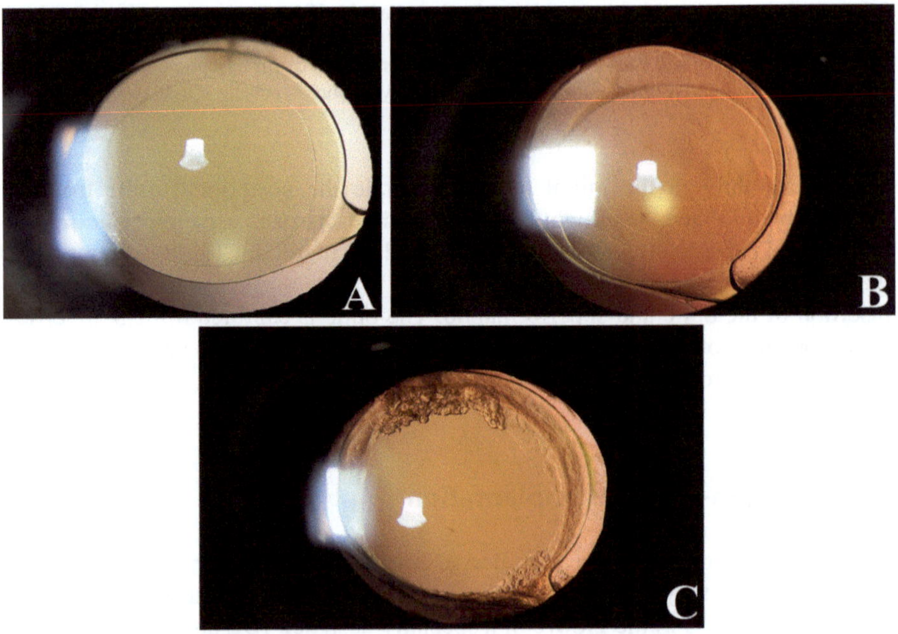

Fig. 24.5 Images of a clear visual axis after congenital cataract surgery with single-piece hydrophobic acrylic IOL: (**a**) 1 month postoperatively, (**b**) 1 year postoperatively, and (**c**) 2 years postoperatively

more "proliferative" than the "fibrous" reaction commonly seen in conjunction with PMMA IOLs. After a primary PCCC and an anterior vitrectomy, VAO rarely occurs in older children with acrylic IOLs. When it does occur, it is usually in a patient operated on in the first year or 2 years of life.

In a 2001 survey conducted by members of the American Society of Cataract and Refractive Surgeons (ASCRS) and the American Association for Pediatric Ophthalmology and Strabismus (AAPOS), 66.8 % of ASCRS respondents and 71.7 % of AAPOS respondents said that they preferred hydrophobic acrylic IOLs for pediatric eyes. A survey by AAPOS members conducted in 2006 showed that for in-the-bag fixation, 93.3 % of the respondents (95.9 and 85.2 % of USA and -non-USA respondents, respectively) preferred hydrophobic acrylic IOLs [58]. AcrySof® hydrophobic acrylic IOLs were preferred by 90.2 % of the overall respondents (94.2 and 77.8 % of USA and non-USA respondents, respectively) [66].

The biocompatibility of IOL materials should be assessed in terms of uveal and capsular biocompatibility. Bioactive materials are those which allow a single LEC layer to bond both with the IOL and the posterior capsule. This produces a sandwich pattern that includes the IOL, the cell monolayer, and the posterior capsule. This sealed sandwich structure might avert further epithelial ingrowth and prevent PCO. Linnola proposed a bioactive-based explanation known as the sandwich theory [67]. This theory suggests that the hydrophobic acrylic material is the most important factor in PCO prevention. Fibronectin and other proteins are theorized to give the acrylic lens a "sticky" property that allows a monolayer of LECs to form between the IOL and the posterior capsule, thereby preventing migration of additional cells behind the IOL [68]. Furthermore, this barrier effect and the inhibition of PCO have been under study for several years.

In one of our studies, we evaluated the combined effect of the surgical technique used and AcrySof® IOL implantation on PCO [31]. When an AcrySof IOL was implanted following PCCC, it was possible to maintain a clear visual axis in 64.3 % of eyes at the end of 3-year follow-up. Visual obscuration in eyes with AcrySof® IOLs is less severe than that in eyes with PMMA IOLs and is therefore less amblyogenic. PCO sets in at a later stage, typically at 14–16 months. On the contrary, with PMMA IOLs, the PCO is fibrous. It is fierce, not only in its appearance, but also in its impact on the development of amblyopia. The opacification sets in rapidly in the first few weeks.

24.7 Clinical Presentation and Characteristics of VAO in Children

Opacification in children is extremely rapid, and therefore, it warrants frequent follow-up, especially in very young children who cannot communicate. When IOLs are implanted in eyes with PCCC, obscuration has been reported to occur as early as 2 weeks after surgery [69]. This opacification progresses with time and a visually significant obscuration can be seen by 6 months [69]. There are several forms of VAO in pediatric eyes apart from PCO such as anterior vitreous face opacification and anterior vitreous reticular response.

24.7.1 Anterior Vitreous Face Opacification

VAO in the form of anterior vitreous face opacification is unique to pediatric eyes. When the AVF is left intact following PCCC, it comes in close apposition to the IOL optic leading to opacification of the vitreous [10]. This is a secondary response of the AVF, which provides a scaffold for the migrating LECs, inflammatory debris, and exudates, resulting in opacification. This opacification is predominantly of the fibrous type with thick, dense, white fibrous tissue producing moderate to severe obscuration of the visual axis. The response depends on the material of the IOL. It was observed within 2 months in eyes implanted with PMMA IOLs, in comparison with its conspicuous absence in eyes implanted with AcrySof IOLs.

24.7.2 Anterior Vitreous Reticular (AVR) Response

In children about 5 years of age, we observed a reticular response with AcrySof® IOLs when the AVF was left intact [70]. We believe that this is a primary response rather than a secondary scaffold response and is different from the visual obscuration that occurs due to the migration of proliferated cells. Morphologically, it appears as a fine reticular meshwork of fibrils with intervening clear spaces and has little impact on visual axis obscuration. Therefore visual acuity is not impaired. The fibrils are so fine that they may be difficult to even document (Fig. 24.6).

24.8 Newer Approaches for PCO Prevention/Reduction

24.8.1 Sutureless Vitrectomy

The 25-gauge as well as the 23-gauge high-speed vitrectomes have found a growing number of applications in pediatric surgery via a pars plana/pars plicata approach or a limbal approach [71, 72]. The 25-gauge system offers several advantages when used in infantile cataract extraction, including sutureless incisions and faster rehabilitation.

24.8.2 Sealed Capsule Irrigation

Maloof and associates have designed a sealed capsule irrigation device (Perfect Capsule™) that can help in selectively irrigating the capsular bag. Pediatric cataract surgeons can use this device to inject certain chemicals thereby eliminating or

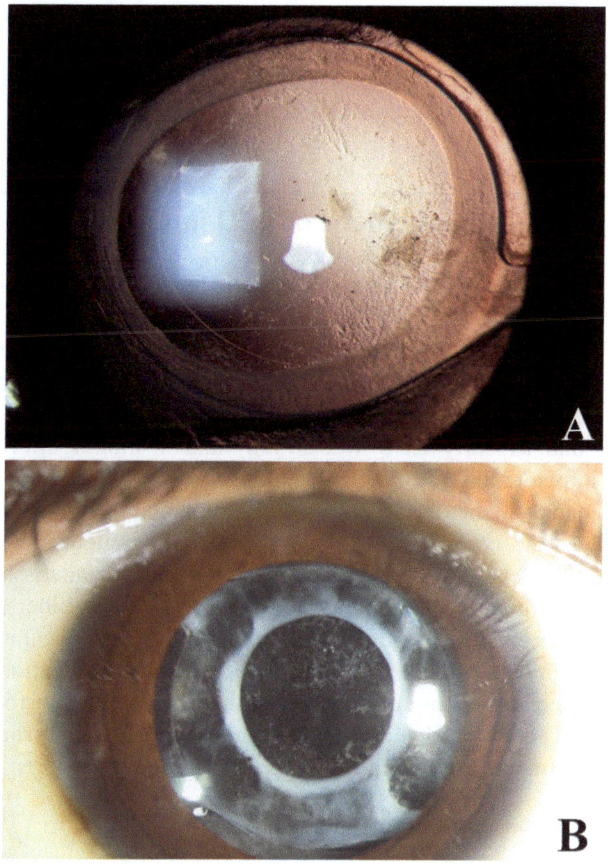

Fig. 24.6 (**a**) "Anterior vitreous reticular response" after PCCC in oblique illumination with single-piece hydrophobic acrylic IOL. (**b**) "Anterior vitreous reticular response" after PCCC in retroillumination with single-piece hydrophobic acrylic IOL at 1.5 years

delaying the onset of VAO [73, 74]. This allows selective and direct intraoperative targeting of LECs with pharmacomodulating agents to control their activity and thereby reduce PCO. However, future studies with long-term follow-up are required to demonstrate the efficacy and safety of drugs used for this purpose in pediatric eyes.

24.8.3 Manual PCCC via the Pars Plicata Approach

Manual PCCC can be performed via the pars plicata [75]. After implantation of the IOL in the capsular bag, all the incisions are sutured with 10–0 nylon and residual ophthalmic viscosurgical device (OVD) is left in the anterior chamber. The pars plicata entry is made 1–1.5 mm behind the limbus. An initial puncture is made in the center of the posterior capsule. Later, a coaxial capsulorhexis forceps is introduced and a flap generated. The edge of the flap is grasped and then regrasped every 2 clock hours fashioning the PCCC in a clockwise manner. This approach is helpful particularly in smaller eyes where the surgeon would like to implant an IOL prior to performing PCCC so as not to jeopardize the stability of the IOL in the capsular bag. It allows a controlled PCCC opening with regular, strong margins, similar to manual PCCC performed through the limbal approach.

24.9 Management of PCO in Children

24.9.1 Nd:YAG Laser Capsulotomy

An Nd:YAG laser capsulotomy is preferred in children over 5 years as they can follow instructions and cooperate during the procedure. The optimal interval for performing this procedure has not been studied. However, this procedure has been done as early as 3 weeks [76] and 1 month after cataract surgery and IOL implantation [77]. Nd:YAG laser capsulotomy is an acceptable option for the management of PCO after IOL implantation in children and only in rare cases leads to complications [62]. Retinal detachment after Nd:YAG capsulotomy for PCO was rare in eyes that had previously undergone uneventful phacoemulsification and IOL implantation [78]. In children older than 4 years, this procedure has been shown to effectively maintain a clear visual axis in 86 % of patients for a mean of 13 months post-Yag [27]·. The re-opacification rate following an Nd:YAG procedure is not influenced by either the delay in performing the primary Nd:YAG procedure or the amount of laser energy used primarily [79]. Repeated laser treatments have been reported in 17–57 % of children.

24.9.2 Secondary Surgical Membranectomy with Vitrectomy

In eyes in which the opacified capsule is too thick and not amenable to Nd:YAG capsulotomy, a secondary surgical membranectomy/capsulotomy is done [80]. The pars plana approach is preferred to avoid damage to the IOL-capsule-zonule complex [81]. This approach also provides access to the vitreous posterior to the margins of the PCCC, which would be difficult to access with the limbal approach. The Vitrectome is introduced through the pars plana, while irrigation is maintained through the limbus. An inadequately performed membranectomy and vitrectomy would act as a scaffold even in the absence of the posterior capsule or intact AVF or both, leading to rapid re-opacification [80].

It has been observed that young patients are usually noncompliant to Nd:YAG capsulotomy. Further young children seem to be at greater risk of requiring repeated interventions for PCO. Hence, this option is not advisable for very young children. The incidence of laser-induced IOL damage is quite high and laser capsulotomy often becomes ineffective in the presence of fibrous PCO. Therefore, for young children, surgical capsulotomy combined with anterior vitrectomy is recommended, rather than Nd:YAG laser capsulotomy.

24.10 Summary

Different patient-related factors, ocular and systemic conditions, and surgical techniques influence the development of PCO. Further, in literature, various alternatives have been suggested to achieve the goal of a clear visual axis. Primary posterior capsulorhexis and anterior vitrectomy should be regarded as mandatory in young children. Optic capture provides an option to reduce or delay VAO and yet avoid anterior vitrectomy. The bag-in-the-lens IOL implantation technique can be considered to avoid re-opacification of the visual axis. Finally the management of the posterior capsule during cataract surgery plays a vital role in ensuring successful technical and functional outcomes. At present, meticulous use of surgical techniques and appropriate IOLs remains the mainstay for retarding the development of PCO in children.

References

1. Lambert SR, Lynn M, Drews-Botsch C, Loupe D, Plager DA, Medow NB, Wilson ME, Buckley EG, Drack AV, Fawcett SL (2001) A comparison of grating visual acuity, strabismus, and reoperation outcomes among children with aphakia and pseudophakia after unilateral cataract surgery during the first six months of life. J AAPOS 5:70–75
2. Trivedi RH, Wilson ME Jr, Bartholomew LR, Lal G, Peterseim MM (2004) Opacification of the visual axis after cataract surgery and single acrylic intraocular lens implantation in the first year of life. J AAPOS 8:156–164
3. Ben Ezra D, Cohen E (1997) Posterior capsulectomy in pediatric cataract surgery: the necessity of a choice. Ophthalmology 104:2168–2174
4. Apple DJ, Solomon KD, Tetz MR, Assia EI, Holland EY, Legler UF, Tsai JC, Castaneda VE, Hoggatt JP, Kostick AM (1992) Posterior capsule opacification. Surv Ophthalmol 37:73–116
5. Faulkner SD, Vorobyov V, Sengpiel F (2006) Visual cortical recovery from reverse occlusion depends on concordant binocular experience. J Neurophysiol 95:1718–1726
6. Kind PC, Mitchell DE, Ahmed B, Blakemore C, Bonhoeffer T, Sengpiel F (2002) Correlated binocular activity guides recovery from monocular deprivation. Nature 416:430–433
7. Mitchell DE, Kind PC, Sengpiel F, Murphy K (2003) Brief daily periods of binocular vision prevent deprivation-induced acuity loss. Curr Biol 13:1704–1708
8. Mitchell DE, Kind PC, Sengpiel F, Murphy K (2006) Short periods of concordant binocular vision prevent the development of deprivation amblyopia. Eur J Neurosci 23:2458–2466
9. Zetterström C, Kugelberg U, Lundgren B, Syrén-Nordqvist S (1996) After-cataract formation in newborn rabbits implanted with intraocular lenses. J Cataract Refract Surg 22:85–88
10. Jones NP, McLeod D, Boulton ME (1995) Massive proliferation of lens epithelial remnants after Nd-YAG laser capsulotomy. Br J Ophthalmol 79:261–263
11. Trivedi RH, Wilson ME (2005) Posterior capsulotomy and anterior vitrectomy for the management of pediatric cataracts. In: Wilson ME, Trivedi RH, Pandey SK (eds) Pediatric cataract surgery: techniques, complications, and management. Lippincott, Williams & Wilkins, Philadelphia, pp 83–92, Chapter 16
12. Tuncer S, Gucukoglu A, Gozum N (2005 Jun) Cataract extraction and primary hydrophobic acrylic intraocular lens implantation in infants. J AAPOS 9(3):250–256
13. Liu CS, Wormstone IM, Duncan G, Marcantonio JM, Webb SF (1996 Apr) A study of human lens cell growth in vitro. A model for posterior capsule opacification. Invest Ophthalmol Vis Sci 37(5):906–914

14. De Groot V, Vrensen GF, Willekens B, Van Tenten Y, Tassignon MJ (2003 May) In vitro study on the closure of posterior capsulorrhexis in the human eye. Invest Ophthalmol Vis Sci 44(5):2076–2083
15. Koch DD, Kohnen T (1997) A retrospective comparison of techniques to prevent secondary cataract formation following posterior chamber intraocular lens implantation in infants and children. Trans Am Ophthalmol Soc 95:351–360, discussion 361–365
16. Buckley EG, Klombers LA, Seaber JH, Scalise-Gordy A, Minzter R (1993) Management of the posterior capsule during pediatric intraocular lens implantation. Am J Ophthalmol 115:722–728
17. Vasavada A, Desai J (1997) Primary posterior capsulorhexis with and without anterior vitrectomy in congenital cataracts. J Cataract Refract Surg 23(Suppl 1):645–651
18. Hollick EJ, Spalton DJ, Ursell PG, Pande MV, Barman SA, Boyce JF, Tilling K (1999) The effect of polymethylmethacrylate, silicone, and polyacrylic intraocular lenses on posterior capsular opacification 3 years after cataract surgery. Ophthalmology 106:49–54, discussion 54–55
19. Peng Q, Apple DJ, Visessook N, Werner L, Pandey SK, Escobar-Gomez M, Schoderbek R, Guindi A (2000) Surgical prevention of posterior capsule opacification. Part 2: enhancement of cortical cleanup by focusing on hydrodissection. J Cataract Refract Surg 26:188–197
20. Apple DJ, Peng Q, Visessook N, Werner L, Pandey SK, Escobar-Gomez M, Ram J, Whiteside SB, Schoderbeck R, Ready EL, Guindi A (2000) Surgical prevention of posterior capsule opacification. Part 1: progress in eliminating this complication of cataract surgery. J Cataract Refract Surg 26:180–187
21. Vasavada AR, Singh R, Apple DJ, Trivedi RH, Pandey SK, Werner L (2002) Effect of hydrodissection on intraoperative performance: randomized study. J Cataract Refract Surg 28:1623–1628
22. Vasavada AR, Trivedi RH, Apple DJ, Ram J, Werner L (2003) Randomized, clinical trial of multiquadrant hydrodissection in pediatric cataract surgery. Am J Ophthalmol 135:84
23. Basti S, Ravishankar U, Gupta S (1996) Results of a prospective evaluation of three methods of management of pediatric cataracts. Ophthalmology 103:713–720
24. O'Keefe M, Fenton S, Lanigan B (2001) Visual outcomes and complications of posterior chamber intraocular lens implantation in the first year of life. J Cataract Refract Surg 27:2006–2011
25. Hosal BM, Biglan AW (2002) Risk factors for secondary membrane formation after removal of pediatric cataract. J Cataract Refract Surg 28:302–309
26. Jensen AA, Basti S, Greenwald MJ, Mets MB (2002) When may the posterior capsule be preserved in pediatric intraocular lens surgery? Ophthalmology 109:324–327, discussion 328
27. Stager DR Jr, Weakley DR Jr, Hunter JS (2002) Long-term rates of PCO following small incision foldable acrylic intraocular lens implantation in children. J Pediatr Ophthalmol Strabismus 39:73–76
28. Müllner-Eidenböck A, Amon M, Moser E, Kruger A, Abela C, Schlemmer Y, Zidek T (2003) Morphological and functional results of AcrySof intraocular lens implantation in children: prospective randomized study of age-related surgical management. J Cataract Refract Surg 29:285–293
29. Raina UK, Mehta DK, Monga S, Arora R (2004) Functional outcomes of acrylic intraocular lenses in pediatric cataract surgery. J Cataract Refract Surg 30:1082–1091
30. Vasavada AR, Trivedi RH, Nath VC (2004) Visual axis opacification after AcrySof intraocular lens implantation in children. J Cataract Refract Surg 30:1073–1081, Erratum in: J Cataract Refract Surg 2004; 30:1826
31. Nihalani BR, Vasavada AR (2006) Single-piece AcrySof intraocular lens implantation in children with congenital and developmental cataract. J Cataract Refract Surg 32:1527–1534
32. Plager DA, Yang S, Neely D, Sprunger D, Sondhi N (2002 Feb) Complications in the first year following cataract surgery with and without IOL in infants and older children. J AAPOS 6 (1):9–14

33. Kugelberg M, Kugelberg U, Bobrova N, Tronina S, Zetterström C (2006 Jun) Implantation of single-piece foldable acrylic IOLs in small children in the Ukraine. Acta Ophthalmol Scand 84 (3):380–387
34. Parks MM (1983) Posterior lens capsulectomy during primary cataract surgery in children. Ophthalmology 90:344–345
35. Knight-Nanan D (1996) O' Keefe M, Bowell R. Outcomes and complications of intraocular lenses in children with cataract. J Cataract Refract Surg 22:730–736
36. BenEzra D, Cohen E (1997) Posterior capsulectomy in pediatric cataract surgery; the necessity of a choice. Ophthalmology 104:2168–2174
37. Dick HB, Schultz T (2013) Femtosecond laser-assisted cataract surgery in infants. J Cataract Refract Surg 39(5):665–668
38. Dholakia SA, Praveen MR, Vasavada AR, Nihalani B (2006) Completion rate of primary posterior continuous curvilinear capsulorhexis and vitreous disturbance during congenital cataract surgery. J AAPOS 10:351–356
39. Praveen MR, Vasavada AR, Koul A, Trivedi RH, Vasavada VA, Vasavada VA (2008) Subtle signs of anterior vitreous face disturbance during posterior capsulorhexis in pediatric cataract surgery. J Cataract Refract Surg 34:163–167
40. Shah SK, Vasavada V, Praveen MR, Vasavada AR, Trivedi RH, Dixit NV (2009) Triamcinolone-assisted vitrectomy in pediatric cataract surgery. J Cataract Refract Surg 35:230–232
41. Guo S, Wagner RS, Caputo A (2004) Management of the anterior and posterior lens capsules and vitreous in pediatric cataract surgery. J Pediatr Ophthalmol Strabismus 41:330–337
42. Chang S (2006) LXII Edward Jackson lecture: open angle glaucoma after vitrectomy. Am J Ophthalmol 141(6):1033–1043
43. Wilson ME Jr, Bartholomew LR, Trivedi RH (2003) Pediatric cataract surgery and intraocular lens implantation: practice styles and preferences of the 2001 ASCRS and AAPOS memberships. J Cataract Refract Surg 29:1811–1820
44. Peng Q, Visessook N, Apple DJ et al (2000) Surgical prevention of posterior capsule opacification. Part-3: intraocular lens optic barrier effect as a second line of defense. J Cataract Refract Surg 26:198
45. Gimbel HV (1997) Posterior continuous curvilinear capsulorhexis and optic capture of the intraocular lens to prevent secondary opacification in pediatric cataract surgery. J Cataract Refract Surg 23(Suppl 1):652–656
46. Gimbel HV, DeBroff BM (1994) Posterior capsulorhexis with optic capture: maintaining a clear visual axis after pediatric cataract surgery. J Cataract Refract Surg 20:658–664
47. Gimbel HV (1996) Posterior capsulorhexis with optic capture in pediatric cataract and intraocular lens surgery. Ophthalmology 103:1871–1875
48. Gimbel HV, DeBroff BM (2004) Intraocular lens optic capture. J Cataract Refract Surg 30:200–206
49. Argento C, Badoza D, Ugrin C (2001) Optic capture of the AcrySof intraocular lens in pediatric cataract surgery. J Cataract Refract Surg 27:1638–1642
50. Vasavada AR, Trivedi RH (2000) Role of optic capture in congenital cataract and intraocular lens surgery in children. J Cataract Refract Surg 26:824–831
51. Vasavada AR, Trivedi RH, Singh R (2001) Necessity of vitrectomy when optic capture is performed in children older than 5 years. J Cataract Refract Surg 27:1185–1193
52. Menapace R (2006) Routine posterior optic buttonholing for eradication of posterior capsule opacification in adults: report of 500 consecutive cases. J Cataract Refract Surg 32:929–943, Erratum in: J Cataract Refract Surg 2006; 32:1410
53. Grieshaber MC, Pienaar A, Stegmann R (2005) Posterior vertical capsulotomy with optic entrapment of the intraocular lens in congenital cataracts–prevention of capsule opacification. J Cataract Refract Surg 31:886–894
54. Tassignon MJ, De Veuster I, Godts D, Kosec D, Van den Dooren K, Gobin L (2007) Bag-in-the-lens intraocular lens implantation in the pediatric eye. J Cataract Refract Surg 33:611–617

55. Leysen I, Coeckelbergh T, Gobin L, Smet H, Daniel Y, De Groot V, Tassignon MJ (2006) Cumulative Nd:YAG laser rate after bag-in-the-lens compared to lens-in-the-bag implantation. J Cataract Refract Surg 32:2085–2090
56. Tassignon MJ, Rozema J, Gobin L (2006) Ring-shaped caliper for better anterior capsulorhexis sizing and centration. J Cataract Refract Surg 32:1253–1255
57. Tassignon M-J, Gobin L, De Veuster I, Godts D (2009) Intérêt de l'implant « bag-in-the-lens » dans la chirurgie de la cataracte de l'enfant. J Fr Ophtalmol 32(7):481–487
58. Zetterström C, Kugelberg U, Oscarson C (1994) Cataract surgery in children with capsulorhexis of anterior and posterior capsules and heparin-surface-modified intraocular lenses. J Cataract Refract Surg 20:599–601
59. Basti S, Aasuri ML, Reddy MK et al (1999) Heparin-surface-modified intraocular lenses in pediatric cataract surgery. Prospective randomized study. J Cataract Refract Surg 25:782–787
60. Wilson ME, Elliott L, Johnson B et al (2001) AcrySof acrylic intraocular lens implantation in children: clinical indications of biocompatibility. J AAPOS 5:377–380
61. Aasuri MK, Fernandes M, Pathan PP (2006) Comparison of acrylic and polymethyl methacrylate lenses in a pediatric population. Indian J Ophthalmol 54:105–109
62. Stager DR Jr, Wang X, Weakley DR Jr, Felius J (2006) The effectiveness of Nd:YAG laser capsulotomy for the treatment of posterior capsule opacification in children with acrylic intraocular lenses. J AAPOS 10:159–163
63. Wilson ME Jr, Trivedi RH, Buckley EG, Granet DB, Lambert SR, Plager DA, Sinskey RM, Vasavada AR (2007) ASCRS white paper. Hydrophobic acrylic intraocular lenses in children. J Cataract Refract Surg 33:1966–1973
64. Ram J, Brar GS, Kaushik S, Gupta A, Gupta A (2003 Aug) Role of posterior capsulotomy with vitrectomy and intraocular lens design and material in reducing posterior capsule opacification after pediatric cataract surgery. J Cataract Refract Surg 29(8):1579–1584
65. Kugelberg M, Zetterstorm C (2002) Pediatric cataract surgery with or without vitrectomy. J Cataract Refract Surg 28:1770–1773
66. Wilson ME Jr, Trivedi RH (2007) Choice of intraocular lens for pediatric cataract surgery: survey of AAPOS members. J Cataract Refract Surg 33:1666–1668
67. Linnola RJ (1997) Sandwich theory: bioactivity-based explanation for posterior capsule opacification. J Cataract Refract Surg 23:1539
68. Linnola RJ, Werner L, Pandey SK et al (2000) Adhesion of fibronectin, vitronectin, laminin, and collagen type IV to intraocular lens materials in pseudophakic human autopsy eyes. Part-1: histological sections. J Cataract Refract Surg 26:1792
69. Raina UK, Gupta V, Arora R, Mehta DK (2002) Posterior continuous curvilinear capsulorhexis with and without optic capture of the posterior chamber intraocular lens in the absence of vitrectomy. J Pediatr Ophthalmol Strabismus 39(5):278–287
70. Vasavada AR, Nath VC, Trivedi RH (2003) Anterior vitreous face behaviour with Acrysof® in paediatric cataract surgery. J AAPOS 7:384–388
71. Moreno-Montañés J, Barrio-Barrio J, García-Layana A (2007) Combined cataract surgery and 25-gauge sutureless vitrectomy for posterior lentiglobus. J Cataract Refract Surg 33:380–382
72. Chee KY, Lam GC (2009) Management of congenital cataract in children younger than 1 year using a 25-gauge vitrectomy system. J Cataract Refract Surg 35:720–724
73. Maloof A, Neilson G, Milverton EJ, Pandey SK (2003) Selective and specific targeting of lens epithelial cells during cataract surgery using sealed-capsule irrigation. J Cataract Refract Surg 29:1566–1568
74. Agarwal A, Agarwal S, Agarwal A, Maloof A (2003) Sealed-capsule irrigation device. J Cataract Refract Surg 29:2274–2276
75. Vasavada AR, Shah SK, Praveen MR, Vasavada VA, Trivedi RH, Karve SJ (2011 Feb) Pars plicata posterior continuous curvilinear capsulorhexis. J Cataract Refract Surg 37(2):221–223
76. Atkinson CS, Hiles DA (1994) Treatment of secondary posterior capsular membranes with the Nd:YAG laser in a pediatric population. Am J Ophthalmol 118:496–501

77. Fenton S, O'Keefe M (1999 Jun) Primary posterior capsulorhexis without anterior vitrectomy in pediatric cataract surgery: longer-term outcome. J Cataract Refract Surg 25(6):763–767
78. Jahn CE, Richter J, Jahn AH, Kremer G, Kron M (2003) Pseudophakic retinal detachment after uneventful phacoemulsification and subsequent neodymium: YAG capsulotomy for capsule opacification. J Cataract Refract Surg 29(5):925–929
79. Hutcheson KA, Drack AV, Ellish NJ, Lambert SR (1999 Oct) Anterior hyaloid face opacification after pediatric Nd:YAG laser capsulotomy. J AAPOS 3(5):303–307
80. Zwaan J, Mullaney PB, Awad A, al-Mesfer S, Wheeler DT (1998) Pediatric intraocular lens implantation. Surgical results and complications in more than 300 patients. Ophthalmology 105(1):112–118
81. Quinn GE, Young TL (1994) Retna and vitreous. In: Isenderg SJ (ed) The Eye in infancy. Mosby-Year Book Inc, St. Louis, p 407

Index

S. Saika et al. (eds.), *Lens Epithelium and Posterior Capsular Opacification*, DOI 10.1007/978-4-431-54300-8, © Springer Japan 2014

MIX
Papier aus verantwortungsvollen Quellen
Paper from responsible sources
FSC® C105338

If you have any concerns about our products,
you can contact us on
ProductSafety@springernature.com

In case Publisher is established outside the EU,
the EU authorized representative is:
Springer Nature Customer Service Center GmbH
Europaplatz 3, 69115 Heidelberg, Germany

Printed by Libri Plureos GmbH
in Hamburg, Germany